Rational Design of Non-precious Metal Oxide Catalysts by Means of Advanced Synthetic and Promotional Routes

Rational Design of Non-precious Metal Oxide Catalysts by Means of Advanced Synthetic and Promotional Routes

Editors

Michalis Konsolakis
Vassilis Stathopoulos

MDPI • Basel • Beijing • Wuhan • Barcelona • Belgrade • Manchester • Tokyo • Cluj • Tianjin

Editors
Michalis Konsolakis
School of Production Engineering and Management
Technical University of Crete
Chania, Crete
Greece

Vassilis Stathopoulos
Department of Agricultural Development, Agrofood and Management of Natural Resources
National and Kapodistrian University of Athens
Psachna, Evia
Greece

Editorial Office
MDPI
St. Alban-Anlage 66
4052 Basel, Switzerland

This is a reprint of articles from the Special Issue published online in the open access journal *Catalysts* (ISSN 2073-4344) (available at: www.mdpi.com/journal/catalysts/special_issues/rational_design).

For citation purposes, cite each article independently as indicated on the article page online and as indicated below:

LastName, A.A.; LastName, B.B.; LastName, C.C. Article Title. *Journal Name* **Year**, *Volume Number*, Page Range.

ISBN 978-3-0365-6164-6 (Hbk)
ISBN 978-3-0365-6163-9 (PDF)

Contents

About the Editors

Michalis Konsolakis

Dr. Konsolakis Michalis is a Full Professor of "Heterogeneous Catalysis & Surface Science"at the School of Production Engineering & Management of the Technical University of Crete, Greece. His research activities are mainly focused in the areas of heterogeneous catalysis and surface science, with particular emphasis on structure–property relationships. Recently, he has focused on the rational design and nano-engineering of metal oxide catalysts by means of advanced synthetic and promotional routes. His published work includes >100 articles in international peer-reviewed journals and >150 articles in conference proceedings. He is a member of the Editorial Board of several international journals in the fields of materials and surface science, also serving as a regular reviewer for >100 scientific journals and research funding agencies.

Vassilis Stathopoulos

Dr. Stathopoulos Vassilis is a Full Professor of "Ceramic, Porous and Catalytic Materials"at the Department of Agricultural Development, Agrofood and Management of Natural Resources, National and Kapodistrial University of Athens. His research activities mainly focus in the areas of heterogeneous catalysis, porous materials and surface engineering. His published work includes >95 articles in international peer-reviewed journals. He is a member of several Editorial Boards in the field of materials and surface science, also serving as a regular reviewer for several scientific journals and research funding agencies.

catalysts

MDPI

Editorial

Rational Design of Non-Precious Metal Oxide Catalysts by Means of Advanced Synthetic and Promotional Routes

Michalis Konsolakis [1,*] and Vassilis N. Stathopoulos [2]

1 School of Production Engineering and Management, Technical University of Crete, 73100 Chania, Greece
2 General Department, National and Kapodistrian University of Athens, Psachna Campus Evia, 34100 Psachna, Greece; vasta@uoa.gr
* Correspondence: mkonsol@pem.tuc.gr; Tel.: +30-28210-37682

Citation: Konsolakis, M.; Stathopoulos, V.N. Rational Design of Non-Precious Metal Oxide Catalysts by Means of Advanced Synthetic and Promotional Routes. *Catalysts* **2021**, *11*, 895. https://doi.org/10.3390/catal11080895

Received: 22 July 2021
Accepted: 23 July 2021
Published: 24 July 2021

Publisher's Note: MDPI stays neutral with regard to jurisdictional claims in published maps and institutional affiliations.

1. Background

Catalysis is an indispensable part of our society, involved in numerous energy and environmental applications, such as the production of value-added chemicals/fuels, hydrocarbons processing, fuel cells applications, abatement of hazardous pollutants, among others. Although, noble metals (NMs)-based catalysts are traditionally employed in various processes, due to their peculiar characteristics and enhanced reactivity, their scarcity and consequently high cost renders them disincentive for practical applications. In this perspective, the rational design and development of earth-abundant NMs-free metal oxides of adequate activity, selectivity and durability constitutes one of the main research pillars in heterogeneous catalysis [1–6]. Towards this direction, however, one crucial question must be answered: Is it possible to fine-tune the local surface chemistry/structure of a single, binary or multicomponent metal oxide in order to be highly efficient-like NMs-in a specific process? Thanks to the huge research progress so far achieved in the fields of (nano) materials synthesis, catalyst tailoring/promotion and surface science, the answer to the aforementioned question is certainly yes.

In specific, the catalytic performance of metal oxides, such as transition metals (TMs)-based mixed metal oxides, spinels, perovskites, hexaaluminates and hydrotalcites can be considerably improved by tailoring the local surface chemistry/structure (e.g., work function, reducibility, oxygen vacancies) and interfacial interactions. The latter can be accomplished by the employment of state-of-the-art nano-synthesis routes towards engineering particle's size and shape (e.g., nanocubes, nanorods) in conjunction to the use of appropriate modifiers (e.g., alkali, graphene oxide) and special pretreatment protocols. This holistic approach can exert a profound influence not only to the surface reactivity of metal sites in its own right, but also to metal-support interfacial activity, offering highly active and stable materials for real-life energy and environmental applications, such as the CO oxidation [7–12], N_2O decomposition [13–15], CO_2 hydrogenation to value-added products [16–20], degradation of organic contaminants [21–33], etc.

2. This Special Issue

In light of the above aspects, the present Special Issue is mainly focused on the fabrication and fine-tuning of NMs-free metal oxide catalysts by means of advanced synthetic and/or promotional routes. It consists of fourteen high quality papers, involving: a comprehensive review article on the recent advances on the rational design and fine-tuning of ceria-based metal oxide catalysts [5]; two articles on the ceria nanoparticles morphological effects on N_2O decomposition [14] and CO oxidation [10] over ceria-based binary oxides; two articles on NO decomposition over K-promoted Co-Mn-Al mixed oxides [34,35]; one article on the effect of ceria synthesis methods on the carbon pathways in dry reforming of methane (DRM) over Ni/CeO_2 catalysts [36]; one article on the impact of synthesis procedure and aliovalent doping in Co-Al spinel-type oxides for lean methane combustion [37];

one article on the support effect on the direct conversion of syngas to higher alcohols over copper-based oxides [38]; one article on nanofibrous Ni/Al_2O_3 catalysts prepared by electrospinning for methane partial oxidation [39]; one article on the influence of precursor compounds on the selective catalytic reduction (SCR) of NO_x by NH_3 over $FeMgO_x$ oxides [40]; one article on the synthesis and active sites determination of ZrO_2-supported WO_x solid acid catalysts [41]; one article on the synthesis and electrocatalytic performance of RuO_2 nanoparticles [42]; one article on the fabrication and photocatalytic performance of mesoporous frameworks of $ZnFe_2O_4$ (ZFO) and $MnFe_2O_4$ (MFO) nanoparticles [43]; and one article on the synthesis of polymer supported catalysts for arylamination reaction [44].

Contribution Highlights

The comprehensive review of M. Konsolakis and M. Lykaki [5] addresses the latest experimental and theoretical advances in the field of the rational design of metal oxide catalysts exemplified by CuO_x/CeO_2 binary system. In particular, it summarizes the general optimization framework that could be followed to fine-tune metal oxide sites and their surrounding environment by means of appropriate synthetic and promotional/modification routes. It was clearly revealed that the modulation of size, shape and electronic state at nanoscale can exert a profound influence not only to the reactivity of metal sites in its own right, but also to metal-support interfacial activity, offering cost-effective and highly active materials for real-life energy and environmental applications.

In view of above aspects, the impact of ceria morphology (nanorods, nanocubes, nanopolyhedra) on the physicochemical properties and the catalytic performance of ceria-based transition metal catalysts was explored by M. Konsolakis and co-workers [10,14]. It was shown that Co_3O_4/CeO_2 of rod-like morphology exhibited the optimum N_2O decomposition performance as compared to other distinct morphologies, due to its abundance in Co^{2+} active sites and Ce^{3+} species in conjunction to its improved reducibility, oxygen kinetics and surface area [14]. Similar conclusions were derived for Fe_2O_3/CeO_2 catalysts for CO oxidation [10]; the rod-shaped sample exhibited the optimum catalytic performance, due to its improved reducibility and abundance in Fe^{2+} species [10]. These findings unambiguously revealed the key role of support morphology towards determining the redox properties and in turn the catalytic performance of reactions following a redox-type mechanism [5,10,14].

Lucie Obalová and co-workers [34,35] systematically explored the impact of preparation parameters and alkali doping on the direct NO decomposition of K-promoted Co-Mn-Al mixed oxides. It was shown that preparation procedure notably affects the physicochemical properties and alkali distribution/stability with great consequences in NO decomposition. Specifically, it was revealed that the presence of potassium promoter notably improves the basicity and reducibility of the catalysts, positively affecting the catalytic activity. However, the calcination time/temperature notably affects the textural characteristics and alkali metal valorization process. The best catalytic performance was achieved for a potassium loading of ca. 1.0 wt.% at a calcination temperature of 700–800 °C. These results clearly revealed the importance of pretreatment conditions in conjunction to surface promotion towards the development of highly active metal oxides.

A.M. Efstathiou and co-workers [36] elegantly designed and conducted transient and isotopic studies to gain insight into the impact of CeO_2 preparation method on the carbon pathways in the dry reforming of methane (DRM) of Ni/CeO_2 catalysts. Among the different preparation methods explored, precipitation led to the lowest amount of carbon deposition. By means of various transient and isotopic studies, it was shown that a large pool of oxygen over precipitated catalysts contributed to the gasification of carbon formed in DRM towards the formation of CO, thus offering an important path for carbon removal.

The impact of different synthetic/modification routes towards enhancing the lean methane combustion of Co_3O_4/Al_2O_3 spinel-type oxides was investigated by Rubén López-Fonseca and co-workers [37]. Three different strategies for enhancing the performance of alumina-supported catalysts were examined: (i) surface protection of the alumina with magnesia prior to the deposition of the cobalt precursor, (ii) coprecipitation of cobalt along with nickel and (iii) surface protection of alumina with ceria. The optimum performance was obtained by the addition of ceria to alumina prior to the deposition of cobalt, which was attributed to the abundance of Co^{3+} species and oxygen vacancies due to the insertion of Ce^{4+} ions into the spinel lattice.

X. Li et al. [38] reported on the impact of support nature (SiO_2, Al_2O_3) and alkali promotion (K) on the synthesis of higher alcohols from CO hydrogenation over Cu-based catalysts. Significant differences on CO conversion and product's selectivity were revealed, attributed to support- and alkali-induced effects on redox and electronic properties.

D. Dong and co-workers [39] investigated nanofibrous Ni/Al_2O_3 catalysts prepared by electrospinning for methane partial oxidation. The impact of different synthesis parameters, such as metal precursor, metal content and calcination temperature were explored. It was shown that by appropriately adjusting the aforementioned parameters highly active and stable catalysts can be obtained.

In a similar manner, L. Xu et al. [40] explored the influence of precursor compounds on the selective catalytic reduction (SCR) of NO_x with NH_3 over Ti-modified $FeMgO_x$ oxides. The key role of precursors towards determining the surface acidity and redox properties and in turn the catalytic performance, was clearly demonstrated. The catalysts derived from $FeSO_4$ and $Mg(NO_3)_2$ precursors exhibited enhanced catalytic activity in the temperature range of ca. 200–400 °C, offering complete NO_x conversion.

R.J. Gorte and co-workers [41] explored thoroughly the reactive sites in WO_x/ZrO_2 catalysts prepared by atomic layer deposition (ALD). By a comparison with a WO_x/ZrO_2 catalyst prepared via conventional impregnation and by employing surface and microscopy techniques three types of catalytic sites were identified, with their concentration varied with the number of ALD cycles. Dehydrogenation sites are associated with ZrO_2, Brønsted-acid sites with monolayer WO_x clusters, while oxidation sites are associated with the WO_x coverage. Such surface chemistry differentiation with the preparation process notably affects acid catalyzed reactions, such as 2-propanol catalytic dehydration.

R. Phul et al. [42] reported on a simple wet chemical route to synthesize ultrafine RuO_2 nanoparticles at controlled temperature as electrocatalysts for oxygen evolution reaction (OER) and oxygen reduction reaction (ORR). These RuO_2 nanoparticles exhibited enhanced bifunctional electrocatalytic performance under different conditions (air, N_2 and O_2 atmosphere), showing excellent potential for electrocatalytic applications. In addition, RuO_2 nanoparticles showed efficient sensing properties rendering them as active nonenzymatic electrochemical sensors for the selective detection of H_2O_2.

The group of G.S. Armatas [43] reported on the preparation of high-surface-area dual component mesoporous frameworks of spinel ferrite $ZnFe_2O_4$ (ZFO) and $MnFe_2O_4$ (MFO) nanoparticles with improved photochemical activity. These mesoporous nanomaterials were synthesized via a polymer-assisted method that allowed the efficient co-assembly of the spinel ferrite colloidal nanoparticles and amphiphilic block-copolymer aggregates. The MFO-ZFO composite materials exhibit excellent performance for photocatalytic reduction of Cr(VI) in aqueous solutions with coexisting organic pollutants (such as phenol, citric acid and EDTA), under UV-vis light irradiation. The enhanced photocatalytic activity of dual component MFO-ZFO mesoporous networks is originated from the combined effect of accessible pore structure, which permits facile diffusion of reactants and products and suitable electronic band structure, which efficiently separates and transports the charge carriers through the ZFO/MFO interface.

B. Chumadathil Pookunoth et al. [44] reported on the immobilization of a 1,3-bis(benzimidazolyl) benzeneCo(II) complex on divinylbenzene cross-linked chloromethylated polystyrene, as an inexpensive polymer matrix. This particular system was tested on the arylamination reaction and showed robustness in the preparation of bioactive adamantanyl-tethered-biphenylamines. Such transition metal-catalyzed cross-coupling reactions between aryl halides and primary/secondary amines to obtain aminated aryl compounds are of particular importance due to the wide field of arylamines applications in the chemicals and pharmaceuticals.

In summary, the aforementioned special issue highlights through the fourteen novel contributions the ongoing importance of the rational design of metal oxide catalysts by means of appropriate synthesis and/or modification routes. It was clearly revealed that the fine-tuning of size, shape and electronic state through appropriate synthetic methods, special pretreatment protocols and surface/structural modification can exert a profound influence on metal's sites reactivity/stability, offering highly active and stable composites for real-life applications.

We are very pleased to serve as Guest Editors on this thematic issue involving fourteen high quality studies. In this regard, we would like to express our gratitude to editorial staff of *Catalysts*, particularly to Assistant Editor, Mrs. Adela Liao, for her efforts and continuous support. Moreover, we are most appreciative to all authors for their contributions and hard work in revising them as well as to all reviewers for their valuable recommendations that assisted authors to upgrade their work to meet high standards of *Catalysts*. We hope that this special issue will be a valuable resource for researchers, students and practitioners, to promote and advance research and applications in the field of the rational design and fabrication of cost-efficient and highly active nano-structured catalysts for energy and environmental applications.

Funding: This research has been co-financed by the European Union and Greek national funds through the Operational Program Competitiveness, Entrepreneurship and Innovation, under the call RESEARCH–CREATE–INNOVATE (project code: T1EDK-00094).

Conflicts of Interest: The authors declare no conflict of interest.

References

1. Konsolakis, M. Surface chemistry and catalysis. *Catalysts* **2016**, *6*, 102. [CrossRef]
2. Konsolakis, M. The role of Copper—Ceria interactions in catalysis science: Recent theoretical and experimental advances. *Appl. Catal. B Environ.* **2016**, *198*, 49–66. [CrossRef]
3. Konsolakis, M. Recent Advances on Nitrous Oxide (N_2O) Decomposition over Non-Noble-Metal Oxide Catalysts: Catalytic Performance, Mechanistic Considerations, and Surface Chemistry Aspects. *ACS Catal.* **2015**, *5*, 6397–6421. [CrossRef]
4. Konsolakis, M.; Stathopoulos, V. Preface on recent advances on surface and interface functionalization in Nano-Catalysis (SUR-INTER-CAT). *Appl. Surf. Sci. Adv.* **2021**, *5*, 100093. [CrossRef]
5. Konsolakis, M.; Lykaki, M. Recent advances on the rational design of non-precious metal oxide catalysts exemplified by CuO_x/CeO_2 binary system: Implications of size, shape and electronic effects on intrinsic reactivity and metal-support interactions. *Catalysts* **2020**, *10*, 160. [CrossRef]
6. Konsolakis, M.; Lykaki, M. Facet-dependent reactivity of ceria nanoparticles exemplified by CeO_2-based transition metal catalysts: A critical review. *Catalysts* **2021**, *11*, 452. [CrossRef]
7. Pandis, P.K.; Perros, D.E.; Stathopoulos, V.N. Doped apatite-type lanthanum silicates in CO oxidation reaction. *Catal. Commun.* **2018**, *114*, 98–103. [CrossRef]
8. Lykaki, M.; Pachatouridou, E.; Carabineiro, S.A.C.; Iliopoulou, E.; Andriopoulou, C.; Kallithrakas-Kontos, N.; Boghosian, S.; Konsolakis, M. Ceria nanoparticles shape effects on the structural defects and surface chemistry: Implications in CO oxidation by Cu/CeO_2 catalysts. *Appl. Catal. B Environ.* **2018**, *230*, 18–28. [CrossRef]
9. Lykaki, M.; Pachatouridou, E.; Iliopoulou, E.; Carabineiro, S.A.C.; Konsolakis, M. Impact of the synthesis parameters on the solid state properties and the CO oxidation performance of ceria nanoparticles. *RSC Adv.* **2017**, *7*, 6160–6169. [CrossRef]
10. Lykaki, M.; Stefa, S.; Carabineiro, S.A.C.; Pandis, P.K.; Stathopoulos, V.N.; Konsolakis, M. Facet-dependent reactivity of Fe_2O_3/CeO_2 nanocomposites: Effect of ceria morphology on CO oxidation. *Catalysts* **2019**, *9*, 371. [CrossRef]
11. Stefa, S.; Lykaki, M.; Binas, V.; Pandis, P.K.; Stathopoulos, V.N.; Konsolakis, M. Hydrothermal Synthesis of ZnO-doped Ceria Nanorods: Effect of ZnO Content on the Redox Properties and the CO Oxidation Performance. *Appl. Sci.* **2020**, *10*, 7605. [CrossRef]
12. Stefa, S.; Lykaki, M.; Fragkoulis, D.; Binas, V.; Pandis, P.K.; Stathopoulos, V.N.; Konsolakis, M. Effect of the preparation method on the physicochemical properties and the CO oxidation performance of nanostructured CeO_2/TiO_2 oxides. *Processes* **2020**, *8*, 847. [CrossRef]
13. Lykaki, M.; Papista, E.; Carabineiro, S.A.C.; Tavares, P.B.; Konsolakis, M. Optimization of N_2O decomposition activity of CuO-CeO_2 mixed oxides by means of synthesis procedure and alkali (Cs) promotion. *Catal. Sci. Technol.* **2018**, *8*, 2312–2322. [CrossRef]
14. Lykaki, M.; Papista, E.; Kaklidis, N.; Carabineiro, S.A.C.; Konsolakis, M. Ceria Nanoparticles' Morphological Effects on the N_2O Decomposition Performance of Co_3O_4/CeO_2 Mixed Oxides. *Catalysts* **2019**, *9*, 233. [CrossRef]

15. Konsolakis, M.; Carabineiro, S.A.C.; Papista, E.; Marnellos, G.E.; Tavares, P.B.; Agostinho Moreira, J. Romaguera-Barcelay, Y.; Figueiredo, J.L. Effect of preparation method on the solid state properties and the deN_2O performance of CuO-CeO_2 oxides. *Catal. Sci. Technol.* **2015**, *5*, 3714–3727. [CrossRef]
16. Díez-Ramírez, J.; Sánchez, P.; Kyriakou, V.; Zafeiratos, S.; Marnellos, G.E.; Konsolakis, M.; Dorado, F. Effect of support nature on the cobalt-catalyzed CO_2 hydrogenation. *J. CO_2 Util.* **2017**, *21*, 562–571. [CrossRef]
17. Konsolakis, M.; Lykaki, M.; Stefa, S.; Carabineiro, S.A.C.; Varvoutis, G.; Papista, E.; Marnellos, G.E. CO_2 Hydrogenation over Nanoceria-Supported Transition Metal Catalysts: Role of Ceria Morphology (Nanorods versus Nanocubes) and Active Phase Nature (Co versus Cu). *Nanomaterials* **2019**, *9*, 1739. [CrossRef]
18. Varvoutis, G.; Lykaki, M.; Papista, E.; Carabineiro, S.A.C.; Psarras, A.C.; Marnellos, G.E.; Konsolakis, M. Effect of alkali (Cs) doping on the surface chemistry and CO_2 hydrogenation performance of CuO/CeO_2 catalysts. *J. CO_2 Util.* **2021**, *44*, 101408. [CrossRef]
19. Varvoutis, G.; Lykaki, M.; Stefa, S.; Binas, V.; Marnellos, G.E.; Konsolakis, M. Deciphering the role of Ni particle size and nickel-ceria interfacial perimeter in the low-temperature CO_2 methanation reaction over remarkably active Ni/CeO_2 nanorods. *Appl. Catal. B Environ.* **2021**, *297*, 120401. [CrossRef]
20. Varvoutis, G.; Lykaki, M.; Stefa, S.; Papista, E.; Carabineiro, S.A.C.; Marnellos, G.E.; Konsolakis, M. Remarkable efficiency of Ni supported on hydrothermally synthesized CeO_2 nanorods for low-temperature CO_2 hydrogenation to methane. *Catal. Commun.* **2020**, *142*, 106036. [CrossRef]
21. Khataee, A.; Gholami, P.; Kalderis, D.; Pachatouridou, E.; Konsolakis, M. Preparation of novel CeO_2-biochar nanocomposite for sonocatalytic degradation of a textile dye. *Ultrason. Sonochem.* **2018**, *41*, 503–513. [CrossRef] [PubMed]
22. Khataee, A.; Kalderis, D.; Gholami, P.; Fazli, A.; Moschogiannaki, M.; Binas, V.; Lykaki, M.; Konsolakis, M. Cu_2O-CuO@biochar composite: Synthesis, characterization and its efficient photocatalytic performance. *Appl. Surf. Sci.* **2019**, *498*, 143846. [CrossRef]
23. Anucha, C.B.; Altin, I.; Bacaksiz, E.; Stathopoulos, V.N.; Polat, I.; Yasar, A.; Yüksel, Ö.F. Silver doped zinc stannate (Ag-$ZnSnO_3$) for the photocatalytic degradation of caffeine under UV irradiation. *Water* **2021**, *13*, 1290. [CrossRef]
24. Konsolakis, M.; Carabineiro, S.A.C.; Marnellos, G.E.; Asad, M.F.; Soares, O.S.G.P.; Pereira, M.F.R.; Órfão, J.J.M.; Figueiredo, J.L. Effect of cobalt loading on the solid state properties and ethyl acetate oxidation performance of cobalt-cerium mixed oxides. *J. Colloid Interface Sci.* **2017**, *496*, 141–149. [CrossRef] [PubMed]
25. Konsolakis, M.; Carabineiro, S.A.C.; Tavares, P.B.; Figueiredo, J.L. Redox properties and VOC oxidation activity of Cu catalysts supported on $Ce_{1-x}Sm_xO_\delta$ mixed oxides. *J. Hazard. Mater.* **2013**, *261*, 512–521. [CrossRef]
26. Khataee, A.; Kalderis, D.; Motlagh, P.Y.; Binas, V.; Stefa, S.; Konsolakis, M. Synthesis of copper (I, II) oxides/hydrochar nanocomposites for the efficient sonocatalytic degradation of organic contaminants. *J. Ind. Eng. Chem.* **2021**, *95*, 73–82. [CrossRef]
27. Khataee, A.; Kayan, B.; Kalderis, D.; Karimi, A.; Akay, S.; Konsolakis, M. Ultrasound-assisted removal of Acid Red 17 using nanosized Fe_3O_4-loaded coffee waste hydrochar. *Ultrason. Sonochem.* **2017**, *35*, 72–80. [CrossRef]
28. Konsolakis, M.; Carabineiro, S.A.C.; Marnellos, G.E.; Asad, M.F.; Soares, O.S.G.P.; Pereira, M.F.R.; Órfão, J.J.M.; Figueiredo, J.L. Volatile organic compounds abatement over copper-based catalysts: Effect of support. *Inorg. Chim. Acta* **2017**, *455*, 473–482. [CrossRef]
29. Carabineiro, S.A.C.; Konsolakis, M.; Marnellos, G.E.N.; Asad, M.F.; Soares, O.S.G.P.; Tavares, P.B.; Pereira, M.F.R.; de Melo Órfão, J.J.; Figueiredo, J.L. Ethyl acetate abatement on copper catalysts supported on ceria doped with rare earth oxides. *Molecules* **2016**, *21*, 644. [CrossRef]
30. Carabineiro, S.A.C.; Chen, X.; Konsolakis, M.; Psarras, A.C.; Tavares, P.B.; Órfão, J.J.M.; Pereira, M.F.R.; Figueiredo, J.L. Catalytic oxidation of toluene on Ce-Co and La-Co mixed oxides synthesized by exotemplating and evaporation methods. *Catal. Today* **2015**, *244*, 161–171. [CrossRef]
31. Bethelanucha, C.; Altin, I.; Bacaksiz, E.; Degirmencioglu, I.; Kucukomeroglu, T.; Yılmaz, S.; Stathopoulos, V.N. Immobilized TiO_2/ZnO Sensitized Copper (II) Phthalocyanine Heterostructure for the Degradation of Ibuprofen under UV Irradiation. *Separations* **2021**, *8*, 24. [CrossRef]
32. Anucha, C.B.; Altin, I.; Biyiklioglu, Z.; Bacaksiz, E.; Polat, I.; Stathopoulos, V.N. Synthesis, characterization, and photocatalytic evaluation of manganese (III) phthalocyanine sensitized $ZnWO_4$ ($ZnWO_4$MnPc) for bisphenol a degradation under uv irradiation. *Nanomaterials* **2020**, *10*, 2139. [CrossRef]
33. Anucha, C.B.; Altin, I.; Bacaksız, E.; Kucukomeroglu, T.; Belay, M.H.; Stathopoulos, V.N. Enhanced photocatalytic activity of $CuWO_4$ doped TiO_2 photocatalyst towards carbamazepine removal under UV irradiation. *Separations* **2021**, *8*, 25. [CrossRef]
34. Jirátová, K.; Pacultová, K.; Balabánová, J.; Karásková, K.; Klegová, A.; Bílková, T.; Jandová, V.; Koštejn, M.; Martaus, A.; Kotarba, A.; et al. Precipitated K-Promoted Co–Mn–Al Mixed Oxides for Direct NO Decomposition: Preparation and Properties. *Catalysts* **2019**, *9*, 592. [CrossRef]
35. Pacultová, K.; Bílková, T.; Klegova, A.; Karásková, K.; Fridrichová, D.; Jirátová, K.; Kiška, T.; Balabánová, J.; Koštejn, M.; Kotarba, A.; et al. Co-Mn-Al mixed oxides promoted by K for direct NO decomposition: Effect of preparation parameters. *Catalysts* **2019**, *9*, 593. [CrossRef]
36. Damaskinos, C.M.; Vasiliades, M.A.; Stathopoulos, V.N.; Efstathiou, A.M. The effect of CeO_2 preparation method on the carbon pathways in the dry reforming of methane on Ni/Ceo_2 studied by transient techniques. *Catalysts* **2019**, *9*, 621. [CrossRef]
37. Choya, A.; de Rivas, B.; Gutiérrez-Ortiz, J.I.; López-Fonseca, R. Comparative study of strategies for enhancing the performance of Co_3O_4/Al_2O_3 catalysts for lean methane combustion. *Catalysts* **2020**, *10*, 757. [CrossRef]

38. Li, X.; Zhang, J.; Zhang, M.; Zhang, W.; Zhang, M.; Xie, H.; Wu, Y.; Tan, Y. The support effects on the direct conversion of syngas to higher alcohol synthesis over copper-based catalysts. *Catalysts* **2019**, *9*, 199. [CrossRef]
39. Ma, Y.; Ma, Y.; Liu, M.; Chen, Y.; Hu, X.; Ye, Z.; Dong, D. Study on nanofibrous catalysts prepared by electrospinning for methane partial oxidation. *Catalysts* **2019**, *9*, 479. [CrossRef]
40. Xu, L.; Yang, Q.; Hu, L.; Wang, D.; Peng, Y.; Shao, Z.; Lu, C.; Li, J. Insights over Titanium Modified $FeMgO_x$ Catalysts for Selective Catalytic Reduction of NO_x with NH_3: Influence of Precursors and Crystalline Structures. *Catalysts* **2019**, *9*, 560. [CrossRef]
41. Wang, C.; Mao, X.; Lee, J.D.; Onn, T.M.; Yeh, Y.H.; Murray, C.B.; Gorte, R.J. A characterization study of reactive sites in ALD-synthesized WO_x/ZrO_2 catalysts. *Catalysts* **2018**, *8*, 292. [CrossRef]
42. Phul, R.; Perwez, M.; Ahmed, J.; Sardar, M.; Alshehri, S.M.; Alhokbany, N.; Majeed Khan, M.A.; Ahmad, T. Efficient multifunctional catalytic and sensing properties of synthesized ruthenium oxide nanoparticles. *Catalysts* **2020**, *10*, 780. [CrossRef]
43. Skliri, E.; Vamvasakis, I.; . Papadas, I.T.; Choulis, S.A.; Armatas, G.S. Mesoporous Composite Networks of Linked $MnFe_2O_4$ and $ZnFe_2O_4$ Nanoparticles as Efficient Photocatalysts for the Reduction of Cr (VI). *Catalysts* **2021**, *11*, 199. [CrossRef]
44. Chumadathil Pookunoth, B.; Eshwar Rao, S.; Deveshegowda, S.N.; Kashinath Metri, P.; Fazl-Ur-Rahman, K.; Periyasamy, G.; Virupaiah, G.; Priya, B.S.; Pandey, V.; Lobie, P.E.; et al. Development of a New Arylamination Reaction Catalyzed by Polymer Bound 1,3-(Bisbenzimidazolyl) Benzene Co(II) Complex and Generation of Bioactive Adamanate Amines. *Catalysts* **2020**, *10*, 1315. [CrossRef]

catalysts
MDPI

Review

Recent Advances on the Rational Design of Non-Precious Metal Oxide Catalysts Exemplified by CuO_x/CeO_2 Binary System: Implications of Size, Shape and Electronic Effects on Intrinsic Reactivity and Metal-Support Interactions

Michalis Konsolakis * and Maria Lykaki

Industrial, Energy and Environmental Systems Lab (IEESL), School of Production Engineering and Management, Technical University of Crete, GR-73100 Chania, Greece; mlykaki@isc.tuc.gr
* Correspondence: mkonsol@pem.tuc.gr; Tel.: +30-28210-37682

Received: 11 December 2019; Accepted: 22 January 2020; Published: 1 February 2020

Abstract: Catalysis is an indispensable part of our society, massively involved in numerous energy and environmental applications. Although, noble metals (NMs)-based catalysts are routinely employed in catalysis, their limited resources and high cost hinder the widespread practical application. In this regard, the development of NMs-free metal oxides (MOs) with improved catalytic activity, selectivity and durability is currently one of the main research pillars in the area of heterogeneous catalysis. The present review, involving our recent efforts in the field, aims to provide the latest advances—mainly in the last 10 years—on the rational design of MOs, i.e., the general optimization framework followed to fine-tune non-precious metal oxide sites and their surrounding environment by means of appropriate synthetic and promotional/modification routes, exemplified by CuO_x/CeO_2 binary system. The fine-tuning of size, shape and electronic/chemical state (e.g., through advanced synthetic routes, special pretreatment protocols, alkali promotion, chemical/structural modification by reduced graphene oxide (rGO)) can exert a profound influence not only to the reactivity of metal sites in its own right, but also to metal-support interfacial activity, offering highly active and stable materials for real-life energy and environmental applications. The main implications of size-, shape- and electronic/chemical-adjustment on the catalytic performance of CuO_x/CeO_2 binary system during some of the most relevant applications in heterogeneous catalysis, such as CO oxidation, N_2O decomposition, preferential oxidation of CO (CO-PROX), water gas shift reaction (WGSR), and CO_2 hydrogenation to value-added products, are thoroughly discussed. It is clearly revealed that the rational design and tailoring of NMs-free metal oxides can lead to extremely active composites, with comparable or even superior reactivity than that of NMs-based catalysts. The obtained conclusions could provide rationales and design principles towards the development of cost-effective, highly active NMs-free MOs, paving also the way for the decrease of noble metals content in NMs-based catalysts.

Keywords: copper-ceria; rational design; size; shape; electronic/chemical functionalization; CO oxidation; N_2O decomposition; preferential oxidation of CO (CO-PROX); water gas shift reaction (WGSR); CO_2 hydrogenation

1. Introduction

The fast growth rate of population in the last decades has led to an unprecedented increase in energy demands. However, the main energy sources fulfilling global demands originate from fossil fuels, rising significant concerns in relation to sources availability and environmental degradation. To this end, the development of emerging energy technologies towards the production of environmentally friendly fuels besides the establishment of cost-effective environmental technologies for climate change mitigation has become a main priority in the scientific and industrial community. Clean and reliable energy supply in conjunction with environmental protection by means of highly- and cost-effective technologies is one of the most significant concerns of the 21st century [1–4].

In view of the above aspects, heterogeneous catalysis is expected to have a key role in the near future towards sustainable development. Heterogeneous catalysis has received considerable attention from both the scientific and industrial community, as it is a field of diverse applications, including the petrochemical industry with the production of high quality chemicals and fuels, the fields of energy conversion and storage, as well as the remediation of the environment through the abatement of hazardous substances, signifying its pivotal role in the world economy [1,3,5–7].

Several types of catalysts have been employed for energy and environmental applications, which can be generally classified into: Noble metal (NMs)-based catalysts and NMs-free metal oxides (MOs), such as bare oxides, mixed metal oxides (MMOs), perovskites, zeolites, hexaaluminates, hydrotalcites, spinels, among others. Among these, NMs-based catalysts have been traditionally used in numerous processes, such as CO oxidation [8–11], nitrous oxide (N_2O) decomposition [12–17], water-gas shift reaction [18–20], carbon dioxide (CO_2) hydrogenation [21–25], etc., exhibiting high activity and selectivity. However, their scarcity and extremely high cost render mandatory the development of highly active, stable and selective catalysts that will be of low cost, nonetheless [26,27].

On the other hand, metal oxides (MOs) prepared from earth-abundant, and inexpensive transition metals have attracted considerable attention as alternatives to rare and expensive NMs, due to their particular features, such as enhanced redox properties, thermal stability and catalytic performance in conjunction to their lower cost [2,3,5,28–43]. The latter is clearly manifested in Figure 1, which schematically depicts the cost of noble metals in comparison with copper (a typical transition metal massively involved in MOs) for the past five-year period. It is evident that the price of noble metals is larger than that of copper by about four orders of magnitude.

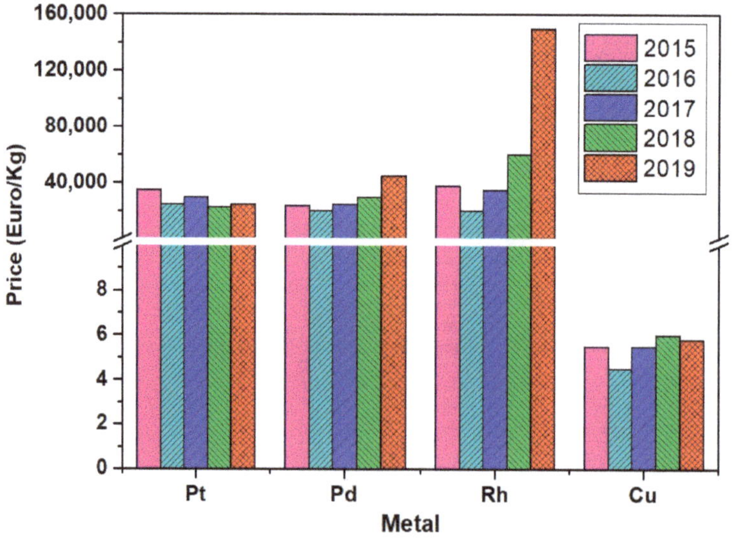

Figure 1. Relative comparison of noble metal and copper metal prices over the past five-year period. Data taken from https://www.infomine.com.

Mixed metal oxides (MMOs) appropriately prepared by admixing two or more single metal oxides in a specific proportion, have lately gained particular attention, since they exhibit unique structural and surface properties, which are completely different from that of parent oxides. Amongst the numerous MOs, transition metal-based oxides have attracted particular attention, due to their peculiar chemisorption capacity, linked to partially filled d-shells [44,45]. For instance, Cu-based oxides can catalyze a variety of reactions following a redox-type mechanism (e.g., photocatalysis), due to the wide range of Cu oxidation states (mainly Cu^0, Cu^I, Cu^{II}), which enables reactivity in multi-electron pathways. On the other hand, reducible oxides, such as ceria, not only provide the basis of active phase dispersion, but could have a profound influence on the intrinsic catalytic activity, through metal-support interactions, as will be further discussed in the sequence. In particular, ceria or cerium oxide (CeO_2) has attracted considerable attention, due to its unique properties, including enhanced thermal stability, high oxygen storage capacity (OSC) and oxygen mobility, as well as superior reducibility driven by the formation of surface/structural defects (e.g., oxygen vacancies) through the rapid interplay between the two oxidation states of cerium (Ce^{3+}/Ce^{4+}) [2,6,38,46–48]. Besides bare ceria's exceptional properties, its combination with transition metals leads to improved catalytic performance, due to the synergy between the metal phase and the support, related to electronic, geometric and bifunctional interactions [40,49–54]. In this regard, the combination of CuO_x and CeO_2 oxides towards the formation of CuO_x/CeO_2 binary oxides, offers catalytic activities comparable or even better to NMs-based catalysts in various applications, such as CO oxidation, N_2O decomposition, preferential oxidation of CO (CO-PROX), as lately reviewed [40].

The peculiar reactivity of CuO_x/CeO_2 system arises not only from the distinct characteristics of individual CuO_x and CeO_2 phases, but mainly from their synergistic interactions. More specifically, the synergistic effects between the different counterparts of MOs can offer unique characteristics (e.g., improved reducibility, abundant structural defects, etc.), reflected then on the catalytic activity [40,55–60]. Various interrelated factors are usually considered under the term "synergy", involving among others:

(i) The superior interfacial reactivity as compared to the reactivity of individual particles;
(ii) The presence of defects (e.g., oxygen vacancies);
(iii) The enhanced reducibility of MOs as compared to single ones;
(iv) The interplay between interfacial redox pairs (e.g., Cu^{2+}/Cu^+ and Ce^{3+}/Ce^{4+}).

In view of the above, very recently, the modulation of metal-support interactions as a tool to enhance the catalytic performance was thoroughly reviewed, disclosing that up to fifteen-fold productivity enhancement can be achieved in reactions related to C1 chemistry by controlling metal-support interactions [61]. However, it is well established today—thanks to the rapid development of sophisticated characterization techniques—that various interrelated factors, such as the composition, the size, the shape, and the electronic state of MOs different counterparts can exert a profound influence on the local surface chemistry and metal-support interactions, and in turn, on the catalytic activity of these multifunctional materials [6,48,51,52,62–74]. In view of this fact, the fine-tuning of MOs towards the development of catalytic materials with the desired cost, activity, selectivity and stability could be considered the "Holy Grail" in the field of catalyst manufacturing. Size, shape, porous, redox and electronic adjustment by means of appropriate synthetic and promotional/modification routes can provide the vehicle to substantially modify not only the reactivity of metal sites in its own right, but also the interfacial activity, offering highly active and stable materials for real-life energy and environmental applications (Figure 2).

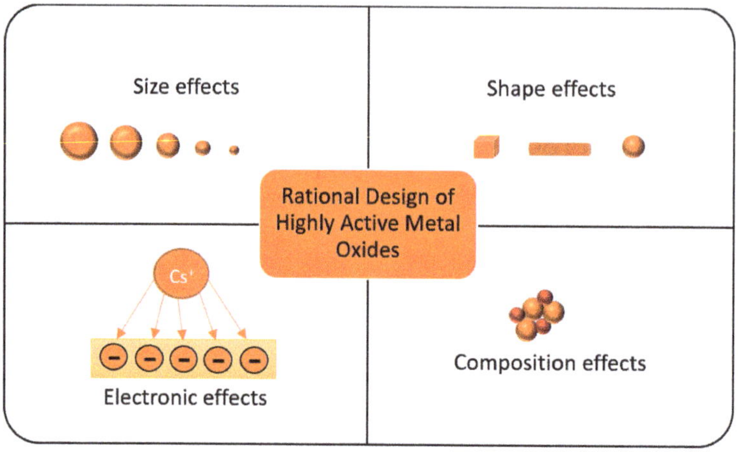

Figure 2. Schematic illustration of metal oxides (MOs) fine-tuning by adjusting their size, shape, composition and electronic/chemical state.

In light of the aforementioned issues, the present review article aims to gain insight into the particular effect of each adjusted parameter on MOs physicochemical characteristics and in turn, on their catalytic performance, towards revealing rigorous structure-property relationships. The basic principles of MOs fine-tuning by modulating the size, shape and electronic/chemical state by means of appropriate synthetic/modification routes will be initially presented. The implications of size, shape and electronic/chemical effects in catalysis will be next exemplified on the basis of state-of-the-art catalytic applications of CuO_x/CeO_2 binary oxide, involving CO oxidation, N_2O decomposition, preferential oxidation of CO (CO-PROX), water gas shift reaction (WGSR), and CO_2 hydrogenation to value-added chemicals/fuels. It should be noted here that the scope of the present article is not to provide a complete survey in relation to the fundamental understanding and practical applications of CuO_x/CeO_2 system, which can be found in various comprehensive reviews [40,49,75–77]. Herein, the CuO_x/CeO_2 system is used as an excellent benchmark to reveal how we can adjust the local surface chemistry and in turn, the catalytic activity of MOs. The obtained conclusions can provide rationales and design principles towards the development of cost-effective, highly active NMs-free MOs of various compositions, paving also the way for the decrease of noble metals content in NMs-based catalysts. The term "CuO_x" instead of CuO is used throughout the text to denote the differentiation of Cu oxidation state depending on synthesis procedure and reaction environment, as discussed in the sequence.

2. Fine-Tuning of Metal Oxides (MOs)

Heterogeneous catalysis traditionally refers to a chemical reaction on the surface of a solid catalyst, involving adsorption and activation of reactant(s) on specific active sites, chemical transformation of adsorbed species and products desorption. Thanks to the rapid development of both in situ and ex situ characterization techniques, it is well acknowledged that the elementary reaction steps are strongly dependent on several parameters involving the size, the shape, the electronic state of individual particles, as well as on their interfacial interactions. Hence, the macroscopic catalytic behavior can be considered as the outcome of interactions between reactants, intermediates and products with the micro(nano)scopic coordination environment of surface atoms, involving geometric arrangements, electronic confinement, and interfacial effects, among others. In view of this fact, the modulation of the above-mentioned parameters can profoundly affect the local surface structure and chemistry with great implications in catalysis. It should be mentioned, however, that due to the interplay between structural and chemical factors, it is quite challenging to disclose the fundamental origin of catalytic performance.

Thus, it is of vital importance to establish reliable structure-property relationships, unveiling the particular role of each factor. The latter could lead to rational design instead of trial-and-error methods by utilizing the fundamental knowledge at the nanoscale.

Moreover, taking into account that the majority of MOs consists of at least two different counterparts, this triggers unique opportunities towards designing various MOs of the same composition, but of different reactivity by adjusting the above-mentioned parameters either in one or both counterparts. For instance, by modulating the size, morphology and electronic/chemical state of ceria carrier in CuO_x/CeO_2 composites, a different extent of metal-support interactions can be attained with implications in catalysis. In a similar manner, the co-adjustment of the shape of both CuO_x and CeO_2 could lead to different synergistic interactions, providing CuO_x/CeO_2 systems of peculiar reactivity. The proposed adjusting approach on the way to fine-tune MOs is schematically illustrated in Figure 3, clearly revealing the unique opportunities in the field of materials synthesis and engineering towards the development of low-cost and highly-effective composites.

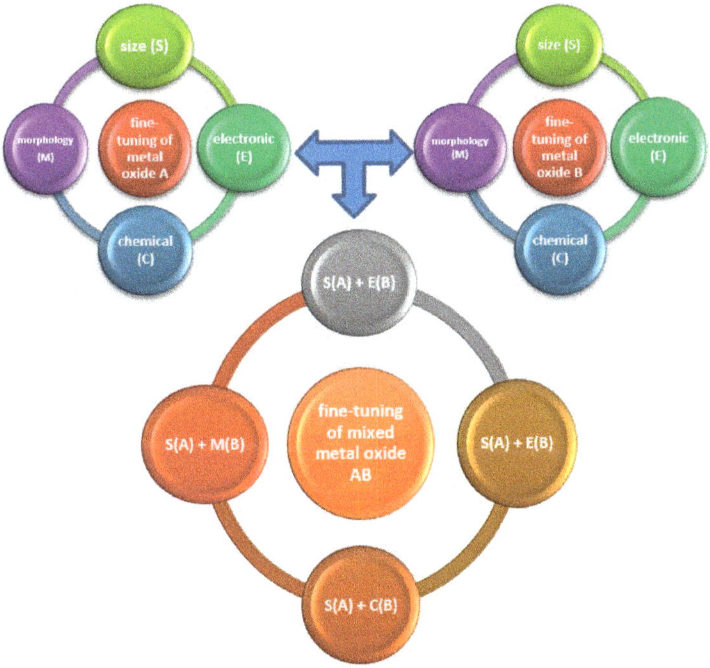

Figure 3. Indicative pathways towards the fine-tuning of a binary metal oxide of the general formula AB by adjusting the size (S), morphology (M), electronic (E) and chemical (C) state of one or both of the individual counterparts A and B. For instance, CuO_x (A)/CeO_2 (B) binary oxides of different reactivity can be obtained by combining the morphology engineering of CeO_2 {denoted as M (B)} with the size engineering of CuO_x {denoted as (S (A)} or by combining both the size and morphology engineering of both counterparts {S(A) + M(A) + S(B) + M (B)}. The scheme is just indicative of the different approaches that can be followed to adjust the local surface chemistry of MOs, without exhausting the margins of all possible combinations.

In the following, the basic principles of size, shape, and electronic/chemical effects are provided in separate sections. It should be stressed, however, that this district presentation does not also mean a district effect of each factor in catalysis. Almost all parameters are interrelated; thus, the discrete role of each one in the catalytic activity of MOs cannot be easily disclosed, as further discussed below.

2.1. *Size Effects*

The rapid development of nanotechnology in the last years enables the fabrication of MOs with tunable size and shape at the nanometer scale. Nowadays, it has been both experimentally and theoretically revealed that the surface, structural and electronic properties of nanoparticles (NPs) differ essentially from the corresponding bulk properties. In general, by decreasing the particle size of metal oxide particles down to few nanometers (e.g., <10 nm), a dramatic increase in activity can be generally obtained, attributed to "size effects". This size-dependent reactivity can be ascribed to different contributions, namely: (i) Quantum size effects, (ii) presence of low coordinated atoms into NPs surface, (iii) electronic state of the surface, (iv) strong interparticle interactions. Hereinafter, the particular effect of every contribution is shortly presented for the sake of following discussions in relation to the fine-tuning of MOs. For additional reading, several comprehensive articles in this topical area are recommended [40,52,62,66,70,71,74,78].

In particular, by decreasing the size of a material down to nanometer scale, the surface-to-volume ratio is largely increased, resulting in an increased population of surface sites, being the active sites in catalysis. Besides the modulation of the fraction of atoms on the topmost surface layer, the number of atoms at corners and edges, being considered more active than those at planes, is considerably increased by decreasing the size. More specifically, size decrease leads to a high density of under-coordinated atoms with exceptional adsorption and catalytic properties [52,62,70,79–83]. Typically, surface sites with low coordination number (CN) demonstrate stronger adsorption ability as compared to those of high CN [66,70,84]; linear relationships between the adsorption energy of various adsorbates and the coordination number have been found for several transition metals, including among others non-precious metals, such as Cu, Ni, and Co [85,86]. Thus, from the geometrical point of view, size decrease has a direct effect on both the number and type of active surface sites reflected then on catalytic activity.

Aside from the "geometric size effects", the electronic state of surface atoms can undergo substantial modifications upon decreasing the particle size down to nanometer scale. In particular, when a bulk material with a continuous electron band is subjected to size decrease down to the nanoscale, the so-called quantum effect or confinement effect is taking place, arising from the presence of discrete electronic states as in the case of molecules [62,70,74,78,87]. For instance, it has been reported that a higher electron density, with a d band close to the Fermi level, can be obtained for Au NPs smaller than ca. 2 nm as compared to bigger ones, with great implications in CO oxidation [88–91].

Recently, thanks to the introduction of new generation sophisticated characterization techniques (e.g., high-angle annular dark-field scanning transmission electron microscopy (HAADF-STEM), extended X-ray absorption fine structure (EXAFS)) and computational methods (e.g., DFT calculations), an indirect size effect linked to the metal-support interactions is clearly revealed. More specifically, even small perturbations between metal nanoparticles and oxide carriers, due to charge transfer between particles, local electric fields, morphological changes, "ligand" effect, etc., can induce a substantial modification in catalytic activity [40,50,55,92–95]. To more accurately describe these phenomena, the term Electronic Metal Support Interactions (EMSI) has recently been proposed by Campbell [96] in contrast to classical Strong Metal Support Interactions (SMSI). In view of this concept, tiny metal clusters composed of a few or even single atoms could play a dominant role in catalysis, despite the fact that they do not account for more than a few percent of the total metal content [40,50]. In view of this fact, it has been shown that by controlling metal (Ni, Pd, Pt) nanocrystal size, the length of metal-ceria interface is appropriately adjusted with significant implications in CO oxidation; normalized reaction rates were dramatically increased with decreasing the size, due to the increased boundary length (Figure 4).

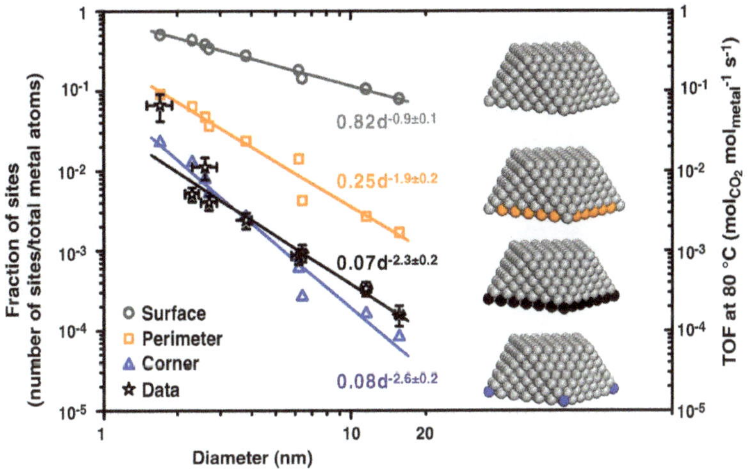

Figure 4. The calculated number of sites with a particular geometry (surface and perimeter or corner atoms in contact with the support) as a function of diameter and the turnover frequency (TOF) at 80 °C for ceria-based metals (Ni, Pd, Pt). Reproduced with permission from Reference [52]. Copyright© 2013, American Association for the Advancement of Science (AAAS).

As an additional implication of size-dependent behavior, the significant effect of particle size on structural defects of reducible carriers (such as ceria) should be mentioned. In fact, a close relationship between the crystal size of ceria and the concentration of oxygen vacancies has been revealed; the large surface-to-volume ratio in conjunction to the exposure of under-coordinated sites can facilitate the formation of oxygen vacancies and the Ce^{3+} fraction in non-stoichiometric $CeO_{2-\delta}$ NPs [71,97–103]. Moreover, an inverse correlation between the lattice parameters of CeO_2 NPs and particle sizes has been established (Figure 5a), attributed to the increase of Ce^{3+} and oxygen vacancies concentration [71]. A similar trend was recorded between the surface-to-bulk oxygen ratio and particle size (Figure 5b).

Closing this part concerning the size effects, it should be noted, that although particle size decrease has, in general, a positive catalytic effect, there is a variation in relation to size-activity relationships depending on catalyst type and reaction environment. For instance, a positive size effect could be obtained if the rate determining step (rds) involves the bond cleavage of a reactant molecule on surface atoms with low coordination number. However, if the same under-coordinated atoms strongly bind dissociated species (e.g., oxygen atoms), this could lead to the poisoning of catalyst surface, and thus, to the negative size effect. In particular, in reactions with no structure sensitivity, the activity remains unaffected by changes in the particle size, while it could decrease with decreasing particle size, referred as negative particle size effect or antipathetic structure sensitivity, or increase as the particle size decreases, referred as positive particle size effect or sympathetic structure sensitivity [79]. Moreover, the activity may reach a maximum when small particles exhibit a negative effect, and larger particles show a positive one [79].

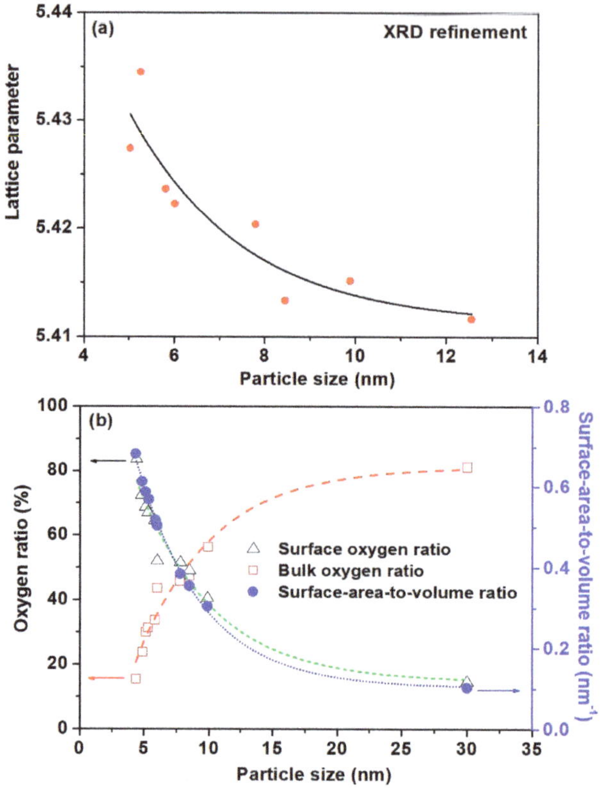

Figure 5. (**a**) Lattice expansion of ceria as a function of size and (**b**) the inverse relationship of surface oxygen to bulk oxygen (TPR) and the correlation of surface oxygen ratio with the theoretical surface to volume ratio. Adapted from Reference [71]. Copyright© 2010, Royal Society of Chemistry.

2.2. Shape Effects

Nanostructured catalysts possess unique properties originating from nanoscale phenomena linked mainly to size effects, commented above, and shape effects. The latter refers to the modification of catalytic activity through the preferential exposure of specific crystallographic facets on the reaction environment, also termed as morphology-dependent nanocatalysis [51,65,66,70,82,104–106]. In particular, the catalytic cycle and hence the reaction efficiency, is determined on reactants adsorption/activation and products desorption processes, being strongly influenced by the surface planes of catalysts particles. In this regard, the simultaneous modulation of size and shape at the nanometer scale can determine the number and the nature of exposed sites, and thus, the catalytic performance. This particular topic is an essential issue within the field of nanocatalysis, aiming to the control of a specific chemical reaction through co-adjusting these parameters at the nanometer scale.

Thanks to the latest advances in materials science, nanostructured catalysts with well-defined crystal facets can be fabricated by precisely controlling nanocrystals nucleation and growth rate [48,63,66,67,78,105]. The obtained crystal morphology is the result of several synthesis parameters, involving temperature, pressure, concentration, and pH, among others. Several reviews have been devoted to the subject [6,38,67,82,97,105,107,108]. Various structures with similar or different dimensions in all directions, such as nanospheres, nanocubes, nanowires, nanorods, nanosheets, etc., could be obtained.

The shape control of ceria and its implications in catalysis is most probably the most extensively investigated system among metal oxides in heterogeneous catalysis [2,6,38,48,51,64,65,73,82,97,104,107,109–112]. The growth rate mechanism of ceria nanocrystals can be affected by various parameters, such as the basicity or polarity of the solvent [113,114], the aging temperature [115,116], the precursor compound [117,118], and the impregnation medium [119]. Regulation of nanocrystals nucleation and growth processes results in specific shapes, such as rods and cubes [48,67]. Moreover, by altering the physicochemical conditions during the synthesis procedure (e.g., by the usage of a capping agent), blocking of certain facets or continuous growth of others may occur.

There are several synthetic approaches for the preparation of ceria nanoparticles, including precipitation [119–121], thermal decomposition [121,122], template or surfactant-assisted method [123–126], microwave-assisted synthesis [127–129], the alcohothermal [124,130] or hydrothermal [121,124,131–135] method, microemulsion [133,136,137], solution combustion [138,139], sol-gel [140–142], sonochemical [143,144], etc. However, not all methods lead to particles of well-ordered size and shape with uniform dispersion on the catalyst surface [145]. Among the different methods, the hydrothermal one has attracted considerable attention, due to the simplicity of the precursor compounds used, the short reaction time, the homogeneity in morphology, as well as the acquisition of various nanostructures, such as rods, polyhedra, cubes, wires [107,117,121,124,131,134,135,146–159]. Ceria nanocrystals have three low-index lattice facets of different activity and stability, namely, {100}, {110}, {111} [48,106], as shown in Figure 6.

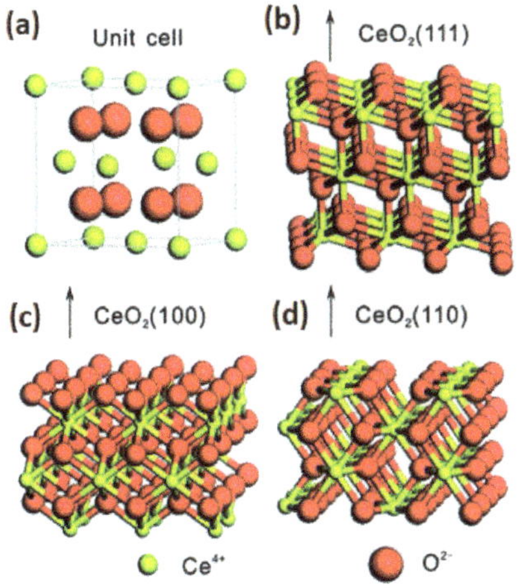

Figure 6. Structural models of CeO_2 (**a**) unit cell, (**b**) (111), (**c**) (100), and (**d**) (110) surfaces without structural optimizations. Reproduced with permission from Reference [106]. Copyright© 2014, Royal Society of Chemistry.

The selective exposure of ceria reactive facets can strongly affect the redox properties of ceria and in turn, its intrinsic characteristics as an active phase or supporting carrier. Popular ceria shapes, mainly, involve nanorods (NR), nanocubes (NC) and nanopolyhedra (NP). Ceria nanorods, mostly, expose the {110} and {100} facets, whereas, nanocubes and nanopolyhedra preferentially expose the {100} and {111} facets, respectively [48,51,106]. By means of both experimental [82,104,109,146,160,161] and theoretical [112,162–166] studies, it was shown that the energy formation of anionic vacancies is dependent on the exposed facets, following the order: {111} > {100} > {110}. In this regard, the reactivity of ceria NR is, in general, increased upon increasing the fraction of {110} and {100} facets [65].

In view of the above, it has been clearly revealed that the activity and selectivity are strongly affected by the exposed crystal planes. For instance, the formation rate of ammonia on Fe crystals follows the sequence: {111} >> {100} > {110} [167]. Similar morphology-dependent effects have been demonstrated for several noble metal [105,168] and metal oxide [51] catalyzed processes.

Very recently, we showed that among ceria nanoparticles of different morphology (nanocubes, nanorods and nanopolyhedra), as shown in Figure 7a, ceria nanorods with {100} and {110} crystal planes, exhibited the optimum redox properties in terms of loosely bound oxygen species population [115,121]. Figure 7b depicts the H_2 temperature-programmed reduction (H_2-TPR) profiles of ceria nanocubes, nanopolyhedra and nanorods, where two main peaks at ca. 550 and 800 °C can be identified, ascribed to the reduction of surface (O_s) and bulk (O_b) oxygen, respectively. Notably, the surface-to-bulk (O_s/O_b) ratio is strongly dependent on ceria morphology following the order: NC (0.71) < NP (0.94) < NR (1.13). It is also worth mentioning that the onset reduction temperature follows the reverse order, i.e.,: NC (589 °C) > NP (555 °C) > NR (545 °C), implying the lower temperature reduction of {110} and {100} surfaces as compared to {111}.

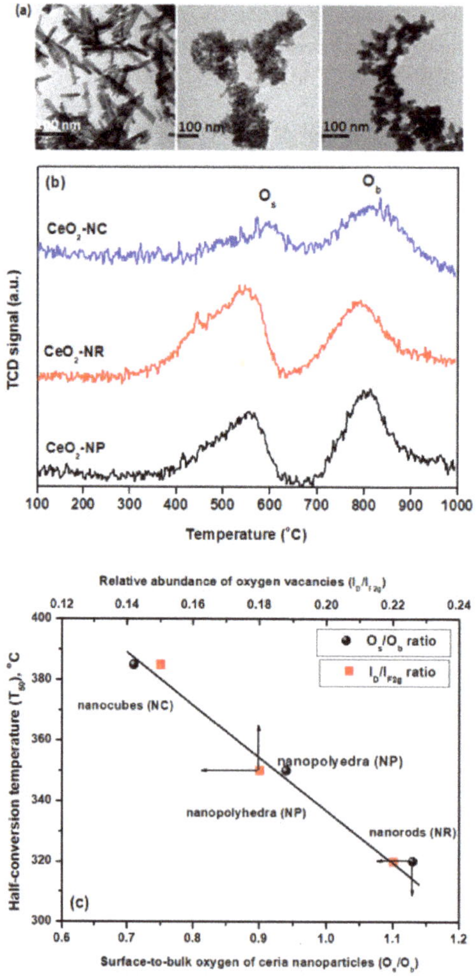

Figure 7. (**a**) TEM images and (**b**) H_2-TPR profiles for ceria nanoparticles of different morphology, i.e., nanorods, nanopolyhedra and nanocubes; (**c**) CO oxidation performance (T_{50}) of ceria nanoparticles as a function of ceria redox properties expressed in terms of surface-to-bulk (O_s/O_b) TPR ratio and I_D/I_{F2g} Raman ratio [115].

More recently, we explored the impact of ceria exposed facets on structural defects by means of in situ Raman spectroscopy following a novel approach involving sequential spectra acquisition under alternating oxidizing and reducing atmospheres [115]. The in situ Raman measurements perfectly corroborated the aforementioned arguments in relation to the impact of crystal planes on the reducibility; the relative abundance of defects and oxygen vacancies exhibited by the I_D/I_{F2g} ratio, as well as the relative reducibility expressed by the detachment of O atoms and the partial $Ce^{4+} \rightarrow Ce^{3+}$ reduction, follow the same trend, i.e., NR > NP > NC.

These results unambiguously indicate that CeO_2-nanorods exhibit the highest concentration of weakly bound oxygen species, linked to enhanced reducibility and oxygen mobility. Interestingly, an almost linear relationship is revealed between the redox properties, expressed either as O_s/O_b or I_D/I_{F2g} ratio, and the CO oxidation performance, in terms of half-conversion temperature (T_{50}), of ceria nanoparticles (Figure 7c), clearly demonstrating the implications of shape modulation in catalysis.

2.3. Electronic Effects

Besides modulating the local surface structure of MOs by size and shape effects, described above, the fine-tuning of electronic structure by appropriate promoters can be considered as an additional modulating tool. Promoters hold a key role in heterogeneous catalysis towards optimizing the catalytic activity, selectivity and stability by modifying the physicochemical features of MOs, and can be classified into two general categories: Structural promoters and electronic promoters. The first category mainly involves the doping of supporting carrier to enhance its structural characteristics and in turn, the stabilization of active phase (e.g., incorporation of rare earth dopants into three-way catalysts [5]). On the other hand, electronic promoters can modify catalysts surface chemistry either directly or indirectly. The former mainly includes the electrostatic interactions between the reactant molecules and the local electric field of promoters. The latter denotes the promoter-induced modifications on metal Fermi level, which is then reflected on the chemisorptive bond strength of reactants and intermediates with great consequences in catalysis. In particular, "promoter effect" is related to the changes in the work function (Φ) of the catalysts surface upon promoter addition, accompanied by substantial modification of its chemisorption properties. The vast majority of electronic promotion over metal oxide catalysts refers to alkali modifiers. It has been well documented that alkali addition can drastically enhance the activity and selectivity of numerous catalytic systems, involving among others Pt-, Pd-, Rh-, Cu-, Fe-based catalysts, in various energy and environmental related reactions (e.g., [169–174]). Various comprehensive studies have been devoted to the role of promoters in heterogeneous catalysis, to which the reader can refer for further reading [175–177].

Figures 8 and 9 depict the "promoter effect" in the case of alkali-doped Co_3O_4 oxides during the N_2O decomposition [178,179]. A close relationship between the catalytic performance (in terms of half-conversion temperature, T_{50}) and the work function (Φ) was disclosed revealing the electronic nature of alkali promotion; electropositive modifiers (such as alkalis) can decrease the work function of the catalyst surface, thus, activating the adsorption/decomposition of electron-acceptor molecules (such as N_2O) [178]. However, at high alkali coverages, depolarization occurs, due to the strengthening of the alkali-alkali bond at the expense of the alkali-surface bond, resulting in a work function increase [180].

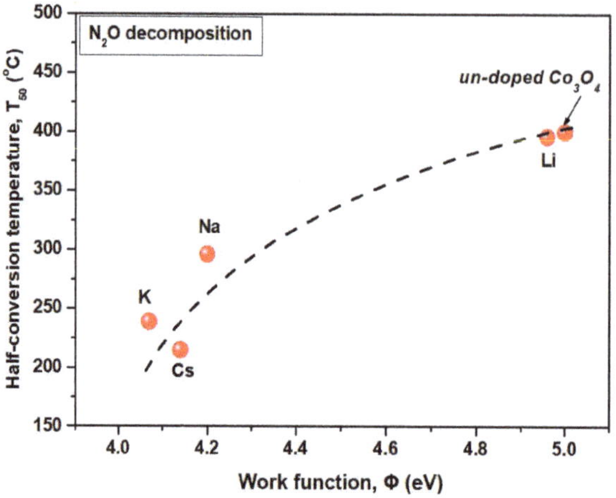

Figure 8. Correlation of half-conversion temperature (T_{50}) with the work function of alkali promoted Co_3O_4. Reaction conditions: 5.0% N_2O; m_{cat} = 300 mg; GHSV = 7000 h^{-1}; alkali coverage = ~2 at/nm^2. Adapted from Reference [178]. Copyright© 2009, Elsevier.

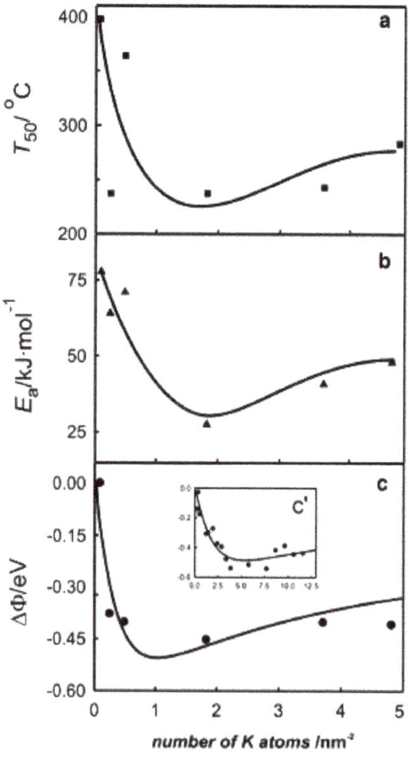

Figure 9. (**a**) The half-conversion temperature of N_2O (T_{50}), (**b**) apparent activation energy (E_a) and (**c**) work function changes ($\Delta\Phi$) as a function of potassium loading (Θ_K) introduced from K_2CO_3 and (c insert) KOH precursor. Reproduced with permission from Reference [179]. Copyright© 2008, Springer Nature.

In this point, it should be mentioned that, depending on the support nature and crystal planes, alkali adsorption may lead to surface reconstruction. This surface reconstruction can be explained by taking into account the structural/electronic perturbations induced by the formation of the alkali-surface bond [181]. As mentioned previously, the crystallographic orientation of the support plays an important role in the diffusion rate of the adsorbed species, as well as in their in-between interaction, resulting in different structural stabilization [181]. For instance, potassium promoter was shown to stabilize certain iron facets in K-promoted iron catalysts, by inducing changes in the crystal growth rate, thus, enabling the formation of small particles with abundance in active facets and affecting the activity and selectivity of the overall system [182]. As shown in Figure 10, by increasing the K/Fe surface atomic ratio, the crystal facets become more stable and the surface energy is decreased [182]. This clearly manifests the pivotal role of alkali addition towards co-adjusting the structural and electronic properties of the catalyst surface, and in turn, the catalytic performance.

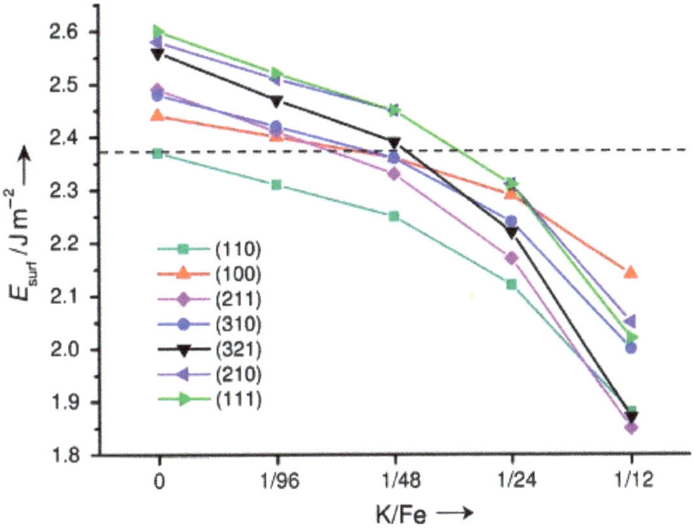

Figure 10. Surface energy variation versus the surface atomic ratio of K to Fe. Reproduced with permission from Reference [182]. Copyright© 2011, John Wiley and Sons.

Besides the alkali-induced modifications on the chemisorption properties, significant alterations on the surface oxygen mobility have been demonstrated; alkali addition could facilitate suprafacial recombination of oxygen towards molecular oxygen desorption, thus, liberating active sites [40]. Both the electronic and redox modifications induced by alkali addition can exert remarkable effects on catalytic activity and selectivity, demonstrating the key role of "promoter effect" as an additional adjusting parameter in catalysis.

It should be noted here, as mentioned in the case of size and shape effects, that the promoter effect is not always positive. The latter strongly depends on reactants type (electron donor or electron acceptor adsorbates) and the work function changes (increase or decrease) induced by the promoter (electropositive or electronegative). Thus, besides the structure-sensitivity, commented above, the electronic-sensitivity of a given reaction should always be taken into account, when attempting to co-adjust the size, shape and electronic state. Although, this is not an easy task, it could be an effective approach towards designing cost-effective and highly active composites, as revealed in the sequence.

2.4. Chemical Modifiers

Besides the extensive use of alkalis or alkaline earths as promoters, numerous other chemical substances can be employed to modulate the local surface chemistry/structure and in turn, the activity, selectivity and long-term stability of parent catalyst (e.g., [74]). In this regard, metal alloys (e.g., Au-Ni alloys as reforming catalysts [183], Pt-Sn alloys for ethanol oxidation [184]) are extensively employed in catalysis towards obtaining highly active and cost-effective catalytic formulations. Several mechanisms are considered responsible for the enhanced performance of bimetallic systems, involving mainly structural (strain effects) and electronic (charge-transfer effects) modifications that can be induced by the interaction between the different counterparts. The latter substantially modifies the binding energy of adsorbates and the path of chemical reactions with major consequences in catalysis [50,74].

In a similar manner, chemical substances with unique physico-chemical properties, such as carbon-based materials, have lately received considerable attention as chemical modifiers or supporting carriers [62,185]. Various carbon materials, such as carbon nanotubes (CNTs), reduced graphene oxide (rGO), ordered mesoporous carbon (OMC), carbon nanofibers (CNFs), and graphitic carbon nitride (g-C_3N_4), have received particular attention in catalysis after the significant progress in controlled synthesis and the fundamental understanding of their properties. In general, nanocarbons (NCs) possess unique physical (large surface area, specific morphology, appropriate pore structure) and chemical (electronic structure, surface acidity/basicity) properties arising from their nanoscale confined structures [185].

The combination of metal nanoparticles (NPs) with carbon materials by means of various synthetic approaches can exert significant modifications on the structural and electronic surrounding of NPs with subsequent implications in catalysis [62,185]. As for example, confined Fe NPs in CNTs exhibit an almost twice yield to C_{5+} hydrocarbons as compared to Fe particles during the syngas conversion to liquid hydrocarbons [186]. The latter was mainly ascribed to the modified structural and redox properties of confined Fe NPs within CNTs [62]. Moreover, the application of graphene in catalysis allows the fabrication of multifunctional materials with distinct heterostructures, which offer quite different properties as compared to individual materials [185,187,188]. In general, carbon materials with exceptional structural and electronic characteristics can be effectively employed either as supporting materials or chemical modifiers, offering unique opportunities towards modulating the intrinsic reactivity of MOs. For instance, it has been found that the homogeneous distribution of copper atoms on the surface of rGO in combination with the outstanding electronic properties of rGO lead to high electrocatalytic activity, due to the synergy between the two components [189].

Metal-organic frameworks (MOFs) are another type of supporting carriers/chemical modifiers consisting of inorganic metal ions or clusters that are bridged with organic ligands in order for one or more dimensional configurations to be formed [190]. These materials exhibit unique properties, such as high surface area and porosity, while their complex network consisting of various channels allows passage in small molecules [191]. The fabrication of MOF-based MOs composites is of great interest, as it results in the development of materials with tunable properties and functionality. Metal nanoparticles regarded as the active centers can be stabilized by MOFs through confinement effects [192]. As for example, Cu, Ni, and Pd nanoparticles encapsulated by MOFs exhibited high catalytic efficiency, ascribed mainly to the synergistic effects of nanoconfinement and electron-donation offered by MOF framework [193–198]. Furthermore, by changing the MOFs functional groups, products distribution may differ, as a consequence of variations induced in the chemical environment of the catalytically active sites [199].

2.5. Pretreatment Effects

Besides the advances that can be induced by adjusting the size, shape and electronic state of MOs, special pretreatment protocols or activation procedures could be applied to further adjust the local surface chemistry of MOs (e.g., [200,201]). In particular, the local surface chemistry of the MOs can be further tailored by appropriate pretreatment protocols, including thermal or chemical pretreatment. According to the pretreatment protocol followed, different properties get affected, resulting in diversified catalytic behavior. By way of illustration, it has been reported that defect engineering by a low-pressure thermal process instead of atmospheric pressure activation, could notably increase the concentration of oxygen vacancy defects and in turn, the CO oxidation activity of ceria nanoparticles, offering an additional tool towards the fine-tuning of MOs [200]. Moreover, it has been documented that the pretreatment protocol (oxidation or reduction) induces significant effects on the local surface structure of cobalt-ceria oxides affecting the dehydroxylation process in ammonia synthesis [202]. In a similar manner, oxidative pretreatment of cobalt-ceria catalysts resulted in an impoverishment of catalyst surface in cobalt species, due to the preferential existence of cerium species on the outer surface, whereas, cobalt and cerium species are uniformly distributed on the catalyst surface through the reduction pretreatment, which gives rise to the formation of oxygen vacancies [33]. In addition, a strong interaction between gold and ceria has been observed after O_2 pretreatment, due to the electron transfer from Au^0 to ceria, giving rise to oxygen vacancy formation, lattice oxygen migration, as well as to the formation of $Au^{\delta+}$-CO and surface bicarbonate species, favoring, thus, the adsorption of CO and the desorption of CO_2 [203]. In terms of T_{100}, CO oxidation performance showed the following order: O_2 pretreatment (74 °C) < N_2 pretreatment (142 °C) < 10% CO/Ar pretreatment (169 °C) [203]. In view of the above short discussion, the pretreatment conditions can affect the facilitation with which certain active species are formed on the catalyst surface, the oxygen mobility or the formation of oxygen defects, with great implications in the catalytic performance.

3. Implications of MOs Fine-Tuning in Catalysis Exemplified by CuO_x/CeO_2 Binary System

In this section, the implications of metal oxides fine-tuning by means of the above-described size, shape and electronic/chemical effects are presented, on the basis of the CuO_x/CeO_2 binary oxide system. This particular catalytic system is selected as representative MOs, taking into account the tremendous fundamental and practical attention lately devoted to the copper-containing cerium oxide materials. More specifically, the abundant availability of copper and ceria and consequently, their lower cost (about four orders of magnitude, Figure 1) render CuO_x/CeO_2 composites strongly competitive. Moreover, their excellent reactivity—linked to peculiar metal-support interactions—in conjunction to their remarkable resistance to various substances, such as carbon dioxide, water and sulfur is of particular fundamental and practical importance [57,76,204]. Remarkably, copper-containing ceria catalysts appropriately adjusted by the aforementioned routes demonstrated catalytic activity similar or even better than NMs-based catalysts in various applications, such as CO oxidation, the decomposition of N_2O and the water-gas shift reaction, among others [115,124,159,205–216].

For instance, the inverse $CeO_x/Cu(111)$ system exhibits superior CO oxidation performance at a relatively low-temperature range (50–100 °C), in which the noble metals do not function well, exhibiting activity values of about one order of magnitude higher than those measured on Pt(100), Pt(111), and Rh(111) [217–219]. The latter has been mainly attributed, on the basis of the most conceptual experimental and theoretical studies, to the existence of Ce^{3+} at the metal-oxide interface which binds O atoms weaker as compared to bulk Ce^{3+} [217,220].

In light of the above aspects, in this section, the main implications of size, shape and electronic/chemical effects on the catalytic performance of CuO_x/CeO_2 system during some of the most relevant applications in heterogeneous catalysis will be discussed. It should be stressed that it is not the aim of this section to provide an extended overview of CuO_x/CeO_2 catalytic applications, which can be found in several comprehensive reviews [3,49,57,75,76,221]. It mainly aims to provide a general optimization framework towards the development of highly active and cost-effective MOs, paving also the way for the decrease of precious metal content in NMs-based catalysts.

3.1. CO Oxidation

CO oxidation is probably the most studied reaction in heterogeneous catalysis, due to its practical and fundamental importance. The catalytic elimination of CO is of great importance in various applications involving, among others, automotive exhaust emissions control and fuel cell systems. More importantly, CO oxidation can serve as a prototype reaction to gain insight into the structure-property relationships.

Regarding, at first, the CO oxidation activity of individual CuO_x phase, it has been clearly revealed that it is strongly dependent on oxidation state, size and morphology. In particular, the following activity order: Cu_2O > metastable cluster CuO > CuO has been revealed, closely related to the ability to release lattice oxygen [222,223]. On the other hand, the exposed crystal planes of CuO_x phase drastically affect the CO oxidation; truncated octahedral Cu_2O with {332} facets displayed better activity than low index {111} and {100} planes [224]. Similarly, CuO with exposed {011} planes is more active that close-packed {111} planes [225]. In view of this fact, it has been found that the CO oxidation activity of CuO mesoporous nanosheets with high-index facets is about 35 times higher than that of the commercial sample [226]. In general, surface vacancies, originated from coordinately unsaturated surface Cu atoms, can easily activate oxygen species towards their reaction with the reducing agent [3].

In a similar manner, theoretical and experimental studies have shown that the energy of anionic vacancies formation over bare ceria follows the order: {111} > {100} > {110}, as previously analyzed [82,104,109,112,146,160–166]. Moreover, a large increase in oxygen vacancies concentration has been found for ceria crystal size lower than ca. 10 nm [98], revealing the interrelation between size and shape effects. In this regard, we recently showed that ceria nanorods with {100} and {110} exposed facets demonstrated the optimum CO oxidation activity amongst ceria samples of different morphology; a close relationship between crystal planes-oxygen exchange kinetics-CO oxidation activity was disclosed [115].

In view of the above aspects, it could be argued that by adjusting the shape and size of individual counterparts of MOs (CuO_x and CeO_x in the case of CuO_x/CeO_2 mixed oxides), significant modifications in their redox and catalytic properties can be obtained. However, in view of the fact that "the whole is more than the sum of its part", the majority of catalytic studies in heterogeneous catalysis is based on CuO_x/CeO_2 mixed oxide than on individual oxides [52,227,228]. The underlying mechanism of this synergistic effect linked to metal-support interactions is the subject of numerous theoretical and experimental studies in catalysis. The latest advances in the field of CuO_x-CeO_2 interactions and their implications in catalysis have been recently reviewed by one of us [40]. In general, the superiority of binary oxides can be ascribed to various interrelated phenomena, involving among others: (i) Electronic perturbations between nanoparticles, (ii) redox interplay between interfacial sites, (iii) facilitation of the formation of structural defects, (iv) improved reducibility and oxygen mobility, (v) unique reactivity of interfacial sites [40]. However, all of these factors are closely related with the intrinsic and extrinsic characteristics of individual oxides, triggering unique opportunities towards the development of highly active MOs by engineering the size and shape of individual oxides and in turn, the interfacial reactivity . Moreover, chemical or electronic effects induced by aliovalent doping can exert a profound influence on the catalytic performance, offering an additional tool towards the rational design of MOs (Figure 2). In the sequence, the optimization of CO oxidation activity of CuO_x/CeO_2 catalysts by means of the above-mentioned approaches is presented, as an indicative example of MOs rational design.

Recently, we thoroughly explored the impact of ceria nanoparticles shape effects on the CO oxidation activity of CuO_x/CeO_2 catalysts. The results clearly revealed the significant role of morphology in the CO oxidation activity, following the order: Nanorods (NR) > nanopolyhedra (NP) > nanocubes (NC), Figure 11. However, more importantly, CuO incorporation to different ceria carriers boosted the catalytic performance, without affecting the order observed for bare CeO_2 (Figure 11), demonstrating the crucial role of support.

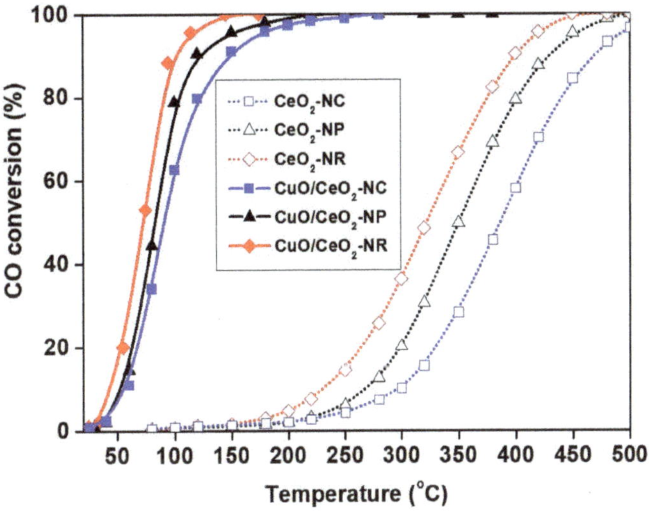

Figure 11. CO conversion as a function of temperature for bare CeO_2 and CuO_x/CeO_2 samples of different morphology (NR, NC and NP, as indicated in each curve). Reaction conditions: 2000 ppm CO, 1 vol.% O_2, GHSV = 39,000 h^{-1} [115].

The CuO_x/CeO_2-NR sample exhibited a half-conversion temperature (T_{50}) of ca. 70 °C, which is much lower to that required for a typical noble metal oxidation catalyst, such as Pt/Al_2O_3 (T_{50} = 230 °C), as shown in Figure 12. Based on a thorough in situ and ex situ characterization study, a perfect relationship between the CO oxidation performance and the following parameters was disclosed: (i) Relative population of Cu^+/Ce^{3+} redox pairs, (ii) abundance of loosely bound oxygen species, expressed in terms of surface-to-bulk oxygen reducibility, (iii) relative concentration of oxygen vacancies, evidenced by the I_D/I_{F2g} Raman ratio (Figure 7c) [115]. Similar conclusions in relation to the key role of ceria morphology in the CO oxidation activity have been reported by several groups [117,134,154,158,210,212,229–231], most of these revealing the superiority of nanorods.

However, it should be noted that similar or even better catalytic activities can be obtained by different morphologies (e.g., [124,211,212]). In this regard, it was recently shown that sub-nanometer copper oxide clusters (1 wt.% Cu loading) deposited on ceria nanospheres (NS) exhibited superior performance as compared to that deposited on nanorods (T_{100} = 122 °C vs. 194 °C) [124]. Extensive characterization investigations revealed that the copper species in nanorods samples existed in both Cu-[O_x]-Ce and CuO_x clusters, while CuO_x clusters dominated in nanospheres. Among these species, only CuO_x clusters could be easily reduced to Cu(I) when they were subjected to interaction with CO, which is considered to be the reason of the enhanced reactivity of CuO_x/CeO_2-NS samples [124].

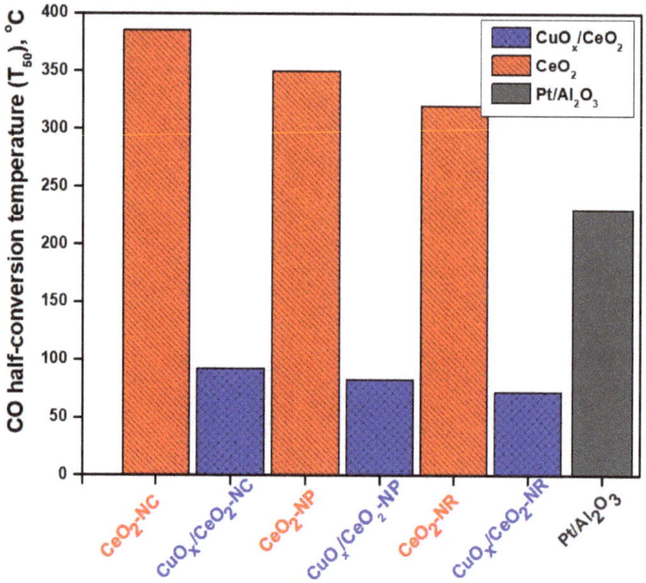

Figure 12. A comparison between bare ceria, copper-ceria and a noble metal catalyst, in terms of half-conversion temperature (T_{50}) for the oxidation of CO. Reaction conditions: 2000 ppm CO, 1 vol.% O_2, GHSV = 39,000 h^{-1}.

So far, numerous synthesis routes and different precursors have been employed to adjust the structural and morphological characteristics of CuO_x/CeO_2 composites, mostly summarized by Prasad and Rattan [76]. For instance, it has been found that the use of Ce(III) instead of Ce(IV) precursors can lead to CuO_x/CeO_2 catalysts with superior reducibility and CO oxidation activity [232]. In particular, it was experimentally shown that Ce(III)-derived samples contained a higher amount of Cu^+ species, through the redox equilibrium $Cu^{2+} + Ce^{3+} \leftrightarrow Cu^+ + Ce^{4+}$, which are responsible for their enhanced oxidation performance [232]. Moreover, CuO_x/CeO_2 samples prepared from copper acetate precursor demonstrated better CO oxidation performance as compared to those prepared from nitrate, chloride and sulfate precursors [233]. Avgouropoulos and co-workers [234,235] recently employed a novel hydrothermal method for the synthesis of atomically dispersed CuO_x/CeO_2 catalysts, offering high CO oxidation performance. By means of various complementary techniques, it was shown that the catalytic activity is mainly related to the nature of highly dispersed copper species rather than the structural/textural characteristics. In a similar manner, Elias et al. [236] reported on the facile synthesis of phase-pure, monodisperse ~3 nm $Cu_{0.1}Ce_{0.9}O_{2-x}$ crystallites via solution-based pyrolysis of heterobimetallic Schiff complexes. An increase of CO oxidation activity by one and three orders of magnitude compared to ceria nanoparticles (3 nm) and microparticles (5 μm), respectively, was attained (Figure 13).

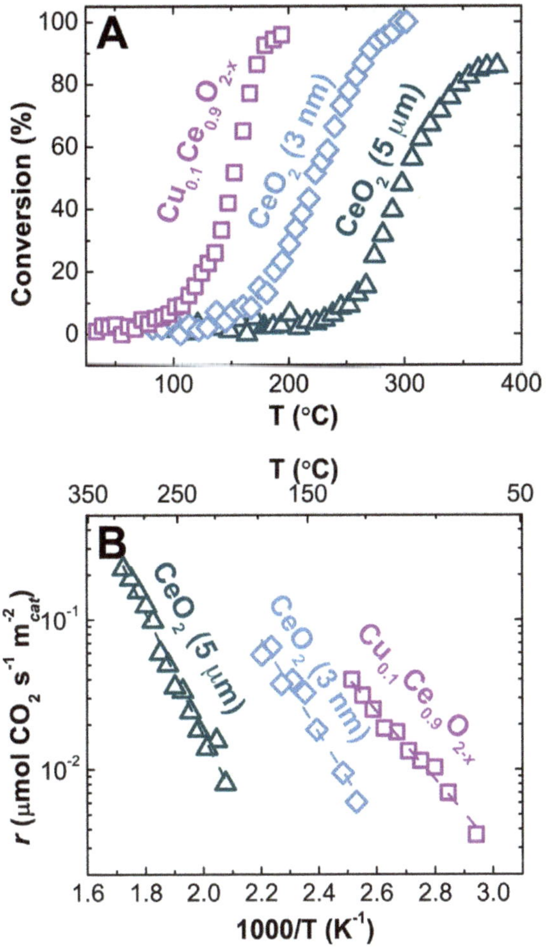

Figure 13. CO oxidation on annealed 3 nm $Cu_{0.1}Ce_{0.9}O_{2-x}$, 3 nm CeO_2 and commercial 5 μm CeO_2 (Sigma-Aldrich). (**A**) "Light off" curves and (**B**) area-normalized Arrhenius plots, measured in 1.0% CO, 2.5% O_2 balanced in He at a flow rate of 1300 mL min^{-1} g_{cat}^{-1} for 20 mg catalyst loadings. Reproduced with permission from Reference [236]. Copyright© 2014, American Chemical Society.

Besides the engineering of shape and size, porous structure engineering could exert a significant influence on the CO oxidation activity of CuO_x/CeO_2 catalysts [123,213]. As for example, three-dimensional CuO_x-doped CeO_2 prepared by a hard template method exhibited complete CO conversion at temperatures as low as 50 °C, due to their improved textural and redox properties [213].

Regarding the influence of CuO_x/CeO_2 composition on the CO oxidation activity, most of the studies revealed an optimum Cu/(Cu+Ce) atomic ratio in the range of 15–30% [212,231,232,237,238]. Within this specific range, the optimum physicochemical characteristics and interfacial interactions can be achieved, reflected then on catalytic activity.

Apart from the above-described approaches that have been put forward to improve the CO oxidation performance of CuO_x/CeO_2 oxides, the addition of aliovalent elements as structural/surface promoters should be mentioned. In view of this fact, it has been found that the modification of ceria support by Mn [239] or Sn [240] can drastically modify the dispersion of CuO_x and the redox interplay between Cu species and support, thus, enhancing the CO oxidation performance. Very recently, the tuning of the interfacial properties of CuO_x/CeO_2 by In_2O_3 doping was also explored [241]. It was found that the CO oxidation performance of In_2O_3-CuO_x/CeO_2 sample greatly exceeds that of parent oxide, offering complete CO conversion at temperatures as low as 100 °C [241]. By means of complementary characterization studies and density functional theory calculations, it was proved that In_2O_3 could modify the geometric structure of CuO_x particles by reducing their size. The latter results in more metal-support interfacial sites and abundant defects. Moreover, the interaction between In and Cu could modify the electronic state of Cu atoms towards the stabilization of partially reduced Cu sites at the interface [241].

Recently, copper-ceria nanosheets were synthesized by using graphene oxide as a sacrificial template, in an attempt to increase the concentration of active interfacial sites [242]. The copper-ceria interaction was further adjusted by appropriate pretreatment, with the catalyst calcined at 600 °C exhibiting complete CO conversion at 90 °C, due to the high concentration in active copper species and oxygen vacancies [242]. Moreover, a sword-like copper-ceria composite derived by a Ce-based MOF with 5 wt.% Cu loading, exhibited superior CO conversion performance (T_{100} = 100 °C) in comparison to other irregular-shaped catalysts, due to the good interfacial contact, which resulted in the abundance of Cu^+ active species and oxygen vacancies [191]. Very recently, triple-shelled CuO_x/CeO_2 hollow nanospheres were synthesized by MOFs, exhibiting high CO conversion performance (T_{100} = 130 °C) [243]. This was mainly ascribed to the porous structure of the triple-shelled morphology, offering an enhanced synergistic interaction between copper and ceria [243].

Table 1 summarizes, at a glance, indicative attempts followed to adjust the interfacial properties and in turn, the CO oxidation performance of CuO_x/CeO_2 binary oxides. It is evident that extremely active composites can be obtained by adjusting the shape, size and electronic/chemical state by means of appropriate synthetic and/or promotional routes. It is of worth pointing out the superiority of finely-tuned CuO_x/CeO_2 samples as compared to noble metal-based catalysts, offering unique opportunities towards the rational design of highly active metal oxide catalysts. Moreover, as further guidance, it would be of particular importance to explore the combining effect of different adjusted parameters (e.g., CuO_x/CeO_2 nanorods co-doped with main-group elements) towards further optimization.

Table 1. Indicative studies towards adjusting the CO oxidation performance of CuO_x/CeO_2 oxides.

Reaction Conditions	Adjusted Parameter (Employed Method)	Optimum System	T_{50} (°C)	Reference
0.2% CO + 1.0% O_2; WHSV = 75,000 mL g^{-1} h^{-1}; GHSV = 39,000 h^{-1}	shape/size (hydrothermal synthesis)	8.5 wt.% Cu/CeO_2-nanorods	72	[115]
1.0% CO + 15.0% O_2; WHSV = 7200 mL g^{-1} h^{-1}	shape/size (hydrothermal synthesis)	15 wt.% Cu/CeO_2-nanorods CeO_2 (15 nm), CuO (6.0 nm)	50	[210]
1.0% CO + 20.0% O_2; WHSV = 80,000 mL g^{-1} h^{-1}	shape/size (alcothermal method)	1.0 wt.% Cu/CeO_2-nanospheres (~130–150 nm spheres comprised of 2–5 nm nanoparticles)	85	[124]
1.0% CO + 2.5% O_2; WHSV = 78,000 mL g^{-1} h^{-1}	size/structure (solution-based pyrolysis of heterobimetallic Schiff complexes)	$Cu_{0.1}Ce_{0.9}O_{2-x}$ monodisperse nanoparticles (~3.0 nm)	150	[236]
1.0% CO + 10.0% O_2; WHSV = 60,000 mL g^{-1} h^{-1}	size/structure (thermolytic decomposition in the presence of capping agent)	9.0 at.% Cu/CeO_2 CeO_2 (3.3 nm)	85	[211]
1.0% CO, air balance; WHSV = 30,000 mL g^{-1} h^{-1}	size/structure (hydrothermal treatment)	Cu/CeO_2-nanospheres (Cu/(Cu+Ce) = 0.33, spherical particles of 300–400 nm diameter composed of nanoparticles of ca. 10 nm)	70	[212]
0.24% CO + 15.0% O_2; WHSV = 60,000 mL g^{-1} h^{-1}	size/structure (hard template method)	10 mol.% Cu/CeO_2-microspheres	150	[123]
1.0% CO, air balance; WHSV = 10,000 mL g^{-1} h^{-1}	size/structure (hard template method)	three-dimensional (3D) Cu/CeO_2 ((Cu/Cu+Ce) = 0.2)	34	[213]
1.0% CO + 21.0% O_2; WHSV = 60,000 mL·g^{-1}·h^{-1}	shape (thermal annealing of CeMOF precursors)	8.0 wt.% Cu/CeO_2-triple-shelled hollow nanospheres	110	[243]
1.0% CO, air balance; WHSV = 52,000 mL g^{-1} h^{-1}	electronic/chemical state (doping by urea combustion method)	5.0 wt.% Cu/$Ce_{0.9}Mn_{0.1}O_2$	120	[239]
2.4% CO + 1.2% O_2; WHSV = 32,000 mL g^{-1} h^{-1}	electronic/chemical state (doping by combustion method)	6.0 wt.% Cu/$Ce_{0.7}Sn_{0.3}O_2$	80	[240]
1.0% CO + 20.0% O_2; WHSV = 60,000 mL g^{-1} h^{-1}	electronic/chemical state (doping by wetness co-impregnation method)	In_2O_3-CuO_x/CeO_2 1.25 wt.% In, 5.0 wt.% Cu	73	[241]
1.0% CO + 1.0% O_2; GHSV = 9600 h^{-1}	-	1.0 wt.% Pt/CeO_2	70	[244]
1.0% CO + 20.0% O_2; WHSV = 60,000 mL g^{-1} h^{-1}	-	3.0 wt.% Pd/CeO_2	120	[241]
0.95% CO + 1.75% O_2; WHSV = 12,000 mL g^{-1} h^{-1}	-	0.2 wt.% Pd/CeO_2	180	[245]

WHSV: Weight hourly space velocity [=] mL g^{-1} h^{-1}; GHSV: Gas hourly space velocity [=] h^{-1}.

3.2. N_2O Decomposition

Nitrous oxide (N_2O) has been lately recognized as one of the most potent greenhouse gases and ozone depleting substances [246]. In view of this fact, the catalytic abatement of N_2O has received particular attention as one of the most promising remediation methods. Although noble metals exhibit satisfactory activity, their high cost and sensitivity to various substances (e.g., O_2, H_2O) hinder widespread applications. Hence, as previously stated, noble metal-free composites have gained particular attention as potential candidates. The recent advances in the field of N_2O decomposition over metal oxides have been recently reviewed by Konsolakis [246]. It was clearly revealed that MOs could be effectively applied for N_2O decomposition, demonstrating comparable or even better catalytic performance compared to NMs-based catalysts. More interestingly, and in line with the aim and scope of the present article, it was shown that very active and stable MOs could be obtained by adjusting their size, shape and electronic state through appropriate synthesis and promotional routes [246].

Herein, we shortly present the main approaches lately followed to improve the deN_2O performance of MOs, exemplified by the CuO_x/CeO_2 system. Table 2 presents indicative studies towards this direction, involving our recent advances in the field [215,247]. It is of worth noticing the comparable or even superior deN_2O performance of finely-tuned CuO_x/CeO_2 samples as compared to noble metal-based catalysts (Table 2).

Recently, we explored the impact of synthesis procedure (impregnation, co-precipitation and exotemplating methods) on the deN_2O performance of CuO_x/CeO_2 mixed oxides [247]. Co-precipitation method resulted in the optimum performance, offering complete N_2O conversion at ca. 550 °C. On the basis of characterization results (XPS, TPR, micro-Raman), the superiority of precipitated samples was ascribed to their enhanced reducibility and the facilitation of Ce^{4+}/Ce^{3+} and Cu^{2+}/Cu^{+} redox cycles. In an attempt to further improve the deN_2O performance of CuO_x/CeO_2 samples, very recently, we explored the potential to further adjust the local surface chemistry and metal-support interactions by means of electronic (alkali) promotion. Notably, the results showed that by co-adjusting the synthesis procedure and the electronic state, highly active deN_2O catalysts could be obtained; the sample with a Cs content of 1.0 at Cs/nm^2 offers a half-conversion temperature (T_{50}) about 200 °C lower as compared to the commercial sample (Figure 14) [215]. The superiority of Cs-doped samples was ascribed to the electronic effect of alkali doping towards stabilizing partially reduced Cu^{+}/Ce^{3+} pairs, which play a pivotal role in the deN_2O process [215,246].

Table 2. Indicative studies followed to adjust the deN_2O performance of CuO_x/CeO_2 oxides.

Reaction Conditions	Adjusted Parameter (Employed Method)	Optimum System	T_{50} (°C)	Reference
0.26% N_2O; GHSV = 19,000 h^{-1}	composition (citrate acid method)	67 mol.% Cu/CeO_2	370	[214]
0.25% N_2O; GHSV = 45,000 h^{-1}	composition (hard template replication)	40 mol.% Cu/CeO_2	440	[248]
0.1% N_2O; WHSV = 90,000 mL g^{-1} h^{-1}	size/structure (various synthesis methods)	20 wt.% Cu/CeO_2 prepared by co-precipitation, CeO_2 (11.8 nm)	465	[247]
0.1% N_2O; WHSV = 90,000 mL g^{-1} h^{-1}	size/structure (co-precipitation method) and electronic state (alkali addition)	Cs-doped (1.0 at/nm^2) Cu/CeO_2 CeO_2 (13.5 nm)	420	[215]
0.2% N_2O; WHSV = 60,000 mL g^{-1} h^{-1}	size/structure (hydrothermal method)	molar ratio Cu/Ce = 1, CeO_2 (7.0 nm), CuO (24 nm)	380	[216]
0.25% N_2O; WHSV = 120,000 mL g^{-1} h^{-1}	shape (hydrothermal method)	4.0 wt.% Cu/CeO_2-nanorods	430	[146]
0.25% N_2O; WHSV = 60,000 mL g^{-1} h^{-1}	shape (glycothermal method)	10 wt.% Cu/CeO_2-nanospheres	380	[205]
0.1% N_2O; WHSV = 60,000 mL g^{-1} h^{-1}	-	0.5 wt% Rh/Al_2O_3	340	[249]
0.1% N_2O; WHSV = 60,000 mL g^{-1} h^{-1}	-	0.5 wt% Pt/Al_2O_3	500	[249]
0.1% N_2O; WHSV = 60,000 mL g^{-1} h^{-1}	-	0.5 wt% Pd/Al_2O_3	>500	[249]

WHSV: Weight hourly space velocity [=] mL g^{-1} h^{-1}; GHSV: Gas hourly space velocity [=] h^{-1}.

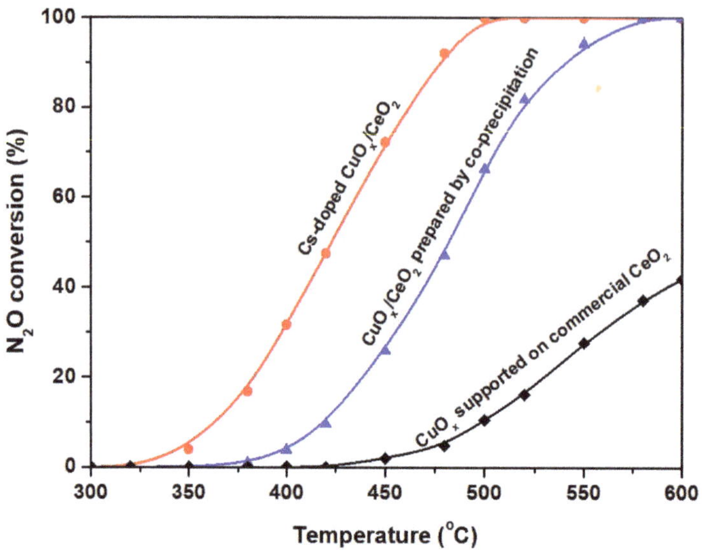

Figure 14. Optimization of deN$_2$O performance of CuO$_x$/CeO$_2$ mixed oxides by co-adjusting synthesis parameters (co-precipitation method) and electronic state (alkali addition). For comparison, the corresponding performance of CuO$_x$ supported on commercial ceria is included. Reaction conditions: 0.1% N$_2$O balanced with He; WHSV = 90,000 mL g^{-1} h^{-1} [215].

The effect of ceria morphology (nanorods, nanocubes, nanopolyhedra) on the deN$_2$O performance of CuO$_x$/CeO$_2$ composites was extensively investigated by Pintar and co-workers [146]. Copper clusters located on {100} and {110} planes—preferentially exposed on ceria nanorods—exhibit a normalized activity ca. 20% higher compared to {111} planes of polyhedra (Figure 15). In terms of conversion performance, the 4.0 wt.% CuO$_x$/Ceria-nanorods exhibited a half-conversion temperature (T_{50}) of about 430 °C compared to 440 °C and 470 °C of nanopolyhedra and nanocubes, respectively. On the basis of a thorough characterization study, it was disclosed that the oxygen mobility and the regeneration of active Cu phase are easier on ceria nanorods, which in turn, facilitates the deN$_2$O activity through oxygen desorption and replenishment of active sites [146]. In a similar manner, CuO$_x$ supported on CeO$_2$ nanospheres exhibited high deN$_2$O performance (T_{50} = 380 °C, Table 2), ascribed mainly to the high population of CuO$_x$ clusters on the high surface area CeO$_2$ nanospheres [205]. These findings clearly demonstrate the significant advances that can be achieved in the deN$_2$O process by engineering the size and shape of metal oxide composites.

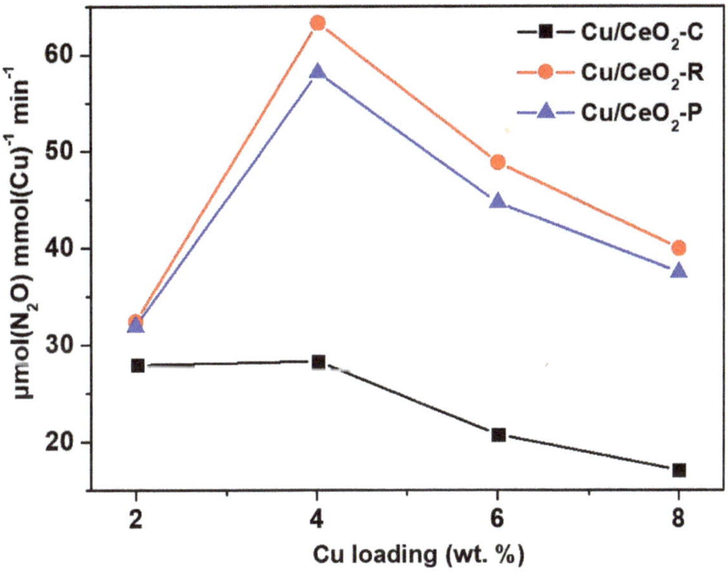

Figure 15. The activity of nanoshaped CuO_x/CeO_2 catalysts measured at T = 375 °C. Adapted from Reference [146]. Copyright© 2015, American Chemical Society.

3.3. Preferential Oxidation of CO (CO-PROX)

The copper-ceria binary oxides are amongst the most widely investigated catalytic systems in the preferential oxidation of carbon monoxide (CO-PROX), a reaction used for the production of highly purified hydrogen and the removal of CO. CuO_x/CeO_2 catalysts have gained particular attention in CO-PROX process, due to their superior performance, which is mainly ascribed to the peculiar properties of copper-ceria interface [40].

In the light of the above-mentioned size, shape and electronic/chemical effects, numerous efforts have been put forward towards optimizing the CO-PROX performance. Indicative approaches followed to fine-tune the CO-PROX performance are summarized in Table 3, and further discussed below.

Table 3. Indicative studies towards adjusting the CO preferential oxidation performance of CuO_x/CeO_2 oxides.

Reaction Conditions	Adjusted Parameter (Employed Method)	Optimum System	Maximum CO Conversion	Reference
1.0% CO + 1.0% O_2 + 50.0% H_2; WHSV = 60,000 mL g^{-1} h^{-1}	composition (hydrothermal method)	5.0 wt.% Cu/CeO_2	99.6% at 130 °C	[250]
1.0% CO + 1.0% O_2 + 50.0% H_2; WHSV = 60,000 mL g^{-1} h^{-1}	composition (co-precipitation method)	30 at.% Cu/CeO_2	~92% at 143 °C	[251]
1.0% CO + 1.0% O_2 + 10.0% H_2O + 15.0% CO_2 + 50.0% H_2; WHSV = 30,000 mL g^{-1} h^{-1}	composition (sol-gel precipitation/ chelating-impregnation)	10 wt.% Cu/CeO_2	~99.5% at 100 °C	[252]
1.0% CO + 1.0% O_2 + 40.0% H_2; WHSV = 30,000 mL g^{-1} h^{-1}	composition (nanocasting method)	7.0 wt.% Cu/CeO_2	100% at 110 °C	[253]
1.2% CO + 1.2% O_2 + 50.0% H_2; WHSV = 20,000 mL g^{-1} h^{-1}	composition/size (freeze-drying method)	6.0 wt.% Cu/CeO_2 CeO_2 (9.9 nm), CuO (10.7 nm)	100% at 90 °C	[254]
1.0% CO + 1.0% O_2 + 50.0% H_2; WHSV = 16,000 mL g^{-1} h^{-1}	composition/size (solvent-free method, cupric nitrate as a copper precursor)	7.5 wt.% Cu/CeO_2 CeO_2 (16.3 nm)	100% at 120 °C	[255]
1.0% CO + 1.0% O_2 + 50.0% H_2 WHSV = 40,000 mL g^{-1} h^{-1}	chemical state (ultrasound-aided impregnation)	$Cu_{0.4}Ce_{0.6}O$/CNTs	100% at 120 °C	[256]
1.25% CO + 1.25% O_2 + 50.0% H_2; WHSV = 20,000 mL g^{-1} h^{-1}	size (Poly(methyl metacrylate) as a template)	6.0 wt.% Cu/CeO_2 CeO_2 (5.6 nm)	100% at 115 °C	[257]
1.0% CO + 1.0% O_2 + 50.0% H_2; WHSV = 36,000 mL g^{-1} h^{-1}	shape (hydrothermal method)	4.0 wt.% Cu/CeO_2-octahedra	95% at 140 °C	[258]
1.0% CO + 1.0% O_2 + 50.0% H_2; WHSV = 60,000 mL g^{-1} h^{-1}	shape (hydrothermal method)	5.0 wt.% Cu/CeO_2-rods/polyhedra	>99.0% at 95/90 °C	[157]
1.0% CO + 1.0% O_2 + 50.0% H_2; WHSV = 40,000 mL g^{-1}·h^{-1}	shape (hydrothermal method)	Cu/CeO_2-spheres CeO_2/CuO = 5	100% at 95 °C	[116]
1.0% CO + 1.0% O_2 + 50.0% H_2; WHSV = 16,000 mL g^{-1}·h^{-1}	shape (alcothermal method)	5.0 wt.% Cu/CeO_2-spheres	100% at 100 °C	[259]
1.0% CO + 1.0% O_2 + 50.0% H_2; WHSV = 40,000 mL g^{-1}·h^{-1}	shape (self-templating method)	Cu/CeO_2-triple-shelled hollow microspheres	100% at 95 °C	[260]
1.0% CO + 1.0% O_2 + 50.0% H_2; WHSV = 120,000 mL g^{-1}·h^{-1}	electronic/chemical state (potassium doping/carbon nanotubes)	Cu/CeO_2/CNTs (2.5 wt.% Cu, 20 wt.% CeO_2, alkali/Cu = 0.68)	100% at 175 °C	[261]
1.0% CO + 1.0% O_2 + 50.0% H_2; WHSV = 60,000 mL g^{-1} h^{-1}	pretreatment (with hydrogen)	10 wt.% Cu/CeO_2	72% at 80 °C	[262]
1.0% CO + 1.0% O_2 + 50.0% H_2; WHSV = 60,000 mL·g^{-1} h^{-1}	pretreatment (with 2 M NaOH and etched with an ionic liquid)	10 wt.% Cu/CeO_2	100% at 150 °C	[263]
1.0% CO + 1.25% O_2 + 50.0% H_2; WHSV = 25,000 mL·g^{-1} h^{-1}	pretreatment (with HNO_3, pH < 4)	7.5 wt.% Cu/CeO_2	100% at 137 °C	[264]

WHSV: Weight hourly space velocity [=] mL g^{-1} h^{-1}.

Several copper-ceria catalytic systems of various copper loadings have been synthesized by different methods, with the optimum Cu loading varying between 5 and 10 wt.% [250,252–254,265]. A further increase in Cu content from 10 to 15 wt.% has been reported to reduce the catalytic activity, due to the large CuO_x agglomerates on the catalyst surface [252]. It was revealed, by means of both ex situ and in situ characterization studies, that the desired CO oxidation process is related to partially reduced Cu^+ species, whereas, highly reduced copper species not strongly associated with CeO_2 favor the undesired H_2 oxidation [40,250,265–267]. In view of this fact, extensive research efforts have been put forward to control the two competitive oxidation processes by appropriately adjusting the geometric and electronic interactions between copper and ceria through the above-described fine-tuning approaches.

Regarding the shape effect, different copper-ceria nanostructures (rods, cubes, spheres, octahedra, spindle or multi-shelled morphologies) have been synthesized and studied for the CO-PROX reaction. It was revealed that the shape-controlled synthesis of ceria nanoparticles has a profound influence on the CO-PROX activity and selectivity. In particular, it was found that rod-shaped and polyhedral copper-ceria systems exhibited higher CO conversion performance (T_{50} = 68 °C) at low-temperatures, as compared to plates (T_{50} = 71 °C) and cubes (T_{50} = 89 °C) [157]. The latter was mainly attributed to the smaller CuO_x clusters subjected to a strong interaction with the ceria carrier, which, in turn, facilitates the formation of Cu^+ sites and oxygen vacancies [157]. More importantly, a close relationship between measurable physicochemical parameters, such as the amount of Cu^+ species and the A_{584}/A_{454} Raman ratio (related to oxygen vacancies) with the catalytic performance was obtained (Figure 16); rod- and polyhedral-shaped samples exhibited the highest values on Cu^+ species and oxygen vacancies, demonstrating, also, the optimum CO-PROX performance [157].

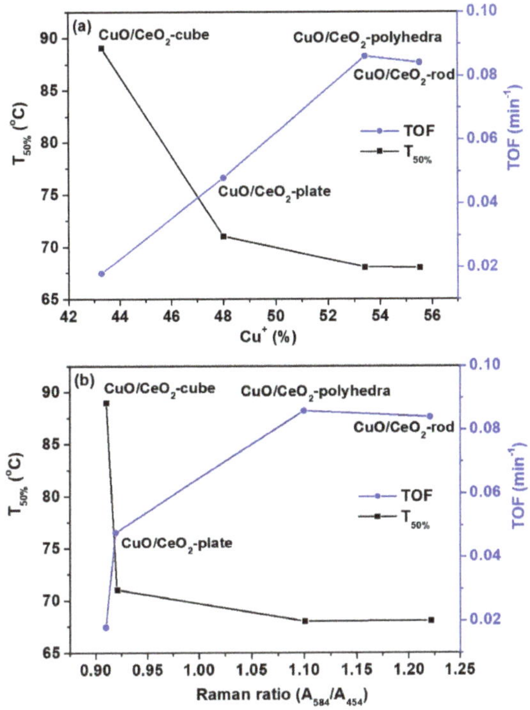

Figure 16. (**a**) TOF (60 °C)/$T_{50\%}$ values versus Cu^+ content and (**b**) TOF (60 °C)/$T_{50\%}$ values versus A_{584}/A_{454} ratios for CuO/CeO_2 catalysts with different morphologies. Adapted from Reference [157]. Copyright© 2016, Royal Society of Chemistry.

In this point, it should be mentioned that in relation to which ceria shape is the most active or selective for the CO-PROX process, inconclusive results are acquired, due to the different reaction conditions applied (see Table 3) in conjunction to the complexity of CO-PROX process, which is affected to a different extent by the various interrelated parameters (e.g., reducibility, metal dispersion, oxygen vacancies, oxidation state, metal-support interactions). Under this perspective, it was reported that copper-ceria nanocubes exhibited higher CO_2 selectivity than copper-ceria nanorods or nanospheres, due to the difficulty of nanocubes to fully reduce the copper oxide species under CO-PROX conditions [133,268], while, at the same time, exhibiting lower CO conversion than nanorods [268] and nanospheres [133]. In a similar manner, CuO_x/CeO_2 spheres and spindles, exposing {111} and {002} facets, showed the highest CO conversion (T_{50} = 69 and 74 °C, respectively), as well as a wide temperature window for total CO conversion (95–195 °C for spheres and 115–215 °C for spindles), in comparison with octahedrons, cubes and rods [116]. Interestingly, in different shaped ceria nanostructures, a close relationship is found between the concentration of oxygen vacancies and the amount of reduced copper species (Figure 17), clearly revealing the key role of exposed facets towards adjusting the catalytic performance. These findings were further substantiated by DFT calculations (Figure 18), showing the high population of oxygen vacancies at the intersection of {111} and {002} facets in opposition to CeO_2 {111} surface [116].

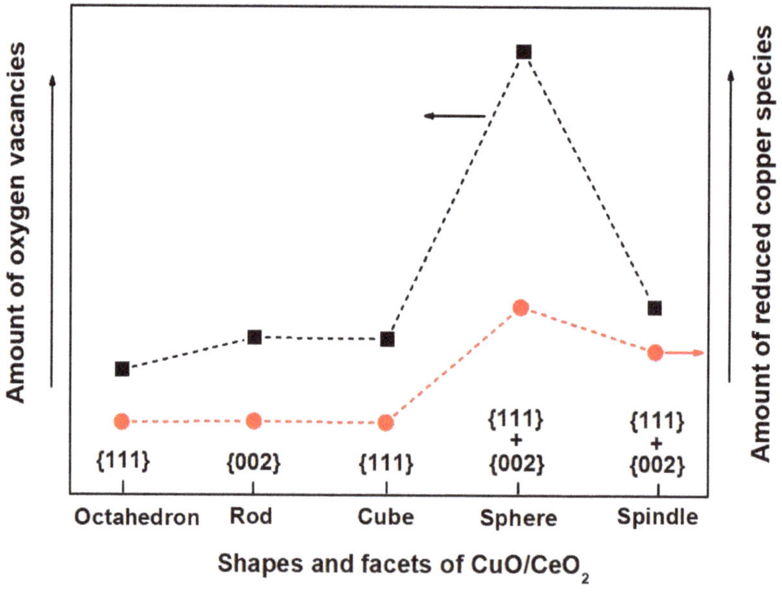

Figure 17. Plots of the number of oxygen vacancies and reduced copper species for the CuO/CeO_2 catalysts with different support shapes. Adapted from Reference [116]. Copyright© 2018, Elsevier.

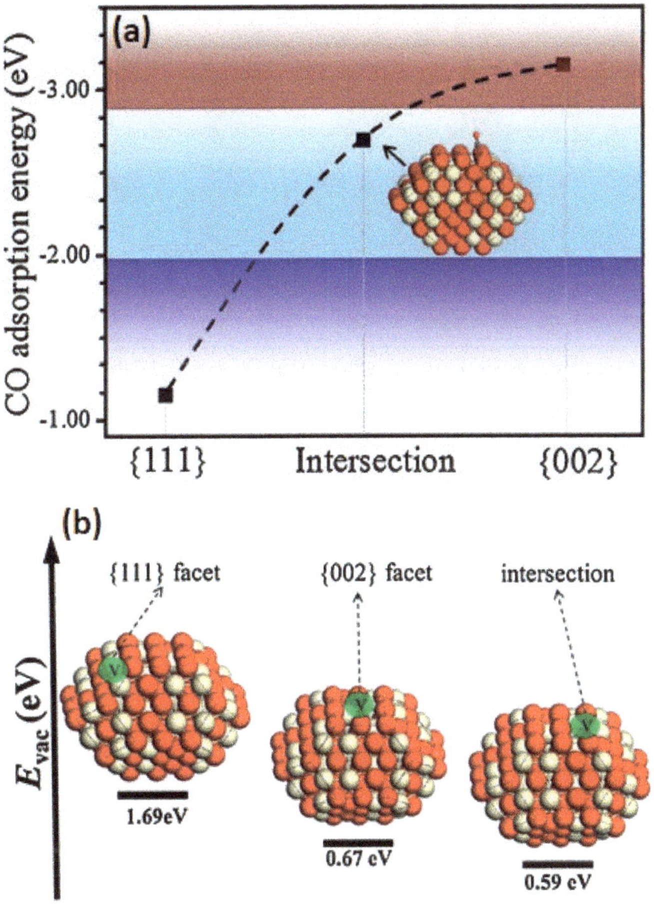

Figure 18. (**a**) DFT-calculated adsorption energy for molecularly adsorbed CO on {111}, {002} facets and their intersection sites of a $Ce_{60}O_{120}$ cluster obtained, (**b**) Models of the optimized geometries of $Ce_{60}O_{119}$ with an oxygen vacancy in the highlighted positions: {111}, {002} facets and their intersection, and values of oxygen vacancy formation energy (E_{vac}) below models. (Color code: O in red, Ce in grey, oxygen vacancy in the highlighted circle labeled "V"). Reproduced with permission from Reference [116]. Copyright© 2018, Elsevier.

In an attempt to optimize the CO-PROX performance through size and shape engineering single- and multi-shelled copper-ceria hollow microspheres were synthesized [260,269]. By tuning the number of shells, the catalytic activity was notably improved, with the triple-shelled structure (Figure 19) exhibiting the highest activity and selectivity (100% CO conversion and 91% CO_2 selectivity at 95 °C), as well as a wide temperature window for complete CO conversion (95–195 °C) [260]. The increase in the number of shells enhances the electronic and geometric interaction between copper and ceria, offering a high population of exposed active sites and an increased space inside the catalyst which facilitates reactants accessibility [260].

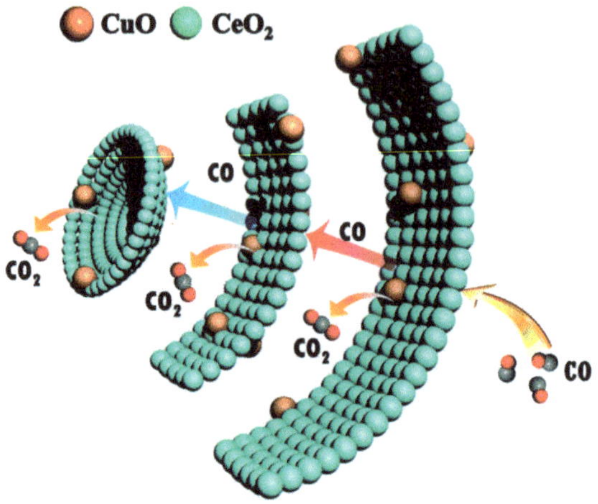

Figure 19. Schematic diagrams for CO-PROX over the CuO_x/CeO_2 hollow microsphere catalysts with triple shells. Reproduced with permission from Reference [260]. Copyright© 2019, Royal Society of Chemistry.

Taking into account the pivotal role of nanoparticles crystallite size/shape and their consequent effect on metal-support interactions, different preparation routes have been investigated for the synthesis of copper-ceria composites, such as the hydrothermal method, the template-assisted method, the solid-state preparation method, sol-gel, co-precipitation, freeze-drying, deposition-precipitation, etc. [250,251,253–255,257,262,270]. For instance, template-assisted synthesis resulted in small ceria crystallite sizes (ca. 5.6 nm), and thus, in a high population of copper-ceria interfacial sites with implications in Cu oxidation state and CO-PROX activity [257]. Moreover, the precursor compounds or the template agent used during the synthesis procedure can affect the pore size and volume or the reducibility of the materials [255,271]. Interestingly, ethanol washing during the preparation of CuO_x/CeO_2 oxides leads to decreased particle sizes, as it affects the dehydration process between precursors particles, resulting in decreased adsorbed water and improved dispersion [272]. Very recently, a novel ultrasound-assisted precipitation method was employed to adjust the defective structure of CeO_2 and in turn, the CO-RPOX activity [273]. By means of characterization techniques and theoretical calculations, it was shown that only two-electron defects on ceria surface (i.e., defects adsorbing oxygen to form peroxides instead of superoxide species which are formed on one-electron defects) were responsible for the formation of Cu^+ and Ce^{3+} species, which were intimately involved in the CO adsorption and oxygen activation processes [273]. In particular, the adsorption of O_2 on two-electron defects resulted in peroxides formation, followed by Cu ions incorporation towards the development of Cu-O-Ce structure. Meanwhile, the two additional electrons in the two-electron defects facilitate the electronic re-dispersion in Cu-O-Ce structure, leading to the creation of Cu^+ (CO adsorption sites) and Ce^{3+} (oxygen activation sites).

Another approach in the direction of catalysts functionalization that has attracted considerable attention in recent years is the preparation of inverse catalytic systems. In particular, the co-existence of Cu^+ and Cu^{2+} ions was observed in star-shaped inverse CeO_2/CuO_x catalysts which exhibited high catalytic activity [274]. Moreover, the alteration of Ce/Cu molar ratio and/or the pH value in the inverse CeO_2/CuO_x catalysts notably affects the morphology and the particle size, which in turn, favor the contact interface between ceria and copper, and thus, the CO oxidation at the expense of H_2 oxidation in PROX process [44]. In addition, a multi-step synthetic approach has been applied for a high concentration of oxygen vacancies to be successfully anchored at the interfaces of the inverse CeO_2/CuO_x system, leading to outstanding CO-PROX activity (~100% CO conversion at a wide temperature window 120–210 °C) and adequate stability [275].

The doping effect on the CO-PROX performance has also been studied in the inverse copper-ceria catalysts [276,277]. It was reported that doping ceria with transition metals (e.g., Fe, Co, Ni) induces changes in the ceria lattice and in the formation of oxygen vacancies [276]. The doping element affects the reducibility of the CeO_2/CuO_x catalysts, while promoting the formation of Ce^{3+} ions and oxygen vacancies, with the NiO-doped CeO_2/CuO_x catalyst exhibiting the highest activity (T_{50} = 68 °C) and the widest temperature window for total CO conversion (115–155 °C) [276]. In the inverse copper-ceria catalysts, it has also been found that the presence of Zn improves the CO-PROX performance, as it has the ability to hinder the CuO reduction to highly reduced copper sites which provide the active sites for the H_2 oxidation [277].

By applying appropriate pretreatment protocols, the CO-PROX performance may also be greatly affected. In particular, the pretreatment of copper-ceria catalysts in an oxidative or reductive atmosphere affects the amount and dispersion of the active species, and consequently, the catalytic performance [262]. The pretreatment with hydrogen led to a breakage of the Cu-[O_x]-Ce structure, which resulted in enhanced catalytic performance, indicating the significance of the highly dispersed CuO_x clusters in the CO-PROX process [262]. Furthermore, the pretreatment in an acidic or a basic environment affects the interaction between the two oxide phases. For instance, the pretreatment of ceria spheres in a basic solvent (2M NaOH), followed by etching in an ionic liquid for the acquisition of ceria nanocubes, resulted in the best catalytic activity at temperatures lower than 150 °C, due to the strong interaction between the highly dispersed CuO_x clusters and ceria support [263]. An acidic treatment with nitric acid in nanorod-shaped CuO_x/CeO_2 catalysts has also been performed by Avgouropoulos and co-workers [264]. It was found that a highly acidic environment (pH < 4) led to an enrichment of catalysts surface in Cu^+ species and to high concentrations of oxygen vacancies and Ce^{3+} species, while facilitating the formation of surface hydroxyls that are considered responsible for controlling the interfacial interactions in the copper-ceria binary system [264]. All the above-mentioned characteristics in conjunction with the better copper dispersion and the improved reducibility of the highly acidic catalysts resulted in enhanced catalytic performance ($T_{50} \simeq 84$ °C) [264]. The same group has also investigated the pretreatment effect of employing ammonia solutions in copper-ceria nanorods [278]. It was revealed that the textural and structural properties of the modified catalysts remained almost unaffected after treatment, whereas, increasing the Cu:NH_3 ratio to 1:4 resulted in higher reducibility and gave rise to Cu^+ and surface lattice oxygen species, leading, thus, to improved catalytic performance [278]. As shown in Figure 20, close relationships between the half-conversion temperature (T_{50}) and the main Raman peak shift or the concentration of Ce^{3+} and oxygen vacancies were observed [278].

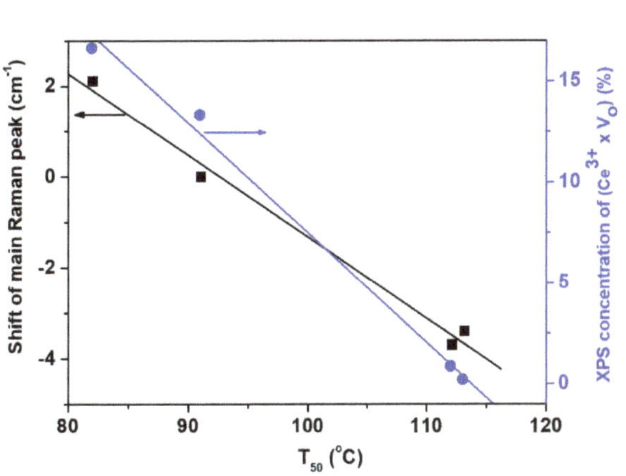

Figure 20. T_{50} vs. (i) shift of the main peak (F_{2g} Raman vibration mode) of fluorite CeO_2 and (ii) surface concentration of Ce^{3+} and oxygen vacancies determined via XPS analysis. Adapted from Reference [278]. Copyright© 2018, John Wiley and Sons.

Another adjusted parameter that can exert a profound influence on the catalytic performance is the electronic promotion mainly induced by alkali modifiers, as it may affect the chemisorption ability of active sites, as well as the copper-ceria interactions. In that context, it was found that the presence of K^+ ions in CuO_x/CeO_2 catalysts has a beneficial effect on CO-PROX process in the presence of both CO_2 and H_2O, since a proper K^+ content was proved to alleviate the CO_2 and H_2O adsorption on the reaction sites and thus, enhancing the catalytic performance [279]. Potassium has also been found to stabilize Cu^+ active species by affecting Cu-Ce interactions [280].

An additional engineering approach towards enhancing the CO-PROX reactivity of CuO_x/CeO_2 oxides involves the employment of chemical substances of specific architecture and textural properties, such as the carbon-based materials (rGO, CNTs, etc.). These materials favor the dispersion of copper and ceria, while affecting the reducibility and the population of oxygen vacancies, thus, resulting in enhanced catalytic performance at low-temperatures [256,281–284]. As for example, the introduction of rGO resulted in abundant Ce^{3+} species and oxygen vacancies, offering high catalytic activity at temperatures below 135 °C and good resistance to CO_2 and H_2O [283].

Interestingly, by combining electronic (alkali promotion) and chemical modification (carbon nanotubes), highly active multifunctional composites can be obtained. In copper-ceria catalysts supported on carbon nanotubes (CNTs) with a specific alkali/Cu atomic ratio, i.e., 0.68, the nature of the alkali metal (Li, Na, K, Cs) has been shown to affect the dispersion of ceria over CNTs and the copper-ceria interaction [261]. K-promoted CuO_x/CeO_2 oxides combined with CNTs exhibited high catalytic activity ($T_{50} \simeq 109$ °C as compared to 175 °C of un-promoted catalyst), attributed to the K-induced modification on redox/electronic properties [261].

3.4. Water-Gas Shift Reaction (WGSR)

The water-gas shift reaction (WGSR) plays a key role in the production of pure hydrogen, through the chemical equilibrium: $CO + H_2O \leftrightarrow CO_2 + H_2$. Among the different catalytic systems, copper-ceria oxides have gained particular attention, due to their low cost and adequate catalytic performance. Moreover, significant efforts have been put forward towards optimizing the low-temperature WGS activity by means of the above discussed methodologies. Regarding CuO_x/CeO_2 catalyzed WGSR, two main reaction mechanisms have been proposed, namely, the redox and the associative mechanism. The first one involves the oxidation of adsorbed CO by oxygen originated by H_2O dissociation. The second one involves the reaction of CO with surface hydroxyl groups towards the formation and subsequent decomposition of various intermediate species, such as formates [161,285].

A thorough study concerning the nature of active species and the role of copper-ceria interface for the low-temperature WGSR has been recently performed by Chen et al. [285]. It was revealed that the activity of copper-ceria catalysts is intrinsically related with the Cu^+ species present at the interfacial perimeter, with the CO molecule being adsorbed on the Cu^+ sites, while water being dissociatively activated on the oxygen vacancies of ceria [285,286]. In a similar manner, Flytzani-Stephanopoulos and co-workers [287] have earlier shown that strongly bound Cu-[O_x]-Ce species, probably associated with oxygen vacancies of ceria, are the active species for the low-temperature WGSR, whereas, the weakly bound copper oxide clusters and CuO_x nanoparticles act as spectators.

Although the distinct role of copper and ceria and their interaction is not well determined, it is generally accepted that the activation of H_2O, linked to copper-ceria interface and oxygen vacancies, is the rate-determining step [285]. Therefore, particular attention has been paid to modulate the interfacial reactivity via the above discussed adjusting approaches. Indicative studies towards modulating the WGSR performance are summarized in Table 4, and further discussed below.

Table 4. Indicative studies towards adjusting the WGSR performance of CuO-CeO_2 oxides.

Reaction Conditions	Adjusted Parameter (Employed Method)	Optimum System	Maximum CO Conversion	Reference
10.0% CO + 12.0% CO_2 + 60.0% H_2; vapor:gas = 1:1; WHSV = 3000 mL g^{-1} h^{-1}	size/structure (precipitation)	20 wt.% Cu/CeO_2	91.7% at 200 °C	[206,207]
2.0% CO + 10.0% H_2O; WHSV = 42,000 mL g^{-1} h^{-1}	size/structure (bulk-nano interfaces by aerosol-spray method)	inverse CeO_2/Cu CeO_2 (2–3 nm)	100% at 350 °C	[288]
1.0% CO + 3.0% H_2O; WHSV = 200,000 mL g^{-1} h^{-1}	shape (microemulsion method)	5.0 wt.% Cu/CeO_2-nanospheres	64% at 350 °C	[136]
10.0% CO + 5.0% CO_2 + 10.0% H_2O + 7.5% H_2; WHSV = 60,000 mL g^{-1} h^{-1}	shape (precipitation method)	Cu/CeO_2-nanoparticles	49% at 400 °C	[119]
3.5% CO + 3.5% CO_2 + 25.0% H_2 + 29.0% H_2O; GHSV = 6000 h^{-1}	shape (hydrothermal method)	10 wt.% Cu/CeO_2-octahedrons	91.3% at 300 °C	[159]
10.0% CO + 5.0% CO_2 + 5.0% H_2; vapor:gas = 1:1; GHSV = 6000 h^{-1}	electronic/chemical state (doping with cobalt by nanocasting)	Cu-Co-CeO_2 (weight ratio of Cu:Co:Ce = 1:2:7)	95% at 300 °C	[289]
10.0% CO + 12.0% CO_2 + 60.0% H_2; vapor:gas = 1:1; WHSV = 2337 mL g^{-1} h^{-1}	electronic/chemical state (doping with yttrium by co-precipitation)	Y-doped Cu/CeO_2 25 wt.% CuO, 2 wt.% Y_2O_3	93.4% at 250 °C	[209]
15.0% CO + 6.0% CO_2 + 55.0% H_2; vapor:gas = 1:1; GHSV = 4500 h^{-1}	pretreatment (with 20 CO_2/2H_2 followed by calcination in O_2)	10 wt.% Cu/CeO_2	86% at 350 °C	[290]

WHSV: Weight hourly space velocity [=] mL g^{-1} h^{-1}. GHSV: Gas hourly space velocity [=] h^{-1}.

The preparation method can affect various characteristics, such as the specific surface area, the total pore volume, the dispersion of the active phase or the crystallite size [206,291]. For instance, copper-ceria catalyst prepared by a hard template method showed higher WGSR activity as compared to the one prepared by co-precipitation (62 vs. 54% CO conversion at 450 °C), due to its larger surface area and higher CuO_x dispersion, while they both exhibited a similar amount of acidic surface sites [291]. Among CuO_x/CeO_2 catalysts synthesized by different precipitation methods, the catalysts prepared by stepwise precipitation showed the highest CO conversion, due to their higher reducibility and oxygen defects [208]. Precipitation was also found to give catalysts with higher WGSR activity, namely, 91.7% CO conversion at 200 °C, in comparison with the hydrothermal (82%) or sol-gel methods (64.5%), due to their abundance in oxygen vacancies, associated with the small CuO_x crystals and large pore volume [206].

The precipitating agent used could also exert a significant impact on the physicochemical properties of CuO_x/CeO_2 catalysts, with the great implication in the catalytic behavior [207,292]. By employing ammonia water instead of ammonium and potassium carbonate, the WGSR activity is notably enhanced (91.7% CO conversion at 200 °C in contrast to 78.3% and 46.2%, respectively), due to the better dispersion of copper species and the stronger copper-ceria interactions [207]. Moreover, the copper precursor compound (nitrate or ammonium ions) and the preparation temperature can notably affect the WGSR activity [292].

Recently, it was found that the dispersion of differently formed copper structures (particles, clusters, layers) on ceria of rod-like morphology is dependent on copper loading, with low copper loadings (1–15 mol.%) exhibiting monolayers and/or bilayers of copper, while a further increase in copper loading up to 28 mol.% results in faceted copper particles and multi-layers of copper [286]. At copper loadings up to 15 mol.%, a linear relationship between the CO conversion and the copper content was observed (Figure 21), indicating that the number of the active interfacial sites (Cu^+-V_o-Ce^{3+}) is significantly increased along with Cu content up to 15 mol.% [286].

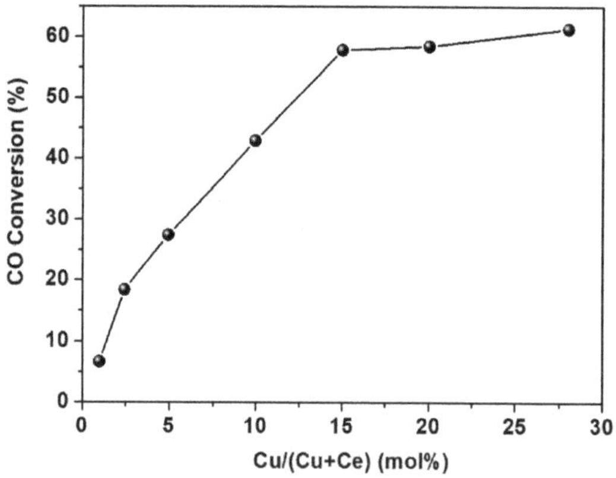

Figure 21. Low-temperature WGS reaction over the Cu/CeO_2 catalysts. CO conversion as a function of the copper content in the respect catalysts. Reaction conditions: 1.0 vol.% CO/3.0 vol.% H_2O/He, 40,000 h^{-1}, 200 °C. Adapted from Reference [286]. Copyright© 2019, Elsevier.

The morphological features of both copper and ceria counterparts notably affect the WGSR activity. In a comprehensive study by Zhang et al. [293], it was reported that Cu cubes exhibit high WGSR activity in contrast to Cu octahedra with the Cu–Cu suboxide (Cu_xO, $x \geq 10$) interface of Cu(100) surface being the active sites. In a similar manner, it was shown that ceria nanoshapes (rods, cubes, octahedra) exhibit different behavior during interaction with CO and H_2O, due to their diverse defect chemistry [294]. Upon CO exposure, ceria nanocubes, exposing {100} planes, favor the formation of oxygen defects at the expense of the existing anti-Frenkel defects, while in nanorods and nanooctahedra (exposing mainly {111} planes) both types of defects are formed [294]. By combining Raman and FTIR results, it was revealed that H_2-reduced ceria rods and octahedra could be further reduced in

CO, resulting in the formation of both defects. In contrast, cubes cannot be further reduced by CO; thus, oxygen is available to form carbonates and bicarbonates by converting Frenkel defects to oxygen vacancies [294]. These findings clearly revealed the key role of both copper and ceria nanoshape on the defect chemistry of individual counterparts. It should be noted, however, that in the binary copper-ceria system, where multifaceted interactions are taking place, the relationships between shape effects and catalytic activity can be rather complex, leading to inconsistent conclusions [119,136,159].

Very recently, Yan et al. [288] reported on a novel structural design approach towards optimizing the WGS activity of CuO_x/CeO_2 system. In particular, inverse copper-ceria catalysts of high efficiency were developed through the fabrication of highly stable bulk-nano interfaces under reaction conditions. Nano-sized ceria particles (2–3 nm) were stabilized on bulk copper resulting in abundant ceria-copper interfaces [288]. This inverse catalyst showed outstanding WGS conversion (T_{100} = 350 °C), due to the high amount of interfacial sites and the strong copper-ceria interaction, which facilitated the dissociation of water and the oxidation of CO [288].

The doping approach has also been employed to enhance the WGSR activity of CuO_x/CeO_2 system [209,295]. For instance, copper-ceria catalysts doped with 2 wt.% yttrium have shown excellent WGSR activity and high thermal stability, as yttrium favored the oxygen vacancy formation on ceria [209]. Recently, Wang et al. [295] performed DFT calculations in order to theoretically investigate the alkali effect on the WGSR activity of Cu(111) and Cu(110) surfaces. It was found that potassium enhances the WGSR activity as it favors the dissociation of H_2O and induces stronger promotion on the (111) surface. With regard to other alkali metals (Na, Rb, Cs), the promoting effect on the dissociation of water differentiates with their electronegativities which induce changes in the work function, i.e., the lower the work function, the stronger the promoting effect of the alkali [295].

Finally, the WGSR activity and the sintering resistance of the CuO_x/CeO_2 catalysts can be further enhanced by improving the metal-support interactions through appropriate pretreatment protocols [290,296]. As for example, the treatment of CuO_x/CeO_2 catalyst in a gas mixture of $20CO_2/2H_2$ led to highly active catalysts, due to the electron enrichment of copper atoms via electronic metal-support interactions [290]. Moreover, ceria pretreatment in different atmosphere (air, vacuum or H_2) affected the WGSR performance of CuO_x/CeO_2 catalysts, with the H_2-pretreated samples exhibiting the highest conversion performance, due to the strong synergism between the two oxide phases, the small CuO_x particle size, and the high concentration in oxygen vacancies [296].

3.5. CO_2 *Hydrogenation*

The hydrogenation of carbon dioxide to value-added chemicals, such as methanol, has received considerable attention, in terms of environmental protection and sustainable energy. The significant role of copper-ceria interfacial sites in the CO_2 hydrogenation process has been confirmed by both theoretical and experimental studies [40,297,298]. In particular, metal-oxide interface plays a key role in CO_2 hydrogenation process, as it could provide the active sites for reactants adsorption, while these interfacial sites may stabilize the key intermediates [299]. In view of this fact, copper-ceria catalysts have shown higher selectivity in methanol than their zirconia-supported counterparts, as the copper-ceria interface favored the dispersion of copper and the oxygen vacancy formation, while the interaction between copper and ceria led to a decrease in copper particle size [300]. The interfaces between the defective CeO_{2-x} and the highly dispersed Cu^+/Cu^0 species are considered the active sites for methanol synthesis in the case of CuO_x/CeO_2 system [301]. Furthermore, the different metal-support interactions between the two catalysts resulted in different reaction intermediates, namely, carbonates for the CuO_x/CeO_2 catalysts and bicarbonates for the zirconia-supported ones, thus, resulting in different selectivity, with the copper-zirconia composites being highly selective in CO [300].

In view of the above aspects, the fine-tuning of the metal-support interface could lead to highly active and selective catalysts. Indicative studies towards adjusting the CO_2 conversion to methanol under similar reaction conditions are summarized in Table 5, and further discussed below.

Table 5. Indicative studies towards adjusting the CO_2 conversion to methanol of CuO_x/CeO_2 oxides.

Reaction Conditions	Adjusted Parameter (Employed Method)	Optimum System	CO_2 Conversion (Methanol Rate or Selectivity) at 260 °C	Reference
CO_2:H_2 = 1:3; P = 3.0 MPa; GHSV = 1200 h^{-1}	electronic/chemical state (doping with zinc and dispersion in SBA-15 by incipient wetness impregnation)	$Cu_{0.5}Zn_{0.4}Ce_{0.1}$/SBA-15	6.5% (33.6 mg g^{-1} h^{-1})	[302]
CO_2:H_2 = 1:3; P = 3.0 MPa; WHSV = 14,400 mL g^{-1} h^{-1}	electronic/chemical state (doping with alumina by co-precipitation)	60 wt.% Cu/AlCeO	17.0% (11.9 mmol g^{-1} h^{-1})	[303]
CO_2:H_2 = 1:3; P = 3.0 MPa; GHSV = 6000 h^{-1}	electronic/chemical state/shape (doping with Ni-Cu by impregnation)	$CuNi_2/CeO_2$-nanotubes Ni/(Cu+Ni) = 2/3	17.8% (18.1 mmol g^{-1} h^{-1})	[304]
CO_2:H_2 = 1:3; P = 3.0 MPa; GHSV = 6000 h^{-1}	electronic/chemical state/shape (doping with Ni-Cu by impregnation)	$CuNi_2/CeO_2$-nanorods 10 wt.% Cu and Ni	18.35% (73.33%)	[305]
CO_2:H_2 = 1:3; P = 2.0 MPa; WHSV = 3000 mL g^{-1} h^{-1}	shape (hydrothermal method)	5.0 wt.% Cu/CeO_2-nanorods	2.5% (71.0%)	[131]

WHSV: Weight hourly space velocity [=] mL g^{-1} h^{-1}; GHSV: Gas hourly space velocity [=] h^{-1}.

The activity and selectivity of the CuO_x/CeO_2 catalysts for methanol synthesis are greatly affected by the support morphology. Copper-ceria nanorods exposing {100} and {110} crystal planes exhibited the highest methanol yield (Table 5) as compared to nanocubes and nanoparticles, due to the strong interaction between the two oxide phases and the high copper dispersion [131]. Copper-ceria nanorods were also found to be more active than nanocubes, while exhibiting similar conversion performance with the nanoparticles, for carbonate (diethyl) hydrogenation [147].

In a similar manner, CuO_x/CeO_2 catalysts led, mainly, to the production of CO at atmospheric pressures through the RWGS reaction, with the nanorod-shaped catalyst exhibiting higher CO_2 conversion (~50% at 450 °C) as compared to nanospheres (~40% at 450 °C), revealing the structure dependence of the RWGS [137]. The active intermediates are preferably formed on the {110} ceria exposed surface of the rod-like morphology, resulting in high catalytic performance [137].

Copper-ceria nanorods of various copper loadings were also investigated in the hydrogenation of carbonate to methanol, with the catalysts of ca. 20 wt.% Cu content exhibiting superior catalytic performance [306]. The copper content can significantly affect the mole fraction of Ce^{3+} and Cu^+ species (Figure 22), and in turn, the methanol yield [306].

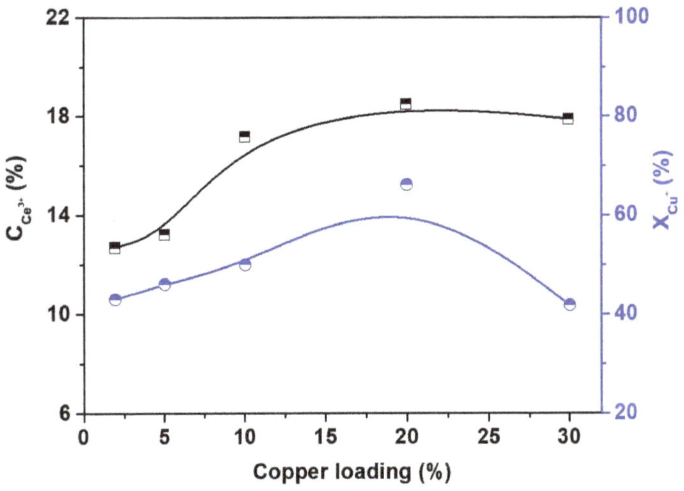

Figure 22. Mole ratio of Ce^{3+} and Cu^+ over the reduced Cu/CeO_2 catalysts as a function of Cu content. Adapted from Reference [306]. Copyright© 2018, John Wiley and Sons.

In addition, a space-confined synthetic approach was applied for the synthesis of highly dispersed copper-ceria catalysts for RWGS, offering 100% CO selectivity at 300 °C and ambient pressure [307]. The enhanced catalytic performance was ascribed to the abundance in interfaces formed among the highly dispersed copper nanoparticles and the Ce^{3+} species, thus, favoring H_2 spillover [307].

The controlled synthesis of multicomponent systems could also be an effective approach for highly active and selective hydrogenation catalysts (e.g., [302–305,308]). For instance, the ternary composite consisting of Cu, ZnO, CeO_x, supported on SBA-15 exhibited high catalytic activity for methanol synthesis (Table 5), due to the peculiar synergistic effects between the different counterparts [302]. Copper was considered to be the active site for hydrogen activation, while ZnO and CeO_x oxides facilitated the CO_2 adsorption and hydrogen spillover on the interfacial sites [302]. In a similar manner, the introduction of alumina to ceria carrier (Cu/AlCe) led to highly active composites (Table 5), mainly ascribed to the high copper dispersion [303]. Interestingly, highly active and selective CO_2 hydrogenation catalysts can be obtained by co-adjusting the composition, structure and shape, in line with the fine-tuning methodology herein proposed (Figures 2 and 3). In this perspective, bimetallic composites (Cu-Ni) incorporated into ceria nanoparticles of specific morphology (e.g., nanotubes, nanorods) could lead to highly active composites for the CO_2 hydrogenation to methanol (Table 5). The enhanced catalytic performance of Cu-Ni/CeO_2 nanoshaped catalysts was mainly interpreted on the basis of the synergistic interaction between Cu and Ni as well as of that between the ceria carrier and Cu-Ni alloy [304,305]. Similarly, the bimetallic Cu-Fe/CeO_2 catalyst has shown enhanced stability in the high-temperature RWGS due mainly to the fact that iron oxide clusters (FeO_x) highly dispersed over ceria act as textural promoters [308]. Finally, copper nanocrystals encapsulated in Zr-based MOFs demonstrated high activity and selectivity for CO_2 hydrogenation to methanol, outperforming the benchmark Cu/ZnO/Al_2O_3 catalyst [197].

4. Outlook and Challenges

In the present review, the copper-ceria binary system has been employed as a reference system to reveal the different approaches that can be followed to modulate the local surface chemistry, and in turn, the catalytic performance of metal oxides (MOs), by means of size, shape and electronic/chemical functionalization. More importantly, the fine-tuning of the above-mentioned parameters can affect not only the reactivity of metal sites in its own right, but also the interfacial activity (e.g., through the formation of oxygen vacancies, facilitation of redox interplay between the different counterparts, etc.) offering a synergistic contribution towards the development of NMs-free highly active and selective composites for several energy and environmental applications.

For instance, the employment of appropriate synthetic routes, such as the hydrothermal method, leads to the development of nanoparticles with specific morphologies, exposing distinct crystal facets of different coordination environments, with great implications in catalysis. Moreover, particles size/shape engineering strongly affects the interfacial reactivity through both geometric and electronic interactions, offering metal oxide systems with the desired properties. In addition, special pretreatment protocols or activation procedures can notably affect the metal dispersion and the population of oxygen defects, with great consequences in the catalytic efficiency. In view of the above, the fine tuning of metal oxides by combining bulk and nano effects or by adjusting the coordination environment could lead to highly efficient catalysts.

Besides the modulation of local surface chemistry by means of size and shape engineering, the electronic/chemical modification (e.g., alkali promotion, incorporation of rGO or g-C_3N_4, employment of MOFs) can be adopted as an additional functionalization tool to regulate the electronic environment and the oxygen exchange kinetics of MOs.

In view of the above aspects, highly active composites, with a comparable or even better performance than that of NMs, have been developed for various processes, such as CO oxidation, N_2O decomposition, preferential oxidation of CO, among others. As for example, the combination of precipitation method with alkali promotion can lead to highly active and oxygen-tolerant CuO_x/CeO_2 catalysts for N_2O decomposition. On the other hand, the modulation of ceria support morphology (nanorods) by the hydrothermal method resulted in CuO_x/CeO_2 composites with superior CO oxidation performance, even better to that of Pt-based catalysts. More importantly, the co-adjustment of different parameters (e.g., the shape of individual counterparts along with the electronic state of metal entities) could lead not only to distinct reactivity of each counterpart, but also to different synergistic interactions, offering mixed metal oxides of unique features. Hence, metal oxides appropriately adjusted by means of suitable synthetic and electronic/chemical modification routes could provide the materials platform for real-life energy and environmental applications.

Another approach towards the fine-tuning of metal oxides could be the employment of computational studies (e.g., DFT calculations) prior to the synthesis of the catalysts, providing, thus, the required feedback that would lead to the focused functionalization of specific parameters. This combinatorial theoretical and experimental approach could result in specific composites with predefined characteristics, while it would save precious time during experimental trials.

Last, but not least, the conclusions drawn from the present survey can provide the design principles for the development of low-loading NMs-based catalysts, paving the way for the decrease of noble metals content in energy and environmental applications in which their use is inevitable. In any case, the fundamental understanding of structure-property relationships is a prerequisite factor towards the rational design of efficient and inexpensive catalytic composites.

Author Contributions: M.L. contributed to paper writing; M.K. contributed to the conception, design and writing of the paper; All authors contributed to the discussion, read and approved the final version of the manuscript. All authors have read and agreed to the published version of the manuscript.

Funding: This research has been co-financed by the European Union and Greek national funds through the Operational Program Competitiveness, Entrepreneurship and Innovation, under the call RESEARCH–CREATE–INNOVATE (project code: T1EDK-00094).

Acknowledgments: The authors would like to express their sincere gratitude to the anonymous reviewers for their constructive remarks that greatly contributed to improving the content and the scientific impact of this review article. The authors also would like to thank the editors for their efforts during the review process.

Conflicts of Interest: The authors declare no conflict of interest.

References

1. Yao, X.; Tang, C.; Gao, F.; Dong, L. Research progress on the catalytic elimination of atmospheric molecular contaminants over supported metal-oxide catalysts. *Catal. Sci. Technol.* **2014**, *4*, 2814–2829. [CrossRef]
2. Melchionna, M.; Fornasiero, P. The role of ceria-based nanostructured materials in energy applications. *Mater. Today* **2014**, *17*, 349–357. [CrossRef]
3. Fang, Y.; Guo, Y. Copper-based non-precious metal heterogeneous catalysts for environmental remediation. *Chin. J. Catal.* **2018**, *39*, 566–582. [CrossRef]
4. Yuan, C.; Wu, H.B.; Xie, Y.; Lou, X.W. Mixed Transition-Metal Oxides: Design, Synthesis, and Energy-Related Applications. *Angew. Chem. Int. Ed.* **2014**, *53*, 1488–1504. [CrossRef]
5. Montini, T.; Melchionna, M.; Monai, M.; Fornasiero, P. Fundamentals and Catalytic Applications of CeO_2-Based Materials. *Chem. Rev.* **2016**, *116*, 5987–6041. [CrossRef] [PubMed]
6. Wu, K.; Sun, L.-D.; Yan, C.-H. Recent Progress in Well-Controlled Synthesis of Ceria-Based Nanocatalysts towards Enhanced Catalytic Performance. *Adv. Energy Mater.* **2016**, *6*, 1600501. [CrossRef]
7. Rodriguez, J.A.; Liu, P.; Graciani, J.; Senanayake, S.D.; Grinter, D.C.; Stacchiola, D.; Hrbek, J.; Fernández-Sanz, J. Inverse Oxide/Metal Catalysts in Fundamental Studies and Practical Applications: A Perspective of Recent Developments. *J. Phys. Chem. Lett.* **2016**, *7*, 2627–2639. [CrossRef] [PubMed]
8. Spezzati, G.; Benavidez, A.D.; DeLaRiva, A.T.; Su, Y.; Hofmann, J.P.; Asahina, S.; Olivier, E.J.; Neethling, J.H.; Miller, J.T.; Datye, A.K.; et al. CO oxidation by Pd supported on CeO_2 (100) and CeO_2 (111) facets. *Appl. Catal. B Environ.* **2019**, *243*, 36–46. [CrossRef]
9. Morfin, F.; Nguyen, T.-S.; Rousset, J.-L.; Piccolo, L. Synergy between hydrogen and ceria in Pt-catalyzed CO oxidation: An investigation on Pt–CeO_2 catalysts synthesized by solution combustion. *Appl. Catal. B Environ.* **2016**, *197*, 2–13. [CrossRef]
10. Gatla, S.; Aubert, D.; Flaud, V.; Grosjean, R.; Lunkenbein, T.; Mathon, O.; Pascarelli, S.; Kaper, H. Facile synthesis of high-surface area platinum-doped ceria for low temperature CO oxidation. *Catal. Today* **2019**, *333*, 105–112. [CrossRef]
11. Gatla, S.; Aubert, D.; Agostini, G.; Mathon, O.; Pascarelli, S.; Lunkenbein, T.; Willinger, M.G.; Kaper, H. Room-Temperature CO Oxidation Catalyst: Low-Temperature Metal-Support Interaction between Platinum Nanoparticles and Nanosized Ceria. *ACS Catal.* **2016**, *6*, 6151–6155. [CrossRef]
12. Parres-Esclapez, S.; Such-Basañez, I.; Illán-Gómez, M.J.; Salinas-Martínez de Lecea, C.; Bueno-López, A. Study by isotopic gases and in situ spectroscopies (DRIFTS, XPS and Raman) of the N_2O decomposition mechanism on Rh/CeO_2 and Rh/γ-Al_2O_3 catalysts. *J. Catal.* **2010**, *276*, 390–401. [CrossRef]
13. Zhu, H.; Li, Y.; Zheng, X. In-situ DRIFTS study of CeO_2 supported Rh catalysts for N_2O decomposition. *Appl. Catal. A Gen.* **2019**, *571*, 89–95. [CrossRef]
14. Zheng, J.; Meyer, S.; Köhler, K. Abatement of nitrous oxide by ruthenium catalysts: Influence of the support. *Appl. Catal. A Gen.* **2015**, *505*, 44–51. [CrossRef]

15. Pachatouridou, E.; Papista, E.; Iliopoulou, E.F.; Delimitis, A.; Goula, G.; Yentekakis, I.V.; Marnellos, G.E.; Konsolakis, M. Nitrous oxide decomposition over Al_2O_3 supported noble metals (Pt, Pd, Ir): Effect of metal loading and feed composition. *J. Environ. Chem. Eng.* **2015**, *3*, 815–821. [CrossRef]
16. Pachatouridou, E.; Papista, E.; Delimitis, A.; Vasiliades, M.A.; Efstathiou, A.M.; Amiridis, M.D.; Alexeev, O.S.; Bloom, D.; Marnellos, G.E.; Konsolakis, M.; et al. N_2O decomposition over ceria-promoted Ir/Al_2O_3 catalysts: The role of ceria. *Appl. Catal. B Environ.* **2016**, *187*, 259–268. [CrossRef]
17. Carabineiro, S.A.C.; Papista, E.; Marnellos, G.E.; Tavares, P.B.; Maldonado-Hódar, F.J.; Konsolakis, M. Catalytic decomposition of N_2O on inorganic oxides: Effect of doping with Au nanoparticles. *Mol. Catal.* **2017**, *436*, 78–89. [CrossRef]
18. Vecchietti, J.; Bonivardi, A.; Xu, W.; Stacchiola, D.; Delgado, J.J.; Calatayud, M.; Collins, S.E. Understanding the Role of Oxygen Vacancies in the Water Gas Shift Reaction on Ceria-Supported Platinum Catalysts. *ACS Catal.* **2014**, *4*, 2088–2096. [CrossRef]
19. Pierre, D.; Deng, W.; Flytzani-Stephanopoulos, M. The Importance of Strongly Bound Pt-CeO_x Species for the Water-gas Shift Reaction: Catalyst Activity and Stability Evaluation. *Top. Catal.* **2007**, *46*, 363–373. [CrossRef]
20. Mei, Z.; Li, Y.; Fan, M.; Zhao, L.; Zhao, J. Effect of the interactions between Pt species and ceria on Pt/ceria catalysts for water gas shift: The XPS studies. *Chem. Eng. J.* **2015**, *259*, 293–302. [CrossRef]
21. Ting, K.W.; Toyao, T.; Siddiki, S.M.A.H.; Shimizu, K.-I. Low-Temperature Hydrogenation of CO_2 to Methanol over Heterogeneous TiO_2-Supported Re Catalysts. *ACS Catal.* **2019**, *9*, 3685–3693. [CrossRef]
22. Wang, F.; He, S.; Chen, H.; Wang, B.; Zheng, L.; Wei, M.; Evans, D.G.; Duan, X. Active Site Dependent Reaction Mechanism over Ru/CeO_2 Catalyst toward CO_2 Methanation. *J. Am. Chem. Soc.* **2016**, *138*, 6298–6305. [CrossRef] [PubMed]
23. Sakpal, T.; Lefferts, L. Structure-dependent activity of CeO_2 supported Ru catalysts for CO_2 methanation. *J. Catal.* **2018**, *367*, 171–180. [CrossRef]
24. Vourros, A.; Garagounis, I.; Kyriakou, V.; Carabineiro, S.A.C.; Maldonado-Hódar, F.J.; Marnellos, G.E.; Konsolakis, M. Carbon dioxide hydrogenation over supported Au nanoparticles: Effect of the support. *J. CO2 Util.* **2017**, *19*, 247–256. [CrossRef]
25. Kyriakou, V.; Vourros, A.; Garagounis, I.; Carabineiro, S.A.C.; Maldonado-Hódar, F.J.; Marnellos, G.E.; Konsolakis, M. Highly active and stable TiO_2-supported Au nanoparticles for CO_2 reduction. *Catal. Commun.* **2017**, *98*, 52–56. [CrossRef]
26. Zhou, Y.; Wang, Z.; Liu, C. Perspective on CO oxidation over Pd-based catalysts. *Catal. Sci. Technol.* **2015**, *5*, 69–81. [CrossRef]
27. Carter, J.H.; Hutchings, G.J. Recent Advances in the Gold-Catalysed Low-Temperature Water–Gas Shift Reaction. *Catalysts* **2018**, *8*, 627. [CrossRef]
28. Konsolakis, M.; Carabineiro, S.A.C.; Marnellos, G.E.; Asad, M.F.; Soares, O.S.G.P.; Pereira, M.F.R.; Órfão, J.J.M.; Figueiredo, J.L. Effect of cobalt loading on the solid state properties and ethyl acetate oxidation performance of cobalt-cerium mixed oxides. *J. Colloid Interface Sci.* **2017**, *496*, 141–149. [CrossRef]
29. Konsolakis, M.; Carabineiro, S.A.C.; Marnellos, G.E.; Asad, M.F.; Soares, O.S.G.P.; Pereira, M.F.R.; Órfão, J.J.M.; Figueiredo, J.L. Volatile organic compounds abatement over copper-based catalysts: Effect of support. *Inorganica Chim. Acta* **2017**, *455*, 473–482. [CrossRef]
30. Konsolakis, M.; Carabineiro, S.A.C.; Tavares, P.B.; Figueiredo, J.L. Redox properties and VOC oxidation activity of Cu catalysts supported on $Ce_{1-x}Sm_xO_\delta$ mixed oxides. *J. Hazard. Mater.* **2013**, *261*, 512–521. [CrossRef]
31. Konsolakis, M.; Ioakeimidis, Z. Surface/structure functionalization of copper-based catalysts by metal-support and/or metal-metal interactions. *Appl. Surf. Sci.* **2014**, *320*, 244–255. [CrossRef]
32. Konsolakis, M.; Ioakimidis, Z.; Kraia, T.; Marnellos, G.E. Hydrogen Production by Ethanol Steam Reforming (ESR) over CeO_2 Supported Transition Metal (Fe, Co, Ni, Cu) Catalysts: Insight into the Structure-Activity Relationship. *Catalysts* **2016**, *6*, 39. [CrossRef]
33. Konsolakis, M.; Sgourakis, M.; Carabineiro, S.A.C. Surface and redox properties of cobalt-ceria binary oxides: On the effect of Co content and pretreatment conditions. *Appl. Surf. Sci.* **2015**, *341*, 48–54. [CrossRef]
34. Kraia, T.; Kaklidis, N.; Konsolakis, M.; Marnellos, G.E. Hydrogen production by H_2S decomposition over ceria supported transition metal (Co, Ni, Fe and Cu) catalysts. *Int. J. Hydrogen Energy* **2019**, *44*, 9753–9762. [CrossRef]

35. Lykaki, M.; Papista, E.; Kaklidis, N.; Carabineiro, S.A.C.; Konsolakis, M. Ceria Nanoparticles' Morphological Effects on the N_2O Decomposition Performance of Co_3O_4/CeO_2 Mixed Oxides. *Catalysts* **2019**, *9*, 233. [CrossRef]
36. Lykaki, M.; Stefa, S.; Carabineiro, S.A.C.; Pandis, P.K.; Stathopoulos, V.N.; Konsolakis, M. Facet-Dependent Reactivity of Fe_2O_3/CeO_2 Nanocomposites: Effect of Ceria Morphology on CO Oxidation. *Catalysts* **2019**, *9*, 371. [CrossRef]
37. Aneggi, E.; Boaro, M.; Colussi, S.; de Leitenburg, C.; Trovarelli, A. Chapter 289—Ceria-Based Materials in Catalysis: Historical Perspective and Future Trends. *Handb. Phys. Chem. Rare Earths* **2016**, *50*, 209–242. [CrossRef]
38. Tang, W.-X.; Gao, P.-X. Nanostructured cerium oxide: preparation, characterization, and application in energy and environmental catalysis. *MRS Commun.* **2016**, *6*, 311–329. [CrossRef]
39. Su, X.; Yang, X.; Zhao, B.; Huang, Y. Designing of highly selective and high-temperature endurable RWGS heterogeneous catalysts: recent advances and the future directions. *J. Energy Chem.* **2017**, *26*, 854–867. [CrossRef]
40. Konsolakis, M. The role of Copper–Ceria interactions in catalysis science: Recent theoretical and experimental advances. *Appl. Catal. B Environ.* **2016**, *198*, 49–66. [CrossRef]
41. Carabineiro, S.A.C.; Chen, X.; Konsolakis, M.; Psarras, A.C.; Tavares, P.B.; Órfão, J.J.M.; Pereira, M.F.R.; Figueiredo, J.L. Catalytic oxidation of toluene on Ce-Co and La-Co mixed oxides synthesized by exotemplating and evaporation methods. *Catal. Today* **2015**, *244*, 161–171. [CrossRef]
42. Carabineiro, S.A.C.; Konsolakis, M.; Marnellos, G.E.N.; Asad, M.F.; Soares, O.S.G.P.; Tavares, P.B.; Pereira, M.F.R.; De Melo Órfão, J.J.; Figueiredo, J.L. Ethyl Acetate Abatement on Copper Catalysts Supported on Ceria Doped with Rare Earth Oxides. *Molecules* **2016**, *21*, 644. [CrossRef]
43. Díez-Ramírez, J.; Sánchez, P.; Kyriakou, V.; Zafeiratos, S.; Marnellos, G.E.; Konsolakis, M.; Dorado, F. Effect of support nature on the cobalt-catalyzed CO_2 hydrogenation. *J. CO2 Util.* **2017**, *21*, 562–571. [CrossRef]
44. Zeng, S.; Zhang, W.; Guo, S.; Su, H. Inverse rod-like CeO_2 supported on CuO prepared by hydrothermal method for preferential oxidation of carbon monoxide. *Catal. Commun.* **2012**, *23*, 62–66. [CrossRef]
45. Rodriguez, J.A.; Liu, P.; Hrbek, J.; Evans, J.; Pérez, M. Water Gas Shift Reaction on Cu and Au Nanoparticles Supported on $CeO_2(111)$ and $ZnO(000\bar{1})$: Intrinsic Activity and Importance of Support Interactions. *Angew. Chem. Int. Ed.* **2007**, *46*, 1329–1332. [CrossRef] [PubMed]
46. Ranga Rao, G.; Mishra, B.G. Structural, redox and catalytic chemistry of ceria based materials. *Bull. Catal. Soc. India* **2003**, *2*, 122–134.
47. Rodriguez, J.A.; Grinter, D.C.; Liu, Z.; Palomino, R.M.; Senanayake, S.D. Ceria-based model catalysts: fundamental studies on the importance of the metal-ceria interface in CO oxidation, the water-gas shift, CO_2 hydrogenation, and methane and alcohol reforming. *Chem. Soc. Rev.* **2017**, *46*, 1824–1841. [CrossRef]
48. Zhang, D.; Du, X.; Shi, L.; Gao, R. Shape-controlled synthesis and catalytic application of ceria nanomaterials. *Dalton Trans.* **2012**, *41*, 14455–14475. [CrossRef]
49. Dong, L.; Yao, X.; Chen, Y. Interactions among supported copper-based catalyst components and their effects on performance: A review. *Chin. J. Catal.* **2013**, *34*, 851–864. [CrossRef]
50. Pacchioni, G. Electronic interactions and charge transfers of metal atoms and clusters on oxide surfaces. *Phys. Chem. Chem. Phys.* **2013**, *15*, 1737–1757. [CrossRef]
51. Zhou, Y.; Li, Y.; Shen, W. Shape Engineering of Oxide Nanoparticles for Heterogeneous Catalysis. *Chem. Asian J.* **2016**, *11*, 1470–1488. [CrossRef] [PubMed]
52. Cargnello, M.; Doan-Nguyen, V.V.T.; Gordon, T.R.; Diaz, R.E.; Stach, E.A.; Gorte, R.J.; Fornasiero, P.; Murray, C.B. Control of Metal Nanocrystal Size Reveals Metal-Support Interface Role for Ceria Catalysts. *Science* **2013**, *341*, 771–773. [CrossRef] [PubMed]
53. Mistry, H.; Behafarid, F.; Reske, R.; Varela, A.S.; Strasser, P.; Roldan Cuenya, B. Tuning Catalytic Selectivity at the Mesoscale via Interparticle Interactions. *ACS Catal.* **2016**, *6*, 1075–1080. [CrossRef]
54. Ahmadi, M.; Mistry, H.; Roldan Cuenya, B. Tailoring the Catalytic Properties of Metal Nanoparticles via Support Interactions. *J. Phys. Chem. Lett.* **2016**, *7*, 3519–3533. [CrossRef] [PubMed]
55. Hermes, E.D.; Jenness, G.R.; Schmidt, J.R. Decoupling the electronic, geometric and interfacial contributions to support effects in heterogeneous catalysis. *Mol. Simul.* **2015**, *41*, 123–133. [CrossRef]
56. Uzunoglu, A.; Zhang, H.; Andreescu, S.; Stanciu, L.A. CeO_2–MO_x (M: Zr, Ti, Cu) mixed metal oxides with enhanced oxygen storage capacity. *J. Mater. Sci.* **2015**, *50*, 3750–3762. [CrossRef]

57. Tang, X.; Zhang, B.; Li, Y.; Xu, Y.; Xin, Q.; Shen, W. CuO/CeO_2 catalysts: Redox features and catalytic behaviors. *Appl. Catal. A Gen.* **2005**, *288*, 116–125. [CrossRef]
58. Lu, Z.; Yang, Z.; He, B.; Castleton, C.; Hermansson, K. Cu-doped ceria: Oxygen vacancy formation made easy. *Chem. Phys. Lett.* **2011**, *510*, 60–66. [CrossRef]
59. Wang, X.; Rodriguez, J.A.; Hanson, J.C.; Gamarra, D.; Martínez-Arias, A.; Fernández-García, M. Unusual Physical and Chemical Properties of Cu in $Ce_{1-x}Cu_xO_2$ oxides. *J. Phys. Chem. B* **2005**, *109*, 19595–19603. [CrossRef]
60. Beckers, J.; Rothenberg, G. Redox properties of doped and supported copper-ceria catalysts. *Dalton Trans.* **2008**, 6573–6578. [CrossRef]
61. Van Deelen, T.W.; Hernández Mejía, C.; De Jong, K.P. Control of metal-support interactions in heterogeneous catalysts to enhance activity and selectivity. *Nat. Catal.* **2019**, *2*, 955–970. [CrossRef]
62. Yang, F.; Deng, D.; Pan, X.; Fu, Q.; Bao, X. Understanding nano effects in catalysis. *Natl. Sci. Rev.* **2015**, *2*, 183–201. [CrossRef]
63. Dinh, C.T.; Nguyen, T.D.; Kleitz, F.; Do, T.O. Chapter 10—Shape-Controlled Synthesis of Metal Oxide Nanocrystals. In *Book Controlled Nanofabrication: Advances and Applications*, 1st ed.; Lui, R.-S., Ed.; Pan Stanford Publishing Pte. Ltd.: Singapore, 2012; pp. 327–367. ISBN 978-981-4316-87-3.
64. Aneggi, E.; Wiater, D.; De Leitenburg, C.; Llorca, J.; Trovarelli, A. Shape-Dependent Activity of Ceria in Soot Combustion. *ACS Catal.* **2014**, *4*, 172–181. [CrossRef]
65. Ta, N.; Liu, J.; Shen, W. Tuning the shape of ceria nanomaterials for catalytic applications. *Chin. J. Catal.* **2013**, *34*, 838–850. [CrossRef]
66. Li, Y.; Shen, W. Morphology-dependent nanocatalysts: Rod-shaped oxides. *Chem. Soc. Rev.* **2014**, *43*, 1543–1574. [CrossRef]
67. Zhou, K.; Li, Y. Catalysis Based on Nanocrystals with Well-Defined Facets. *Angew. Chem. Int. Ed.* **2012**, *51*, 602–613. [CrossRef]
68. Liu, L.; Corma, A. Metal Catalysts for Heterogeneous Catalysis: From Single Atoms to Nanoclusters and Nanoparticles. *Chem. Rev.* **2018**, *118*, 4981–5079. [CrossRef]
69. Capdevila-Cortada, M.; Vilé, G.; Teschner, D.; Pérez-Ramírez, J.; López, N. Reactivity descriptors for ceria in catalysis. *Appl. Catal. B Environ.* **2016**, *197*, 299–312. [CrossRef]
70. Cao, S.; Tao, F.; Tang, Y.; Li, Y.; Yu, J. Size- and shape-dependent catalytic performances of oxidation and reduction reactions on nanocatalysts. *Chem. Soc. Rev.* **2016**, *45*, 4747–4765. [CrossRef]
71. Xu, J.; Harmer, J.; Li, G.; Chapman, T.; Collier, P.; Longworth, S.; Tsang, S.C. Size dependent oxygen buffering capacity of ceria nanocrystals. *Chem. Commun.* **2010**, *46*, 1887–1889. [CrossRef]
72. Puigdollers, A.R.; Schlexer, P.; Tosoni, S.; Pacchioni, G. Increasing Oxide Reducibility: The Role of Metal/Oxide Interfaces in the Formation of Oxygen Vacancies. *ACS Catal.* **2017**, *7*, 6493–6513. [CrossRef]
73. Sayle, T.X.T.; Caddeo, F.; Zhang, X.; Sakthivel, T.; Das, S.; Seal, S.; Ptasinska, S.; Sayle, D.C. Structure-Activity Map of Ceria Nanoparticles, Nanocubes, and Mesoporous Architectures. *Chem. Mater.* **2016**, *28*, 7287–7295. [CrossRef]
74. Roldan Cuenya, B. Synthesis and catalytic properties of metal nanoparticles: Size, shape, support, composition, and oxidation state effects. *Thin Solid Films* **2010**, *518*, 3127–3150. [CrossRef]
75. Martínez-Arias, A.; Gamarra, D.; Hungría, A.B.; Fernández-García, M.; Munuera, G.; Hornés, A.; Bera, P.; Conesa, J.C.; Cámara, A.L. Characterization of Active Sites/Entities and Redox/Catalytic Correlations in Copper-Ceria-Based Catalysts for Preferential Oxidation of CO in H_2-Rich Streams. *Catalysts* **2013**, *3*, 378–400. [CrossRef]
76. Prasad, R.; Rattan, G. Preparation Methods and Applications of CuO-CeO_2 Catalysts: A Short Review. *Bull. Chem. React. Eng. Catal.* **2010**, *5*, 7–30. [CrossRef]
77. Beckers, J.; Rothenberg, G. Sustainable selective oxidations using ceria-based materials. *Green Chem.* **2010**, *12*, 939–948. [CrossRef]
78. Philippot, K.; Serp, P. Chapter 1—Concepts in Nanocatalysis. In *Book Nanomaterials in Catalysis*, 1st ed.; Serp, P., Philippot, K., Eds.; Wiley-VCH Verlag GmbH & Co. KGaA.: Weinheim, Germany, 2013; pp. 1–54. [CrossRef]
79. Che, M.; Bennett, C.O. The Influence of Particle Size on the Catalytic Properties of Supported Metals. *Adv. Catal.* **1989**, *36*, 55–172. [CrossRef]

80. Hvolbæk, B.; Janssens, T.V.W.; Clausen, B.S.; Falsig, H.; Christensen, C.H.; Nørskov, J.K. Catalytic activity of Au nanoparticles. *Nano Today* **2007**, *2*, 14–18. [CrossRef]
81. Tao, F.; Dag, S.; Wang, L.-W.; Liu, Z.; Butcher, D.R.; Bluhm, H.; Salmeron, M.; Somorjai, G.A. Break-Up of Stepped Platinum Catalyst Surfaces by High CO Coverage. *Science* **2010**, *327*, 850–853. [CrossRef]
82. Qiao, Z.-A.; Wu, Z.; Dai, S. Shape-Controlled Ceria-Based Nanostructures for Catalysis Applications. *ChemSusChem* **2013**, *6*, 1821–1833. [CrossRef]
83. Vinod, C.P. Surface science as a tool for probing nanocatalysis phenomena. *Catal. Today* **2010**, *154*, 113–117. [CrossRef]
84. Somorjai, G.A.; Li, Y. *Introduction to Surface Chemistry and Catalysis*, 2nd ed.; John Wiley & Sons: Hoboken, NJ, USA, 2010; ISBN 978-0-470-50823-7.
85. Calle-Vallejo, F.; Loffreda, D.; Koper, M.T.M.; Sautet, P. Introducing structural sensitivity into adsorption-energy scaling relations by means of coordination numbers. *Nat. Chem.* **2015**, *7*, 403–410. [CrossRef] [PubMed]
86. Brodersen, S.H.; Grønbjerg, U.; Hvolbæk, B.; Schiøtz, J. Understanding the catalytic activity of gold nanoparticles through multi-scale simulations. *J. Catal.* **2011**, *284*, 34–41. [CrossRef]
87. Hamid, S.B.A.; Schlögl, R. The Impact of Nanoscience in Heterogeneous Catalysis. In *Book Nano-Micro Interface Bridging Micro Nano Worlds*, 2nd ed.; Van de Voorde, M., Werner, M., Fecht, H.-J., Eds.; Wiley-VCH Verlag GmbH & Co. KGaA.: Hoboken, NJ, USA, 2015; Volume 2, pp. 405–430. [CrossRef]
88. Cleveland, C.L.; Landman, U.; Schaaff, T.G.; Shafigullin, M.N.; Stephens, P.W.; Whetten, R.L. Structural Evolution of Smaller Gold Nanocrystals: The Truncated Decahedral Motif. *Phys. Rev. Lett.* **1997**, *79*, 1873–1876. [CrossRef]
89. Chen, M.; Goodman, D.W. Catalytically active gold on ordered titania supports. *Chem. Soc. Rev.* **2008**, *37*, 1860–1870. [CrossRef]
90. Van Bokhoven, J.A.; Louis, C.; Miller, J.T.; Tromp, M.; Safonova, O.V.; Glatzel, P. Activation of Oxygen on Gold/Alumina Catalysts: In Situ High-Energy-Resolution Fluorescence and Time-Resolved X-ray Spectroscopy. *Angew. Chem. Int. Ed.* **2006**, *45*, 4651–4654. [CrossRef]
91. Walsh, M.J.; Yoshida, K.; Kuwabara, A.; Pay, M.L.; Gai, P.L.; Boyes, E.D. On the Structural Origin of the Catalytic Properties of Inherently Strained Ultrasmall Decahedral Gold Nanoparticles. *Nano Lett.* **2012**, *12*, 2027–2031. [CrossRef]
92. Senanayake, S.D.; Rodriguez, J.A.; Stacchiola, D. Electronic Metal-Support Interactions and the Production of Hydrogen Through the Water-Gas Shift Reaction and Ethanol Steam Reforming: Fundamental Studies with Well-Defined Model Catalysts. *Top. Catal.* **2013**, *56*, 1488–1498. [CrossRef]
93. Hu, P.; Huang, Z.; Amghouz, Z.; Makkee, M.; Xu, F.; Kapteijn, F.; Dikhtiarenko, A.; Chen, Y.; Gu, X.; Tang, X. Electronic Metal-Support Interactions in Single-Atom Catalysts. *Angew. Chem. Int. Ed.* **2014**, *53*, 3418–3421. [CrossRef]
94. Han, Z.-K.; Zhang, L.; Liu, M.; Ganduglia-Pirovano, M.V.; Gao, Y. The Structure of Oxygen Vacancies in the Near-Surface of Reduced CeO_2 (111) Under Strain. *Front. Chem.* **2019**, *7*, 436. [CrossRef]
95. Murugan, B.; Ramaswamy, A.V. Defect-Site Promoted Surface Reorganization in Nanocrystalline Ceria for the Low-Temperature Activation of Ethylbenzene. *J. Am. Chem. Soc.* **2007**, *129*, 3062–3063. [CrossRef] [PubMed]
96. Campbell, C.T. Catalyst–support interactions: Electronic perturbations. *Nat. Chem.* **2012**, *4*, 597–598. [CrossRef] [PubMed]
97. Sun, C.; Li, H.; Chen, L. Nanostructured ceria-based materials: Synthesis, properties, and applications. *Energy Environ. Sci.* **2012**, *5*, 8475–8505. [CrossRef]
98. Zhou, X.-D.; Huebner, W. Size-induced lattice relaxation in CeO_2 nanoparticles. *Appl. Phys. Lett.* **2001**, *79*, 3512–3514. [CrossRef]
99. Dutta, P.; Pal, S.; Seehra, M.S.; Shi, Y.; Eyring, E.M.; Ernst, R.D. Concentration of Ce^{3+} and Oxygen Vacancies in Cerium Oxide Nanoparticles. *Chem. Mater.* **2006**, *18*, 5144–5146. [CrossRef]
100. Hailstone, R.K.; DiFrancesco, A.G.; Leong, J.G.; Allston, T.D.; Reed, K.J. A Study of Lattice Expansion in CeO_2 Nanoparticles by Transmission Electron Microscopy. *J. Phys. Chem. C* **2009**, *113*, 15155–15159. [CrossRef]
101. Migani, A.; Vayssilov, G.N.; Bromley, S.T.; Illas, F.; Neyman, K.M. Dramatic reduction of the oxygen vacancy formation energy in ceria particles: a possible key to their remarkable reactivity at the nanoscale. *J. Mater. Chem.* **2010**, *20*, 10535–10546. [CrossRef]

102. Bruix, A.; Neyman, K.M. Modeling Ceria-Based Nanomaterials for Catalysis and Related Applications. *Catal. Lett.* **2016**, *146*, 2053–2080. [CrossRef]
103. Sk, M.A.; Kozlov, S.M.; Lim, K.H.; Migani, A.; Neyman, K.M. Oxygen vacancies in self-assemblies of ceria nanoparticles. *J. Mater. Chem. A* **2014**, *2*, 18329–18338. [CrossRef]
104. Trovarelli, A.; Llorca, J. Ceria Catalysts at Nanoscale: How Do Crystal Shapes Shape Catalysis? *ACS Catal.* **2017**, *7*, 4716–4735. [CrossRef]
105. Li, Y.; Liu, Q.; Shen, W. Morphology-dependent nanocatalysis: metal particles. *Dalt. Trans.* **2011**, *40*, 5811–5826. [CrossRef] [PubMed]
106. Huang, W.; Gao, Y. Morphology-dependent surface chemistry and catalysis of CeO_2 nanocrystals. *Catal. Sci. Technol.* **2014**, *4*, 3772–3784. [CrossRef]
107. Datta, S.; Torrente-Murciano, L. Nanostructured faceted ceria as oxidation catalyst. *Curr. Opin. Chem. Eng.* **2018**, *20*, 99–106. [CrossRef]
108. Munnik, P.; De Jongh, P.E.; De Jong, K.P. Recent Developments in the Synthesis of Supported Catalysts. *Chem. Rev.* **2015**, *115*, 6687–6718. [CrossRef]
109. Vilé, G.; Colussi, S.; Krumeich, F.; Trovarelli, A.; Pérez-Ramírez, J. Opposite Face Sensitivity of CeO_2 in Hydrogenation and Oxidation Catalysis. *Angew. Chem. Int. Ed.* **2014**, *53*, 12069–12072. [CrossRef]
110. Yuan, Q.; Duan, H.-H.; Li, L.-L.; Sun, L.-D.; Zhang, Y.-W.; Yan, C.-H. Controlled synthesis and assembly of ceria-based nanomaterials. *J. Colloid Interface Sci.* **2009**, *335*, 151–167. [CrossRef]
111. Mullins, D.R. The surface chemistry of cerium oxide. *Surf. Sci. Rep.* **2015**, *70*, 42–85. [CrossRef]
112. Paier, J.; Penschke, C.; Sauer, J. Oxygen Defects and Surface Chemistry of Ceria: Quantum Chemical Studies Compared to Experiment. *Chem. Rev.* **2013**, *113*, 3949–3985. [CrossRef]
113. Yang, W.; Wang, X.; Song, S.; Zhang, H. Syntheses and Applications of Noble-Metal-free CeO_2-Based Mixed-Oxide Nanocatalysts. *Chem* **2019**, *5*, 1743–1774. [CrossRef]
114. Castanet, U.; Feral-Martin, C.; Demourgues, A.; Neale, R.L.; Sayle, D.C.; Caddeo, F.; Flitcroft, J.M.; Caygill, R.; Pointon, B.J.; Molinari, M.; et al. Controlling the {111}/{110} Surface Ratio of Cuboidal Ceria Nanoparticles. *ACS Appl. Mater. Interfaces* **2019**, *11*, 11384–11390. [CrossRef]
115. Lykaki, M.; Pachatouridou, E.; Carabineiro, S.A.C.; Iliopoulou, E.; Andriopoulou, C.; Kallithrakas-Kontos, N.; Boghosian, S.; Konsolakis, M. Ceria nanoparticles shape effects on the structural defects and surface chemistry: Implications in CO oxidation by Cu/CeO_2 catalysts. *Appl. Catal. B Environ.* **2018**, *230*, 18–28. [CrossRef]
116. Xie, Y.; Wu, J.; Jing, G.; Zhang, H.; Zeng, S.; Tian, X.; Zou, X.; Wen, J.; Su, H.; Zhong, C.-J.; et al. Structural origin of high catalytic activity for preferential CO oxidation over CuO/CeO_2 nanocatalysts with different shapes. *Appl. Catal. B Environ.* **2018**, *239*, 665–676. [CrossRef]
117. He, H.; Yang, P.; Li, J.; Shi, R.; Chen, L.; Zhang, A.; Zhu, Y. Controllable synthesis, characterization, and CO oxidation activity of CeO_2 nanostructures with various morphologies. *Ceram. Int.* **2016**, *42*, 7810–7818. [CrossRef]
118. Chang, H.; Ma, L.; Yang, S.; Li, J.; Chen, L.; Wang, W.; Hao, J. Comparison of preparation methods for ceria catalyst and the effect of surface and bulk sulfates on its activity toward NH_3-SCR. *J. Hazard. Mater.* **2013**, *262*, 782–788. [CrossRef] [PubMed]
119. Gawade, P.; Mirkelamoglu, B.; Ozkan, U.S. The Role of Support Morphology and Impregnation Medium on the Water Gas Shift Activity of Ceria-Supported Copper Catalysts. *J. Phys. Chem. C* **2010**, *114*, 18173–18181. [CrossRef]
120. Liu, Y.H.; Zuo, J.C.; Ren, X.F.; Yong, L. Synthesis and character of cerium oxide (CeO_2) nanoparticles by the precipitation method. *Metalurgija* **2014**, *53*, 463–465.
121. Lykaki, M.; Pachatouridou, E.; Iliopoulou, E.; Carabineiro, S.A.C.; Konsolakis, M. Impact of the synthesis parameters on the solid state properties and the CO oxidation performance of ceria nanoparticles. *RSC Adv.* **2017**, *7*, 6160–6169. [CrossRef]
122. Shang, H.; Zhang, X.; Xu, J.; Han, Y. Effects of preparation methods on the activity of CuO/CeO_2 catalysts for CO oxidation. *Front. Chem. Sci. Eng.* **2017**, *11*, 603–612. [CrossRef]
123. Zhou, L.; Li, X.; Yao, Z.; Chen, Z.; Hong, M.; Zhu, R.; Liang, Y.; Zhao, J. Transition-Metal Doped Ceria Microspheres with Nanoporous Structures for CO Oxidation. *Sci. Rep.* **2016**, *6*, 23900. [CrossRef]
124. Wang, W.-W.; Yu, W.-Z.; Du, P.-P.; Xu, H.; Jin, Z.; Si, R.; Ma, C.; Shi, S.; Jia, C.-J.; Yan, C.-H. Crystal Plane Effect of Ceria on Supported Copper Oxide Cluster Catalyst for CO Oxidation: Importance of Metal-Support Interaction. *ACS Catal.* **2017**, *7*, 1313–1329. [CrossRef]

125. Nakagawa, K.; Ohshima, T.; Tezuka, Y.; Katayama, M.; Katoh, M.; Sugiyama, S. Morphological effects of CeO_2 nanostructures for catalytic soot combustion of CuO/CeO_2. *Catal. Today* **2015**, *246*, 67–71. [CrossRef]
126. Chaudhary, S.; Sharma, P.; Kumar, R.; Mehta, S.K. Nanoscale surface designing of Cerium oxide nanoparticles for controlling growth, stability, optical and thermal properties. *Ceram. Int.* **2015**, *41*, 10995–11003. [CrossRef]
127. Miyazaki, H.; Kato, J.-I.; Sakamoto, N.; Wakiya, N.; Ota, T.; Suzuki, H. Synthesis of CeO_2 nanoparticles by rapid thermal decomposition using microwave heating. *Adv. Appl. Ceram.* **2010**, *109*, 123–127. [CrossRef]
128. Yang, H.; Huang, C.; Tang, A.; Zhang, X.; Yang, W. Microwave-assisted synthesis of ceria nanoparticles. *Mater. Res. Bull.* **2005**, *40*, 1690–1695. [CrossRef]
129. Zawadzki, M. Preparation and characterization of ceria nanoparticles by microwave-assisted solvothermal process. *J. Alloys Compd.* **2008**, *454*, 347–351. [CrossRef]
130. Zhang, Y.-W.; Si, R.; Liao, C.-S.; Yan, C.-H.; Xiao, C.-X.; Kou, Y. Facile Alcohothermal Synthesis, Size-Dependent Ultraviolet Absorption, and Enhanced CO Conversion Activity of Ceria Nanocrystals. *J. Phys. Chem. B* **2003**, *107*, 10159–10167. [CrossRef]
131. Ouyang, B.; Tan, W.; Liu, B. Morphology effect of nanostructure ceria on the Cu/CeO_2 catalysts for synthesis of methanol from CO_2 hydrogenation. *Catal. Commun.* **2017**, *95*, 36–39. [CrossRef]
132. Mai, H.-X.; Sun, L.-D.; Zhang, Y.-W.; Si, R.; Feng, W.; Zhang, H.-P.; Liu, H.-C.; Yan, C.-H. Shape-Selective Synthesis and Oxygen Storage Behavior of Ceria Nanopolyhedra, Nanorods, and Nanocubes. *J. Phys. Chem. B* **2005**, *109*, 24380–24385. [CrossRef]
133. Gamarra, D.; López Cámara, A.; Monte, M.; Rasmussen, S.B.; Chinchilla, L.E.; Hungría, A.B.; Munuera, G.; Gyorffy, N.; Schay, Z.; Cortés Corberán, V.; et al. Preferential oxidation of CO in excess H_2 over CuO/CeO_2 catalysts: Characterization and performance as a function of the exposed face present in the CeO_2 support. *Appl. Catal. B Environ.* **2013**, *130–131*, 224–238. [CrossRef]
134. Piumetti, M.; Bensaid, S.; Andana, T.; Dosa, M.; Novara, C.; Giorgis, F.; Russo, N.; Fino, D. Nanostructured Ceria-Based Materials: Effect of the Hydrothermal Synthesis Conditions on the Structural Properties and Catalytic Activity. *Catalysts* **2017**, *7*, 174. [CrossRef]
135. Araiza, D.G.; Gómez-Cortés, A.; Díaz, G. Partial oxidation of methanol over copper supported on nanoshaped ceria for hydrogen production. *Catal. Today* **2017**, *282*, 185–194. [CrossRef]
136. Yao, S.Y.; Xu, W.Q.; Johnston-Peck, A.C.; Zhao, F.Z.; Liu, Z.Y.; Luo, S.; Senanayake, S.D.; Martínez-Arias, A.; Liu, W.J.; Rodriguez, J.A. Morphological effects of the nanostructured ceria support on the activity and stability of CuO/CeO_2 catalysts for the water-gas shift reaction. *Phys. Chem. Chem. Phys.* **2014**, *16*, 17183–17195. [CrossRef] [PubMed]
137. Lin, L.; Yao, S.; Liu, Z.; Zhang, F.; Li, N.; Vovchok, D.; Martínez-Arias, A.; Castañeda, R.; Lin, J.; Senanayake, S.D.; et al. In Situ Characterization of Cu/CeO_2 Nanocatalysts for CO_2 Hydrogenation: Morphological Effects of Nanostructured Ceria on the Catalytic Activity. *J. Phys. Chem. C* **2018**, *122*, 12934–12943. [CrossRef]
138. Andana, T.; Piumetti, M.; Bensaid, S.; Veyre, L.; Thieuleux, C.; Russo, N.; Fino, D.; Quadrelli, E.A.; Pirone, R. CuO nanoparticles supported by ceria for NO_x-assisted soot oxidation: insight into catalytic activity and sintering. *Appl. Catal. B Environ.* **2017**, *216*, 41–58. [CrossRef]
139. Miceli, P.; Bensaid, S.; Russo, N.; Fino, D. Effect of the morphological and surface properties of CeO_2-based catalysts on the soot oxidation activity. *Chem. Eng. J.* **2015**, *278*, 190–198. [CrossRef]
140. Liu, J.; Li, Y.; Zhang, J.; He, D. Glycerol carbonylation with CO_2 to glycerol carbonate over CeO_2 catalyst and the influence of CeO_2 preparation methods and reaction parameters. *Appl. Catal. A Gen.* **2016**, *513*, 9–18. [CrossRef]
141. He, H.-W.; Wu, X.-Q.; Ren, W.; Shi, P.; Yao, X.; Song, Z.-T. Synthesis of crystalline cerium dioxide hydrosol by a sol-gel method. *Ceram. Int.* **2012**, *38S*, S501–S504. [CrossRef]
142. Phonthammachai, N.; Rumruangwong, M.; Gulari, E.; Jamieson, A.M.; Jitkarnka, S.; Wongkasemjit, S. Synthesis and rheological properties of mesoporous nanocrystalline CeO_2 via sol-gel process. *Colloids Surf. A Physicochem. Eng. Asp.* **2004**, *247*, 61–68. [CrossRef]
143. Pinjari, D.V.; Pandit, A.B. Room temperature synthesis of crystalline CeO_2 nanopowder: Advantage of sonochemical method over conventional method. *Ultrason. Sonochem.* **2011**, *18*, 1118–1123. [CrossRef]
144. Yin, L.; Wang, Y.; Pang, G.; Koltypin, Y.; Gedanken, A. Sonochemical Synthesis of Cerium Oxide Nanoparticles—Effect of Additives and Quantum Size Effect. *J. Colloid Interface Sci.* **2002**, *246*, 78–84. [CrossRef]

145. Roldan Cuenya, B.; Behafarid, F. Nanocatalysis: size- and shape-dependent chemisorption and catalytic reactivity. *Surf. Sci. Rep.* **2015**, *70*, 135–187. [CrossRef]
146. Zabilskiy, M.; Djinović, P.; Tchernychova, E.; Tkachenko, O.P.; Kustov, L.M.; Pintar, A. Nanoshaped CuO/CeO_2 Materials: Effect of the Exposed Ceria Surfaces on Catalytic Activity in N_2O Decomposition Reaction. *ACS Catal.* **2015**, *5*, 5357–5365. [CrossRef]
147. Cui, Y.; Dai, W.-L. Support morphology and crystal plane effect of Cu/CeO_2 nanomaterial on the physicochemical and catalytic properties for carbonate hydrogenation. *Catal. Sci. Technol.* **2016**, *6*, 7752–7762. [CrossRef]
148. Liu, L.; Yao, Z.; Deng, Y.; Gao, F.; Liu, B.; Dong, L. Morphology and Crystal-Plane Effects of Nanoscale Ceria on the Activity of CuO/CeO_2 for NO Reduction by CO. *ChemCatChem* **2011**, *3*, 978–989. [CrossRef]
149. López, J.M.; Gilbank, A.L.; García, T.; Solsona, B.; Agouram, S.; Torrente-Murciano, L. The prevalence of surface oxygen vacancies over the mobility of bulk oxygen in nanostructured ceria for the total toluene oxidation. *Appl. Catal. B Environ.* **2015**, *174–175*, 403–412. [CrossRef]
150. Araiza, D.G.; Gómez-Cortés, A.; Díaz, G. Reactivity of methanol over copper supported on well-shaped CeO_2: A TPD-DRIFTS study. *Catal. Sci. Technol.* **2017**, *7*, 5224–5235. [CrossRef]
151. Sudarsanam, P.; Hillary, B.; Amin, M.H.; Rockstroh, N.; Bentrup, U.; Brückner, A.; Bhargava, S.K. Heterostructured Copper-Ceria and Iron-Ceria Nanorods: Role of Morphology, Redox, and Acid Properties in Catalytic Diesel Soot Combustion. *Langmuir* **2018**, *34*, 2663–2673. [CrossRef]
152. Piumetti, M.; Bensaid, S.; Russo, N.; Fino, D. Nanostructured ceria-based catalysts for soot combustion: Investigations on the surface sensitivity. *Appl. Catal. B Environ.* **2015**, *165*, 742–751. [CrossRef]
153. Wang, S.; Zhao, L.; Wang, W.; Zhao, Y.; Zhang, G.; Ma, X.; Gong, J. Morphology control of ceria nanocrystals for catalytic conversion of CO_2 with methanol. *Nanoscale* **2013**, *5*, 5582–5588. [CrossRef]
154. Wu, Z.; Li, M.; Overbury, S.H. On the structure dependence of CO oxidation over CeO_2 nanocrystals with well-defined surface planes. *J. Catal.* **2012**, *285*, 61–73. [CrossRef]
155. Kovacevic, M.; Mojet, B.L.; Van Ommen, J.G.; Lefferts, L. Effects of Morphology of Cerium Oxide Catalysts for Reverse Water Gas Shift Reaction. *Catal. Lett.* **2016**, *146*, 770–777. [CrossRef]
156. Piumetti, M.; Andana, T.; Bensaid, S.; Russo, N.; Fino, D.; Pirone, R. Study on the CO Oxidation over Ceria-Based Nanocatalysts. *Nanoscale Res. Lett.* **2016**, *11*, 165. [CrossRef] [PubMed]
157. Guo, X.; Zhou, R. A new insight into the morphology effect of ceria on CuO/CeO_2 catalysts for CO selective oxidation in hydrogen-rich gas. *Catal. Sci. Technol.* **2016**, *6*, 3862–3871. [CrossRef]
158. Mock, S.A.; Sharp, S.E.; Stoner, T.R.; Radetic, M.J.; Zell, E.T.; Wang, R. CeO_2 nanorods-supported transition metal catalysts for CO oxidation. *J. Colloid Interface Sci.* **2016**, *466*, 261–267. [CrossRef]
159. Ren, Z.; Peng, F.; Li, J.; Liang, X.; Chen, B. Morphology-Dependent Properties of Cu/CeO_2 Catalysts for the Water-Gas Shift Reaction. *Catalysts* **2017**, *7*, 48. [CrossRef]
160. Li, Y.; Wei, Z.; Gao, F.; Kovarik, L.; Peden, C.H.F.; Wang, Y. Effects of CeO_2 support facets on VO_x/CeO_2 catalysts in oxidative dehydrogenation of methanol. *J. Catal.* **2014**, *315*, 15–24. [CrossRef]
161. Mudiyanselage, K.; Senanayake, S.D.; Feria, L.; Kundu, S.; Baber, A.E.; Graciani, J.; Vidal, A.B.; Agnoli, S.; Evans, J.; Chang, R.; et al. Importance of the Metal-Oxide Interface in Catalysis: In Situ Studies of the Water-Gas Shift Reaction by Ambient-Pressure X-ray Photoelectron Spectroscopy. *Angew. Chem. Int. Ed.* **2013**, *52*, 5101–5105. [CrossRef]
162. Mayernick, A.D.; Janik, M.J. Methane Activation and Oxygen Vacancy Formation over CeO_2 and Zr, Pd Substituted CeO_2 Surfaces. *J. Phys. Chem. C* **2008**, *112*, 14955–14964. [CrossRef]
163. Nolan, M.; Grigoleit, S.; Sayle, D.C.; Parker, S.C.; Watson, G.W. Density functional theory studies of the structure and electronic structure of pure and defective low index surfaces of ceria. *Surf. Sci.* **2005**, *576*, 217–229. [CrossRef]
164. Nolan, M.; Parker, S.C.; Watson, G.W. The electronic structure of oxygen vacancy defects at the low index surfaces of ceria. *Surf. Sci.* **2005**, *595*, 223–232. [CrossRef]
165. Sayle, T.X.T.; Parker, S.C.; Sayle, D.C. Oxidising CO to CO_2 using ceria nanoparticles. *Phys. Chem. Chem. Phys.* **2005**, *7*, 2936–2941. [CrossRef] [PubMed]
166. Sayle, T.X.T.; Cantoni, M.; Bhatta, U.M.; Parker, S.C.; Hall, S.R.; Möbus, G.; Molinari, M.; Reid, D.; Seal, S.; Sayle, D.C. Strain and Architecture-Tuned Reactivity in Ceria Nanostructures; Enhanced Catalytic Oxidation of CO to CO_2. *Chem. Mater.* **2012**, *24*, 1811–1821. [CrossRef]
167. Spencer, N.D.; Schoonmaker, R.C.; Somorjai, G.A. Iron Single Crystals as Ammonia Synthesis Catalysts: Effect of Surface Structure on Catalyst Activity. *J. Catal.* **1982**, *74*, 129–135. [CrossRef]

168. Xie, S.; Choi, S.-I.; Xia, X.; Xia, Y. Catalysis on faceted noble-metal nanocrystals: both shape and size matter. *Curr. Opin. Chem. Eng.* **2013**, *2*, 142–150. [CrossRef]
169. Zhai, Y.; Pierre, D.; Si, R.; Deng, W.; Ferrin, P.; Nilekar, A.U.; Peng, G.; Herron, J.A.; Bell, D.C.; Saltsburg, H.; et al. Alkali-Stabilized Pt-OH_x Species Catalyze Low-Temperature Water-Gas Shift Reactions. *Science* **2010**, *329*, 1633–1636. [CrossRef] [PubMed]
170. Stakheev, A.Y.; Kustov, L.M. Effects of the support on the morphology and electronic properties of supported metal clusters: modern concepts and progress in 1990s. *Appl. Catal. A Gen.* **1999**, *188*, 3–35. [CrossRef]
171. Lee, D.W.; Yoo, B.R. Advanced metal oxide (supported) catalysts: Synthesis and applications. *J. Ind. Eng. Chem.* **2014**, *20*, 3947–3959. [CrossRef]
172. Yentekakis, I.V.; Konsolakis, M.; Lambert, R.M.; MacLeod, N.; Nalbantian, L. Extraordinarily effective promotion by sodium in emission control catalysis: NO reduction by propene over Na-promoted Pt/γ-Al_2O_3. *Appl. Catal. B Environ.* **1999**, *22*, 123–133. [CrossRef]
173. Konsolakis, M.; Vrontaki, M.; Avgouropoulos, G.; Ioannides, T.; Yentekakis, I.V. Novel doubly-promoted catalysts for the lean NO_x reduction by H_2 + CO: Pd(K)/Al_2O_3-(TiO_2). *Appl. Catal. B Environ.* **2006**, *68*, 59–67. [CrossRef]
174. Konsolakis, M.; Aligizou, F.; Goula, G.; Yentekakis, I.V. N_2O decomposition over doubly-promoted Pt(K)/Al_2O_3-(CeO_2-La_2O_3) structured catalysts: On the combined effects of promotion and feed composition. *Chem. Eng. J.* **2013**, *230*, 286–295. [CrossRef]
175. Janek, J.C.G.; Vayenas, S.; Bebelis, C.; Pliangos, S.; Brosda, D. Tsiplakides: Electrochemical activation of catalysis: promotion, electrochemical promotion, and metal-support interaction. *J. Solid State Electrochem.* **2002**, *7*, 60–61. [CrossRef]
176. De Lucas-Consuegra, A. New Trends of Alkali Promotion in Heterogeneous Catalysis: Electrochemical Promotion with Alkaline Ionic Conductors. *Catal. Surv. Asia* **2015**, *19*, 25–37. [CrossRef]
177. Ertl, G.; Knözinger, H.; Weitkamp, J. *Handbook of Heterogeneous Catalysis*; VCH Verlagsgesellschaft mbH: Weinheim, Germany, 1997; ISBN 9783527619474.
178. Stelmachowski, P.; Maniak, G.; Kotarba, A.; Sojka, Z. Strong electronic promotion of Co_3O_4 towards N_2O decomposition by surface alkali dopants. *Catal. Commun.* **2009**, *10*, 1062–1065. [CrossRef]
179. Zasada, F.; Stelmachowski, P.; Maniak, G.; Paul, J.-F.; Kotarba, A.; Sojka, Z. Potassium Promotion of Cobalt Spinel Catalyst for N_2O Decomposition-Accounted by Work Function Measurements and DFT Modelling. *Catal. Lett.* **2009**, *127*, 126–131. [CrossRef]
180. Kiskinova, M.P. Chapter 7—Theoretical Approaches to the Description of the Modifier Effects. In *Book Studies in Surface Science and Catalysis—Poisoning and Promotion in Catalysis based on Surface Science Concepts and Experiments*, 1st ed.; Kiskinova, M.P., Delmon, B., Yates, J.T., Eds.; Elsevier Science Publishers B.V.: Amsterdam, The Netherlands, 1991; Volume 70, pp. 285–307. [CrossRef]
181. Kiskinova, M.P. Chapter 4—Interaction of Atomic Adsorbates, Acting As Promoters Or Poisons With Single Crystal Metal Surfaces. In *Book Studies in Surface Science and Catalysis—Poisoning and Promotion in Catalysis Based on Surface Science Concepts and Experiments*, 1st ed.; Kiskinova, M.P., Delmon, B., Yates, J.T., Eds.; Elsevier Science Publishers B.V.: Amsterdam, The Netherlands, 1991; Volume 70, pp. 19–68. [CrossRef]
182. Huo, C.-F.; Wu, B.-S.; Gao, P.; Yang, Y.; Li, Y.-W.; Jiao, H. The Mechanism of Potassium Promoter: Enhancing the Stability of Active Surfaces. *Angew. Chem. Int. Ed.* **2011**, *50*, 7403–7406. [CrossRef]
183. Besenbacher, F.; Chorkendorff, I.; Clausen, B.S.; Hammer, B.; Molenbroek, A.M.; Nørskov, J.K.; Stensgaard, I. Design of a Surface Alloy Catalyst for Steam Reforming. *Science* **1998**, *279*, 1913–1915. [CrossRef]
184. Lamy, C.; Belgsir, E.M.; Léger, J.-M. Electrocatalytic oxidation of aliphatic alcohols: Application to the direct alcohol fuel cell (DAFC). *J. Appl. Electrochem.* **2001**, *31*, 799–809. [CrossRef]
185. Liang, Y.N.; Oh, W.-D.; Li, Y.; Hu, X. Nanocarbons as platforms for developing novel catalytic composites: overview and prospects. *Appl. Catal. A Gen.* **2018**, *562*, 94–105. [CrossRef]
186. Chen, W.; Fan, Z.; Pan, X.; Bao, X. Effect of Confinement in Carbon Nanotubes on the Activity of Fischer-Tropsch Iron Catalyst. *J. Am. Chem. Soc.* **2008**, *130*, 9414–9419. [CrossRef]
187. Geim, A.K.; Grigorieva, I.V. Van der Waals heterostructures. *Nature* **2013**, *499*, 419–425. [CrossRef]
188. Fu, Q.; Bao, X. Catalysis on a metal surface with a graphitic cover. *Chin. J. Catal.* **2015**, *36*, 517–519. [CrossRef]
189. Ania, C.O.; Seredych, M.; Rodriguez-Castellon, E.; Bandosz, T.J. New copper/GO based material as an efficient oxygen reduction catalyst in an alkaline medium: The role of unique Cu/rGO architecture. *Appl. Catal. B Environ.* **2015**, *163*, 424–435. [CrossRef]

190. Safaei, M.; Foroughi, M.M.; Ebrahimpoor, N.; Jahani, S.; Omidi, A.; Khatami, M. A review on metal-organic frameworks: Synthesis and applications. *TrAC Trends Anal. Chem.* **2019**, *118*, 401–425. [CrossRef]
191. Wang, Y.; Yang, Y.; Liu, N.; Wang, Y.; Zhang, X. Sword-like CuO/CeO_2 composites derived from a Ce-BTC metal-organic framework with superior CO oxidation performance. *RSC Adv.* **2018**, *8*, 33096–33102. [CrossRef]
192. Yang, Q.; Xu, Q.; Jiang, H.-L. Metal-organic frameworks meet metal nanoparticles: synergistic effect for enhanced catalysis. *Chem. Soc. Rev.* **2017**, *46*, 4774–4808. [CrossRef] [PubMed]
193. Chen, L.; Chen, H.; Luque, R.; Li, Y. Metal–organic framework encapsulated Pd nanoparticles: towards advanced heterogeneous catalysts. *Chem. Sci.* **2014**, *5*, 3708–3714. [CrossRef]
194. Chen, L.; Xu, Q. Metal-Organic Framework Composites for Catalysis. *Matter* **2019**, *1*, 57–89. [CrossRef]
195. Li, P.-Z.; Aranishi, K.; Xu, Q. ZIF-8 immobilized nickel nanoparticles: highly effective catalysts for hydrogen generation from hydrolysis of ammonia borane. *Chem. Commun.* **2012**, *48*, 3173–3175. [CrossRef]
196. Abdel-Mageed, A.M.; Rungtaweevoranit, B.; Parlinska-Wojtan, M.; Pei, X.; Yaghi, O.M.; Jürgen Behm, R. Highly Active and Stable Single-Atom Cu Catalysts Supported by a Metal-Organic Framework. *J. Am. Chem. Soc.* **2019**, *141*, 5201–5210. [CrossRef]
197. Rungtaweevoranit, B.; Baek, J.; Araujo, J.R.; Archanjo, B.S.; Choi, K.M.; Yaghi, O.M.; Somorjai, G.A. Copper Nanocrystals Encapsulated in Zr-based Metal-Organic Frameworks for Highly Selective CO_2 Hydrogenation to Methanol. *Nano Lett.* **2016**, *16*, 7645–7649. [CrossRef]
198. Ye, J.-Y.; Liu, C.-J. $Cu_3(BTC)_2$: CO oxidation over MOF based catalysts. *Chem. Commun.* **2011**, *47*, 2167–2169. [CrossRef] [PubMed]
199. Li, X.; Goh, T.W.; Li, L.; Xiao, C.; Guo, Z.; Zeng, X.C.; Huang, W. Controlling Catalytic Properties of Pd Nanoclusters through Their Chemical Environment at the Atomic Level Using Isoreticular Metal-Organic Frameworks. *ACS Catal.* **2016**, *6*, 3461–3468. [CrossRef]
200. Lawrence, N.J.; Brewer, J.R.; Wang, L.; Wu, T.-S.; Wells-Kingsbury, J.; Ihrig, M.M.; Wang, G.; Soo, Y.-L.; Mei, W.-N.; Cheung, C.L. Defect Engineering in Cubic Cerium Oxide Nanostructures for Catalytic Oxidation. *Nano Lett.* **2011**, *11*, 2666–2671. [CrossRef] [PubMed]
201. Mock, S.A.; Zell, E.T.; Hossain, S.T.; Wang, R. Effect of Reduction Treatment on CO Oxidation with CeO_2 Nanorod-Supported CuO_x Catalysts. *ChemCatChem* **2018**, *10*, 311–319. [CrossRef]
202. Lin, B.; Qi, Y.; Wei, K.; Lin, J. Effect of pretreatment on ceria-supported cobalt catalyst for ammonia synthesis. *RSC Adv.* **2014**, *4*, 38093–38102. [CrossRef]
203. Ren, Y.; Tang, K.; Wei, J.; Yang, H.; Wei, H.; Yang, Y. Pretreatment Effect on Ceria-Supported Gold Nanocatalysts for CO Oxidation: Importance of the Gold–Ceria Interaction. *Energy Technol.* **2018**, *6*, 379–390. [CrossRef]
204. Gawande, M.B.; Goswami, A.; Felpin, F.-X.; Asefa, T.; Huang, X.; Silva, R.; Zou, X.; Zboril, R.; Varma, R.S. Cu and Cu-Based Nanoparticles: Synthesis and Applications in Catalysis. *Chem. Rev.* **2016**, *116*, 3722–3811. [CrossRef]
205. Zabilskiy, M.; Djinović, P.; Erjavec, B.; Dražić, G.; Pintar, A. Small CuO clusters on CeO_2 nanospheres as active species for catalytic N_2O decomposition. *Appl. Catal. B Environ.* **2015**, *163*, 113–122. [CrossRef]
206. Li, L.; Zhan, Y.; Zheng, Q.; Zheng, Y.; Chen, C.; She, Y.; Lin, X.; Wei, K. Water-Gas Shift Reaction over CuO/CeO_2 Catalysts: Effect of the Thermal Stability and Oxygen Vacancies of CeO_2 Supports Previously Prepared by Different Methods. *Catal. Lett.* **2009**, *130*, 532–540. [CrossRef]
207. Li, L.; Song, L.; Wang, H.; Chen, C.; She, Y.; Zhan, Y.; Lin, X.; Zheng, Q. Water-gas shift reaction over CuO/CeO_2 catalysts: Effect of CeO_2 supports previously prepared by precipitation with different precipitants. *Int. J. Hydrogen Energy* **2011**, *36*, 8839–8849. [CrossRef]
208. Li, L.; Song, L.; Chen, C.; Zhang, Y.; Zhan, Y.; Lin, X.; Zheng, Q.; Wang, H.; Ma, H.; Ding, L.; et al. Modified precipitation processes and optimized copper content of CuO-CeO_2 catalysts for water-gas shift reaction. *Int. J. Hydrogen Energy* **2014**, *39*, 19570–19582. [CrossRef]
209. She, Y.; Li, L.; Zhan, Y.; Lin, X.; Zheng, Q.; Wei, K. Effect of yttrium addition on water-gas shift reaction over CuO/CeO_2 catalysts. *J. Rare Earths* **2009**, *27*, 411–417. [CrossRef]
210. Dong, F.; Meng, Y.; Han, W.; Zhao, H.; Tang, Z. Morphology effects on surface chemical properties and lattice defects of Cu/CeO_2 catalysts applied for low-temperature CO oxidation. *Sci. Rep.* **2019**, *9*, 12056. [CrossRef] [PubMed]

211. Fotopoulos, A.; Arvanitidis, J.; Christofilos, D.; Papaggelis, K.; Kalyva, M.; Triantafyllidis, K.; Niarchos, D.; Boukos, N.; Basina, G.; Tzitzios, V. One Pot Synthesis and Characterization of Ultra Fine CeO_2 and Cu/CeO_2 Nanoparticles. Application for Low Temperature CO Oxidation. *J. Nanosci. Nanotechnol.* **2011**, *11*, 8593–8598. [CrossRef]
212. Qin, J.; Lu, J.; Cao, M.; Hu, C. Synthesis of porous CuO-CeO_2 nanospheres with an enhanced low-temperature CO oxidation activity. *Nanoscale* **2010**, *2*, 2739–2743. [CrossRef] [PubMed]
213. Su, Y.; Dai, L.; Zhang, Q.; Li, Y.; Peng, J.; Wu, R.; Han, W.; Tang, Z.; Wang, Y. Fabrication of Cu-Doped CeO_2 Catalysts with Different Dimension Pore Structures for CO Catalytic Oxidation. *Catal. Surv. Asia* **2016**, *20*, 231–240. [CrossRef]
214. Zhou, H.; Huang, Z.; Sun, C.; Qin, F.; Xiong, D.; Shen, W.; Xu, H. Catalytic decomposition of N_2O over $Cu_xCe_{1\text{-}x}O_y$ mixed oxides. *Appl. Catal. B Environ.* **2012**, *125*, 492–498. [CrossRef]
215. Lykaki, M.; Papista, E.; Carabineiro, S.A.C.; Tavares, P.B.; Konsolakis, M. Optimization of N_2O decomposition activity of CuO-CeO_2 mixed oxides by means of synthesis procedure and alkali (Cs) promotion. *Catal. Sci. Technol.* **2018**, *8*, 2312–2322. [CrossRef]
216. Liu, Z.; He, C.; Chen, B.; Liu, H. CuO-CeO_2 mixed oxide catalyst for the catalytic decomposition of N_2O in the presence of oxygen. *Catal. Today* **2017**, *297*, 78–83. [CrossRef]
217. Yang, F.; Graciani, J.; Evans, J.; Liu, P.; Hrbek, J.; Sanz, J.F.; Rodriguez, J.A. CO oxidation on Inverse CeO_x/Cu(111) Catalysts: High Catalytic Activity and Ceria-Promoted Dissociation of O_2. *J. Am. Chem. Soc.* **2011**, *133*, 3444–3451. [CrossRef]
218. Berlowitz, P.J.; Peden, C.H.F.; Wayne Goodman, D. Kinetics of carbon monoxide oxidation on single-crystal palladium, platinum, and iridium. *J. Phys. Chem.* **1988**, *92*, 5213–5221. [CrossRef]
219. Rodriguez, J.A.; Wayne Goodman, D. High-pressure catalytic reactions over single-crystal metal surfaces. *Surf. Sci. Rep.* **1991**, *14*, 1–107. [CrossRef]
220. Gamarra, D.; Belver, C.; Fernández-García, M.; Martínez-Arias, A. Selective CO Oxidation in Excess H_2 over Copper-Ceria Catalysts: Identification of Active Entities/Species. *J. Am. Chem. Soc.* **2007**, *129*, 12064–12065. [CrossRef] [PubMed]
221. Senanayake, S.D.; Stacchiola, D.; Rodriguez, J.A. Unique Properties of Ceria Nanoparticles Supported on Metals: Novel Inverse Ceria/Copper Catalysts for CO Oxidation and the Water-Gas Shift Reaction. *Acc. Chem. Res.* **2013**, *46*, 1702–1711. [CrossRef] [PubMed]
222. Pillai, U.R.; Deevi, S. Room temperature oxidation of carbon monoxide over copper oxide catalyst. *Appl. Catal. B Environ.* **2006**, *64*, 146–151. [CrossRef]
223. Zhao, D.; Tu, C.-M.; Hu, X.-J.; Zhang, N. Notable in situ surface transformation of Cu_2O nanomaterials leads to dramatic activity enhancement for CO oxidation. *RSC Adv.* **2017**, *7*, 37596–37603. [CrossRef]
224. Wang, X.; Liu, C.; Zheng, B.; Jiang, Y.; Zhang, L.; Xie, Z.; Zheng, L. Controlled synthesis of concave Cu_2O microcrystals enclosed by {hhl} high-index facets and enhanced catalytic activity. *J. Mater. Chem. A* **2013**, *1*, 282–287. [CrossRef]
225. Zhou, K.; Wang, R.; Xu, B.; Li, Y. Synthesis, characterization and catalytic properties of CuO nanocrystals with various shapes. *Nanotechnology* **2006**, *17*, 3939–3943. [CrossRef]
226. Huang, H.; Zhang, L.; Wu, K.; Yu, Q.; Chen, R.; Yang, H.; Peng, X.; Ye, Z. Hetero-metal cation control of CuO nanostructures and their high catalytic performance for CO oxidation. *Nanoscale* **2012**, *4*, 7832–7841. [CrossRef]
227. Jia, A.-P.; Jiang, S.-Y.; Lu, J.-Q.; Luo, M.-F. Study of Catalytic Activity at the CuO–CeO_2 Interface for CO Oxidation. *J. Phys. Chem. C* **2010**, *114*, 21605–21610. [CrossRef]
228. Vayssilov, G.N.; Lykhach, Y.; Migani, A.; Staudt, T.; Petrova, G.P.; Tsud, N.; Skála, T.; Bruix, A.; Illas, F.; Prince, K.C.; et al. Support nanostructure boosts oxygen transfer to catalytically active platinum nanoparticles. *Nat. Mater.* **2011**, *10*, 310–315. [CrossRef] [PubMed]
229. Hossain, S.T.; Almesned, Y.; Zhang, K.; Zell, E.T.; Bernard, D.T.; Balaz, S.; Wang, R. Support structure effect on CO oxidation: A comparative study on SiO_2 nanospheres and CeO_2 nanorods supported CuO_x catalysts. *Appl. Surf. Sci.* **2018**, *428*, 598–608. [CrossRef]
230. Yao, S.; Mudiyanselage, K.; Xu, W.; Johnston-Peck, A.C.; Hanson, J.C.; Wu, T.; Stacchiola, D.; Rodriguez, J.A.; Zhao, H.; Beyer, K.A.; et al. Unraveling the Dynamic Nature of a CuO/CeO_2 Catalyst for CO Oxidation in Operando: A Combined Study of XANES (Fluorescence) and Drifts. *ACS Catal.* **2014**, *4*, 1650–1661. [CrossRef]

231. Hossain, S.T.; Azeeva, E.; Zhang, K.; Zell, E.T.; Bernard, D.T.; Balaz, S.; Wang, R. A comparative study of CO oxidation over Cu-O-Ce solid solutions and CuO/CeO_2 nanorods catalysts. *Appl. Surf. Sci.* **2018**, *455*, 132–143. [CrossRef]
232. Qi, L.; Yu, Q.; Dai, Y.; Tang, C.; Liu, L.; Zhang, H.; Gao, F.; Dong, L.; Chen, Y. Influence of cerium precursors on the structure and reducibility of mesoporous $CuO\text{-}CeO_2$ catalysts for CO oxidation. *Appl. Catal. B Environ.* **2012**, *119–120*, 308–320. [CrossRef]
233. Sun, S.; Mao, D.; Yu, J.; Yang, Z.; Lu, G.; Ma, Z. Low-temperature CO oxidation on CuO/CeO_2 catalysts: the significant effect of copper precursor and calcination temperature. *Catal. Sci. Technol.* **2015**, *5*, 3166–3181. [CrossRef]
234. Kappis, K.; Papadopoulos, C.; Papavasiliou, J.; Vakros, J.; Georgiou, Y.; Deligiannakis, Y.; Avgouropoulos, G. Tuning the Catalytic Properties of Copper-Promoted Nanoceria via a Hydrothermal Method. *Catalysts* **2019**, *9*, 138. [CrossRef]
235. Papadopoulos, C.; Kappis, K.; Papavasiliou, J.; Vakros, J.; Kuśmierz, M.; Gac, W.; Georgiou, Y.; Deligiannakis, Y.; Avgouropoulos, G. Copper-promoted ceria catalysts for CO oxidation reaction. *Catal. Today* **2019**, in press. [CrossRef]
236. Elias, J.S.; Risch, M.; Giordano, L.; Mansour, A.N.; Shao-Horn, Y. Structure, Bonding, and Catalytic Activity of Monodisperse, Transition-Metal-Substituted CeO_2 Nanoparticles. *J. Am. Chem. Soc.* **2014**, *136*, 17193–17200. [CrossRef]
237. Li, Y.; Cai, Y.; Xing, X.; Chen, N.; Deng, D.; Wang, Y. Catalytic activity for CO oxidation of $Cu\text{-}CeO_2$ composite nanoparticles synthesized by a hydrothermal method. *Anal. Methods* **2015**, *7*, 3238–3245. [CrossRef]
238. Ma, J.; Jin, G.; Gao, J.; Li, Y.; Dong, L.; Huang, M.; Huang, Q.; Li, B. Catalytic effect of two-phase intergrowth and coexistence $CuO\text{-}CeO_2$. *J. Mater. Chem. A* **2015**, *3*, 24358–24370. [CrossRef]
239. Zhao, F.; Gong, M.; Zhang, G.; Li, J. Effect of the loading content of CuO on the activity and structure of CuO/Ce-Mn-O catalysts for CO oxidation. *J. Rare Earths* **2015**, *33*, 604–610. [CrossRef]
240. Lin, R.; Luo, M.-F.; Zhong, Y.-J.; Yan, Z.-L.; Liu, G.-Y.; Liu, W.-P. Comparative study of $CuO/Ce_{0.7}Sn_{0.3}O_2$, CuO/CeO_2 and CuO/SnO_2 catalysts for low-temperature CO oxidation. *Appl. Catal. A Gen.* **2003**, *255*, 331–336. [CrossRef]
241. Zhang, X.-M.; Tian, P.; Tu, W.; Zhang, Z.; Xu, J.; Han, Y.-F. Tuning the Dynamic Interfacial Structure of Copper-Ceria Catalysts by Indium Oxide during CO Oxidation. *ACS Catal.* **2018**, *8*, 5261–5275. [CrossRef]
242. Li, W.; Shen, X.; Zeng, R.; Chen, J.; Xiao, W.; Ding, S.; Chen, C.; Zhang, R.; Zhang, N. Constructing copper-ceria nanosheets with high concentration of interfacial active sites for enhanced performance in CO oxidation. *Appl. Surf. Sci.* **2019**, *492*, 818–825. [CrossRef]
243. Song, X.-Z.; Su, Q.-F.; Li, S.-J.; Liu, S.-H.; Zhang, N.; Meng, Y.-L.; Chen, X.; Tan, Z. Triple-shelled CuO/CeO_2 hollow nanospheres derived from metal–organic frameworks as highly efficient catalysts for CO oxidation. *New J. Chem.* **2019**, *43*, 16096–16102. [CrossRef]
244. Liu, H.-H.; Wang, Y.; Jia, A.-P.; Wang, S.-Y.; Luo, M.-F.; Lu, J.-Q. Oxygen vacancy promoted CO oxidation over Pt/CeO_2 catalysts: A reaction at $Pt\text{-}CeO_2$ interface. *Appl. Surf. Sci.* **2014**, *314*, 725–734. [CrossRef]
245. Meng, L.; Jia, A.-P.; Lu, J.-Q.; Luo, L.-F.; Huang, W.-X.; Luo, M.-F. Synergetic Effects of PdO Species on CO Oxidation over $PdO\text{-}CeO_2$ Catalysts. *J. Phys. Chem. C* **2011**, *115*, 19789–19796. [CrossRef]
246. Konsolakis, M. Recent Advances on Nitrous Oxide (N_2O) Decomposition over Non-Noble-Metal Oxide Catalysts: Catalytic Performance, Mechanistic Considerations, and Surface Chemistry Aspects. *ACS Catal.* **2015**, *5*, 6397–6421. [CrossRef]
247. Konsolakis, M.; Carabineiro, S.A.C.; Papista, E.; Marnellos, G.E.; Tavares, P.B.; Agostinho Moreira, J.; Romaguera-Barcelay, Y.; Figueiredo, J.L. Effect of preparation method on the solid state properties and the deN_2O performance of $CuO\text{-}CeO_2$ oxides. *Catal. Sci. Technol.* **2015**, *5*, 3714–3727. [CrossRef]
248. Zabilskiy, M.; Erjavec, B.; Djinović, P.; Pintar, A. Ordered mesoporous $CuO\text{-}CeO_2$ mixed oxides as an effective catalyst for N_2O decomposition. *Chem. Eng. J.* **2014**, *254*, 153–162. [CrossRef]
249. Parres-Esclapez, S.; Illán-Gómez, M.J.; De Lecea, C.S.M.; Bueno-López, A. On the importance of the catalyst redox properties in the N_2O decomposition over alumina and ceria supported Rh, Pd and Pt. *Appl. Catal. B Environ.* **2010**, *96*, 370–378. [CrossRef]
250. Zhu, P.; Li, J.; Zuo, S.; Zhou, R. Preferential oxidation properties of CO in excess hydrogen over $CuO\text{-}CeO_2$ catalyst prepared by hydrothermal method. *Appl. Surf. Sci.* **2008**, *255*, 2903–2909. [CrossRef]

251. Du, P.-P.; Wang, W.-W.; Jia, C.-J.; Song, Q.-S.; Huang, Y.-Y.; Si, R. Effect of strongly bound copper species in copper-ceria catalyst for preferential oxidation of carbon monoxide. *Appl. Catal. A Gen.* **2016**, *518*, 87–101. [CrossRef]
252. Wu, Z.; Zhu, H.; Qin, Z.; Wang, H.; Ding, J.; Huang, L.; Wang, J. CO preferential oxidation in H_2-rich stream over a CuO/CeO_2 catalyst with high H_2O and CO_2 tolerance. *Fuel* **2013**, *104*, 41–45. [CrossRef]
253. Jampa, S.; Wangkawee, K.; Tantisriyanurak, S.; Changpradit, J.; Jamieson, A.M.; Chaisuwan, T.; Luengnaruemitchai, A.; Wongkasemjit, S. High performance and stability of copper loading on mesoporous ceria catalyst for preferential oxidation of CO in presence of excess of hydrogen. *Int. J. Hydrogen Energy* **2017**, *42*, 5537–5548. [CrossRef]
254. Arango-Díaz, A.; Moretti, E.; Talon, A.; Storaro, L.; Lenarda, M.; Núñez, P.; Marrero-Jerez, J.; Jiménez-Jiménez, J.; Jiménez-López, A.; Rodríguez-Castellón, E. Preferential CO oxidation (CO-PROX) catalyzed by CuO supported on nanocrystalline CeO_2 prepared by a freeze-drying method. *Appl. Catal. A Gen.* **2014**, *477*, 54–63. [CrossRef]
255. Wang, J.; Pu, H.; Wan, G.; Chen, K.; Lu, J.; Lei, Y.; Zhong, L.; He, S.; Han, C.; Luo, Y. Promoted the reduction of Cu^{2+} to enhance $CuO–CeO_2$ catalysts for CO preferential oxidation in H_2-rich streams: Effects of preparation methods and copper precursors. *Int. J. Hydrogen Energy* **2017**, *42*, 21955–21968. [CrossRef]
256. Shi, L.; Zhang, G.; Wang, Y. Tailoring catalytic performance of carbon nanotubes confined $CuO–CeO_2$ catalysts for CO preferential oxidation. *Int. J. Hydrogen Energy* **2018**, *43*, 18211–18219. [CrossRef]
257. Cecilia, J.A.; Arango-Díaz, A.; Marrero-Jerez, J.; Núñez, P.; Moretti, E.; Storaro, L.; Rodríguez-Castellón, E. Catalytic Behaviour of $CuO-CeO_2$ Systems Prepared by Different Synthetic Methodologies in the CO-PROX Reaction under CO_2-H_2O Feed Stream. *Catalysts* **2017**, *7*, 160. [CrossRef]
258. Han, J.; Kim, H.J.; Yoon, S.; Lee, H. Shape effect of ceria in Cu/ceria catalysts for preferential CO oxidation. *J. Mol. Catal. A Chem.* **2011**, *335*, 82–88. [CrossRef]
259. Zou, Q.; Zhao, Y.; Jin, X.; Fang, J.; Li, D.; Li, K.; Lu, J.; Luo, Y. Ceria-nano supported copper oxide catalysts for CO preferential oxidation: Importance of oxygen species and metal-support interaction. *Appl. Surf. Sci.* **2019**, *494*, 1166–1176. [CrossRef]
260. Yu, X.; Wu, J.; Zhang, A.; Xue, L.; Wang, Q.; Tian, X.; Shan, S.; Zhong, C.-J.; Zeng, S. Hollow copper-ceria microspheres with single and multiple shells for preferential CO oxidation. *CrystEngComm* **2019**, *21*, 3619–3626. [CrossRef]
261. Dongil, A.B.; Bachiller-Baeza, B.; Castillejos, E.; Escalona, N.; Guerrero-Ruiz, A.; Rodríguez-Ramos, I. Promoter effect of alkalis on CuO/CeO_2/carbon nanotubes systems for the PROx reaction. *Catal. Today* **2018**, *301*, 141–146. [CrossRef]
262. Wang, W.-W.; Du, P.-P.; Zou, S.-H.; He, H.-Y.; Wang, R.-X.; Jin, Z.; Shi, S.; Huang, Y.-Y.; Si, R.; Song, Q.-S.; et al. Highly Dispersed Copper Oxide Clusters as Active Species in Copper-Ceria Catalyst for Preferential Oxidation of Carbon Monoxide. *ACS Catal.* **2015**, *5*, 2088–2099. [CrossRef]
263. Gong, X.; Liu, B.; Kang, B.; Xu, G.; Wang, Q.; Jia, C.; Zhang, J. Boosting Cu-Ce interaction in Cu_xO/CeO_2 nanocube catalysts for enhanced catalytic performance of preferential oxidation of CO in H_2-rich gases. *Mol. Catal.* **2017**, *436*, 90–99. [CrossRef]
264. Papavasiliou, J.; Vakros, J.; Avgouropoulos, G. Impact of acid treatment of $CuO-CeO_2$ catalysts on the preferential oxidation of CO reaction. *Catal. Commun.* **2018**, *115*, 68–72. [CrossRef]
265. Borchers, C.; Martin, M.L.; Vorobjeva, G.A.; Morozova, O.S.; Firsova, A.A.; Leonov, A.V.; Kurmaev, E.Z.; Kukharenko, A.I.; Zhidkov, I.S.; Cholakh, S.O. $Cu–CeO_2$ nanocomposites: mechanochemical synthesis, physico-chemical properties, CO-PROX activity. *J. Nanopart. Res.* **2016**, *18*, 344. [CrossRef]
266. Martínez-Arias, A.; Gamarra, D.; Fernández-García, M.; Hornés, A.; Belver, C. Spectroscopic Study on the Nature of Active Entities in Copper-Ceria CO-PROX Catalysts. *Top. Catal.* **2009**, *52*, 1425–1432. [CrossRef]
267. Lu, J.; Wang, J.; Zou, Q.; He, D.; Zhang, L.; Xu, Z.; He, S.; Luo, Y. Unravelling the Nature of the Active Species as well as the Doping Effect over Cu/Ce-Based Catalyst for Carbon Monoxide Preferential Oxidation. *ACS Catal.* **2019**, *9*, 2177–2195. [CrossRef]
268. Monte, M.; Gamarra, D.; López Cámara, A.; Rasmussen, S.B.; Gyorffy, N.; Schay, Z.; Martínez-Arias, A.; Conesa, J.C. Preferential oxidation of CO in excess H_2 over CuO/CeO_2 catalysts: Performance as a function of the copper coverage and exposed face present in the CeO_2 support. *Catal. Today* **2014**, *229*, 104–113. [CrossRef]

269. Li, W.; Hu, Y.; Jiang, H.; Yang, S.; Li, C. Facile synthesis of multi-shelled hollow Cu–CeO_2 microspheres with promoted catalytic performance for preferential oxidation of CO. *Mater. Chem. Phys.* **2019**, *226*, 158–168. [CrossRef]
270. Firsova, A.A.; Morozova, O.S.; Vorob'eva, G.A.; Leonov, A.V.; Kukharenko, A.I.; Cholakh, S.O.; Kurmaev, E.Z.; Korchak, V.N. Mechanochemical Activation of Cu–CeO_2 Mixture as a Promising Technique for the Solid-State Synthesis of Catalysts for the Selective Oxidation of CO in the Presence of H_2. *Kinet. Catal.* **2018**, *59*, 160–173. [CrossRef]
271. Yen, H.; Seo, Y.; Kaliaguine, S.; Kleitz, F. Tailored Mesostructured Copper/Ceria Catalysts with Enhanced Performance for Preferential Oxidation of CO at Low Temperature. *Angew. Chem. Int. Ed.* **2012**, *51*, 12032–12035. [CrossRef] [PubMed]
272. Liu, Z.; Chen, J.; Zhou, R.; Zheng, X. Influence of Ethanol Washing in Precursor on CuO-CeO_2 Catalysts in Preferential Oxidation of CO in Excess Hydrogen. *Catal. Lett.* **2008**, *123*, 102–106. [CrossRef]
273. Xia, Y.; Lao, J.; Ye, J.; Cheng, D.-G.; Chen, F.; Zhan, X. Role of Two-Electron Defects on the CeO_2 Surface in CO Preferential Oxidation over CuO/CeO_2 Catalysts. *ACS Sustain. Chem. Eng.* **2019**, *7*, 18421–18433. [CrossRef]
274. Xie, Y.; Yin, Y.; Zeng, S.; Gao, M.; Su, H. Coexistence of Cu^+ and Cu^{2+} in star-shaped CeO_2/Cu_xO catalyst for preferential CO oxidation. *Catal. Commun.* **2017**, *99*, 110–114. [CrossRef]
275. Chen, S.; Li, L.; Hu, W.; Huang, X.; Li, Q.; Xu, Y.; Zuo, Y.; Li, G. Anchoring High-Concentration Oxygen Vacancies at Interfaces of CeO_{2-x}/Cu toward Enhanced Activity for Preferential CO Oxidation. *ACS Appl. Mater. Interfaces* **2015**, *7*, 22999–23007. [CrossRef]
276. Zhang, L.; Chen, T.; Zeng, S.; Su, H. Effect of doping elements on oxygen vacancies and lattice oxygen in CeO_2-CuO catalysts. *J. Environ. Chem. Eng.* **2016**, *4*, 2785–2794. [CrossRef]
277. López Cámara, C.A.; Cortés Corberán, V.; Barrio, L.; Zhou, G.; Si, R.; Hanson, J.C.; Monte, M.; Conesa, J.C.; Rodriguez, J.A.; Martínez-Arias, A. Improving the CO-PROX Performance of Inverse CeO_2/CuO Catalysts: Doping of the CuO Component with Zn. *J. Phys. Chem. C* **2014**, *118*, 9030–9041. [CrossRef]
278. Papavasiliou, J.; Rawski, M.; Vakros, J.; Avgouropoulos, G. A Novel Post-Synthesis Modification of CuO-CeO_2 Catalysts: Effect on Their Activity for Selective CO Oxidation. *ChemCatChem* **2018**, *10*, 2096–2106. [CrossRef]
279. Liu, Z.; Zhou, R.; Zheng, X. Influence of residual K^+ on the catalytic performance of CuO-CeO_2 catalysts in preferential oxidation of CO in excess hydrogen. *Int. J. Hydrogen Energy* **2008**, *33*, 791–796. [CrossRef]
280. Dongil, A.B.; Bachiller-Baeza, B.; Castillejos, E.; Escalona, N.; Guerrero-Ruiz, A.; Rodríguez-Ramos, I. The promoter effect of potassium in CuO/CeO_2 systems supported on carbon nanotubes and graphene for the CO-PROX reaction. *Catal. Sci. Technol.* **2016**, *6*, 6118–6127. [CrossRef]
281. Ding, J.; Li, L.; Li, H.; Chen, S.; Fang, S.; Feng, T.; Li, G. Optimum Preferential Oxidation Performance of CeO_2-CuO_x-RGO Composites through Interfacial Regulation. *ACS Appl. Mater. Interfaces* **2018**, *10*, 7935–7945. [CrossRef] [PubMed]
282. Zhang, H.; Xu, C.; Ding, J.; Su, H.; Zeng, S. RGO/MWCNTs/Cu_xO-CeO_2 ternary nanocomposites for preferential CO oxidation in hydrogen-rich streams. *Appl. Surf. Sci.* **2017**, *426*, 50–55. [CrossRef]
283. Xu, C.; Zeng, S.; Zhang, H.; Xie, Y.; Zhang, A.; Jing, G.; Su, H. Facile hydrothermal procedure to synthesize sheet-on-sheet reduced graphene oxide (RGO)/Cu_xO-CeO_2 nanocomposites for preferential oxidation of carbon monoxide. *Int. J. Hydrogen Energy* **2017**, *42*, 14133–14143. [CrossRef]
284. Wang, Q.; Zhang, H.; Wu, J.; Tuya, N.; Zhao, Y.; Liu, S.; Dong, Y.; Li, P.; Xu, Y.; Zeng, S. Experimental and computational studies on copper-cerium catalysts supported on nitrogen-doped porous carbon for preferential oxidation of CO. *Catal. Sci. Technol.* **2019**, *9*, 3023–3035. [CrossRef]
285. Chen, A.; Yu, X.; Zhou, Y.; Miao, S.; Li, Y.; Kuld, S.; Sehested, J.; Liu, J.; Aoki, T.; Hong, S.; et al. Structure of the catalytically active copper–ceria interfacial perimeter. *Nat. Catal.* **2019**, *2*, 334–341. [CrossRef]
286. Ning, J.; Zhou, Y.; Chen, A.; Li, Y.; Miao, S.; Shen, W. Dispersion of copper on ceria for the low-temperature water-gas shift reaction. *Catal. Today* **2019**, in press. [CrossRef]
287. Si, R.; Raitano, J.; Yi, N.; Zhang, L.; Chan, S.-W.; Flytzani-Stephanopoulos, M. Structure sensitivity of the low-temperature water-gas shift reaction on Cu-CeO_2 catalysts. *Catal. Today* **2012**, *180*, 68–80. [CrossRef]
288. Yan, H.; Yang, C.; Shao, W.-P.; Cai, L.-H.; Wang, W.-W.; Jin, Z.; Jia, C.-J. Construction of stabilized bulk-nano interfaces for highly promoted inverse CeO_2/Cu catalyst. *Nat. Commun.* **2019**, *10*, 3470. [CrossRef]
289. Wang, X.; Mi, J.; Lin, Z.; Lin, Y.; Jiang, L.; Cao, Y. Efficient fabrication of mesoporous active Cu-Co-CeO_2 catalysts for water-gas shift. *Mater. Lett.* **2016**, *162*, 214–217. [CrossRef]

290. Wang, X.; Liu, Y.; Peng, X.; Lin, B.; Cao, Y.; Jiang, L. Sacrificial Adsorbate Strategy Achieved Strong Metal-Support Interaction of Stable Cu Nanocatalysts. *ACS Appl. Energy Mater.* **2018**, *1*, 1408–1414. [CrossRef]
291. Djinović, P.; Batista, J.; Levec, J.; Pintar, A. Comparison of water-gas shift reaction activity and long-term stability of nanostructured CuO-CeO_2 catalysts prepared by hard template and co-precipitation methods. *Appl. Catal. A Gen.* **2009**, *364*, 156–165. [CrossRef]
292. Chen, C.; Zhan, Y.; Li, D.; Zhang, Y.; Lin, X.; Jiang, L.; Zheng, Q. Preparation of CuO/CeO_2 Catalyst with Enhanced Catalytic Performance for Water-Gas Shift Reaction in Hydrogen Production. *Energy Technol.* **2018**, *6*, 1096–1103. [CrossRef]
293. Zhang, Z.; Wang, S.-S.; Song, R.; Cao, T.; Luo, L.; Chen, X.; Gao, Y.; Lu, J.; Li, W.-X.; Huang, W. The most active Cu facet for low-temperature water gas shift reaction. *Nat. Commun.* **2017**, *8*, 488. [CrossRef] [PubMed]
294. Agarwal, S.; Zhu, X.; Hensen, E.J.M.; Mojet, B.L.; Lefferts, L. Surface-Dependence of Defect Chemistry of Nanostructured Ceria. *J. Phys. Chem. C* **2015**, *119*, 12423–12433. [CrossRef]
295. Wang, Y.-X.; Wang, G.-C. A Systematic Theoretical Study of Water Gas Shift Reaction on Cu(111) and Cu(110): Potassium Effect. *ACS Catal.* **2019**, *9*, 2261–2274. [CrossRef]
296. Chen, C.; Zhan, Y.; Zhou, J.; Li, D.; Zhang, Y.; Lin, X.; Jiang, L.; Zheng, Q. Cu/CeO_2 Catalyst for Water-Gas Shift Reaction: Effect of CeO_2 Pretreatment. *ChemPhysChem* **2018**, *19*, 1448–1455. [CrossRef]
297. Graciani, J.; Mudiyanselage, K.; Xu, F.; Baber, A.E.; Evans, J.; Senanayake, S.D.; Stacchiola, D.J.; Liu, P.; Hrbek, J.; Fernández Sanz, J.; et al. Highly active copper-ceria and copper-ceria-titania catalysts for methanol synthesis from CO_2. *Science* **2014**, *345*, 546–550. [CrossRef]
298. Rodriguez, J.A.; Liu, P.; Stacchiola, D.J.; Senanayake, S.D.; White, M.G.; Chen, J.G. Hydrogenation of CO_2 to Methanol: Importance of Metal-Oxide and Metal-Carbide Interfaces in the Activation of CO_2. *ACS Catal.* **2015**, *5*, 6696–6706. [CrossRef]
299. Kattel, S.; Liu, P.; Chen, J.G. Tuning Selectivity of CO_2 Hydrogenation Reactions at the Metal/Oxide Interface. *J. Am. Chem. Soc.* **2017**, *139*, 9739–9754. [CrossRef]
300. Wang, W.; Qu, Z.; Song, L.; Fu, Q. CO_2 hydrogenation to methanol over Cu/CeO_2 and Cu/ZrO_2 catalysts: Tuning methanol selectivity via metal-support interaction. *J. Energy Chem.* **2020**, *40*, 22–30. [CrossRef]
301. Van de Water, L.G.A.; Wilkinson, S.K.; Smith, R.A.P.; Watson, M.J. Understanding methanol synthesis from CO/H_2 feeds over Cu/CeO_2 catalysts. *J. Catal.* **2018**, *364*, 57–68. [CrossRef]
302. Hu, X.; Zhao, C.; Guan, Q.; Hu, X.; Li, W.; Chen, J. Selective hydrogenation of CO_2 over a Ce promoted Cu-based catalyst confined by SBA-15. *Inorg. Chem. Front.* **2019**, *6*, 1799–1812. [CrossRef]
303. Li, S.; Guo, L.; Ishihara, T. Hydrogenation of CO_2 to methanol over Cu/AlCeO catalyst. *Catal. Today* **2020**, *339*, 352–361. [CrossRef]
304. Tan, Q.; Shi, Z.; Wu, D. CO_2 Hydrogenation to Methanol over a Highly Active Cu-Ni/CeO_2-Nanotube Catalyst. *Ind. Eng. Chem. Res.* **2018**, *57*, 10148–10158. [CrossRef]
305. Tan, Q.; Shi, Z.; Wu, D. CO_2 hydrogenation over differently morphological CeO_2-supported Cu-Ni catalysts. *Int. J. Energy Res.* **2019**, *43*, 5392–5404. [CrossRef]
306. Li, H.; Cui, Y.; Liu, Q.; Dai, W.-L. Insight into the Synergism between Copper Species and Surface Defects Influenced by Copper Content over Copper/Ceria Catalysts for the Hydrogenation of Carbonate. *ChemCatChem* **2018**, *10*, 619–624. [CrossRef]
307. Yang, S.-C.; Pang, S.H.; Sulmonetti, T.P.; Su, W.-N.; Lee, J.-F.; Hwang, B.-J.; Jones, C.W. Synergy between Ceria Oxygen Vacancies and Cu Nanoparticles Facilitates the Catalytic Conversion of CO_2 to CO under Mild Conditions. *ACS Catal.* **2018**, *8*, 12056–12066. [CrossRef]
308. Lin, L.; Yao, S.; Rui, N.; Han, L.; Zhang, F.; Gerlak, C.A.; Liu, Z.; Cen, J.; Song, L.; Senanayake, S.D.; et al. Conversion of CO_2 on a highly active and stable Cu/FeO_x/CeO_2 catalyst: tuning catalytic performance by oxide-oxide interactions. *Catal. Sci. Technol.* **2019**, *9*, 3735–3742. [CrossRef]

catalysts

MDPI

Article

Facet-Dependent Reactivity of Fe_2O_3/CeO_2 Nanocomposites: Effect of Ceria Morphology on CO Oxidation

Maria Lykaki [1], Sofia Stefa [1], Sónia A. C. Carabineiro [2], Pavlos K. Pandis [3], Vassilis N. Stathopoulos [3] and Michalis Konsolakis [1,*]

[1] Industrial, Energy and Environmental Systems Lab (IEESL), School of Production Engineering and Management, Technical University of Crete, GR-73100 Chania, Greece; mlykaki@isc.tuc.gr (M.L.); sstefa@isc.tuc.gr (S.S.)

[2] Laboratório de Catálise e Materiais (LCM), Laboratório Associado LSRE-LCM, Faculdade de Engenharia, Universidade do Porto, 4200-465 Porto, Portugal; scarabin@fe.up.pt

[3] Laboratory of Chemistry and Materials Technology, General Department, School of Sciences, National and Kapodistrian University of Athens, GR-34400 Psachna Campus, Evia, Greece; ppandis@teiste.gr (P.K.P.); vasta@uoa.gr (V.N.S.)

* Correspondence: mkonsol@pem.tuc.gr; Tel.: +30-28210-37682

Received: 19 March 2019; Accepted: 15 April 2019; Published: 19 April 2019

Abstract: Ceria has been widely studied either as catalyst itself or support of various active phases in many catalytic reactions, due to its unique redox and surface properties in conjunction to its lower cost, compared to noble metal-based catalytic systems. The rational design of catalytic materials, through appropriate tailoring of the particles' shape and size, in order to acquire highly efficient nanocatalysts, is of major significance. Iron is considered to be one of the cheapest transition metals while its interaction with ceria support and their shape-dependent catalytic activity has not been fully investigated. In this work, we report on ceria nanostructures morphological effects (cubes, polyhedra, rods) on the textural, structural, surface, redox properties and, consequently, on the CO oxidation performance of the iron-ceria mixed oxides (Fe_2O_3/CeO_2). A full characterization study involving N_2 adsorption at −196 °C, X-ray diffraction (XRD), transmission electron microscopy (TEM), scanning electron microscopy-energy dispersive X-ray spectroscopy (SEM-EDS), temperature programmed reduction (TPR), and X-ray photoelectron spectroscopy (XPS) was performed. The results clearly revealed the key role of support morphology on the physicochemical properties and the catalytic behavior of the iron-ceria binary system, with the rod-shaped sample exhibiting the highest catalytic performance, both in terms of conversion and specific activity, due to its improved reducibility and oxygen mobility, along with its abundance in Fe^{2+} species.

Keywords: ceria morphology; facet dependence; Fe_2O_3/CeO_2 mixed oxides; CO oxidation

---

1. Introduction

Ceria (CeO_2), or cerium oxide, has been extensively used in a variety of catalytic applications such as oxidation processes, steam reforming, water-gas shift reaction, reduction of NOx, among others [1–3]. Ceria's unique properties, such as improved thermal stability, high oxygen storage capacity (OSC), and oxygen mobility render it an exceptional component for ceria-based catalytic materials [2,4–6]. Actually, its ability to switch between the oxidation states Ce^{3+} and Ce^{4+}, supplemented by the formation of surface defects, such as oxygen vacancies, is accounted for its enhanced redox properties [2,3,7,8]. Such a behavior is identified in other oxide systems such as manganese oxides [9–11] and perovskites [12,13], but ceria still remains one of the best redox materials [14].

Recent studies [14–23] have focused on the synthesis of nanostructured materials with well-defined morphology. By tailoring particles' shape and size, certain crystal facets can be exposed, leading to different structural and redox properties, hence, resulting in improved catalytic activity [8,15,17,19]. In fact, it has been shown that the anionic vacancies energy formation strongly depends on the exposed facets, following the order: {111} > {100} > {110}.

In addition, as the crystallite size decreases, materials in the nanoscale exhibit a plethora of oxygen defects and enhanced catalytic performance [7]. It has been revealed that there is a strong dependence between support morphology and catalytic activity, a fact showing the significance of the fine-tuning of ceria with predefined textural and structural characteristics [17,18,21,22].

Several reports [22,24–29] have shown that the introduction of transition metal oxides into the ceria lattice improves the catalytic performance of the mixed oxides compared to bare ones because of the "synergistic" effect induced by the interactions between the oxide phases, a phenomenon that has not been fully comprehended. In this regard, it is of great importance to develop cost-effective and highly efficient catalysts based on iron oxide (Fe_2O_3), which is considered to be one of the cheapest metal oxides [30]. Among the various catalytic systems, iron-ceria mixed oxides have been studied in several catalytic reactions, such as oxidation processes [30–34], reduction processes [35,36], decomposition reactions [37,38], soot combustion [39,40], etc.

However, the Fe_2O_3/CeO_2 binary system has not been extensively investigated in relation to the support shape dependence of catalytic activity, which is a highly engaging topic. To fully address this, most of the studies concern the effect of iron content [39,41], calcination temperature [42], or the distribution of iron ions in solid solutions [43] of Fe-doped CeO_2 catalysts. Moreover, others have focused on the improvement of OSC and oxygen mobility of iron-doped ceria catalysts [44]. In a similar manner, Bao et al. [31] investigated the oxidation of CO in Fe_2O_3/CeO_2 catalysts from the perspective of solid solution formation and surface oxygen vacancies.

Only a few studies have focused on the Fe_2O_3/CeO_2 system from the perspective of ceria shape effect. For instance, Torrente-Murciano et al. [45] have shown that the control of support morphology in the hydrogenation of CO_2 to hydrocarbons leads to the exposure of different crystal facets, resulting in enhanced metal-support interactions, a fact corroborated by the reducibility studies. The structure-dependent catalytic performance of iron-ceria nanorods and nanopolyhedra for NO reduction has been studied, and this dependence was mainly attributed to the exposed facets of ceria nanoshapes, along with the synergistic effect between iron and ceria [35]. Furthermore, the support role on the decomposition of sulfuric acid has been studied through the development of a series of supported iron oxide-based catalysts [37]. Also, Sudarsanam et al. [40] studied the role of support morphology in copper-ceria and iron-ceria nanorods for diesel soot combustion, and revealed an abundance in oxygen vacancies in both catalytic systems, with copper-ceria nanorods, however, exhibiting the best catalytic performance, which was ascribed to the high reducibility of ceria and the large amount of acid sites.

Despite the various studies regarding the iron-ceria mixed oxides, there is still plenty of room to elaborate on the morphology dependence of catalytic activity. In the present work, three different ceria nanostructures (nanorods (NR), nanopolyhedra (NP), nanocubes (NC)) were hydrothermally synthesized and used as supports for the iron oxide phase (Fe_2O_3/CeO_2). For comparison purposes, two additional samples were investigated: bare iron oxide sample prepared by thermal decomposition (Fe_2O_3-D) and iron-ceria mixed oxide prepared by a physical mixture of Fe_2O_3-D and CeO_2-NR (Fe_2O_3-D + CeO_2-NR). Several characterization techniques, namely N_2 adsorption at −196 °C (Brunauer–Emmett–Teller (BET) method), X-ray diffraction (XRD), transmission electron microscopy (TEM), temperature programmed reduction using H_2 (H_2-TPR), and X-ray photoelectron spectroscopy (XPS), have been employed in order to gain insight into the effect of support morphology on the textural, structural, redox, surface properties and, consequently, on the catalytic performance of the iron-ceria binary system. The oxidation of CO was employed as probe reaction in order to disclose structure–activity relationships.

2. Results and Discussion

2.1. Textural and Structural Characteristics (BET/XRD)

The main textural/structural characteristics of bare ceria and Fe_2O_3/CeO_2 samples are summarized in Table 1. On the basis of the BET surface area, the following order was obtained for the bare ceria samples: CeO_2-NP (87.9 m^2/g) > CeO_2-NR (79.3 m^2/g) > CeO_2-NC (37.3 m^2/g). A decrease in the surface area was observed by the addition of iron into the ceria support, with the Fe_2O_3/CeO_2-NR sample exhibiting the highest BET surface area (68.6 m^2/g), accompanied by Fe_2O_3/CeO_2-NP (64.2 m^2/g) and Fe_2O_3/CeO_2-NC (32.2 m^2/g).

Table 1. Textural/structural characteristics of bare CeO_2, Fe_2O_3-D, and Fe_2O_3/CeO_2 samples.

Sample	BET Analysis			XRD Analysis	
	BET Surface Area (m^2/g)	Pore Volume (cm^3/g)	Average Pore Diameter (nm)	Crystallite Size (nm), D_{XRD} [1]	
				CeO_2	Fe_2O_3
CeO_2-NR	79.3	0.48	24.2	15.0	-
CeO_2-NP	87.9	0.17	7.9	11.0	-
CeO_2-NC	37.3	0.26	27.4	27.0	-
Fe_2O_3/CeO_2-NR	68.6	0.19	11.3	9.7	7.2
Fe_2O_3/CeO_2-NP	64.2	0.12	7.6	8.5	16.5
Fe_2O_3/CeO_2-NC	32.2	0.19	23.3	16.8	52.3
Fe_2O_3-D	27.0	0.15	22.3	-	23.3

[1] Calculated applying the Williamson–Hall plot after Rietveld refinement of diffractograms.

Figure 1a shows the Barrett–Joyner–Halenda (BJH) desorption pore size distribution (PSD) of as-prepared samples. The corresponding adsorption–desorption isotherms are depicted in Figure 1b. In all cases, maxima at pore diameters higher than 3 nm are obtained, implying the presence of mesopores which can be further corroborated by the existence of adsorption–desorption isotherms of type IV with hysteresis loop at a relative pressure above 0.5 (Figure 1b) [33,35,46,47]. As observed in Table 1 and Figure 1a, the incorporation of iron into the ceria lattice results in a decrease in pore volume and average pore diameter, with the sample of rod-like morphology exhibiting the highest reduction percentages (60% and 53%, respectively), while nanocubes and nanopolyhedra show a much lower percentage decrease, namely 27% and 29% in pore volume and 15% and 3.8% in average pore diameter, respectively.

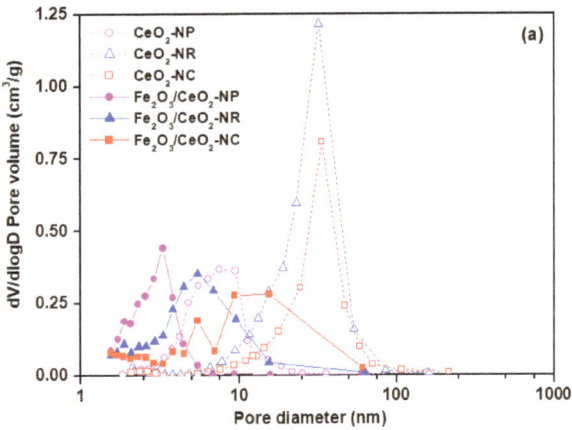

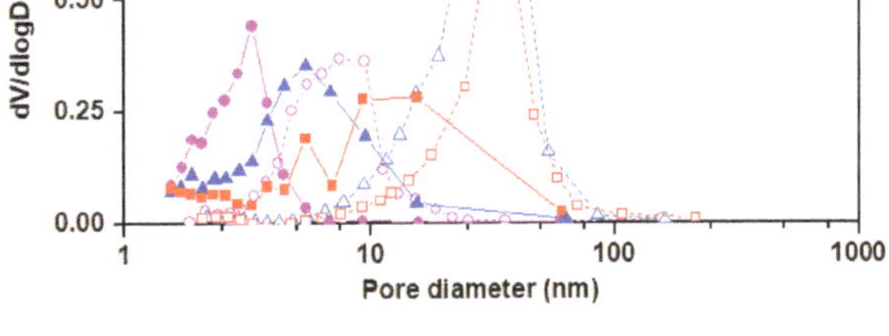

Figure 1. *Cont.*

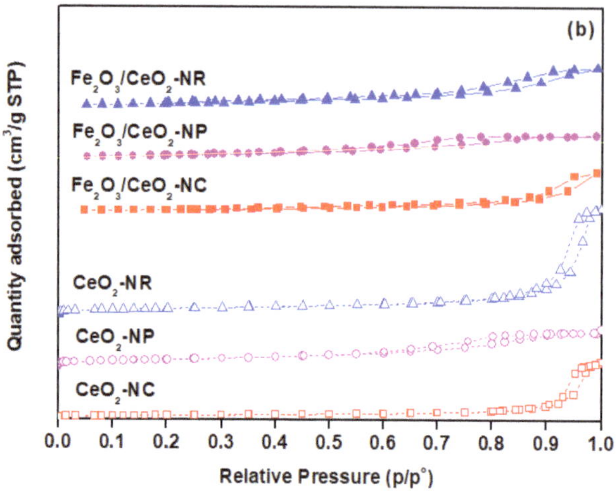

Figure 1. (**a**) Barrett–Joyner–Halenda (BJH) desorption pore size distribution (PSD) and (**b**) adsorption–desorption isotherms of bare CeO_2 and Fe_2O_3/CeO_2 samples.

The obtained order in BET surface can be mainly interpreted on the basis of the different pore volume and pore size distribution of ceria and iron-ceria samples. In particular, CeO_2-NR and CeO_2-NP possess a pore volume of 0.48 and 0.17 cm^3/g, respectively, exhibiting, however, a completely different pore size distribution (mean pore size at 24.2 and 7.9 nm, respectively, Figure 1a). On the other hand, CeO_2-NC exhibits an intermediate pore volume (0.26 cm^3/g) along with a higher mean pore size (27.4 nm). These differences in pore volume and pore size distribution, linked to different ceria morphologies, can be mainly accounted for the observed variations in BET surface area. Moreover, it is worth noticing that the BET surface area follows, in general, the reverse order of the crystallite size of both CeO_2 and Fe_2O_3 phases (Table 1), i.e., the larger the crystallite size the lower the BET surface area.

The XRD patterns of all samples are demonstrated in Figure 2. The main peaks can be indexed to (111), (200), (220), (311), (222), (400), (331), (420), (422), (511), and (440) planes which are attributed to ceria face-centered cubic fluorite structure (Fm3m symmetry, no. 225) [48–50]. In the Fe_2O_3-D sample, the peaks observed correspond to (012), (104), (110), (006), (113), (202), (024), (116), (018), (122), (214), (300), (208), (119), (220), (036), (312), (134), (226), (042), (232), (324), and (410) planes of the hematite phase [51]. There are two very small peaks at 2θ values ~35.96 and 54.26° in Fe_2O_3/CeO_2 samples which correspond to the hematite (Fe_2O_3) phase (JCPDS card 33-0664) [40,52]. However, the peak at 2θ 41.03° present in the Fe_2O_3-D sample, which is attributed to the hematite phase [40], is not observed in the mixed oxides. The peaks are characteristic of the cerianite and hematite phases in accordance with the nominal composition of the mixed oxides (Table 1). By applying the Williamson–Hall plot after Rietveld refinement of diffractograms, the average crystallite sizes of cerianite and hematite phases were calculated (Table 1). In particular, the CeO_2 crystallite size is 27.0, 15.0, and 11.0 nm for CeO_2-NC, CeO_2-NR, and CeO_2-NP, respectively, following the same order in Fe_2O_3/CeO_2 samples, i.e., 16.8, 9.7, and 8.5 nm for Fe_2O_3/CeO_2-NC, Fe_2O_3/CeO_2-NR, and Fe_2O_3/CeO_2-NP, respectively. Taking into account the crystallite size of iron oxide phase, the following order was obtained: Fe_2O_3/CeO_2-NC (52.3 nm) > Fe_2O_3/CeO_2-NP (16.5 nm) > Fe_2O_3/CeO_2-NR (7.2 nm). The lower crystallite size of ceria in Fe_2O_3/CeO_2 samples as compared to bare ceria samples should be mentioned. The addition of iron ions through the wet impregnation method and the subsequent calcination of iron-ceria composites could be considered responsible for the decrease of ceria crystallite size. In a similar manner, the high dispersibility of the iron-ceria mixed oxides and/or the formation of solid solutions have been considered as contributing factors to this decrease in ceria crystallite size [53–57].

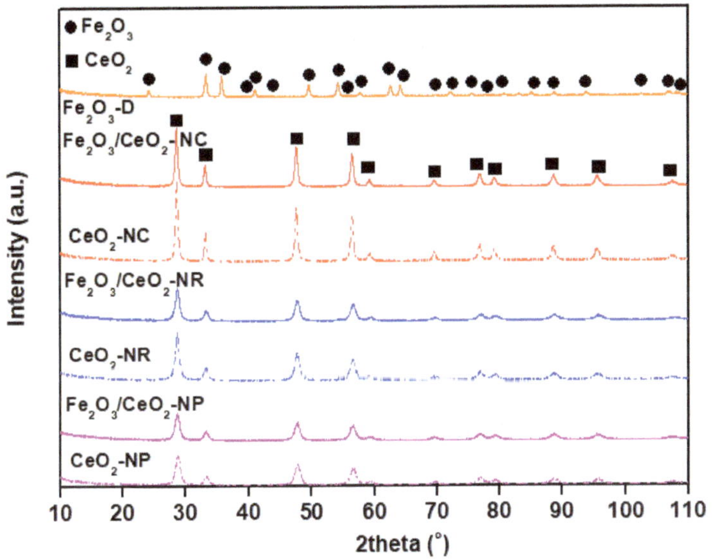

Figure 2. XRD patterns of bare CeO_2, Fe_2O_3-D, and Fe_2O_3/CeO_2 samples.

2.2. *Morphological Characteristics (TEM, SEM-EDS)*

Transmission electron microscopy analyses were performed so as to further investigate the morphological features of the various ceria nanostructures. The TEM images of bare ceria samples as well as those of iron-ceria mixed oxides are presented in Figures 3a–c and 3d–f, respectively. CeO_2-NR displays ceria in the rod-like morphology (Figure 3a) with 25–200 nm in length. Nanopolyhedra of irregular shapes and cubes are shown in Figure 3b and c, respectively. As it can be observed by the TEM images of iron-ceria mixed oxides (Figure 3d–f), the support morphology remains unaffected by the addition of iron to the ceria carrier. Scanning electron microscopy analyses along with energy dispersive X-ray spectrometry (EDS) were performed in addition to obtain the elemental mapping images of the Fe_2O_3/CeO_2 samples (Figure 4). The SEM images of Fe_2O_3/CeO_2-NR, Fe_2O_3/CeO_2-NP, and Fe_2O_3/CeO_2-NC are depicted in Figure 4a,e,i, respectively, while the corresponding elemental mapping images are shown in Figure 4b–d, Figure 4f–h, and Figure 4j–l, respectively. As it is obvious from SEM-EDS analysis, there is a uniform distribution of all elements (Ce, Fe, O) in the iron-ceria mixed oxides. Noteworthy, the Fe_2O_3/CeO_2-NR sample (Figure 4b–c) exhibits a higher amount of cerium than compared to iron, while iron-ceria nanopolyhedra (Figure 4f–g) and iron-ceria nanocubes (Figure 4j–k) exhibit a larger population in iron. To further gain insight into the surface elemental composition of the iron-ceria samples, XPS analysis was performed in addition, which corroborated the above findings (see below).

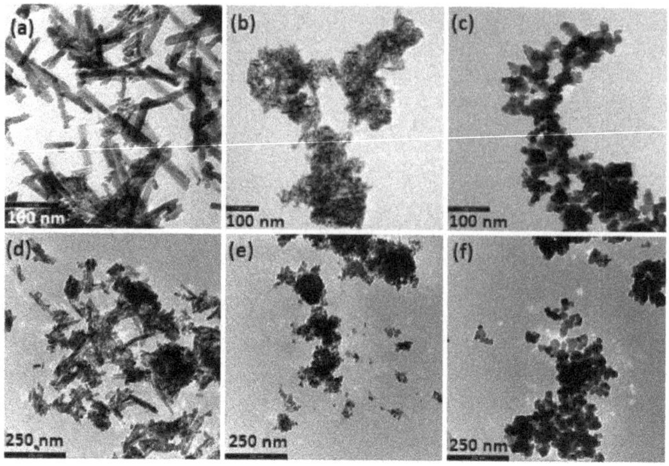

Figure 3. TEM images of the samples: (**a**) CeO_2-NR, (**b**) CeO_2-NP, (**c**) CeO_2-NC, (**d**) Fe_2O_3/CeO_2-NR, (**e**) Fe_2O_3/CeO_2-NP, and (**f**) Fe_2O_3/CeO_2-NC.

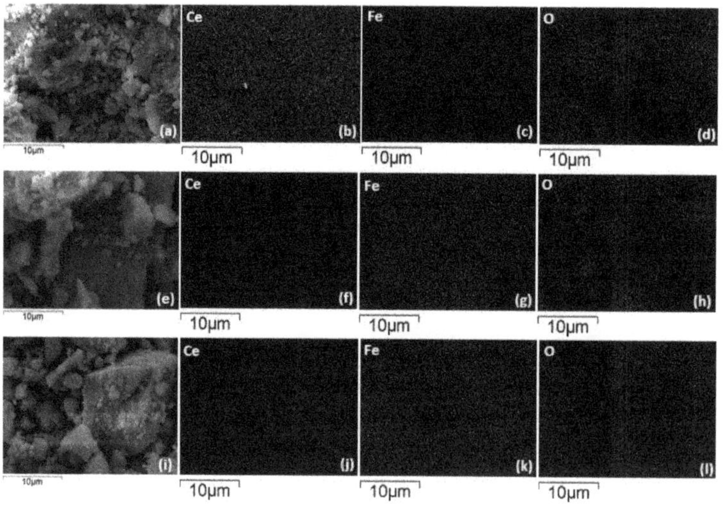

Figure 4. SEM-EDS elemental mapping images of the samples: (**a–d**) Fe_2O_3/CeO_2-NR, (**e–h**) Fe_2O_3/CeO_2-NP, (i-l) Fe_2O_3/CeO_2-NC.

2.3. Redox Properties (H_2-TPR)

The effect of support morphology on the reducibility of the samples was investigated by TPR experiments. The reduction profiles of bare ceria samples are depicted in Figure 5 and they include two broad peaks centered at 526–551 °C and 789–813 °C, which are attributed to the surface oxygen (O_s) and bulk oxygen (O_b) ceria reduction, respectively [58,59]. As observed in Figure 5, in ceria nanocubes, the O_s peak is smaller in comparison with ceria nanorods and nanopolyhedra due to the smaller amount of easily reducible oxygen available in the cubic sample, a fact closely related to the exposed crystal facets, as it has been reported in previous studies [45,60]. In Table 2, the hydrogen consumption that corresponds to ceria surface and bulk oxygen reduction is presented. Regarding the ratio of surface to bulk oxygen for the bare ceria samples, the following order is obtained: CeO_2-NR (1.13) > CeO_2-NP (0.94) > CeO_2-NC (0.71), indicating the higher reducibility of the rod-shaped sample which possesses the largest population of weakly bound oxygen species. It is of note that the catalytic conversion follows the same trend with the surface-to-bulk ratio, signifying the interrelationship between the reducibility and the catalytic performance, as will be further discussed in the sequence.

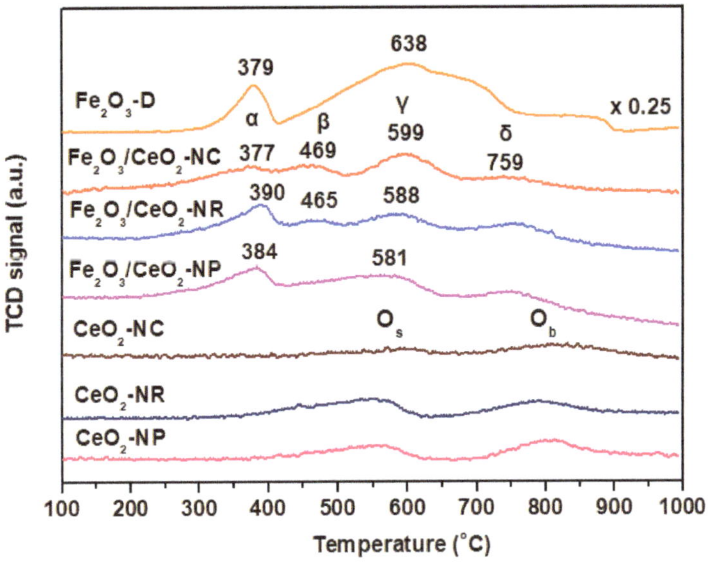

Figure 5. H_2-TPR profiles of bare CeO_2, Fe_2O_3-D, and Fe_2O_3/CeO_2 samples.

Table 2. Redox properties of bare CeO_2, Fe_2O_3-D, and Fe_2O_3/CeO_2 samples.

Sample	H_2 consumption (mmol H_2/g) [a]			O_s/O_b Ratio	Peak Temperature (°C)	
	O_s Peak	O_b Peak	Total		O_s Peak	O_b Peak
CeO_2-NR	0.59	0.52	1.11	1.13	545	788
CeO_2-NP	0.48	0.51	0.99	0.94	555	804
CeO_2-NC	0.41	0.58	0.99	0.71	589	809
	H_2 Consumption (mmol H_2/g) [a]	H_2 Excess (mmol H_2/g) [b]	Peak α	Peak β	Peak γ	Peak δ
Fe_2O_3/CeO_2-NR	3.42	1.42	390	465	588	759
Fe_2O_3/CeO_2-NP	2.93	0.93	384	-	581	759
Fe_2O_3/CeO_2-NC	2.47	0.47	377	469	599	759
Fe_2O_3-D	17.59	-	379	-	638	-

[a] Estimated by the area of the corresponding TPR peaks, which is calibrated against a known amount of CuO standard sample. [b] Estimated by the subtraction of H_2 amount required for Fe_2O_3 reduction in 7.5 wt.% Fe_2O_3/CeO_2 samples (~2 mmol/g) from the total H_2 consumption.

Figure 5 also shows the reduction profiles of Fe_2O_3/CeO_2 samples, as well as that of bare Fe_2O_3-D material. The main TPR peaks accompanied by the corresponding H_2 consumption (mmol H_2/g) are presented in Table 2. The Fe_2O_3-D sample exhibits a sharp peak at 379 °C along with a broader peak at ~638 °C, consisting of two overlapping bands, which are ascribed to the stepwise reduction of hematite to magnetite and magnetite to metallic iron, respectively, i.e., $Fe_2O_3 \rightarrow Fe_3O_4$ and $Fe_3O_4 \rightarrow Fe^0$ [44,58]. However, the number of steps (two or three) involved in the reduction of pure iron oxide has not been fully clarified [61], as it is considered to be dependent on the amount ratio of hydrogen to water present in the reduction process, an issue addressed by Zielinski et al. [62]. It should be also mentioned that the reduction of FeO (wustite) to Fe^0 cannot be easily observed because of the metastable nature of FeO as well as its disproportion to Fe_3O_4 and metallic iron ($4FeO \rightarrow Fe_3O_4 + Fe$) at a temperature below 619 °C [38].

The Fe_2O_3/CeO_2-NC and Fe_2O_3/CeO_2-NR samples exhibit four reduction peaks in the range of ~377–390 °C (peak α), 465–469 °C (peak β), 588–599 °C (peak γ), and 759 °C (peak δ), whereas Fe_2O_3/CeO_2-NP shows three reduction peaks centered at ~384 °C (peak α), 581 °C (peak γ), and 759 °C (peak δ), as the second and third peaks could have been merged, therefore justifying the absence of peak β [63]. The peaks at 465–469 °C (peak β) and 759 °C (peak δ) are attributed to the ceria surface oxygen and bulk oxygen reduction, respectively, while the peaks at ~377–390 °C (peak α) and 581–599 °C (peak γ) are ascribed to the iron species reduction, namely $Fe_2O_3 \rightarrow Fe_3O_4$ and $Fe_3O_4 \rightarrow Fe^0$, respectively [40,44]. In particular, the two aforementioned peaks could be referring to dispersed and clustered Fe_2O_3, accordingly [64]. Interestingly, the high-temperature peak δ (759 °C) that corresponds to ceria bulk oxygen reduction remains unaffected by the addition of iron, which can be attributed to the preferred existence of iron at the outermost shell of the nanoparticles [50]. It should also be noted that the addition of iron to ceria supports results into a downward shift of surface ceria TPR peaks, implying the pronounced effect of iron-ceria interactions on the reducibility [35,38].

According to the hydrogen consumed (Table 2), iron-ceria nanorods exhibit the highest value of H_2 consumption (3.42 mmol H_2/g) followed by nanopolyhedra (2.93 mmol H_2/g) and nanocubes (2.47 mmol H_2/g) perfectly matched to the catalytic conversion order (see below). It is also worth noticing that the amount of H_2 required for the reduction of Fe_2O_3/CeO_2 samples always surpasses the theoretical amount of H_2 for the complete reduction of Fe_2O_3 to Fe (~2 mmol H_2/g, on the basis of a Fe nominal loading of 7.5 wt.%). The latter reveals the facilitation of ceria capping oxygen reduction in the presence of iron, further corroborating the above findings and the synergistic function of metal and support. The H_2 excess uptake (mmol/g, Table 2), reflecting the extent of ceria oxygen reduction, follows the sequence Fe_2O_3/CeO_2-NR (1.42) > Fe_2O_3/CeO_2-NP (0.93) > Fe_2O_3/CeO_2-NC (0.47), in line to supports' reducibility.

2.4. *Surface Properties (XPS)*

In order to further investigate the impact of support morphology on the chemical composition and oxidation state of the samples, XPS analysis was performed. The Ce 3d XPS spectra of the samples are shown in Figure 6a. The Ce3d core level spectra were deconvoluted into eight components consisting of three pairs of spin-orbit doublets of Ce^{4+} and two peaks corresponding to Ce^{3+} [44,59,65]. In particular, the Ce $3d_{3/2}$ spin-obit components represented by the u lines include three characteristic peaks labeled as u (900.7 eV), u″ (907.6 eV), and u‴ (916.4 eV). The Ce $3d_{5/2}$ spin-orbit components represented by the v lines, contain three peaks labeled as v (882.2 eV), v″ (888.8 eV), and v‴ (898.2 eV). The aforementioned three pairs of peaks are ascribed to Ce^{4+} while the two lines denoted as u′ (902.1 eV) and v′ (883.8 eV) are attributed to Ce^{3+}. The proportion of Ce^{3+} ions with regard to the total cerium is calculated from the area ratio of the sum of the Ce^{3+} species to that of the total cerium species [66]. Table 3 summarizes the results derived by XPS analysis for all the samples. Bare ceria supports show similar amounts of the Ce^{3+} species between 23.3 and 25.3%. With regard to the iron-ceria samples, the population of Ce^{3+} ions is slightly higher than bare ceria samples, varying between 25.3 and 28.5%, without, however, exhibiting significant alterations between the samples of different morphology. In a similar manner, it has been shown that the relative amount of reduced Ce^{3+} species is similar among CuO/CeO_2 samples of different morphology [67,68].

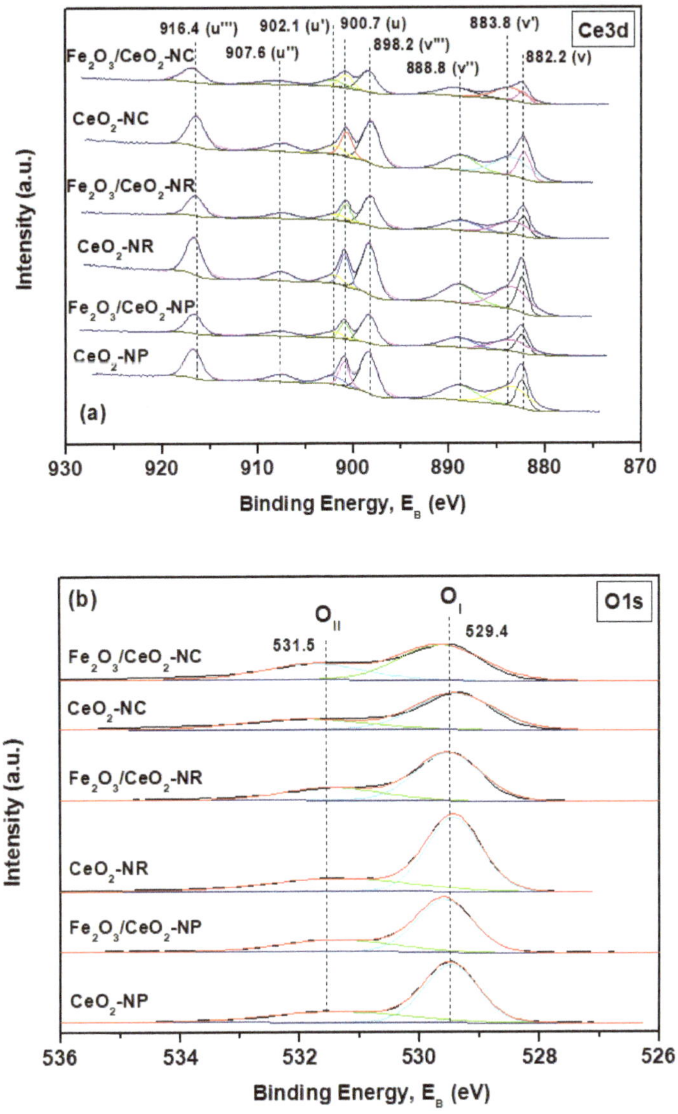

Figure 6. *Cont.*

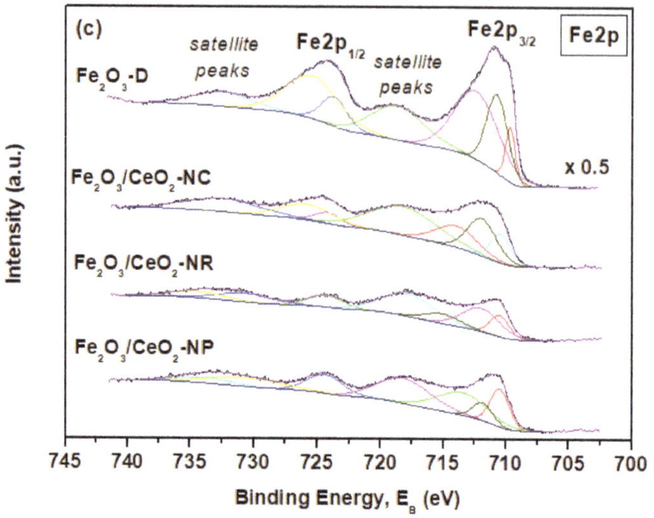

Figure 6. XPS spectra of (**a**) Ce 3d, (**b**) O 1s, and (**c**) Fe 2p of bare CeO_2, Fe_2O_3-D, and Fe_2O_3/CeO_2 samples.

Figure 6b shows the O 1s XPS spectra of the samples, which consist of two peaks. In general, the peak at lower binding energy (529.4 eV) corresponds to lattice oxygen (O_I) of the metal oxide phases, such as O^{2-}, whereas the peak at higher binding energy (531.5 eV) is ascribed to chemisorbed oxygen (O_{II}) including adsorbed oxygen (O^-/O_2^{2-}), adsorbed water, hydroxyl (OH^-), and carbonate (CO_3^{2-}) species [40,69,70]. Upon iron addition, a slight shift in the O_I band (Figure 5b) to higher binding energy occurs, which can be attributed to the electronegative effect of iron on the environment surrounding the cerium-oxygen bond [40,44]. As can be observed from Table 3, the following order, in terms of O_I/O_{II} ratio, is obtained for bare ceria samples: CeO_2-NR (2.13) > CeO_2-NP (2.04) > CeO_2-NC (1.99), which is well-correlated with the catalytic performance order (see below). The latter indicates the key role of lattice oxygen on the CO oxidation, as discussed in the sequence. Interestingly, exactly the same trend is obtained for iron-ceria samples, i.e., Fe_2O_3/CeO_2-NR (2.52) > Fe_2O_3/CeO_2-NP (2.25) > Fe_2O_3/CeO_2-NC (1.84) (Table 3), indicating the key role of support on the O_I/O_{II} ratio. These results, in conjunction to TPR studies, demonstrate that the samples with the rod-like morphology exhibit the highest concentration of easily reduced oxygen species, offering the optimum reducibility and oxygen kinetics.

Figure 6c shows the Fe 2p XPS spectra of Fe_2O_3-D and Fe_2O_3/CeO_2 samples. All samples exhibit two main peaks around 710.8 eV and 724.1 eV, corresponding to Fe $2p_{3/2}$ and Fe $2p_{1/2}$, respectively, as well as two satellite peaks at 717.9 and 732.6 eV which indicate the existence of Fe^{3+}, in agreement with the XRD results (Figure 2) [40]. However, it is worth noticing that the peak observed at 709.6-710.5 eV could be assigned to Fe^{2+} species [71], which is suggested to be formed by the interaction between the two oxide phases, through an interfacial redox process: $xFe_2O_3 + (2 - y)CeO_{2-x} \rightarrow xFe_2O_{3-y} + (2 - y)CeO_2$ [72–74]. Taking into account the Fe^{2+} (%) amount (Table 3), calculated by curve fitting including the satellite peaks (Figure 6c), the following order is obtained for the Fe_2O_3/CeO_2 samples of different morphology: Fe_2O_3/CeO_2-NR (14.4) > Fe_2O_3/CeO_2-NP (13.3) > Fe_2O_3/CeO_2-NC (13.1), which again coincides with the catalytic conversion order as described in the sequence and it is indicative of the interfacial interaction between the two oxide phases [72]. It should be pointed out that the aforementioned Fe^{2+} (%) amount order is in full compliance with the order of O_s/O_b ratio and the reducibility of the mixed oxides (Table 2), disclosing the interrelationship between electronic and redox properties induced by iron-ceria interactions.

The surface atomic ratio Fe/(Fe + Ce) of the Fe_2O_3/CeO_2 samples is presented in Table 3. Obviously, the surface atomic ratio of the rod-shaped sample is near to the nominal one (0.2), indicating a uniform distribution of iron and cerium species over the entire sample. However, nanopolyhedra and nanocubes exhibit higher values of surface atomic ratio than the nominal composition, namely 0.34 and 0.43, respectively, indicating an enrichment of the catalyst's surface in iron species or equally an impoverishment of catalyst's surface to cerium species. These results further corroborate the SEM-EDS analysis (Figure 4), as previously discussed.

Table 3. XPS results of bare CeO_2 and Fe_2O_3/CeO_2 samples.

Sample	Fe^{2+} (%)	Ce^{3+} (%)	O_I/O_{II}	Fe/(Fe + Ce)
CeO_2-NR	-	24.3	2.13	-
CeO_2-NP	-	25.3	2.04	-
CeO_2-NC	-	23.3	1.99	-
Fe_2O_3/CeO_2-NR	14.4	25.3	2.52	0.28
Fe_2O_3/CeO_2-NP	13.3	26.5	2.25	0.34
Fe_2O_3/CeO_2-NC	13.1	28.5	1.84	0.43

2.5. *Catalytic Evaluation Studies*

In order to gain insight into the morphological effect of ceria support on the catalytic performance of the Fe_2O_3/CeO_2 binary system, the oxidation of CO was investigated, as a model reaction. Figure 7 shows the CO conversion as a function of temperature of bare ceria carriers of different morphology as well as of the corresponding Fe_2O_3/CeO_2 samples. For comparison purposes, the catalytic performance of bare Fe_2O_3 as well as of a mechanical mixture of Fe_2O_3-D + CeO_2-NR (see experimental section) was investigated in parallel to reveal the individual or synergistic effect of catalyst's counterparts. Bare ceria carriers demonstrate inferior performance with, however, significant alterations between the samples of different morphology. In particular, in terms of half-conversion temperature (T_{50}), the following trend is obtained for bare ceria samples: CeO_2-NR (307 °C) > CeO_2-NP (323 °C) > CeO_2-NC (369 °C). Notably, the incorporation of iron into the ceria lattice clearly enhances the catalytic performance without, however, affecting the CO conversion order of bare ceria carriers: Fe_2O_3/CeO_2-NR (166 °C) > Fe_2O_3/CeO_2-NP (182 °C) > Fe_2O_3/CeO_2-NC (219 °C) > Fe_2O_3-D + CeO_2-NR (272 °C) > Fe_2O_3-D (277 °C). It is also worth noticing that the preparation method significantly affects the catalytic performance. Specifically, the hydrothermal method, which results in the development of ceria nanoparticles of different morphology, in conjunction to the addition of iron through the wet impregnation method, leads to highly active iron-ceria composites, as compared to the iron-ceria mixed oxide prepared by mechanical mixing (Fe_2O_3-D + CeO_2-NR, green line in Figure 7). More specifically, the conversion profile of Fe_2O_3/CeO_2-NR has been shifted by more than 100 °C to lower temperatures as compared to that of bare Fe_2O_3-D, CeO_2-NR, and Fe_2O_3-D + CeO_2-NR mechanical mixture, clearly revealing the synergistic interaction between CeO_2 and Fe_2O_3 induced by the preparation procedure followed. At this point it should be noted that a stable conversion performance (~80%) was attained at 200 °C during a short term (24 h) stability experiment over the most active Fe_2O_3/CeO_2-NR sample (not shown for brevity).

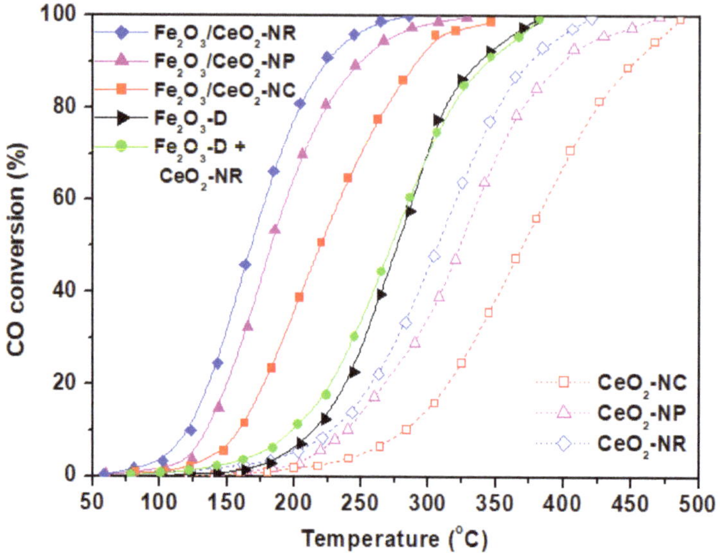

Figure 7. CO conversion as a function of temperature for bare CeO_2 and Fe_2O_3/CeO_2 samples of different morphology (NR, NC, and NP, as indicated in each curve). Reaction conditions: 2000 ppm CO, 1 vol.% O_2, Gas Hourly Space Velocity (GHSV) = 40,000 h^{-1}.

To more closely gain insight into the intrinsic reactivity of as-prepared samples, the reaction rate under differential conditions (Gas Hourly Space Velocity (GHSV) = 40,000 h^{-1}, m_{cat} = 100 mg, $X_{CO} < 15\%$) was obtained in the form of Arrhenius plots (Figure 8), and the corresponding activation energies (E_a) are summarized in Table 4. The superiority of Fe_2O_3/CeO_2 samples as compared to bare CeO_2 and Fe_2O_3 at a given temperature is again obvious (Figure 8), clearly revealing the synergistic iron-ceria interactions. Moreover, the same activity order to the CO conversion performance (Figure 7) is obtained, further validating the aforementioned structure–activity relationships. In relation to the activation energies, the bare nanorod sample (CeO_2-NR) exhibits the lowest activation energy (44.2 kJ·mol^{-1}), followed by nanopolyhedra (46.7 kJ·mol^{-1}) and nanocubes (49.8 kJ·mol^{-1}), a trend identical to their CO conversion performance. A similar trend is shown for the mixed oxides, with Fe_2O_3/CeO_2-NR showing the lowest activation energy among Fe_2O_3 supported samples, followed by the Fe_2O_3/CeO_2-NP and Fe_2O_3/CeO_2-NC samples. On the other hand, the bare Fe_2O_3 sample exhibits a much higher activation energy (63.3 kJ mol^{-1}) compared to bare ceria and iron-ceria samples, indicating a higher energy barrier for CO oxidation over Fe_2O_3. In view of this fact, the higher activation energy of iron supported ceria samples (53.5–58.6 kJ mol^{-1}) compared to bare ceria samples (44.2–49.8 kJ mol^{-1}) can receive a consistent explanation. However, it is worth noticing that iron-ceria catalysts demonstrate the highest activity (Figure 8, Table 4), despite their higher E_a, implying a more facile reaction path most probably induced by the iron-ceria interfacial sites. These findings clearly reveal the pivotal role of ceria support morphology in conjunction with iron-ceria interface towards determining the activation energy and the catalytic activity of Fe_2O_3/CeO_2 samples.

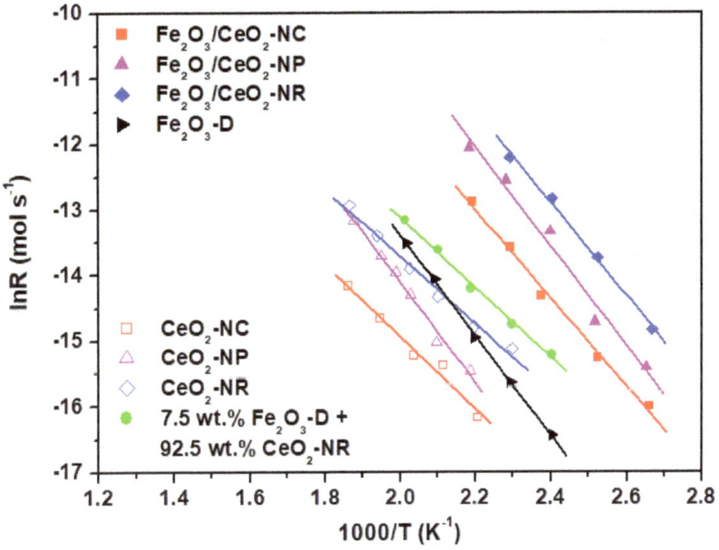

Figure 8. Arrhenius plots for CO conversion as a function of temperature for Ceria and Fe_2O_3/Ceria samples of different morphology (NR, NC, and NP, as indicated). Reaction conditions: 2000 ppm CO, 1 vol.% O_2, GHSV = 40,000 h^{-1}.

Table 4. Activation Energies (E_a) for the CO oxidation reaction over bare CeO_2 and Fe_2O_3/CeO_2 samples.

Sample	Ea (kJ mol^{-1}) (± 0.1)	R^2 (from Fitting Procedure)
CeO_2-NR	44.2	0.9943
	46.7	0.9906
CeO_2-NC	49.8	0.9916
Fe_2O_3/CeO_2-NR	53.5	0.9875
Fe_2O_3/CeO_2-NP	55.7	0.9853
Fe_2O_3/CeO_2-NC	58.6	0.9959
Fe_2O_3-D	63.3	0.9995
Fe_2O_3-D + CeO_2-NR	48.1	0.9983

The enhanced textural properties (BET surface area and pore volume) of the Fe_2O_3/CeO_2-NR sample in comparison with the Fe_2O_3/CeO_2-NP and Fe_2O_3/CeO_2-NC samples should be also mentioned, which could be further accounted for its enhanced CO conversion performance. In order to further elucidate the relationship between CO oxidation activity and the aforementioned textural properties, the specific activity, normalized both per unit of catalyst mass (nmol g^{-1} s^{-1}) and surface area (nmol m^{-2} s^{-1}) was calculated under differential reaction conditions ($X_{CO} < 15\%$, T = 125 °C, GHSV = 40,000 h^{-1}). These specific parameters can reflect more accurately the impact of intrinsic properties of the Fe_2O_3/CeO_2 mixed oxides on the catalytic performance. The results are summarized in Table 5. It is evident that Fe_2O_3/CeO_2-NR exhibits the best catalytic performance (both in terms of conversion and specific activity) as compared to the other two iron-ceria polymorphs, revealing the pivotal role of the exposed crystal planes and the redox characteristics, rather than that of textural characteristics, on the CO oxidation performance. Similar conclusions have been already drawn in our previous work on CO oxidation over CuO/CeO_2 mixed oxides [18].

Table 5. CO conversion and specific activity of Fe_2O_3/CeO_2 samples at 125 °C. Reaction conditions: 2000 ppm CO, 1 vol.% O_2, GHSV = 40,000 h^{-1}.

Sample	CO Conversion (%)	Specific Activity	
		r (nmol g^{-1} s^{-1})	r (nmol m^{-2} s^{-1})
Fe_2O_3/CeO_2-NC	2.40	26	0.80
Fe_2O_3/CeO_2-NP	4.46	48	0.75
Fe_2O_3/CeO_2-NR	11.69	126	1.84

Summarizing, the nanorod-shaped sample (Fe_2O_3/CeO_2-NR) exhibits the best catalytic performance, followed by nanopolyhedra (Fe_2O_3/CeO_2-NP) and nanocubes (Fe_2O_3/CeO_2-NC). The same order, however much inferior, is observed for the bare ceria samples as well, demonstrating the fundamental role of support morphology. This enhanced catalytic behavior of the nanorod sample can be mainly attributed, on the basis of the present findings, to its improved reducibility and oxygen mobility, enforced by the strong interaction between ceria nanorods and iron species, along with the abundance of ceria nanorods in oxygen vacancies. The latter has been clearly established by in situ Raman spectroscopy, revealing the following trend in the relative abundance of structural defects: CeO_2-NR > CeO_2-NP > CeO_2-NC [18].

More specifically, the oxidation of CO over ceria-based materials follows a Mars-van Krevelen-type of mechanism, in which the reaction includes alternating reduction/oxidation steps that result in the formation of surface oxygen vacancies, which is considered to be a reactivity descriptor on doped ceria surfaces [75], and their subsequent refill by gas phase oxygen [31]. In view of this fact, it has been shown that there is a close relationship between the fundamental steps of the above mentioned mechanism and the oxygen species of ceria's {111} crystal plane [75] and this could also be the case for the nanorod sample, which shows abundance in lattice oxygen (Table 3) and easily reduced (Table 2) species. The aforementioned mechanism reveals the significance of the support's redox properties, as the synergism observed in the Fe_2O_3/CeO_2 samples is mainly due to the Ce^{4+}/Ce^{3+} and Fe^{3+}/Fe^{2+} redox couples [50]. In a similar manner, Luo et al. [74] have shown that high ratios of Fe^{2+}/Fe^{3+} and Ce^{3+}/Ce^{4+} favor the conversion of CO in the inverse CeO_2-Fe_2O_3 catalysts. In general, the Fe^{2+}/Fe^{3+} pairs are considered catalytic centers for the oxidation of CO as the coexistence of Fe^{2+}/Fe^{3+} and Ce^{4+}/Ce^{3+} pairs facilitate the electron transfer between the mixed sites Fe^{2+}-Ce^{4+} and/or Fe^{3+}-Ce^{3+}, leading to enhanced CO oxidation activity [74,76], while the presence of Fe^{2+} and Ce^{3+} ions is related to the formation of oxygen vacancies [74]. Moreover, Chen et al. [77] have shown by DFT computational studies of CO oxidation over iron-modified ceria {111} surfaces that the CO molecule is adsorbed to the iron adatom which is then oxidized by the lattice oxygen, resulting in the desorption of CO_2 and the formation of oxygen vacancies, which is considered to be the rate-determining step.

In conclusion, the enhanced catalytic behavior of the Fe_2O_3/CeO_2 sample of rod-like morphology could be mainly attributed to its superior reducibility and oxygen kinetics, closely related to the abundance in weakly bound oxygen species (Table 2) and Fe^{2+} ions (Table 3). In view of this fact, the above statements regarding structure–function relationships can be further corroborated by the perfect relationship between the catalytic performance and the surface-to-bulk ratio (O_s/O_b) (Figure 9), as it has been similarly reported by our group [18] for CO oxidation in copper-ceria samples. Furthermore, it is worth pointing out that the CO oxidation performance follows the same order, namely nanorods > nanopolyhedra > nanocubes, regardless of the active phase used, as the above mentioned trend was also observed in CuO/CeO_2 nanoparticles of the same morphology, clearly reflecting the key role of support morphology on the catalytic behavior [18].

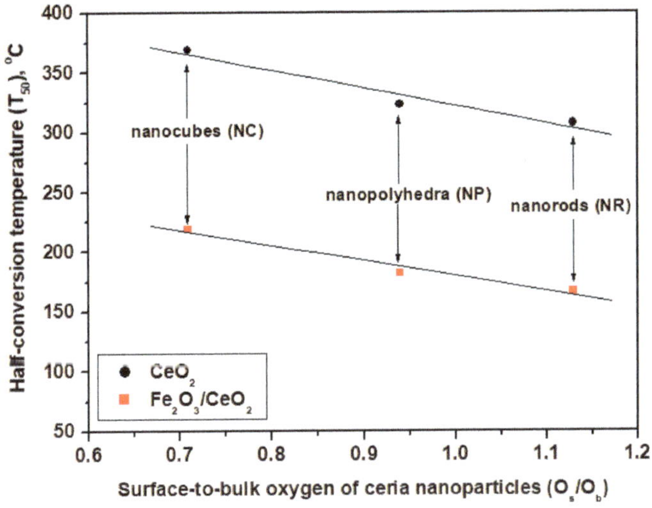

Figure 9. Relationship between the half-conversion temperature (T_{50}) and the surface-to-bulk oxygen ratio (O_s/O_b).

3. Materials and Methods

3.1. Materials Synthesis

In the present work, all chemicals were of analytical reagent grade. $Ce(NO_3)_3{\cdot}6H_2O$ (purity ≥ 99.0%, Fluka, Bucharest, Romania) and $Fe(NO_3)_3{\cdot}9H_2O$ (≥ 98%, Sigma-Aldrich, St. Louis, MO, USA) were the precursors used for the synthesis of bare ceria and iron-ceria samples. NaOH (purity ≥ 98%, Sigma-Aldrich, St. Louis, MO, USA) and absolute ethanol (≥ 98%, Honeywell, Charlotte, North Carolina, USA) were also used during preparation procedures. Firstly, the bare ceria samples were prepared by the hydrothermal method as thoroughly described in our previous work [18]. In brief, appropriate amounts of $Ce(NO_3)_3{\cdot}6H_2O$ and NaOH were dissolved in double deionized water, mixed under vigorous stirring for 1 h and aged for 24 h, at 90 °C for ceria nanorods and nanopolyhedra and at 180 °C for ceria nanocubes. Subsequently, centrifugation was used for the recovery of the solid products which were extensively washed with double deionized water until pH 7 for the removal of any co-precipitated salts, as well as absolute ethanol in order to avoid the agglomeration of the nanoparticles. Finally, the resulting precipitate was dried at 90 °C for 12 h and calcined at 500 °C for 2 h under air flow (heating ramp 5 °C/min). The bare ceria samples are denoted as CeO_2-NX, where NX corresponds to NR–nanorods, NP–nanopolyhedra, and NC–nanocubes.

The Fe_2O_3/CeO_2-NX mixed catalysts were prepared by the wet impregnation method, employing an aqueous solution of $Fe(NO_3)_3{\cdot}9H_2O$, in order to obtain an atomic ratio Fe/(Fe+Ce) of 0.2, which corresponds to a Fe loading of 7.5 wt.%. This particular ratio was dictated from our previous studies over a series of MO_x/CeO_2 catalysts where the M/(M + Ce) atomic content is always kept constant to 0.2 [18,78]. Moreover, relevant literature studies over FeO_x/CeO_2 catalysts revealed that optimum redox/surface properties can be obtained at similar metal contents, e.g., [40,50]. Heating of the resulting suspensions under stirring until water evaporation, drying at 90 °C for 12 h, and final calcination at 500 °C for 2 h under air flow (heating ramp 5 °C/min) occurred.

For comparison purposes, two additional samples were synthesized. A bare iron oxide sample denoted as Fe_2O_3-D was prepared by thermal decomposition of $Fe(NO_3)_3{\cdot}9H_2O$ at 500 °C for 2 h, and an iron-ceria mixed oxide was prepared as a physical mixture of 7.5 wt.% Fe_2O_3-D and 92.5 wt.% CeO_2-NR, as it was mixed in agate by hand.

3.2. *Materials Characterization*

The porosity of the materials was evaluated by the N_2-adsorption isotherms at –196 °C, using an ASAP 2010 (Micromeritics, Norcross, GA, USA) apparatus (from ReQuimTe Analyses Laboratory, Universidade Nova de Lisboa, Portugal). Samples were previously degassed at 300 °C for 6 h. The specific surface area was calculated by the Brunauer-Emmett-Teller (BET) equation [79].

Structural characterization was carried out by means of XRD in a PAN'alytical X'Pert MPD (PANanalytical, Almelo, Netherlands) equipped with a X'Celerator detector and secondary monochromator (Cu Kα λ = 0.154 nm, 50 kV, 40 mA; data recorded at a 0.017° step size, 100 s/step) in University of Trás-os-Montes e Alto Douro. The collected spectra were analyzed by Rietveld refinement using PowderCell software (by Werner Kraus and Gert Nolze, http://www.ccp14.ac.uk), allowing the determination of crystallite sizes by means of the Williamson–Hall plot.

The samples were imaged by TEM. The analyses were performed on a Leo 906E apparatus (Austin, TX, USA), at 120 kV in University of Trás-os-Montes e Alto Douro. Samples were prepared ultrasonically dispersion and a 400 mesh formvar/carbon copper grid (Agar Scientific, Essex, UK) was dipped into the solution for the TEM analysis.

The surface morphology was also investigated by Scanning Electron Microscopy (SEM, JEOL JSM-6390LV, JEOL Ltd., Akishima, Tokyo, Japan), operating at 20 keV, equipped with an energy dispersive X-ray spectrometry (EDS) system. The powders were placed on double-sided adhesive tape and sputtered with Au for 39 seconds in order to create a coating with a thickness of 10 nm, approximately. The specimens were observed under two different detection modes: secondary and backscattered electrons.

The redox properties were assessed by TPR experiments in an AMI-200 Catalyst Characterization Instrument (Altamira Instruments, Pittsburgh, PA, USA), employing H_2 as a reducing agent. In a typical H_2-TPR experiment, 50 mg of the sample (grain size 180–355 μm) was heated up to 1100 °C (10 °C/min), under H_2 flow (1.5 cm^3) balanced with He (29 cm^3). The amount of H_2 consumed (mmol g^{-1}) was calculated by taking into account the integrated area of TPR peaks, calibrated against a known amount of CuO standard sample [80,81].

The surface composition and the chemical state of each element were determined by XPS analyses, performed on a VG Scientific ESCALAB 200A spectrometer using Al Kα radiation (1486.6 eV) in CEMUP. The charge effect was corrected using the C1s peak as a reference (binding energy of 285 eV). The CASAXPS software (http://www.casaxps.com/) was used for data analysis.

3.3. *Catalytic Evaluation Studies*

Catalytic oxidation of CO was performed in a quartz fixed-bed tubular microreactor (12.95 mm i.d.) at atmospheric pressure, loaded with 0.10 g of catalyst. The reaction stream consisted of 2000 ppm of CO and 1 vol.% O_2 balanced with He in a total feed stream of 80 mL min^{-1} which was controlled by EL-Flow Bronkhorst Mass Flow controllers (Bronkhorst High-Tech B.V., Ruurlo, Netherlands) and homogenized by a mixing chamber. The catalyst temperature was recorded using a K-Type thermocouple placed in the catalyst bed and the gas hour space velocity (GHSV) of the feed stream was 40,000 h^{-1}.

Prior to catalytic experiments, all samples were treated under a 20 cm^3/min flow of 20 vol.% O_2 in He heating up to 480 °C with 10 degrees/min. Samples remained at 480 °C for 30 min and then the temperature was decreased to 25 °C with the same rate. Final purge with He flow was carried out in order to remove physiosorbed species.

Catalytic evaluation measurements were carried out every 20 degrees up to 500 °C. CO and CO_2 in the effluent gas were analyzed by gas chromatography (GC) using a fast response micro GC Varian CP-4900 equipped with 2 fully equipped channels with separated TCD detectors, injectors and 2 capillary columns (Molecular Sieve 5X and PoraPlot Q), all provided from Varian B.V., Middelburg,

Netherlands. The conversion of CO (X_{CO}) was calculated from the difference in CO concentration between the inlet and outlet gas streams, according to the equation:

$$X_{CO}(\%) = \frac{[CO]_{in} - [CO]_{out}}{[CO]_{in}} \times 100 \quad (1)$$

The specific reaction rate (r, mol m^{-2} s^{-1}) of the CO conversion was also estimated under differential reaction conditions (X_{CO} < 15%, T=125°C, W/F = 0.075 g s cm^{-3}) using the following formula:

$$r\left(\text{mol m}^{-2}\,\text{s}^{-1}\right) = \frac{X_{CO} \times [CO]_{in} \times F\left(\frac{cm^3}{min}\right)}{100 \times 60\left(\frac{s}{min}\right) \times V_m\left(\frac{cm^3}{mol}\right) \times m_{cat}(g) \times S_{BET}\left(\frac{m^2}{g}\right)} \quad (2)$$

where F and V_m are the total flow rate and gas molar volume, respectively, at standard ambient temperature and pressure conditions (298 K and 1 bar), m_{cat} is the catalyst's mass, and S_{BET} is the surface area.

4. Conclusions

In the present work, ceria nanostructures of different morphology (cubes, polyhedra, rods) were synthesized by the hydrothermal method and used as carriers for the iron oxide phase. The bare ceria samples, as well as the iron-ceria mixed oxides, were catalytically evaluated in the oxidation of CO. Regarding the CO conversion performance, the same trend was observed in both bare ceria and Fe_2O_3/CeO_2 mixed oxides, namely nanorods > nanopolyhedra > nanocubes, demonstrating the crucial role of support morphology, with iron addition, however, significantly enhancing the catalytic behavior. In terms of half-conversion temperature (T_{50}), iron-ceria nanorods (Fe_2O_3/CeO_2-NR) exhibit the best catalytic performance (T_{50} = 166 °C), attributed mainly to their enhanced reducibility and oxygen kinetics linked to their abundance in loosely bound oxygen species, their highest amount in lattice oxygen, and their largest amount of Fe^{2+}. The present findings demonstrate that the co-doping of cheap and abundant transition metals (such as iron) on a reducible carrier (such as ceria), along with the fine-tuning of their morphology, should be considered as a novel approach towards developing highly active, noble metal-free, catalysts.

Author Contributions: M.L. and S.S contributed to materials synthesis and characterization, results interpretation and paper writing; S.A.C.C. contributed to materials characterization; P.K.P. and V.N.S. contributed to catalytic evaluation studies, results interpretation and paper writing; M.K. contributed to the conception, design, results interpretation and writing of the paper; All authors contributed to the discussion, read and approved the final version of the manuscript.

Funding: This research has been co-financed by the European Union and Greek national funds through the Operational Program Competitiveness, Entrepreneurship and Innovation, under the call RESEARCH–CREATE–INNOVATE (project code: T1EDK-00094). This work was also financially supported by Associate Laboratory LSRE-LCM—UID/EQU/50020/2019—funded by national funds through FCT/MCTES (PIDDAC). S.A.C.C. acknowledges Fundação para a Ciência e a Tecnologia (Portugal) for Investigador FCT program (IF/01381/2013/CP1160/CT0007), with financing from the European Social Fund and the Human Potential Operational Program.

Acknowledgments: We are grateful to Carlos Sá (CEMUP) for the assistance with the XPS measurements, Pedro Tavares (UTAD) for the TEM and XRD analyses and Nuno Costa (ReQuimTe) for the N_2 adsorption results.

Conflicts of Interest: The authors declare no conflict of interest.

References

1. Paier, J.; Penschke, C.; Sauer, J. Oxygen defects and surface chemistry of ceria: Quantum chemical studies compared to experiment. *Chem. Rev.* **2013**, *113*, 3949–3985. [CrossRef]
2. Montini, T.; Melchionna, M.; Monai, M.; Fornasiero, P. Fundamentals and Catalytic Applications of CeO_2-Based Materials. *Chem. Rev.* **2016**, *116*, 5987–6041. [CrossRef] [PubMed]

3. Tang, W.-X.; Gao, P.-X. Nanostructured cerium oxide: preparation, characterization, and application in energy and environmental catalysis. *MRS Commun.* **2016**, *6*, 311–329. [CrossRef]
4. Sun, C.; Xue, D. Size-dependent oxygen storage ability of nano-sized ceria. *Phys. Chem. Chem. Phys.* **2013**, *15*, 14414–14419. [CrossRef] [PubMed]
5. Nolan, M. *Surface Effects in the Reactivity of Ceria: A First Principles Perspective*; Elsevier Inc.: Amsterdam, The Netherlands, 2015; ISBN 9780128013403.
6. Devaiah, D.; Reddy, L.H.; Park, S.E.; Reddy, B.M. Ceria–zirconia mixed oxides: Synthetic methods and applications. *Catal. Rev. Sci. Eng.* **2018**, *60*, 177–277. [CrossRef]
7. Zhang, D.; Du, X.; Shi, L.; Gao, R. Shape-controlled synthesis and catalytic application of ceria nanomaterials. *Dalton Trans.* **2012**, *41*, 14455–14475. [CrossRef]
8. Melchionna, M.; Fornasiero, P. The role of ceria-based nanostructured materials in energy applications. *Mater. Today* **2014**, *17*, 349–357. [CrossRef]
9. Stathopoulos, V.N.; Costa, C.N.; Pomonis, P.J.; Efstathiou, A.M. The $CH_4/NO/O_2$ "lean-deNO*x*" reaction on mesoporous Mn-based mixed oxides. *Top. Catal.* **2001**, *16*, 231–235. [CrossRef]
10. Stathopoulos, V.N.; Petrakis, D.E.; Hudson, M.; Falaras, P.; Neofytides, S.G.; Pomonis, P.J. Novel Mn-based Mesoporous Mixed Oxidic Solids. *Stud. Surf. Sci. Catal.* **2000**, *128*, 593–602. [CrossRef]
11. Stathopoulos, V.N.; Belessi, V.C.; Costa, C.N.; Neophytides, S.; Falaras, P.; Efstathiou, A.M.; Pomonis, P.J. Catalytic activity of high surface area mesoporous Mn-based mixed oxides for the deep oxidation of methane and lean-NOx reduction. *Stud. Surf. Sci. Catal.* **2000**, *130*, 1529–1534. [CrossRef]
12. Stathopoulos, V.N.; Belessi, V.C.; Bakas, T.V.; Neophytides, S.G.; Costa, C.N.; Pomonis, P.J.; Efstathiou, A.M. Comparative study of La–Sr–Fe–O perovskite-type oxides prepared by ceramic and surfactant methods over the CH_4 and H_2 lean-deNOx. *Appl. Catal. B Environ.* **2009**, *93*, 1–11. [CrossRef]
13. Stathopoulos, V.N.; Belessi, V.C.; Ladavos, A.K. Samarium based high surface area perovskite type oxides $SmFe_{1-X}Al_XO_3$ (X = 0.00, 0.50, 0.95). Part II, Catalytic combustion of CH_4. *React. Kinet. Catal. Lett.* **2001**, *72*, 49–55. [CrossRef]
14. Corberán, V.C.; Rives, V.; Stathopoulos, V. Chapter 7—Recent Applications of Nanometal Oxide Catalysts in Oxidation Reactions. In *Advanced Nanomaterials for Catalysis and Energy; Synthesis, Characterization and Applications*; Elsevier: Amsterdam, The Netherlands, 2019; pp. 227–293. ISBN 9780128148075.
15. Qiao, Z.A.; Wu, Z.; Dai, S. Shape-controlled ceria-based nanostructures for catalysis applications. *ChemSusChem* **2013**, *6*, 1821–1833. [CrossRef] [PubMed]
16. Matte, L.P.; Kilian, A.S.; Luza, L.; Alves, M.C.M.; Morais, J.; Baptista, D.L.; Dupont, J.; Bernardi, F. Influence of the CeO_2 Support on the Reduction Properties of Cu/CeO_2 and Ni/CeO_2 Nanoparticles. *J. Phys. Chem. C* **2015**, *119*, 26459–26470. [CrossRef]
17. Zabilskiy, M.; Djinović, P.; Tchernychova, E.; Tkachenko, O.P.; Kustov, L.M.; Pintar, A. Nanoshaped CuO/CeO_2 Materials: Effect of the Exposed Ceria Surfaces on Catalytic Activity in N_2O Decomposition Reaction. *ACS Catal.* **2015**, *5*, 5357–5365. [CrossRef]
18. Lykaki, M.; Pachatouridou, E.; Carabineiro, S.A.C.; Iliopoulou, E.; Andriopoulou, C.; Kallithrakas-Kontos, N.; Boghosian, S.; Konsolakis, M. Ceria nanoparticles shape effects on the structural defects and surface chemistry: Implications in CO oxidation by Cu/CeO_2 catalysts. *Appl. Catal. B Environ.* **2018**, *230*, 18–28. [CrossRef]
19. Wu, Z.; Li, M.; Overbury, S.H. On the structure dependence of CO oxidation over CeO_2 nanocrystals with well-defined surface planes. *J. Catal.* **2012**, *285*, 61–73. [CrossRef]
20. Vilé, G.; Colussi, S.; Krumeich, F.; Trovarelli, A.; Pérez-Ramírez, J. Opposite face sensitivity of CeO_2 in hydrogenation and oxidation catalysis. *Angew. Chem. Int. Ed.* **2014**, *53*, 12069–12072. [CrossRef] [PubMed]
21. Aneggi, E.; Wiater, D.; De Leitenburg, C.; Llorca, J.; Trovarelli, A. Shape-dependent activity of ceria in soot combustion. *ACS Catal.* **2014**, *4*, 172–181. [CrossRef]
22. Cargnello, M.; Doan-Nguyen, V.V.T.; Gordon, T.R.; Diaz, R.E.; Stach, E.A.; Gorte, R.J.; Fornasiero, P.; Murray, C.B. Control of Metal Nanocrystal Size Reveals Metal-Support Interface Role for Ceria Catalysts. *Science* **2013**, *341*, 771–773. [CrossRef] [PubMed]
23. Ahmadi, M.; Mistry, H.; Roldan Cuenya, B. Tailoring the Catalytic Properties of Metal Nanoparticles via Support Interactions. *J. Phys. Chem. Lett.* **2016**, *7*, 3519–3533. [CrossRef] [PubMed]
24. Qiu, N.; Zhang, J.; Wu, Z. Peculiar surface–interface properties of nanocrystalline ceria–cobalt oxides with enhanced oxygen storage capacity. *Phys. Chem. Chem. Phys.* **2014**, *16*, 22659–22664. [CrossRef] [PubMed]

25. Konsolakis, M.; Carabineiro, S.A.C.; Papista, E.; Marnellos, G.E.; Tavares, P.B.; Agostinho Moreira, J.; Romaguera-Barcelay, Y.; Figueiredo, J.L. Effect of preparation method on the solid state properties and the deN_2O performance of CuO-CeO_2 oxides. *Catal. Sci. Technol.* **2015**, *5*, 3714–3727. [CrossRef]
26. Konsolakis, M. The role of Copper–Ceria interactions in catalysis science: Recent theoretical and experimental advances. *Appl. Catal. B Environ.* **2016**, *198*, 49–66. [CrossRef]
27. Wu, K.; Sun, L.D.; Yan, C.H. Recent Progress in Well-Controlled Synthesis of Ceria-Based Nanocatalysts towards Enhanced Catalytic Performance. *Adv. Energy Mater.* **2016**, *6*, 1600501. [CrossRef]
28. Devaiah, D.; Thrimurthulu, G.; Smirniotis, P.G.; Reddy, B.M. Nanocrystalline alumina-supported ceria-praseodymia solid solutions: Structural characteristics and catalytic CO oxidation. *RSC Adv.* **2016**, *6*, 44826–44837. [CrossRef]
29. Devaiah, D.; Tsuzuki, T.; Thirupathi, B.; Smirniotis, P.G.; Reddy, B.M. $Ce_{0.80}M_{0.12}Sn_{0.08}O_{2-\delta}$ (M = Hf, Zr, Pr, and La) ternary oxide solid solutions with superior properties for CO oxidation. *RSC Adv.* **2015**, *5*, 30275–30285. [CrossRef]
30. Zhu, X.; Du, Y.; Wang, H.; Wei, Y.; Li, K.; Sun, L. Chemical interaction of Ce-Fe mixed oxides for methane selective oxidation. *J. Rare Earths* **2014**, *32*, 824–830. [CrossRef]
31. Bao, H.; Chen, X.; Fang, J.; Jiang, Z.; Huang, W. Structure-activity Relation of Fe_2O_3–CeO_2 Composite Catalysts in CO Oxidation. *Catal. Lett.* **2008**, *125*, 160–167. [CrossRef]
32. Wang, W.; Li, W.; Guo, R.; Chen, Q.; Wang, Q.; Pan, W.; Hu, G. A $CeFeO_x$ catalyst for catalytic oxidation of NO to NO_2. *J. Rare Earths* **2016**, *34*, 876–881. [CrossRef]
33. Tan, L.; Tao, Q.; Gao, H.; Li, J.; Jia, D.; Yang, M. Preparation and catalytic performance of mesoporous ceria-base composites CuO/CeO_2, Fe_2O_3/CeO_2 and La_2O_3/CeO_2. *J. Porous Mater.* **2017**, *24*, 795–803. [CrossRef]
34. Massa, P.; Dafinov, A.; Cabello, F.M.; Fenoglio, R. Catalytic wet peroxide oxidation of phenolic solutions over Fe_2O_3/CeO_2 and WO_3/CeO_2 catalyst systems. *Catal. Commun.* **2008**, *9*, 1533–1538. [CrossRef]
35. Han, J.; Meeprasert, J.; Maitarad, P.; Nammuangruk, S.; Shi, L.; Zhang, D. Investigation of the Facet-Dependent Catalytic Performance of Fe_2O_3/CeO_2 for the Selective Catalytic Reduction of NO with NH_3. *J. Phys. Chem. C* **2016**, *120*, 1523–1533. [CrossRef]
36. Prieto-Centurion, D.; Eaton, T.R.; Roberts, C.A.; Fanson, P.T.; Notestein, J.M. Catalytic reduction of NO with H_2 over redox-cycling Fe on CeO_2. *Appl. Catal. B Environ.* **2015**, *168–169*, 68–76. [CrossRef]
37. Nadar, A.; Banerjee, A.M.; Pai, M.R.; Pai, R.V.; Meena, S.S.; Tewari, R.; Tripathi, A.K. Catalytic properties of dispersed iron oxides Fe_2O_3/MO_2 (M = Zr, Ce, Ti and Si) for sulfuric acid decomposition reaction: Role of support. *Int. J. Hydrogen Energy* **2018**, *43*, 37–52. [CrossRef]
38. Perez-Alonso, F.J.; Melián-Cabrera, I.; López Granados, M.; Kapteijn, F.; Fierro, J.L.G. Synergy of $Fe_xCe_{1-x}O_2$ mixed oxides for N_2O decomposition. *J. Catal.* **2006**, *239*, 340–346. [CrossRef]
39. Shen, Q.; Lu, G.; Du, C.; Guo, Y.; Wang, Y.; Guo, Y.; Gong, X. Role and reduction of NOx in the catalytic combustion of soot over iron—ceria mixed oxide catalyst. *Chem. Eng. J.* **2013**, *218*, 164–172. [CrossRef]
40. Sudarsanam, P.; Hillary, B.; Amin, M.H.; Rockstroh, N.; Bentrup, U.; Brückner, A.; Bhargava, S.K. Heterostructured Copper-Ceria and Iron-Ceria Nanorods: Role of Morphology, Redox, and Acid Properties in Catalytic Diesel Soot Combustion. *Langmuir* **2018**, *34*, 2663–2673. [CrossRef]
41. Lu, F.; Jiang, B.B.; Wang, J.; Huang, Z.; Liao, Z.; Yang, Y. Insights into the improvement effect of Fe doping into the CeO_2 catalyst for vapor phase ketonization of carboxylic acids. *Mol. Catal.* **2018**, *444*, 22–33. [CrossRef]
42. Li, K.; Wang, H.; Wei, Y.; Yan, D. Transformation of methane into synthesis gas using the redox property of Ce-Fe mixed oxides: Effect of calcination temperature. *Int. J. Hydrogen Energy* **2011**, *36*, 3471–3482. [CrossRef]
43. Bao, H.; Qian, K.; Fang, J.; Huang, W. Fe-doped CeO_2 solid solutions: Substituting-site doping versus interstitial-site doping, bulk doping versus surface doping. *Appl. Surf. Sci.* **2017**, *414*, 131–139. [CrossRef]
44. Reddy, A.S.; Chen, C.Y.; Chen, C.C.; Chien, S.H.; Lin, C.J.; Lin, K.H.; Chen, C.L.; Chang, S.C. Synthesis and characterization of Fe/CeO_2 catalysts: Epoxidation of cyclohexene. *J. Mol. Catal. A Chem.* **2010**, *318*, 60–67. [CrossRef]
45. Torrente-Murciano, L.; Chapman, R.S.L.; Narvaez-Dinamarca, A.; Mattia, D.; Jones, M.D. Effect of nanostructured ceria as support for the iron catalysed hydrogenation of CO_2 into hydrocarbons. *Phys. Chem. Chem. Phys.* **2016**, *18*, 15496–15500. [CrossRef]
46. Kumar, S.; Sharma, C. Synthesis, characterization and catalytic wet air oxidation property of mesoporous $Ce_{1-x}Fe_xO_2$ mixed oxides. *Mater. Chem. Phys.* **2015**, *155*, 223–231. [CrossRef]

47. Trens, P.; Stathopoulos, V.; Hudson, M.J.; Pomonis, P. Synthesis and characterization of packed mesoporous tungsteno-silicates: application to the catalytic dehydrogenation of 2-propanol. *Appl. Catal. A Gen.* **2004**, *263*, 103–108. [CrossRef]
48. Farahmandjou, M.; Zarinkamar, M. Synthesis of nano-sized ceria (CeO_2) particles via a cerium hydroxy carbonate precursor and the effect of reaction temperature on particle morphology. *J. Ultrafine Grained Nanostructured Mater.* **2015**, *48*, 5–10. [CrossRef]
49. Sharma, V.; Eberhardt, K.M.; Sharma, R.; Adams, J.B.; Crozier, P.A. A spray drying system for synthesis of rare-earth doped cerium oxide nanoparticles. *Chem. Phys. Lett.* **2010**, *495*, 280–286. [CrossRef]
50. Sahoo, T.R.; Armandi, M.; Arletti, R.; Piumetti, M.; Bensaid, S.; Manzoli, M.; Panda, S.R.; Bonelli, B. Pure and Fe-doped CeO_2 nanoparticles obtained by microwave assisted combustion synthesis: Physico-chemical properties ruling their catalytic activity towards CO oxidation and soot combustion. *Appl. Catal. B Environ.* **2017**, *211*, 31–45. [CrossRef]
51. Carabineiro, S.A.C.; Papista, E.; Marnellos, G.E.; Tavares, P.B.; Maldonado-Hódar, F.J.; Konsolakis, M. Catalytic decomposition of N_2O on inorganic oxides: Effect of doping with Au nanoparticles. *Mol. Catal.* **2017**, *436*, 78–89. [CrossRef]
52. Trpkov, D.; Panjan, M.; Kopanja, L.; Tadić, M. Hydrothermal synthesis, morphology, magnetic properties and self-assembly of hierarchical α-Fe_2O_3 (hematite) mushroom-, cube- and sphere-like superstructures. *Appl. Surf. Sci.* **2018**, *457*, 427–438. [CrossRef]
53. Zhu, X.; Wei, Y.; Wang, H.; Li, K. Ce-Fe oxygen carriers for chemical-looping steam methane reforming. *Int. J. Hydrogen Energy* **2013**, *38*, 4492–4501. [CrossRef]
54. Ma, S.; Chen, S.; Zhu, M.; Zhao, Z.; Hu, J.; Wu, M.; Toan, S.; Xiang, W. Enhanced sintering resistance of Fe_2O_3/CeO_2 oxygen carrier for chemical looping hydrogen generation using core-shell structure. *Int. J. Hydrogen Energy* **2019**, *44*, 6491–6504. [CrossRef]
55. Li, K.; Wang, H.; Wei, Y.; Yan, D. Syngas production from methane and air via a redox process using Ce-Fe mixed oxides as oxygen carriers. *Appl. Catal. B Environ.* **2010**, *97*, 361–372. [CrossRef]
56. Channei, D.; Inceesungvorn, B.; Wetchakun, N.; Phanichphant, S.; Nakaruk, A.; Koshy, P.; Sorrell, C.C. Photocatalytic activity under visible light of Fe-doped CeO_2 nanoparticles synthesized by flame spray pyrolysis. *Ceram. Int.* **2013**, *39*, 3129–3134. [CrossRef]
57. Hosseini, S.Y.; Khosravi-Nikou, M.R.; Shariati, A. Production of hydrogen and syngas using chemical looping technology via cerium-iron mixed oxides. *Chem. Eng. Process. Process Intensif.* **2019**, *139*, 23–33. [CrossRef]
58. Qiao, D.; Lu, G.; Liu, X.; Guo, Y.; Wang, Y.; Guo, Y. Preparation of $Ce_{1-x}Fe_xO_2$ solid solution and its catalytic performance for oxidation of CH_4 and CO. *J. Mater. Sci.* **2011**, *46*, 3500–3506. [CrossRef]
59. Li, D.; Li, K.; Xu, R.; Wang, H.; Tian, D.; Wei, Y.; Zhu, X.; Zeng, C.; Zeng, L. $Ce_{1-x}Fe_xO_{2-\delta}$ catalysts for catalytic methane combustion: Role of oxygen vacancy and structural dependence. *Catal. Today* **2018**, *318*, 73–85. [CrossRef]
60. Désaunay, T.; Bonura, G.; Chiodo, V.; Freni, S.; Couzinié, J.P.; Bourgon, J.; Ringuedé, A.; Labat, F.; Adamo, C.; Cassir, M. Surface-dependent oxidation of H_2 on CeO_2 surfaces. *J. Catal.* **2013**, *297*, 193–201. [CrossRef]
61. Alvarez, P.; Araya, P.; Rojas, R.; Guerrero, S.; Aguila, G. Activity of alumina supported Fe catalysts for N_2O decomposition: effects of the iron content and thermal treatment. *J. Chil. Chem. Soc.* **2017**, *62*, 3752–3759. [CrossRef]
62. Zieliński, J.; Zglinicka, I.; Znak, L.; Kaszkur, Z. Reduction of Fe_2O_3 with hydrogen. *Appl. Catal. A Gen.* **2010**, *381*, 191–196. [CrossRef]
63. Pudukudy, M.; Yaakob, Z.; Jia, Q.; Takriff, M.S. Catalytic decomposition of methane over rare earth metal (Ce and La) oxides supported iron catalysts. *Appl. Surf. Sci.* **2019**, *467*, 236–248. [CrossRef]
64. Cheng, X.; Zhang, X.; Su, D.; Wang, Z.; Chang, J.; Ma, C. NO reduction by CO over copper catalyst supported on mixed CeO_2 and Fe_2O_3: Catalyst design and activity test. *Appl. Catal. B Environ.* **2018**, *239*, 485–501. [CrossRef]
65. Zhang, X.; Wang, J.; Song, Z.; Zhao, H.; Xing, Y.; Zhao, M.; Zhao, J.; Ma, Z.; Zhang, P.; Tsubaki, N. Promotion of surface acidity and surface species of doped Fe and SO_4^{2-} over CeO_2 catalytic for NH_3-SCR reaction. *Mol. Catal.* **2019**, *463*, 1–7. [CrossRef]
66. Wang, C.; Cheng, Q.; Wang, X.; Ma, K.; Bai, X.; Tan, S.; Tian, Y.; Ding, T.; Zheng, L.; Zhang, J.; et al. Enhanced catalytic performance for CO preferential oxidation over CuO catalysts supported on highly defective CeO_2 nanocrystals. *Appl. Surf. Sci.* **2017**, *422*, 932–943. [CrossRef]

67. Piumetti, M.; Andana, T.; Bensaid, S.; Russo, N.; Fino, D.; Pirone, R. Study on the CO Oxidation over Ceria-Based Nanocatalysts. *Nanoscale Res. Lett.* **2016**, *11*, 165. [CrossRef]
68. Wu, Z.; Li, M.; Howe, J.; Meyer, H.M.; Overbury, S.H. Probing defect sites on CeO_2 nanocrystals with well-defined surface planes by Raman spectroscopy and O_2 adsorption. *Langmuir* **2010**, *26*, 16595–16606. [CrossRef]
69. Hillary, B.; Sudarsanam, P.; Amin, M.H.; Bhargava, S.K. Nanoscale Cobalt-Manganese Oxide Catalyst Supported on Shape-Controlled Cerium Oxide: Effect of Nanointerface Configuration on Structural, Redox, and Catalytic Properties. *Langmuir* **2017**, *33*, 1743–1750. [CrossRef]
70. Andana, T.; Piumetti, M.; Bensaid, S.; Veyre, L.; Thieuleux, C.; Russo, N.; Fino, D.; Quadrelli, E.A.; Pirone, R. CuO nanoparticles supported by ceria for NO_x-assisted soot oxidation: insight into catalytic activity and sintering. *Appl. Catal. B Environ.* **2017**, *216*, 41–58. [CrossRef]
71. Konsolakis, M.; Ioakimidis, Z.; Kraia, T.; Marnellos, G.E. Hydrogen Production by Ethanol Steam Reforming (ESR) over CeO_2 Supported Transition Metal (Fe, Co, Ni, Cu) Catalysts: Insight into the Structure-Activity Relationship. *Catalysts* **2016**, *6*, 39. [CrossRef]
72. Wang, H.; Jin, B.; Wang, H.; Ma, N.; Liu, W.; Weng, D.; Wu, X.; Liu, S. Study of Ag promoted Fe_2O_3@CeO_2 as superior soot oxidation catalysts: The role of Fe_2O_3 crystal plane and tandem oxygen delivery. *Appl. Catal. B Environ.* **2018**, *237*, 251–262. [CrossRef]
73. Liu, F.; Wang, Z.; Wang, D.; Chen, D.; Chen, F.; Li, X. Morphology and Crystal-Plane Effects of Fe/W-CeO_2 for Selective Catalytic Reduction of NO with NH_3. *Catalysts* **2019**, *9*, 288. [CrossRef]
74. Luo, Y.; Chen, R.; Peng, W.; Tang, G.; Gao, X. Inverse CeO_2-Fe_2O_3 catalyst for superior low-temperature CO conversion efficiency. *Appl. Surf. Sci.* **2017**, *416*, 911–917. [CrossRef]
75. Kim, K.; Han, J.W. Mechanistic study for enhanced CO oxidation activity on (Mn,Fe) co-doped CeO_2 (111). *Catal. Today* **2017**, *293–294*, 82–88. [CrossRef]
76. Said, A.E.A.A.; Abd El-Wahab, M.M.M.; Goda, M.N. Synthesis and characterization of pure and (Ce, Zr, Ag) doped mesoporous CuO-Fe_2O_3 as highly efficient and stable nanocatalysts for CO oxidation at low temperature. *Appl. Surf. Sci.* **2016**, *390*, 649–665. [CrossRef]
77. Chen, H.T.; Chang, J.G. Computational investigation of CO adsorption and oxidation on iron-modified cerium oxide. *J. Phys. Chem. C* **2011**, *115*, 14745–14753. [CrossRef]
78. Lykaki, M.; Papista, E.; Kaklidis, N.; Carabineiro, S.A.C.; Konsolakis, M. Ceria Nanoparticles' Morphological Effects on the N_2O Decomposition Performance of Co_3O_4/CeO_2 Mixed Oxides. *Catalysts* **2019**, *9*, 233. [CrossRef]
79. Brunauer, S.; Emmett, P.H.; Teller, E. Adsorption of Gases in Multimolecular Layers. *J. Am. Chem. Soc.* **1938**, *60*, 309–319. [CrossRef]
80. Xu, J.; Harmer, J.; Li, G.; Chapman, T.; Collier, P.; Longworth, S.; Tsang, S.C. Size dependent oxygen buffering capacity of ceria nanocrystals. *Chem. Commun.* **2010**, *46*, 1887–1889. [CrossRef]
81. Barthos, R.; Hegyessy, A.; Klébert, S.; Valyon, J. Vanadium dispersion and catalytic activity of Pd/VO_x/SBA-15 catalysts in the Wacker oxidation of ethylene. *Microporous Mesoporous Mater.* **2015**, *207*, 1–8. [CrossRef]

catalysts

MDPI

Article

Ceria Nanoparticles' Morphological Effects on the N_2O Decomposition Performance of Co_3O_4/CeO_2 Mixed Oxides

Maria Lykaki [1], Eleni Papista [2], Nikolaos Kaklidis [2], Sónia A. C. Carabineiro [3] and Michalis Konsolakis [1,*]

1 School of Production Engineering and Management, Technical University of Crete, 73100 Chania, Greece; mlykaki@isc.tuc.gr

2 Department of Mechanical Engineering, University of Western Macedonia, GR-50100 Kozani, Greece; epapista@uowm.gr (E.P.); nkaklidis@uowm.gr (N.K.)

3 Laboratório de Catálise e Materiais (LCM), Laboratório Associado LSRE-LCM, Faculdade de Engenharia, Universidade do Porto, 4200-465 Porto, Portugal; scarabin@fe.up.pt

* Correspondence: mkonsol@pem.tuc.gr; Tel.: +30-28210-37682

Received: 15 January 2019; Accepted: 18 February 2019; Published: 3 March 2019

Abstract: Ceria-based oxides have been widely explored recently in the direct decomposition of N_2O (deN_2O) due to their unique redox/surface properties and lower cost as compared to noble metal-based catalysts. Cobalt oxide dispersed on ceria is among the most active mixed oxides with its efficiency strongly affected by counterpart features, such as particle size and morphology. In this work, the morphological effect of ceria nanostructures (nanorods (NR), nanocubes (NC), nanopolyhedra (NP)) on the solid-state properties and the deN_2O performance of the Co_3O_4/CeO_2 binary system is investigated. Several characterization methods involving N_2 adsorption at −196 °C, X-ray diffraction (XRD), temperature programmed reduction (TPR), X-ray photoelectron spectroscopy (XPS) and transmission electron microscopy (TEM) were carried out to disclose structure–property relationships. The results revealed the importance of support morphology on the physicochemical properties and the N_2O conversion performance of bare ceria samples, following the order nanorods (NR) > nanopolyhedra (NP) > nanocubes (NC). More importantly, Co_3O_4 impregnation to different carriers towards the formation of Co_3O_4/CeO_2 mixed oxides greatly enhanced the deN_2O performance as compared to bare ceria samples, without, however, affecting the conversion sequence, implying the pivotal role of ceria support. The Co_3O_4/CeO_2 sample with the rod-like morphology exhibited the best deN_2O performance (100% N_2O conversion at 500 °C) due to its abundance in Co^{2+} active sites and Ce^{3+} species in conjunction to its improved reducibility, oxygen kinetics and surface area.

Keywords: ceria nanoparticles; morphological effects; Co_3O_4/CeO_2 mixed oxides, deN_2O process

1. Introduction

Nitrous oxide (N_2O) is one of the most significant greenhouse gases contributing to the depletion of the ozone layer. N_2O has a much higher global warming potential (GWP) compared to CO_2 (310 times higher) and a long atmospheric lifetime (114 years). The emissions of N_2O are derived by both natural and anthropogenic sources. The main anthropogenic sources for N_2O emissions involve agriculture (use of fertilizers), chemical industry (adipic and nitric acid production), the combustion of fossil fuels, as well as biomass burning, etc. [1–4].

The abatement of N_2O emissions is of paramount importance and the direct catalytic decomposition of nitrous oxide to molecular nitrogen and oxygen (deN_2O process) is considered to be a highly efficient remediation method. Thus far, several catalytic systems, such as supported noble metals [5–7], perovskites [8–10], hexaaluminates [11–14], spinels [15–18], zeolites [19–22] and mixed oxides [23–27], have been used for N_2O decomposition. Although noble metals exhibit satisfactory activity for the deN_2O process, their high cost and the deterioration of their catalytic efficiency from gases present in the exhaust gas stream (e.g., O_2) act as inhibiting factors for practical applications [1,28]. Hence, research efforts have focused on the development of noble metal-free mixed oxides of high activity, stability and low cost, as recently reviewed [1].

Among the different transition metal oxides, cobalt spinel shows unique physicochemical characteristics, such as thermal stability and high reducibility, making it an excellent candidate for the deN_2O process [15,23,29,30]. However, the high cost of cobalt renders mandatory its dispersion to high surface area supports like ceria, magnesia, etc. [31,32]. Among the various supports investigated, ceria exhibits unique redox properties associated with its high oxygen storage capacity (OSC), rendering this material highly effective in many catalytic processes [23,33–35]. Furthermore, the synergistic effects induced by strong metal–ceria interactions, in nanoscale, can modify the surface chemistry of the materials through geometric or/and electronic perturbations, leading to improved redox properties and catalytic activity [36–40].

However, the catalytic efficiency of transition metal oxides, involving ceria-based mixed oxides, can be considerably affected by the different counterpart characteristics, such as particle size and morphology. In this regard, engineering the particle size and shape (e.g., nanorods and nanocubes) through the employment of advanced nano-synthesis paths has lately received particular attention [33,41–43]. Interestingly, the support morphology greatly affects the redox properties, oxygen mobility and, subsequently, the catalytic activity of the mixed oxides. For instance, Lin et al. [44] prepared Co_3O_4/CeO_2 catalysts with three different support morphologies, namely polyhedra, nanorods and hexagonal shapes, with polyhedra exhibiting the highest catalytic activity for ammonia synthesis. In a similar manner, by tailoring the support morphology, CuO/CeO_2 nanoshaped materials of enhanced reducibility and deN_2O performance can be obtained [45]. Andrade-Martínez et al. [46] investigated the catalytic reduction of N_2O over CuO/SiO_2 catalysts, revealing the key role of the spherical ordered mesoporous support, along with its functionalization through copper addition, on the improved catalytic activity and stability, making this material comparable to noble metal-reported systems. Different support morphologies (rods, plates and cubes) have also been employed for the low temperature CH_2Br_2 oxidation revealing the superiority of cobalt-ceria nanorods in the catalytic performance [47]. Moreover, cobalt oxide supported on ceria of different morphology (nanoparticles, nanorods and nanocubes) has been investigated for the catalytic oxidation of toluene with the nanoparticles exhibiting the highest catalytic activity due to the synergism at the interface between the two oxide phases, which leads to an improved reducibility [48]. Very recently, the influence of support morphology (nanorods, nanocubes and nanopolyhedra) on the surface and structural properties of CuO/CeO_2 mixed oxide has been thoroughly explored through both *in situ* and *ex situ* characterization techniques. The results disclosed the significance of the ceria morphology on the reducibility and oxygen kinetics, revealing the order nanorods > nanopolyhedra > nanocubes [49].

In this work, ceria structures of various morphologies (nanopolyhedra, nanorods and nanocubes) were hydrothermally prepared, and then cobalt was impregnated into the above ceria supports. The

and the deN_2O performance of Co_3O_4/CeO_2 mixed oxides. The results clearly revealed that support morphology can exert a profound influence on the N_2O decomposition, paving the way toward the rational design of highly efficient deN_2O catalysts.

2. Results and Discussion

2.1. *Textural/Structural Analysis (BET and XRD)*

The main textural and structural characteristics of bare ceria samples and Co_3O_4/CeO_2 mixed oxides (hereinafter denoted as Co/CeO_2) are summarized in Table 1. According to the BET surface area, the following order is acquired: CeO_2-NP (88 m^2 g^{-1}) > CeO_2-NR (79 m^2 g^{-1}) > CeO_2-NC (37 m^2 g^{-1}). The addition of cobalt into CeO_2 decreases the surface area, without, however, significantly affecting the order obtained for bare ceria samples. The Co/CeO_2-NR sample exhibits the highest value (72 m^2 g^{-1}) succeeded by Co/CeO_2-NP (71 m^2 g^{-1}) and Co/CeO_2-NC (28 m^2 g^{-1}). Regarding the average pore diameter and pore volume, they both decreased upon the addition of Co to ceria nanorods and nanocubes. However, concerning ceria nanopolyhedra, the addition of cobalt leads to a small increase in the average pore diameter, whereas the pore volume is not significantly affected (Table 1).

Table 1. The textural and structural properties of bare CeO_2 and Co/CeO_2 samples.

Sample	BET Analysis			XRD Analysis	
	BET Surface Area (m^2 g^{-1})	Pore Volume (cm^3/g)	Average Pore Diameter (nm)	Crystallite Size (nm), D_{XRD} [1]	
				CeO_2	Co_3O_4
CeO_2-NC	37	0.26	27.4	27 ± 1	-
CeO_2-NR	79	0.48	24.2	15 ± 1	-
CeO_2-NP	88	0.17	7.9	11 ± 1	-
Co/CeO_2-NC	28	0.15	22.6	24 ± 1	19 ± 1
Co/CeO_2-NR	72	0.31	17.4	14 ± 1	16 ± 1
Co/CeO_2-NP	71	0.17	9.8	11 ± 1	15 ± 1

[1] Calculated applying the Williamson–Hall plot after the Rietveld refinement of diffractograms.

Figure 1a shows the Barret-Joyner-Halenda (BJH) desorption pore size distributions (PSD) of bare ceria and Co/CeO_2 catalysts. According to the pore size distribution, all the samples have their maxima at a pore diameter more than 3 nm, designating the presence of mesopores [50]. It is obvious that bare ceria samples with the nanocube (CeO_2-NC) and nanorod morphology (CeO_2-NR) exhibit similar particle size distributions, whereas in ceria nanopolyhedra (CeO_2-NP), a narrower PSD is observed. Noteworthily, PSD remains practically unaffected by the addition of cobalt in all cases. As it can be observed in Figure 1b which shows the adsorption–desorption isotherms, all samples demonstrate type IV isotherms with a hysteresis loop at a relative pressure > 0.5, further corroborating the mesoporous structure of the materials [51,52].

The XRD patterns of the samples are shown in Figure 2. The main peaks can be indexed to (111), (200), (220), (311), (222), (400), (331), (420), (422), (511) and (440) planes which are attributed to ceria face-centered cubic fluorite structure (Fm3m symmetry, no. 225) [53,54]. There are three small peaks at 2θ values of approx. 36, 44 and 64° which are typical of Co_3O_4 [33]. These three diffraction peaks correspond to the (311), (400) and (440) planes of Co_3O_4, respectively. The average crystallite diameter of the oxide phases (CeO_2 and Co_3O_4) was assessed by an XRD analysis by means of the Williamson–Hall plot (Table 1). The CeO_2 crystallite size measurements showed 24, 14 and 11 nm for Co/CeO_2-NC, Co/CeO_2-NR and Co/CeO_2-NP, respectively. As it is obvious from Table 1, there is a small decrease in the ceria crystallite size for nanocubes and nanorods, whereas no changes were observed for nanopolyhedra, indicating that the structural characteristics of ceria supports do not get significantly affected upon cobalt addition, as it will be further corroborated by a TEM analysis (see below). In a similar manner, the BET analysis (Table 1) indicates no significant modifications on the pore characteristics of ceria nanopolyhedra upon cobalt addition, which could be ascribed to their irregular morphology. It should be also noted that the samples with nanocube morphology exhibit the smallest BET surface area and the largest CeO_2 and Co_3O_4 crystallite sizes in comparison to nanorods and nanopolyhedra. As for the crystallite size of cobalt oxide phase, the following sequence was obtained: Co/CeO_2-NC (19 nm) > Co/CeO_2-NR (16 nm) > Co/CeO_2-NP (15 nm), which perfectly matches the order obtained for CeO_2.

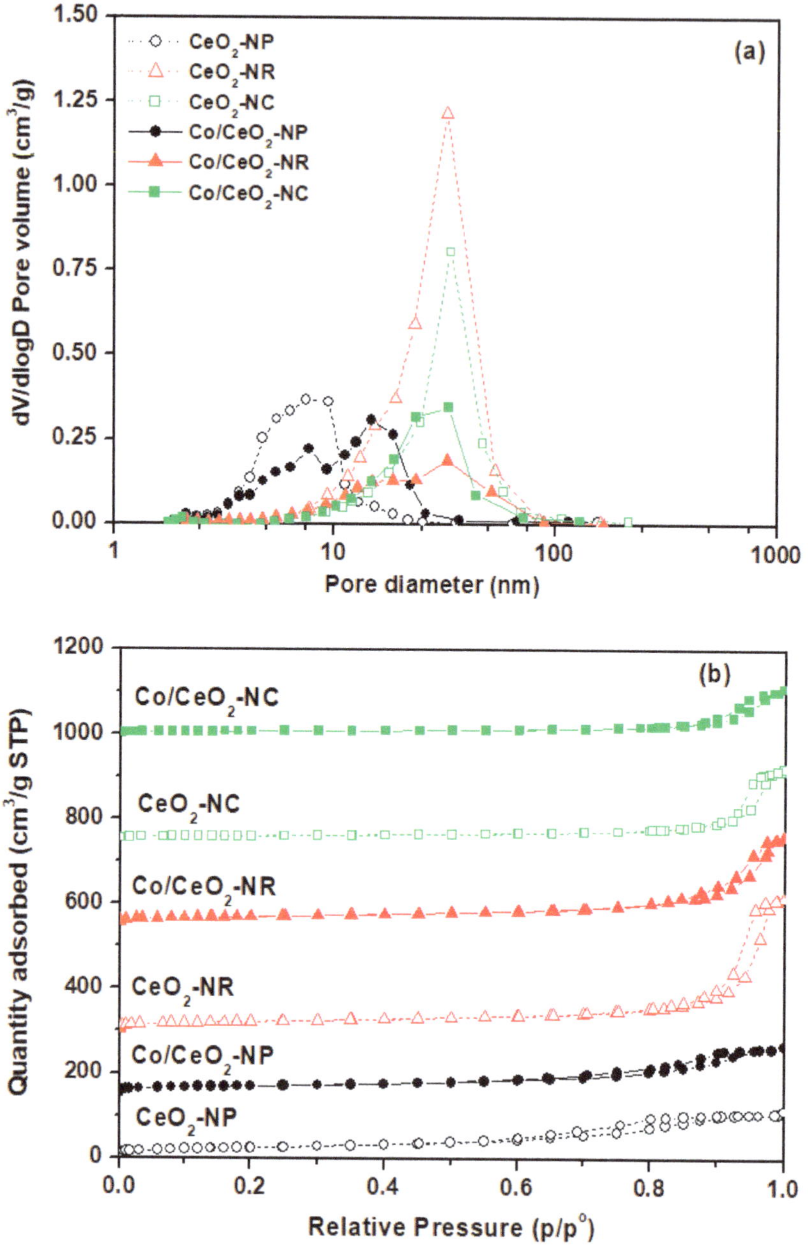

Figure 1. (**a**) The BJH (Barret-Joyner-Halenda) desorption pore size distribution (PSD) and (**b**) the adsorption–desorption isotherms of CeO_2 and Co/CeO_2 samples.

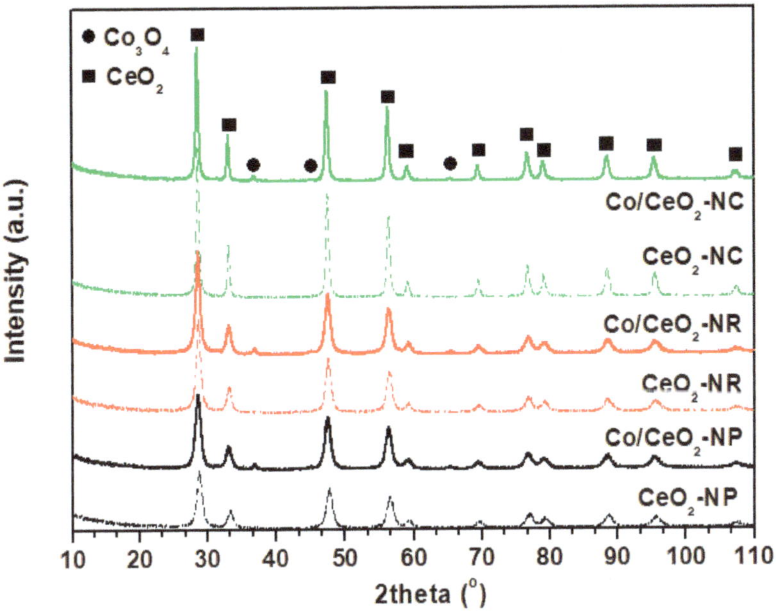

Figure 2. The XRD patterns of the CeO_2 and Co/CeO_2 samples.

2.2. *Morphological Characterization (TEM)*

Transmission electron microscopy (TEM) has been applied so as to examine the morphological differences among the materials. Figure 3a–c shows the TEM images of ceria supports. The CeO_2-NR sample (Figure 3a) exhibits a rod-shaped morphology with the length varying between 25 and 200 nm. Figure 3b and c demonstrates mainly irregular-shaped nanopolyhedra and cubes, respectively. Figure 3d–f illustrates the images derived by TEM analysis for the Co/CeO_2 mixed oxides. Evidently, the morphology is not affected by cobalt addition to the ceria carrier.

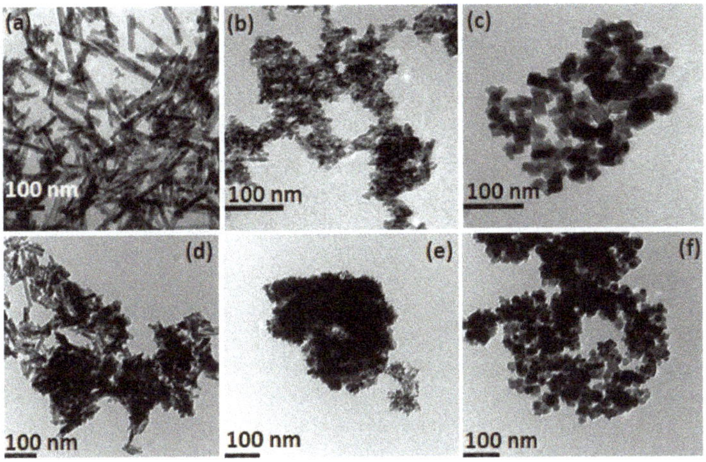

Figure 3. The transmission electron microscopy images of CeO_2 (**a**–**c**) and Co/CeO_2 (**d**–**f**) samples: (**a**) CeO_2-NR, (**b**) CeO_2-NP, (**c**) CeO_2-NC, (**d**) Co/CeO_2-NR, (**e**) Co/CeO_2-NP and (**f**) Co/CeO_2-NC.

2.3. Redox Properties (H_2-Temperature Programmed Reduction (TPR))

H_2-TPR experiments took place to investigate the ceria shape effect on the redox properties of as-prepared samples. Figure 4a shows the TPR profiles of bare ceria samples, consisting of two wide-ranging peaks which are centred at 526–551 °C and 789–813 °C. These peaks are attributed to ceria surface oxygen (O_s) and bulk oxygen (O_b) reduction, respectively [33,49,55]. In Table 2, the hydrogen consumption corresponding to surface oxygen as well as to bulk oxygen reduction is presented. Based on the ratio of surface-to-bulk oxygen (O_s/O_b), the following order was acquired: CeO_2-NR (1.13) > CeO_2-NP (0.94) > CeO_2-NC (0.71). This indicates the superior reducibility of the rod-shaped sample as it exhibits the highest amount of loosely bound oxygen species. The latter is expected to notably affect the deN_2O process, where the desorption of adsorbed oxygen species mainly determines the reaction rate (*vide infra*).

Table 2. The redox properties of the bare CeO_2 and Co/CeO_2 samples.

Sample	H_2 Consumption (mmol H_2 g^{-1}) [a]			O_s/O_b Ratio	Peak Temperature (°C)	
	O_s Peak	O_b Peak	Total		O_s Peak	O_b Peak
CeO_2-NP	0.48	0.51	0.99	0.94	555	804
CeO_2-NR	0.59	0.52	1.11	1.13	545	788
CeO_2-NC	0.41	0.58	0.99	0.71	589	809
	Peaks a+b	CeO_2 Peak	Total		Peak a	Peak b
Co/CeO_2-NP	2.40	0.61	3.01		333	388
Co/CeO_2-NR	2.37	0.62	2.99		318	388
Co/CeO_2-NC	2.05	0.32	2.37		335	405

[a] Estimated by the area of the corresponding temperature programmed reduction (TPR) peaks, which is calibrated against a known amount of CuO standard sample.

The reduction profiles of the Co/CeO_2 samples as well as the one of a Co_3O_4 reference are shown in Figure 4b. Table 2 summarizes the main TPR peaks along with the hydrogen consumption (mmol H_2 g^{-1}). Pure Co_3O_4 shows two reduction peaks (a and b) in much lower temperatures than those of bare ceria samples, namely 305 °C and 415 °C. They are ascribed to the stepwise reduction of $Co_3O_4 \rightarrow CoO \rightarrow Co$, respectively [44,56–58].

On the other hand, Co/CeO_2 samples exhibit two main peaks at the temperature range of 318–335 °C (peak a) and 388–405 °C (peak b), ascribed to the reduction of Co^{3+} to Co^{2+} and Co^{2+} to Co^0, respectively [33,59,60]. Obviously, the cobalt addition facilitates the reduction of ceria surface oxygen, shifting the peaks centered at 526–551 °C to a lower temperature (comparison of Figure 4a,b). They also exhibit a broad peak above 800 °C, attributed to the ceria subsurface oxygen reduction, while the capping oxygen reduction overlaps with the reduction of CoO [33,56,61]. Apparently, the reduction of the mixed oxides takes place in lower temperatures compared to the bare ceria samples, demonstrating the beneficial effect of cobalt on the surface oxygen reduction of ceria. In fact, the interaction between the two oxide phases could be considered responsible for the improved reducibility and oxygen mobility, as thoroughly discussed in previous studies [48,49,62]. According to the consumption of hydrogen in the low-temperature range (Table 2), which could be related to the cobalt species reduction along with the ceria surface oxygen reduction, the Co/CeO_2-NP and Co/CeO_2-NR samples exhibit a similar H_2 uptake (about 2.40 mmol H_2 g^{-1}) while the sample of nanocube morphology exhibits a much lower value (2.05 mmol H_2 g^{-1}). This trend is well-matched to the catalytic results (*vide infra*), revealing the key role of redox ability on the deN_2O process.

Moreover, the Co/CeO_2-NR sample exhibits the lowest reduction temperature (peak at 318 °C) in comparison with the other samples (peak ca. 335 °C), indicating the facilitation of Co^{3+} species reduction over ceria nanorods. Noteworthily, the theoretical amount of hydrogen for the complete reduction of Co_3O_4 to Co (approx. 1.76 mmol H_2 g^{-1}, based on a 7.8 wt. % nominal loading of Co) is always surpassed by the hydrogen amount required for the reduction of Co/CeO_2 samples (Table 2).

The latter reveals the facilitation of ceria capping oxygen reduction in the presence of cobalt, further corroborating the above findings.

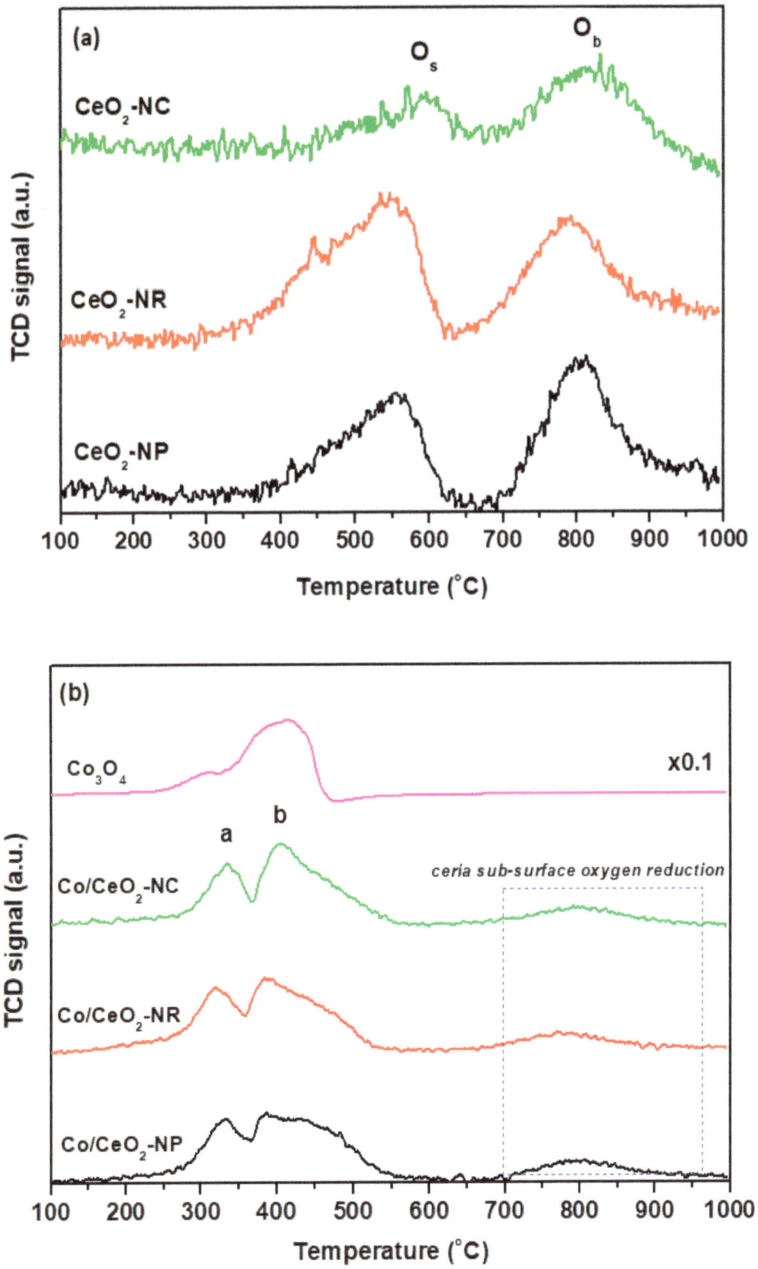

Figure 4. The H_2-TPR profiles of (**a**) bare CeO_2 and (**b**) Co_3O_4, Co/CeO_2 samples.

2.4. *Surface Analysis (X-ray Photoelectron Spectroscopy (XPS))*

An XPS analysis was performed in order to investigate the effect of ceria morphology on the elemental chemical states and surface composition of Co/CeO_2 mixed oxides. Figure 5a shows the Ce3d XPS spectra of ceria nanoparticles of different morphology and the corresponding Co/CeO_2 samples, which can be deconvoluted into eight components [63–65], with the assignment of the characteristic peaks having been thoroughly described in our previous work [49]. In brief, the three pairs of peaks labeled as u, v; u″, v″; and u‴, v‴ are ascribed to Ce^{4+}, whereas the residual u′ and v′ peaks are ascribed to Ce^{3+} species.

The corresponding O 1s spectra of the samples are depicted in Figure 5b. The low binding energy peak at 529.3 eV is attributed to the lattice oxygen (O_I) of Co_3O_4 and CeO_2 phases, and the high binding energy peak at 531.3 eV corresponds to the chemisorbed oxygen (O_{II}) such as adsorbed oxygen (O^-/O_2^{2-}) and water, carbonate as well as hydroxyl species [23,56].

The proportion of Ce^{3+} (%) as well as the O_{II}/O_I ratio for all samples is summarized in Table 3. Bare ceria supports exhibit a similar amount of Ce^{3+} ranging from 23.3 to 25.3%. Regarding, the cobalt-ceria samples, the population of Ce^{3+} species is slightly higher, varying between 26.1 and 28.5%. In particular, the Co/CeO_2-NR sample exhibits the highest amount (28.5%) followed by Co/CeO_2-NP (26.7%) and Co/CeO_2-NC (26.1%), indicating the abundance of the nanorod samples in oxygen vacancies. Interestingly, the relative ratio of adsorbed to lattice oxygen (O_{II}/O_I) and the Ce^{3+} (%) follow the same order, namely, Co/CeO_2-NR (0.60) > Co/CeO_2-NP (0.53) > Co/CeO_2-NC (0.51), perfectly matched to the order obtained for the catalytic performance, as it will be discussed in the sequence. It should be also noted that Co addition to CeO_2-NR enhances both the population of reduced Ce^{3+} species and the O_{II}/O_I ratio, revealing the synergistic interactions between cerium and cobalt oxides toward the formation of highly reducible composites, in agreement with the TPR results.

Table 3. The XPS results of bare CeO_2 and Co/CeO_2 samples.

Sample	Co^{2+}/Co^{3+}	Ce^{3+} (%)	O_{II}/O_I
CeO_2-NC	-	23.3	0.50
CeO_2-NR	-	24.3	0.47
CeO_2-NP	-	25.3	0.49
Co/CeO_2-NC	1.06	26.1	0.51
Co/CeO_2-NR	1.32	28.5	0.60
Co/CeO_2-NP	0.94	26.7	0.53

Figure 6 depicts the Co 2p XPS spectra of Co/CeO_2 samples along with the spectrum obtained for the Co_3O_4 reference sample for comparison purposes. The samples exhibit two major peaks of $Co2p_{3/2}$ (780 eV) and $Co2p_{1/2}$ (795 eV). According to peaks' positions and shapes, the structure of the cobalt spinel is formed [23,66,67]. The Co^{2+}/Co^{3+} ratio of Co/CeO_2 samples derived by the deconvolution of the $Co2p_{1/2}$ and $Co2p_{3/2}$ peaks is included in Table 3. The nanorod sample, which offers the best deN_2O performance (*vide infra*), exhibits the highest Co^{2+}/Co^{3+} ratio (1.32), followed by nanocubes (1.06) and nanopolyhedra (0.94). In view of this fact, it has been reported that samples with a high Co^{2+}/Co^{3+} ratio exhibit better deN_2O performance [3,20,22,43,59], further corroborating the present findings.

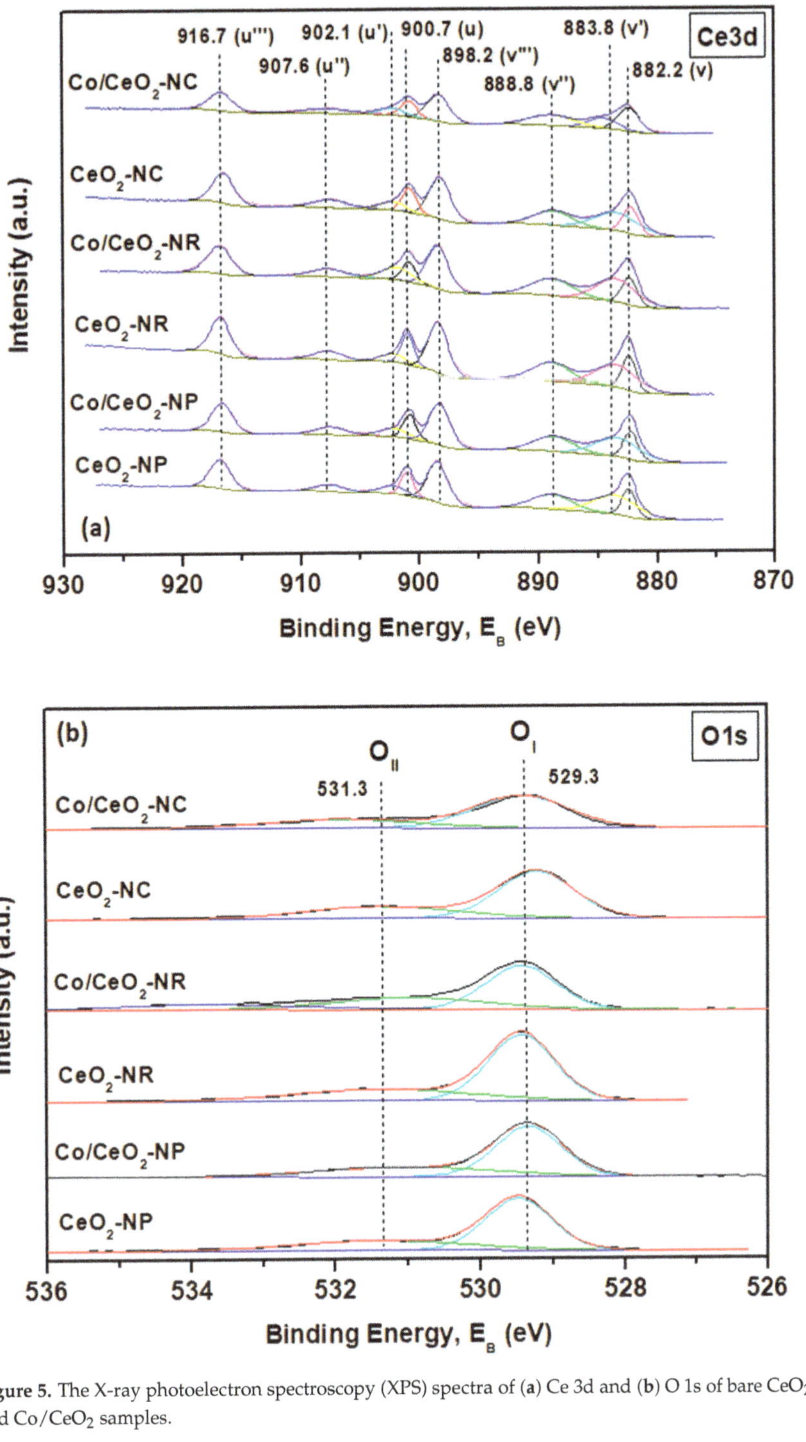

Figure 5. The X-ray photoelectron spectroscopy (XPS) spectra of (**a**) Ce 3d and (**b**) O 1s of bare CeO_2 and Co/CeO_2 samples.

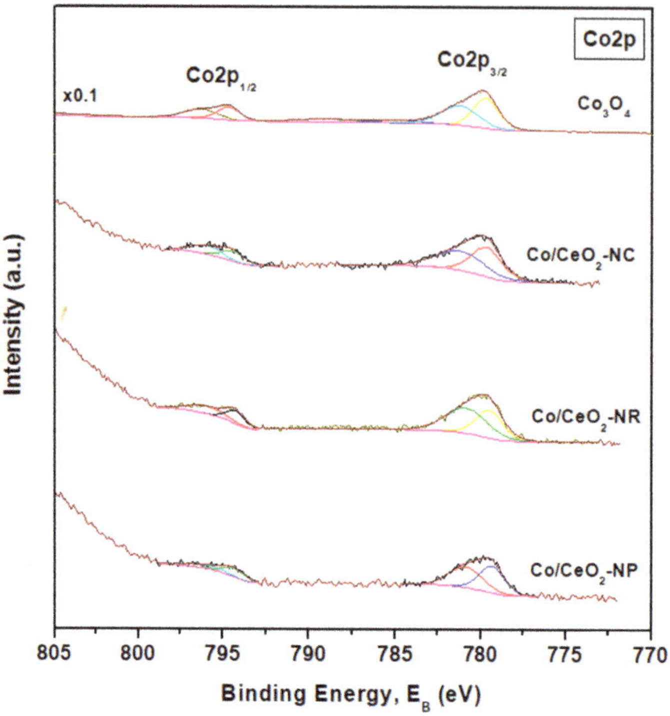

Figure 6. The Co 2p XPS spectra of the Co_3O_4 and Co/CeO_2 samples: The Co 2p XPS spectra of Co/CeO_2 samples have been magnified.

2.5. Catalytic Evaluation Studies

The impact of ceria morphology on the catalytic decomposition of N_2O under oxygen deficient and oxygen excess conditions was next examined. Figure 7a,b shows the N_2O conversion profiles as a temperature function for bare ceria as well as Co/CeO_2 samples in the absence and presence of oxygen, respectively. The Co/CeO_2-NR sample exhibits the best catalytic performance, both in the absence and presence of oxygen in the gas stream. Apparently, the addition of cobalt in the ceria lattice enormously enhances the catalytic efficiency without, however, affecting the catalytic order of bare ceria samples, suggesting the pivotal role of ceria morphology on the deN_2O performance. In terms of the half-conversion temperature (T_{50}), the following order is obtained for the mixed oxides in the absence of oxygen: Co/CeO_2-NR (449 °C) > Co/CeO_2-NP (458 °C) > Co/CeO_2-NC (464 °C). The same trend is observed in the presence of oxygen as well, although in slightly higher temperatures, due to its competitive sorption on the catalyst surface. In this point, it should be noted that the un-catalyzed reaction shows nearly zero N_2O conversion in the temperature range investigated, as previously reported [29,46,68].

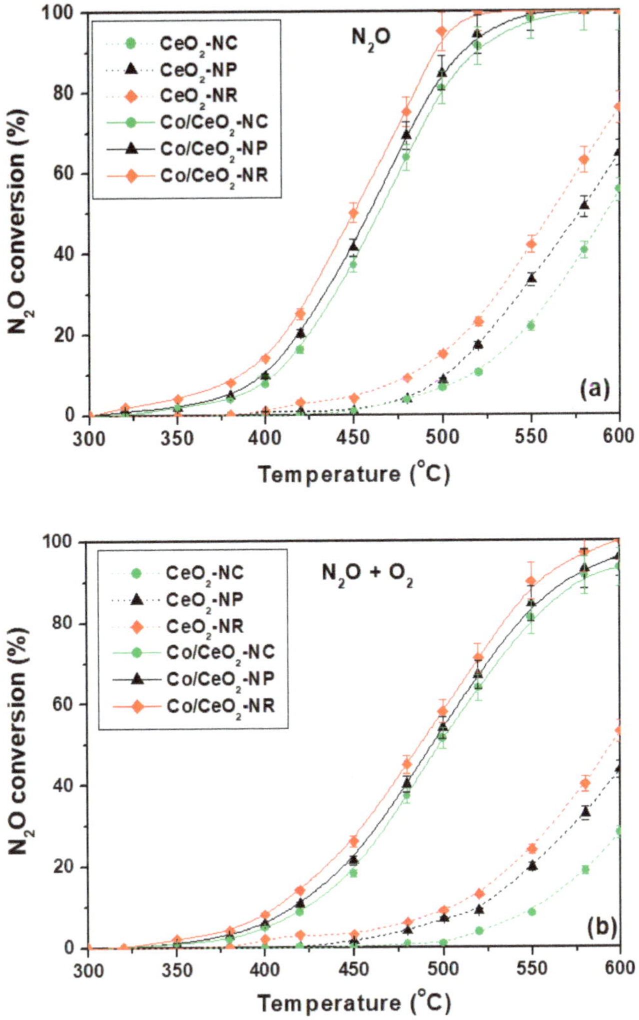

Figure 7. N_2O conversion as a function of temperature for CeO_2 and Co/CeO_2 samples of different morphology (**a**) in the absence and (**b**) in the presence of oxygen: The reaction conditions are 1000 ppm N_2O, 0 or 2% O_2 and Gas Hour Space Velocity (GHSV) = 40,000 h^{-1}.

The above findings can be well-interpreted by taking into account a redox-type mechanism for the decomposition of N_2O over cobalt spinel oxides [4,23,24,30,59,69–73]:

$$Co^{2+} + N_2O \rightarrow Co^{3+}\text{-}O^- + N_2 \tag{1}$$

$$Co^{3+}\text{-}O^- + N_2O \rightarrow Co^{3+}\text{-}O_2{}^- + N_2 \tag{2}$$

$$Co^{3+}\text{-}O_2{}^- \rightarrow Co^{2+} + O_2 \tag{3}$$

In this mechanistic sequence, N_2O is initially chemisorbed on the Co^{2+} sites (Equation (1)) which are considered as the active centres for initiating the N_2O dissociative adsorption. Then, the regeneration of the active sites is taking place through the Co^{3+}/Co^{2+} redox cycle, involving the combination of O^- into O_2^- (Equation (2)) and the desorption of molecular oxygen (Equation (3)), which finally leads to the regeneration of those sites [69].

However, in the case of Co_3O_4/CeO_2 mixed oxides, the excellent redox characteristics of ceria, such as oxygen storage capacity and oxygen mobility, can be further accounted for the regeneration of active sites through the following steps [69]:

$$Co^{3+}\text{-}O^- + Ce^{3+}\text{-}O_{vac} \rightarrow Co^{2+} + Ce^{4+}\text{-}O^- \quad (4)$$

$$2Ce^{4+}\text{-}O^- \leftrightarrow Ce^{4+}\text{-}O_2^{2-}\text{-}Ce^{4+} \quad (5)$$

$$Ce^{4+}\text{-}O^- + N_2O \rightarrow Ce^{3+}\text{-}O_{vac} + N_2 + O_2 \quad (6)$$

Based on the above mechanistic scheme, the superiority of the Co/CeO_2 sample with a rod-like morphology can receive a consistent explanation. More specifically, nanorod-shaped ceria with (110) and (100) reactive planes exhibit enhanced oxygen kinetics and reducibility as it has the highest population of loosely bound oxygen species (Table 2), which is a decisive factor in terms of deN_2O activity. In other words, the high amount of weakly bound oxygen species present in the Co_3O_4/CeO_2 samples of rod-like morphology, linked directly to oxygen vacancy formation and oxygen mobility, could be considered responsible for the formation and the consequent regeneration of active sites. In this regard, a perfect interrelation between the catalytic performance (in terms of the half-conversion temperature, T_{50}) and the redox properties (in terms of the ratio of surface oxygen to bulk oxygen, O_s/O_b) is disclosed, as illustrated in Figure 8. This clearly justifies the key role of redox properties on the deN_2O process. In a similar manner, Liu et al. [28] have pointed out that the synergistic interaction between the two oxide phases in a CuO–CeO_2 mixed oxide enhances the reducibility and consequently the deN_2O efficiency as the surface-adsorbed oxygen species is easily desorbed and the active sites' regeneration is enabled.

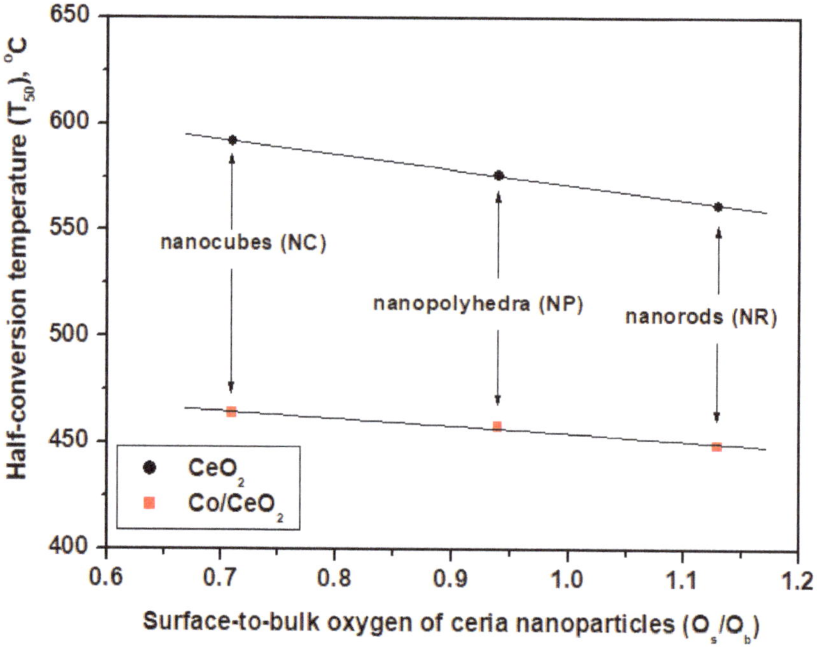

Figure 8. The half-conversion temperature (T_{50}) as a function of the TPR surface-to-bulk oxygen ratio (O_s/O_b).

More interestingly, the deN_2O performance of CeO_2 as well as the Co_3O_4/CeO_2 samples totally coincides, indicating the significance of the ceria carrier. However, the superiority of the mixed oxides in comparison to the bare ceria samples is evident, reflecting the synergistic interactions between cobalt and cerium oxides. The latter is manifested by the improved redox properties (in terms of H_2 consumption and TPR onset temperature) of Co_3O_4/CeO_2 mixed oxides as compared to bare ceria (Table 2). In a similar manner, the incorporation of cobalt into the ceria lattice increases both the amount of the adsorbed oxygen species (O^-/O_2^{2-}) and Ce^{3+} (Table 3), related with the surface oxygen reduction and the abundance in oxygen vacancies (O_{vac}).

Moreover, ceria nanorods facilitate the reduction of Co^{3+} to Co^{2+} active sites (Table 3), further contributing to the superior catalytic performance of the Co/CeO_2-NR sample. Along the same line, it has been recently reported that ceria nanorods stabilize the partial oxidation state of Co in CoO_X/CeO_2 catalysts *via* the facilitation of oxygen transfer at the metal-support interface [74]. It should be, therefore, deduced that ceria nanorods with the exposed (110) and (100) facets show the highest surface-to-bulk oxygen ratio resulting in improved reducibility and oxygen kinetics while exhibiting the highest amount of weakly bound oxygen species which is a decisive factor in the deN_2O process. Upon cobalt addition, the nanorod sample exhibits in addition the highest population in Ce^{3+}/Co^{2+} redox pairs, indicative of abundant oxygen vacancies, which, along with its enhanced reducibility, leads to a superior deN_2O performance.

In this point, the enhanced textural characteristics (BET area and pore volume) of Co/CeO_2-NR as compared to Co/CeO_2-NC should be also mentioned, which could be further accounted for its enhanced deN_2O performance. Thus, by taking into account the specific activity normalized per unit of surface area (nmol m^{-2} s^{-1}) instead of mass unit (nmol g^{-1} s^{-1}), an inferior performance is observed for Co/CeO_2-NR compared to Co/CeO_2-NC (Table 4). On the other hand, Co_3O_4/CeO_2-NR exhibits a superior deN_2O performance (both in terms of conversion and specific activity) as compared to Co_3O_4/CeO_2-NP despite their similar structural (crystallite size) and textural (surface area) properties (Table 1). The latter clearly reveals the importance of exposed facets and redox properties on the deN_2O process, as it has been similarly reported by Zabilskiy et al. [45] for CuO/CeO_2 nanostructures of different morphology. Therefore, on the basis of the present findings, it can be deduced that the enhanced N_2O conversion performance of Co_3O_4/CeO_2-NR mixed oxides could be attributed to a compromise between redox and textural characteristics.

Table 4. The N_2O conversion and specific activity of Co/CeO_2 samples at 420 °C: The reaction conditions are 1000 ppm N_2O, 0 or 2 vol. % O_2 and GHSV = 40,000 h^{-1}.

Sample	N_2O Conversion (%)		Specific Activity			
	O_2 Absence	O_2 Presence	O_2 Absence		O_2 Presence	
			r (nmol g^{-1} s^{-1})	r (nmol m^{-2} s^{-1})	r (nmol g^{-1} s^{-1})	r (nmol m^{-2} s^{-1})
Co/CeO_2-NC	16.2	8.6	166	5.9	88	3.1
Co/CeO_2-NP	20.2	10.7	207	2.9	109	1.5
Co/CeO_2-NR	25	14	256	3.6	143	2.0

3. Materials and Methods

3.1. Materials Synthesis

In the present work, the chemicals that were used were of analytical reagent grade. $Ce(NO_3)_3 \cdot 6H_2O$ (Fluka, Bucharest, Romania, purity ≥99.0%) and $Co(NO_3)_2 \cdot 6H_2O$ (Sigma-Aldrich, Taufkirchen, Germany, purity ≥98%) were employed as precursor compounds for the preparation of bare ceria as well as Co/CeO_2 mixed oxides. Also, NaOH (Sigma-Aldrich, Taufkirchen, Germany, purity ≥98%) and ethanol (ACROS Organics, Geel, Belgium, purity 99.8%,) were used during materials synthesis. Initially, the hydrothermal method was applied for the preparation of bare ceria nanoparticles, as described in detail in our previous work [49]. In brief, ceria nanorods (CeO_2-NR) were synthesized by dissolving NaOH (36.7 M) in double deionized water and then adding an appropriate amount of an aqueous solution of $Ce(NO_3)_3 \cdot 6H_2O$ (0.13 M) under vigorous stirring. Next, the transfer of the final slurry into a Teflon bottle and its aging at 90 °C for 24 h occurred. For the synthesis of ceria nanopolyhedra (CeO_2-NP), a similar procedure was followed, employing, however, a lower amount of NaOH (6 M). In order to synthesize ceria nanocubes (CeO_2-NC), the same procedure as the one described above for the synthesis of ceria nanorods was followed, with the obtained slurry to be aged at 180 °C instead of 90 °C. In all cases, centrifugation was used for the recovery of the solid products that were thoroughly washed with double deionized water until a neutral pH was reached and finally washed with ethanol so as to avoid the nanoparticles' agglomeration. Afterwards, drying of the precipitate at 90 °C for 12 h followed by calcination at 500 °C for 2 h under air flow (heating ramp 5 °C/min) was carried out.

The Co/CeO_2-NX catalysts where NX stands for NP: nanopolyhedra, NR: nanorods and NC: nanocubes were prepared by wet impregnation, employing an aqueous solution of $Co(NO_3)_2 \cdot 6H_2O$, in order to achieve an atomic ratio Co/(Co+Ce) of 0.2 which corresponds to 7.8 wt. % of Co loading. Heating under stirring of the obtained suspensions until complete water evaporation occurred, followed by drying at 90 °C for 12 h and final calcination at 500 °C for 2 h under air flow (heating ramp 5 °C/min).

3.2. *Materials Characterization*

The porosity of the materials was evaluated by the N_2-adsorption isotherms at −196 °C, using an ASAP 2010 (Micromeritics, Norcross, GA, USA) apparatus (from ReQuimTe Analyses Laboratory, Universidade Nova de Lisboa, Lisboa, Portugal). The samples were previously degassed at 300 °C for 6 h. The specific surface area was calculated by the Brunauer–Emmett–Teller (BET) equation [75].

Structural characterization was carried out by means of X-ray diffraction (XRD) in a PAN'alytical X'Pert MPD equipped with a X'Celerator detector and secondary monochromator (Cu Kα λ = 0.154 nm, 50 kV, 40 mA; data recorded at a 0.017° step size, 100 s/step) in the University of Trás-os-Montes e Alto Douro. The collected spectra were analyzed by Rietveld refinement using PowderCell software, allowing the determination of crystallite sizes by means of the Williamson–Hall plot.

The redox properties were assessed by Temperature Programmed Reduction (TPR) experiments in an AMI-200 Catalyst Characterization Instrument (Altamira Instruments, Pittsburgh, PA, USA), employing H_2 as a reducing agent. In a typical H_2-TPR experiment, 50 mg of the sample (grain size 180–355 μm) was heated up to 1100 °C (10 °C/min) under H_2 flow (1.5 cm^3) balanced with He (29 cm^3). The amount of H_2 consumed (mmol g^{-1}) was calculated by taking into account the integrated area of TPR peaks, calibrated against a known amount of CuO standard sample [76,77].

The surface composition and the chemical state of each element were determined by X-ray photoelectron spectroscopy (XPS) analyses, performed on a VG Scientific ESCALAB 200A spectrometer using Al Kα radiation (1486.6 eV) in CEMUP. The charge effect was corrected using the C1s' peak as a reference (binding energy of 285 eV). The CASAXPS software was used for data analysis.

The samples were imaged by transmission electron microscopy (TEM). The analyses were performed on a Leo 906E apparatus (Austin, TX, USA), at 120 kV in the University of Trás-os-Montes e Alto Douro. The samples were prepared by ultrasonic dispersion, and a 400 mesh formvar/carbon copper grid (Agar Scientific, Essex, UK) was dipped into the solution for the TEM analysis.

3.3. *Catalytic Performance Evaluation*

The catalytic studies for the N_2O decomposition took place in a quartz fixed-bed U-shaped reactor (0.8 cm i.d.) with 100 mg of catalyst loading (grain size 180–355 μm). The feed gas (1000 ppm N_2O and 0 or 2 vol. % O_2) total flow rate was 150 cm^3/min which corresponds to a Gas Hour Space Velocity (GHSV) of 40,000 h^{-1}. The analysis of the gases was performed by a gas chromatograph (SHIMADZU 14B). The apparatus is equipped with a TCD detector and two separation columns (Molecular Sieve 5A for O_2, N_2 measurements and Porapack QS for N_2O measurement). Prior to the catalytic activity measurements, the materials under consideration were subjected to further processing under He flow (100 cm^3/min) at 400 °C. In order to minimize the external and internal diffusion limitations, preliminary experiments concerning the influence of particle size and W/F ratio on deN_2O catalytic performance were carried out. Based on these experiments, a catalyst particle size in the range of 180–355 μm was selected, in addition to a W/F ratio of 0.04 g s cm^{-3}. The conversion of N_2O (X_{N2O}) was calculated from the difference in N_2O concentration between the inlet and outlet gas streams, according to the equation

$$X_{N_2O}(\%) = \frac{[N_2O]_{in} - [N_2O]_{out}}{[N_2O]_{in}} \times 100 \quad (7)$$

The specific reaction rate (r, mol m^{-2} s^{-1}) of the N_2O decomposition was also estimated using the following formula:

$$r\left(mol\, m^{-2}\, s^{-1}\right) = \frac{X_{N_2O} \times [N_2O]_{in} \times F\left(\frac{cm^3}{min}\right)}{100 \times 60\left(\frac{s}{min}\right) \times V_m\left(\frac{cm^3}{mol}\right) \times m_{cat}(g) \times S_{BET}\left(\frac{m^2}{g}\right)} \quad (8)$$

where F and V_m are the total flow rate and gas molar volume, respectively, at standard ambient temperature and pressure conditions (298 K and 1 bar), m_{cat} is the catalyst's mass and S_{BET} is the surface area.

4. Conclusions

In this work, three different ceria nanoshaped materials (nanorods, nanocubes and nanopolyhedra) were hydrothermally synthesized and used as supports for the cobalt oxide phase. Both single CeO_2 and Co/CeO_2 mixed oxides were catalytically assessed during the decomposition of N_2O in the presence and absence of oxygen. For bare ceria samples, the following deN_2O order was obtained: CeO_2-NR (nanorods) > CeO_2-NP (nanopolyhedra) > CeO_2-NC (nanocubes). Most importantly, cobalt addition to the CeO_2 carriers greatly enhances the N_2O decomposition, not affecting at all the order obtained for the bare ceria supports and clearly reflecting the key role of support. The present results clearly reveal the key role of support morphology on the textural, structural and redox properties, reflected then on the catalytic performance of Co_3O_4/CeO_2 mixed oxides. Among the different samples investigated, the cobalt-ceria nanorods (Co/CeO_2-NR) exposing {100} and {110} facets showed the best deN_2O performance, ascribed mainly to their abundance in Co^{2+} active sites in conjunction to their enhanced redox and textural properties.

Author Contributions: M.L. contributed to materials synthesis, results interpretation and paper writing; E.P. and N.K. contributed to catalytic evaluation studies; S.A.C.C. contributed to the materials characterization; M.K. contributed to the conception, design, results interpretation and writing of the paper; all authors contributed to the discussion and read and approved the final version of the manuscript.

Acknowledgments: The research work was supported by the Hellenic Foundation for Research and Innovation (HFRI) and the General Secretariat for Research and Technology (GSRT), under the HFRI PhD Fellowship grant (GA. no. 34252). This research has been cofinanced by the European Union and Greek national funds through the Operational Program Competitiveness, Entrepreneurship and Innovation, under the call RESEARCH—CREATE—INNOVATE (project code: T1EDK-00094). This work was also financially supported by Associate Laboratory LSRE-LCM—UID/EQU/50020/2019—funded by national funds through FCT/MCTES (PIDDAC). S.A.C.C. acknowledges Fundação para a Ciência e a Tecnologia (Portugal) for Investigador FCT program (IF/01381/2013/CP1160/CT0007), with financing from the European Social Fund and the Human Potential Operational Program. We are grateful to Carlos Sá (CEMUP) for the assistance with the XPS measurements, Pedro Tavares (UTAD) for the TEM and XRD analyses and Nuno Costa (ReQuimTe) for the N_2 adsorption results.

Conflicts of Interest: The authors declare no conflict of interest.

References

1. Konsolakis, M. Recent Advances on Nitrous Oxide (N_2O) Decomposition over Non-Noble-Metal Oxide Catalysts: Catalytic Performance, Mechanistic Considerations and Surface Chemistry Aspects. *ACS Catal.* **2015**, *5*, 6397–6421. [CrossRef]
2. Liu, Z.; He, F.; Ma, L.; Peng, S. Recent Advances in Catalytic Decomposition of N_2O on Noble Metal and Metal Oxide Catalysts. *Catal. Surv. Asia* **2016**, *20*, 121–132. [CrossRef]
3. Kim, M.J.; Lee, S.J.; Ryu, I.S.; Jeon, M.W.; Moon, S.H.; Roh, H.S.; Jeon, S.G. Catalytic decomposition of N_2O over cobalt based spinel oxides: The role of additives. *Mol. Catal.* **2017**, *442*, 202–207. [CrossRef]
4. Rutkowska, M. Catalytic Decomposition of N_2O Using a New Generation of Functionalized Microporous and Mesoporous Inorganic Materials. *Wiadomości Chemiczne* **2015**, *69*, 297–315.
5. Piumetti, M.; Hussain, M.; Fino, D.; Russo, N. Mesoporous silica supported Rh catalysts for high concentration N_2O decomposition. *Appl. Catal. B Environ.* **2015**, *165*, 158–168. [CrossRef]
6. Kim, S.S.; Lee, S.J.; Hong, S.C. Effect of CeO_2 addition to Rh/Al_2O_3 catalyst on N_2O decomposition. *Chem. Eng. J.* **2011**, *169*, 173–179. [CrossRef]
7. Hussain, M.; Fino, D.; Russo, N. Development of modified KIT-6 and SBA-15-spherical supported Rh catalysts for N_2O abatement: From powder to monolith supported catalysts. *Chem. Eng. J.* **2014**, *238*, 198–205. [CrossRef]
8. Wu, Y.; Cordier, C.; Berrier, E.; Nuns, N.; Dujardin, C.; Granger, P. Surface reconstructions of $LaCo_{1-x}Fe_xO_3$ at high temperature during N_2O decomposition in realistic exhaust gas composition: Impact on the catalytic properties. *Appl. Catal. B Environ.* **2013**, *140–141*, 151–163. [CrossRef]
9. Ivanov, D.V.; Pinaeva, L.G.; Isupova, L.A.; Sadovskaya, E.M.; Prosvirin, I.P.; Gerasimov, E.Y.; Yakovleva, I.S. Effect of surface decoration with $LaSrFeO_4$ on oxygen mobility and catalytic activity of $La_{0.4}Sr_{0.6}FeO_{3-\delta}$ in high-temperature N_2O decomposition, methane combustion and ammonia oxidation. *Appl. Catal. A Gen.* **2013**, *457*, 42–51. [CrossRef]
10. Kumar, S.; Vinu, A.; Subrt, J.; Bakardjieva, S.; Rayalu, S.; Teraoka, Y.; Labhsetwar, N. Catalytic N_2O decomposition on $Pr_{0.8}Ba_{0.2}MnO_3$ type perovskite catalyst for industrial emission control. *Catal. Today* **2012**, *198*, 125–132. [CrossRef]
11. Zhang, Y.; Wang, X.; Zhu, Y.; Zhang, T. Stabilization mechanism and crystallographic sites of Ru in Fe-promoted barium hexaaluminate under high-temperature condition for N_2O decomposition. *Appl. Catal. B Environ.* **2013**, *129*, 382–393. [CrossRef]
12. Zhang, Y.; Wang, X.; Zhu, Y.; Hou, B.; Yang, X.; Liu, X.; Wang, J.; Li, J.; Zhang, T. Characterization of Fe substitution into La-hexaaluminate systems and the effect on N_2O catalytic decomposition. *J. Phys. Chem. C* **2014**, *118*, 1999–2010. [CrossRef]
13. Zhang, Y.; Wang, X.; Zhu, Y.; Liu, X.; Zhang, T. Thermal evolution crystal structure and Fe crystallographic sites in $LaFe_xAl_{12-x}O_{19}$ hexaaluminates. *J. Phys. Chem. C* **2014**, *118*, 10792–10804. [CrossRef]
14. Pérez-Ramírez, J.; Santiago, M. Metal-substituted hexaaluminates for high-temperature N_2O abatement. *Chem. Commun.* **2007**, *2*, 619–621. [CrossRef] [PubMed]
15. Stelmachowski, P.; Maniak, G.; Kaczmarczyk, J.; Zasada, F.; Piskorz, W.; Kotarba, A.; Sojka, Z. Mg and Al substituted cobalt spinels as catalysts for low temperature deN_2O-Evidence for octahedral cobalt active sites. *Appl. Catal. B Environ.* **2014**, *146*, 105–111. [CrossRef]
16. Grzybek, G.; Stelmachowski, P.; Gudyka, S.; Duch, J.; Ćmil, K.; Kotarba, A.; Sojka, Z. Insights into the twofold role of Cs doping on deN_2O activity of cobalt spinel catalyst-towards rational optimization of the precursor and loading. *Appl. Catal. B Environ.* **2015**, *168–169*, 509–514. [CrossRef]
17. Zasada, F.; Piskorz, W.; Janas, J.; Gryboś, J.; Indyka, P.; Sojka, Z. Reactive Oxygen Species on the (100) Facet of Cobalt Spinel Nanocatalyst and their Relevance in $^{16}O_2/^{18}O_2$ Isotopic Exchange, deN_2O, and deCH_4 Processes-A Theoretical and Experimental Account. *ACS Catal.* **2015**, *5*, 6879–6892. [CrossRef]
18. Amrousse, R.; Chang, P.-J.; Choklati, A.; Friche, A.; Rai, M.; Bachar, A.; Follet-Houttemane, C.; Hori, K. Catalytic decomposition of N_2O over Ni and Mg-magnetite catalysts. *Catal. Sci. Technol.* **2013**, *3*, 2288. [CrossRef]
19. Zou, W.; Xie, P.; Hua, W.; Wang, Y.; Kong, D.; Yue, Y.; Ma, Z.; Yang, W.; Gao, Z. Catalytic decomposition of N_2O over Cu-ZSM-5 nanosheets. *J. Mol. Catal. A Chem.* **2014**, *394*, 83–88. [CrossRef]

20. Zhang, X.; Shen, Q.; He, C.; Ma, C.; Cheng, J.; Liu, Z.; Hao, Z. Decomposition of nitrous oxide over Co-zeolite catalysts: role of zeolite structure and active site. *Catal. Sci. Technol.* **2012**, *2*, 1249–1258. [CrossRef]
21. Rutkowska, M.; Piwowarska, Z.; Micek, E.; Chmielarz, L. Hierarchical Fe-, Cu- and Co-Beta zeolites obtained by mesotemplate-free method. Part I: Synthesis and catalytic activity in N_2O decomposition. *Microporous Mesoporous Mater.* **2015**, *209*, 54–65. [CrossRef]
22. Xie, P.; Luo, Y.; Ma, Z.; Wang, L.; Huang, C.; Yue, Y.; Hua, W.; Gao, Z. CoZSM-11 catalysts for N_2O decomposition: Effect of preparation methods and nature of active sites. *Appl. Catal. B Environ.* **2015**, *170–171*, 34–42. [CrossRef]
23. Grzybek, G.; Stelmachowski, P.; Gudyka, S.; Indyka, P.; Sojka, Z.; Guillén-Hurtado, N.; Rico-Pérez, V.; Bueno-López, A.; Kotarba, A. Strong dispersion effect of cobalt spinel active phase spread over ceria for catalytic N_2O decomposition: The role of the interface periphery. *Appl. Catal. B Environ.* **2016**, *180*, 622–629. [CrossRef]
24. Chromčáková, Ž.; Obalová, L.; Kovanda, F.; Legut, D.; Titov, A.; Ritz, M.; Fridrichová, D.; Michalik, S.; Kuśtrowski, P.; Jirátová, K. Effect of precursor synthesis on catalytic activity of Co_3O_4 in N_2O decomposition. *Catal. Today* **2015**, *257*, 18–25. [CrossRef]
25. Franken, T.; Palkovits, R. Investigation of potassium doped mixed spinels $Cu_xCo_{3-x}O_4$ as catalysts for an efficient N_2O decomposition in real reaction conditions. *Appl. Catal. B Environ.* **2015**, *176–177*, 298–305. [CrossRef]
26. Zabilskiy, M.; Erjavec, B.; Djinović, P.; Pintar, A. Ordered mesoporous CuO-CeO_2 mixed oxides as an effective catalyst for N_2O decomposition. *Chem. Eng. J.* **2014**, *254*, 153–162. [CrossRef]
27. Xue, L.; He, H.; Liu, C.; Zhang, C.; Zhang, B. Promotion Effects and Mechanism of Alkali Metals and Alkaline Earth Metals on Cobalt - Cerium Composite Oxide Catalysts for N_2O Decomposition. *Environ. Sci. Technol.* **2009**, *43*, 890–895. [CrossRef] [PubMed]
28. Liu, Z.; He, C.; Chen, B.; Liu, H. CuO-CeO_2 mixed oxide catalyst for the catalytic decomposition of N_2O in the presence of oxygen. *Catal. Today* **2017**, *297*, 78–83. [CrossRef]
29. Russo, N.; Fino, D.; Saracco, G.; Specchia, V. N_2O catalytic decomposition over various spinel-type oxides. *Catal. Today* **2007**, *119*, 228–232. [CrossRef]
30. Xue, L.; Zhang, C.; He, H.; Teraoka, Y. Catalytic decomposition of N_2O over CeO_2 promoted Co_3O_4 spinel catalyst. *Appl. Catal. B Environ.* **2007**, *75*, 167–174. [CrossRef]
31. Yang, J.; Lukashuk, L.; Akbarzadeh, J.; Stöger-Pollach, M.; Peterlik, H.; Föttinger, K.; Rupprechter, G.; Schubert, U. Different Synthesis Protocols for Co_3O_4-CeO_2 Catalysts-Part 1: Influence on the Morphology on the Nanoscale. *Chem. Eur. J.* **2015**, *21*, 885–892. [CrossRef] [PubMed]
32. Shen, Q.; Li, L.; Li, J.; Tian, H.; Hao, Z. A study on N_2O catalytic decomposition over Co/MgO catalysts. *J. Hazard. Mater.* **2009**, *163*, 1332–1337. [CrossRef] [PubMed]
33. Luo, J.-Y.; Meng, M.; Li, X.; Li, X.-G.; Zha, Y.-Q.; Hu, T.-D.; Xie, Y.-N.; Zhang, J. Mesoporous Co_3O_4–CeO_2 and Pd/Co_3O_4–CeO_2 catalysts: Synthesis, characterization and mechanistic study of their catalytic properties for low-temperature CO oxidation. *J. Catal.* **2008**, *254*, 310–324. [CrossRef]
34. Wang, H.; Ye, J.L.; Liu, Y.; Li, Y.D.; Qin, Y.N. Steam reforming of ethanol over Co_3O_4/CeO_2 catalysts prepared by different methods. *Catal. Today* **2007**, *129*, 305–312. [CrossRef]
35. Qiu, N.; Zhang, J.; Wu, Z. Peculiar surface-interface properties of nanocrystalline ceria-cobalt oxides with enhanced oxygen storage capacity. *Phys. Chem. Chem. Phys.* **2014**, *16*, 22659–22664. [CrossRef] [PubMed]
36. Vinod, C.P. Surface science as a tool for probing nanocatalysis phenomena. *Catal. Today* **2010**, *154*, 113–117. [CrossRef]
37. Hu, P.; Huang, Z.; Amghouz, Z.; Makkee, M.; Xu, F.; Kapteijn, F.; Dikhtiarenko, A.; Chen, Y.; Gu, X.; Tang, X. Electronic Metal-Support Interactions in Single-Atom Catalysts. *Angew. Chem.* **2014**, *126*, 3486–3489. [CrossRef]
38. Elias, J.S.; Risch, M.; Giordano, L.; Mansour, A.N.; Shao-Horn, Y. Structure, Bonding, and Catalytic Activity of Monodisperse, Transition-Metal-Substituted CeO_2 Nanoparticles. *J. Am. Chem. Soc.* **2014**, *136*, 17193–17200. [CrossRef] [PubMed]
39. Cargnello, M.; Fornasiero, P.; Gorte, R.J. Opportunities for tailoring catalytic properties through metal-support interactions. *Catal. Lett.* **2012**, *142*, 1043–1048. [CrossRef]
40. Díez-Ramírez, J.; Sánchez, P.; Kyriakou, V.; Zafeiratos, S.; Marnellos, G.E.; Konsolakis, M.; Dorado, F. Effect of support nature on the cobalt-catalyzed CO_2 hydrogenation. *J. CO2 Util.* **2017**, *21*, 562–571. [CrossRef]

41. Cargnello, M.; Doan-Nguyen, V.V.T.; Gordon, T.R.; Diaz, R.E.; Stach, E.A.; Gorte, R.J.; Fornasiero, P.; Murray, C.B. Control of Metal Nanocrystal Size Reveals Metal-Support Interface Role for Ceria Catalysts. *Science* **2013**, *341*, 771–773. [CrossRef] [PubMed]
42. Vayssilov, G.N.; Lykhach, Y.; Migani, A.; Staudt, T.; Petrova, G.P.; Tsud, N.; Skála, T.; Bruix, A.; Illas, F.; Prince, K.C.; et al. Support nanostructure boosts oxygen transfer to catalytically active platinum nanoparticles. *Nat. Mater.* **2011**, *10*, 310–315. [CrossRef] [PubMed]
43. Mahammadunnisa, S.K.; Akanksha, T.; Krushnamurty, K.; Subrahmanyam, C. Catalytic decomposition of N_2O over CeO_2 supported Co_3O_4 catalysts. *J. Chem. Sci.* **2016**, *128*, 1795–1804. [CrossRef]
44. Lin, B.; Liu, Y.; Heng, L.; Ni, J.; Lin, J.; Jiang, L. Effect of ceria morphology on the catalytic activity of Co/CeO_2 catalyst for ammonia synthesis. *Catal. Commun.* **2017**, *101*, 15–19. [CrossRef]
45. Zabilskiy, M.; Djinović, P.; Tchernychova, E.; Tkachenko, O.P.; Kustov, L.M.; Pintar, A. Nanoshaped CuO/CeO_2 Materials: Effect of the Exposed Ceria Surfaces on Catalytic Activity in N_2O Decomposition Reaction. *ACS Catal.* **2015**, *5*, 5357–5365. [CrossRef]
46. Andrade-Martínez, J.; Ortega-Zarzosa, G.; Gómez-Cortés, A.; Rodríguez-González, V. N_2O catalytic reduction over different porous SiO_2 materials functionalized with copper. *Powder Technol.* **2015**, *274*, 305–312. [CrossRef]
47. Mei, J.; Ke, Y.; Yu, Z.; Hu, X.; Qu, Z.; Yan, N. Morphology-dependent properties of Co_3O_4/CeO_2 catalysts for low temperature dibromomethane (CH_2Br_2) oxidation. *Chem. Eng. J.* **2017**, *320*, 124–134. [CrossRef]
48. Hu, F.; Peng, Y.; Chen, J.; Liu, S.; Song, H.; Li, J. Low content of CoO_x supported on nanocrystalline CeO_2 for toluene combustion: The importance of interfaces between active sites and supports. *Appl. Catal. B Environ.* **2019**, *240*, 329–336. [CrossRef]
49. Lykaki, M.; Pachatouridou, E.; Carabineiro, S.A.C.; Iliopoulou, E.; Andriopoulou, C.; Kallithrakas-Kontos, N.; Boghosian, S.; Konsolakis, M. Ceria nanoparticles shape effects on the structural defects and surface chemistry: Implications in CO oxidation by Cu/CeO_2 catalysts. *Appl. Catal. B Environ.* **2018**, *230*, 18–28. [CrossRef]
50. Allwar, A.; Md Noor, A.; Nawi, M. Textural Characteristics of Activated Carbons Prepared from Oil Palm Shells Activated with $ZnCl_2$ and Pyrolysis Under Nitrogen and Carbon Dioxide. *J. Phys. Sci.* **2008**, *19*, 93–104.
51. Kumar, S.; Sharma, C. Synthesis, characterization and catalytic wet air oxidation property of mesoporous $Ce_{1-x}Fe_xO_2$ mixed oxides. *Mater. Chem. Phys.* **2015**, *155*, 223–231.
52. Tan, L.; Tao, Q.; Gao, H.; Li, J.; Jia, D.; Yang, M. Preparation and catalytic performance of mesoporous ceria-base composites CuO/CeO_2, Fe_2O_3/CeO_2 and La_2O_3/CeO_2. *J. Porous Mater.* **2017**, *24*, 795–803. [CrossRef]
53. Farahmandjou, M.; Zarinkamar, M. Synthesis of nano-sized ceria (CeO_2) particles via a cerium hydroxy carbonate precursor and the effect of reaction temperature on particle morphology. *J. Ultrafine Grained Nanostructured Mater.* **2015**, *48*, 5–10.
54. Sharma, V.; Eberhardt, K.M.; Sharma, R.; Adams, J.B.; Crozier, P.A. A spray drying system for synthesis of rare-earth doped cerium oxide nanoparticles. *Chem. Phys. Lett.* **2010**, *495*, 280–286. [CrossRef]
55. Liu, J.; Zhao, Z.; Wang, J.; Xu, C.; Duan, A.; Jiang, G.; Yang, Q. The highly active catalysts of nanometric CeO_2-supported cobalt oxides for soot combustion. *Appl. Catal. B Environ.* **2008**, *84*, 185–195. [CrossRef]
56. Yu, S.W.; Huang, H.H.; Tang, C.W.; Wang, C.B. The effect of accessible oxygen over Co_3O_4-CeO_2 catalysts on the steam reforming of ethanol. *Int. J. Hydrogen Energy* **2014**, *39*, 20700–20711. [CrossRef]
57. Konsolakis, M.; Carabineiro, S.A.C.; Marnellos, G.E.; Asad, M.F.; Soares, O.S.G.P.; Pereira, M.F.R.; Órfão, J.J.M.; Figueiredo, J.L. Effect of cobalt loading on the solid state properties and ethyl acetate oxidation performance of cobalt-cerium mixed oxides. *J. Colloid Interface Sci.* **2017**, *496*, 141–149. [CrossRef] [PubMed]
58. Mock, S.A.; Sharp, S.E.; Stoner, T.R.; Radetic, M.J.; Zell, E.T.; Wang, R. CeO_2 nanorods-supported transition metal catalysts for CO oxidation. *J. Colloid Interface Sci.* **2016**, *466*, 261–267. [CrossRef] [PubMed]
59. Wang, Y.; Hu, X.; Zheng, K.; Zhang, H.; Zhao, Y. Effect of precipitants on the catalytic activity of Co–Ce composite oxide for N_2O catalytic decomposition. *Reac. Kinet. Mech. Cat.* **2018**, *123*, 707–721. [CrossRef]
60. Wang, L.; Liu, H. Mesoporous Co-CeO_2 catalyst prepared by colloidal solution combustion method for reverse water-gas shift reaction. *Catal. Today* **2018**, *316*, 155–161. [CrossRef]

61. Kumar Megarajan, S.; Rayalu, S.; Teraoka, Y.; Labhsetwar, N. High NO oxidation catalytic activity on non-noble metal based cobalt-ceria catalyst for diesel soot oxidation. *J. Mol. Catal. A Chem.* **2014**, *385*, 112–118. [CrossRef]
62. Konsolakis, M. The role of Copper–Ceria interactions in catalysis science: Recent theoretical and experimental advances. *Appl. Catal. B Environ.* **2016**, *198*, 49–66. [CrossRef]
63. Cui, Y.; Dai, W.-L. Support morphology and crystal plane effect of Cu/CeO_2 nanomaterial on the physicochemical and catalytic properties for carbonate hydrogenation. *Catal. Sci. Technol.* **2016**, *6*, 7752–7762. [CrossRef]
64. Guo, X.; Zhou, R. A new insight into the morphology effect of ceria on CuO/CeO_2 catalysts for CO selective oxidation in hydrogen-rich gas. *Catal. Sci. Technol.* **2016**, *6*, 3862–3871. [CrossRef]
65. Wang, C.; Cheng, Q.; Wang, X.; Ma, K.; Bai, X.; Tan, S.; Tian, Y.; Ding, T.; Zheng, L.; Zhang, J.; et al. Enhanced catalytic performance for CO preferential oxidation over CuO catalysts supported on highly defective CeO_2 nanocrystals. *Appl. Surf. Sci.* **2017**, *422*, 932–943. [CrossRef]
66. Dou, J.; Tang, Y.; Nie, L.; Andolina, C.M.; Zhang, X.; House, S.; Li, Y.; Yang, J.; Tao, F.F. Complete Oxidation of Methane on Co_3O_4/CeO_2 Nanocomposite: A Synergic Effect. *Catal. Today* **2018**, *311*, 48–55. [CrossRef]
67. Konsolakis, M.; Sgourakis, M.; Carabineiro, S.A.C. Surface and redox properties of cobalt-ceria binary oxides: On the effect of Co content and pretreatment conditions. *Appl. Surf. Sci.* **2015**, *341*, 48–54. [CrossRef]
68. Ma, J.; Rodriguez, N.M.; Vannice, M.A.; Baker, R.T.K. Nitrous oxide decomposition and reduction over copper catalysts supported on various types of carbonaceous materials. *Top. Catal.* **2000**, *10*, 27–38. [CrossRef]
69. You, Y.; Chang, H.; Ma, L.; Guo, L.; Qin, X.; Li, J.; Li, J. Enhancement of N_2O decomposition performance by N_2O pretreatment over Ce-Co-O catalyst. *Chem. Eng. J.* **2018**, *347*, 184–192. [CrossRef]
70. Iwanek, E.; Krawczyk, K.; Petryk, J.; Sobczak, J.W.; Kaszkur, Z. Direct nitrous oxide decomposition with CoO_x-CeO_2 catalysts. *Appl. Catal. B Environ.* **2011**, *106*, 416–422. [CrossRef]
71. Piskorz, W.; Zasada, F.; Stelmachowski, P.; Kotarba, A.; Sojka, Z. Decomposition of N_2O over the surface of cobalt spinel: A DFT account of reactivity experiments. *Catal. Today* **2008**, *137*, 418–422. [CrossRef]
72. Abu-Zied, B.M.; Soliman, S.A.; Abdellah, S.E. Pure and Ni-substituted Co_3O_4 spinel catalysts for direct N_2O decomposition. *Chin. J. Catal.* **2014**, *35*, 1105–1112. [CrossRef]
73. Stelmachowski, P.; Maniak, G.; Kotarba, A.; Sojka, Z. Strong electronic promotion of Co_3O_4 towards N_2O decomposition by surface alkali dopants. *Catal. Commun.* **2009**, *10*, 1062–1065. [CrossRef]
74. Savereide, L.; Nauert, S.L.; Roberts, C.A.; Notestein, J.M. The effect of support morphology on CoO_X/CeO_2 catalysts for the reduction of NO by CO. *J. Catal.* **2018**, *366*, 150–158. [CrossRef]
75. Brunauer, S.; Emmett, P.H.; Teller, E. Adsorption of Gases in Multimolecular Layers. *J. Am. Chem. Soc.* **1938**, *60*, 309–319. [CrossRef]
76. Barthos, R.; Hegyessy, A.; Klébert, S.; Valyon, J. Vanadium dispersion and catalytic activity of $Pd/VO_x/SBA$-15 catalysts in the Wacker oxidation of ethylene. *Microporous Mesoporous Mater.* **2015**, *207*, 1–8. [CrossRef]
77. Xu, J.; Harmer, J.; Li, G.; Chapman, T.; Collier, P.; Longworth, S.; Tsang, S.C. Size dependent oxygen buffering capacity of ceria nanocrystals. *Chem. Commun.* **2010**, *46*, 1887–1889. [CrossRef] [PubMed]

catalysts

MDPI

Article

Precipitated K-Promoted Co–Mn–Al Mixed Oxides for Direct NO Decomposition: Preparation and Properties

Květa Jirátová [1,*], Kateřina Pacultová [2], Jana Balabánová [1], Kateřina Karásková [2], Anna Klegová [2], Tereza Bílková [2], Věra Jandová [1], Martin Koštejn [1], Alexandr Martaus [2], Andrzej Kotarba [3] and Lucie Obalová [2]

1 Institute of Chemical Process Fundamentals of the CAS, v.v.i., Rozvojová 135, CZ- 165 02 Praha 6- Suchdol, Czech Republic

2 Institute of Environmental Technology, VŠB- TU Ostrava, 17. listopadu 15/2172, CZ- 708 00 Ostrava- Poruba, Czech Republic

3 Faculty of Chemistry, Jagiellonian University, Gronostajowa 2, PL- 30387 Krakow, Poland

* Correspondence: jiratova@icpf.cas.cz

Received: 28 May 2019; Accepted: 5 July 2019; Published: 9 July 2019

Abstract: Direct decomposition of nitric oxide (NO) proceeds over Co–Mn–Al mixed oxides promoted by potassium. In this study, answers to the following questions have been searched: Do the properties of the K-promoted Co–Mn–Al catalysts prepared by different methods differ from each other? The K-precipitated Co–Mn–Al oxide catalysts were prepared by the precipitation of metal nitrates with a solution of K_2CO_3/KOH, followed by the washing of the precipitate to different degrees of residual K amounts, and by cthe alcination of the precursors at 500 °C. The properties of the prepared catalysts were compared with those of the best catalyst prepared by the K-impregnation of a wet cake of Co–Mn–Al oxide precursors. The solids were characterized by chemical analysis, DTG, XRD, N_2 physisorption, FTIR, temperature programmed reduction (H_2-TPR), temperature programmed CO_2 desorption (CO_2-TPD), X-ray photoelectron spectrometry (XPS), and the species-resolved thermal alkali desorption method (SR-TAD). The washing of the K-precipitated cake resulted in decreasing the K amount in the solid, which affected the basicity, reducibility, and non-linearly catalytic activity in NO decomposition. The highest activity was found at ca 8 wt.% of K, while that of the best K-impregnated wet cake catalyst was at about 2 wt.% of K. The optimization of the cake washing conditions led to a higher catalytic activity.

Keywords: NO decomposition; Co–Mn–Al mixed oxides; catalyst preparation; potassium promoter

---

1. Introduction

Nitric oxide (NO) is considered a harmful gas, as it negatively affects the environment by attenuating the ozone layer and contributing to the formation of acid rain and smog [1]. NO is formed by natural processes and anthropogenic activity. This paper is focused on studies of the catalytic processes devoted to the removal of nitrogen oxide from stationary exhaust sources. In principle, primary or secondary methods can be used for NOx emissions abatement. Primary methods aim to prevent NO_x formation during combustion processes. Secondary methods reduce the concentration of the already formed NO_x. The most commonly used secondary methods include selective non-catalytic reduction (SNCR) and selective catalytic reduction (SCR) [2,3]. The SNCR proceeds without the presence of a catalyst, but requires high reaction temperatures (above 850 °C) to meet the high efficiency of the process. Such conditions can be achieved in laboratories, but it is not easy to meet them at real industrial conditions, because there are many factors influencing them, and, thus, it is difficult to maintain the

SNCR process constant [4]. On the other hand, SCR proceeds at much lower temperatures (200–450 °C) but needs the presence of a catalyst. The disadvantage of the SCR method is the need to use a reducing agent, most often ammonia and urea [5]. The use of a reducing agent increases operating costs, and the process may be accompanied by unwanted emissions of unreacted reducing agents or other reaction by-products. Moreover, in the case of ammonia storage, strict safety measures are required.

The direct decomposition of nitric oxide has been studied and described in some publications [6]. Apart from metals (Pt, Pd, Ag, Rh, Ni, Cu, Mo, Co, Au), metal oxides like Co_3O_4, Fe_2O_3, NiO, CuO, and ZrO_2; lanthanides; perovskites; and mixed oxides were also studied for direct NO decomposition [7–10]. A Co_3O_4 catalyst alkalized by a small amount of Li, Na, K, Rb, or Cs showed higher NO conversion than non-promoted Co_3O_4. Alkali metals were added by the impregnation method, and from all investigated promoters, potassium showed the best results [11].

Alkali metals are often used in the preparation of heterogeneous catalysts for their ability to modify structural and/or the electronic properties of metal oxides [12,13]. It is known [14] that potassium adsorption on metal surfaces results in the dominant lowering of the surface potential on sites adjacent to a potassium atom and a small, but still significant, lowering of the potential on sites located further away.

As shown in [5], the method of catalyst preparation affects its activity and stability. Pacultova et al. [15] found out that potassium modified Co–Mn–Al mixed oxide catalysts prepared from hydrotalcite-like precursors could efficiently catalyze the decomposition reaction, as it proceeds successfully in the temperature region of 600–700 °C. The durability of the activity of the catalyst can be influenced by stability of potassium in the catalysts at the applied reaction conditions. The co-precipitation of metal nitrates by a Na_2CO_3/NaOH water solution and subsequent impregnation of the resulting wet cake by KNO_3 solution (the K-impregnation of wet cake is labeled as BP) led to a higher activity and stability of the catalyst than the other examined methods, like the impregnation of the calcined Co–Mn–Al precursor with a solution of KNO_3 or the calcination of corresponding metal nitrates. A rather complicated preparation procedure is main disadvantage of the preparation method consisting in metal salt co-precipitation by solutions of Na compounds and following the impregnation of a cake with K salts.

In the paper of Yongzhao Wang et al. [16], cobalt–cerium mixed oxides modified by K were prepared by the co-precipitation of a metal nitrates solution with an aqueous solution of K_2CO_3 and/or KOH, and the catalysts were used in a deN_2O process.

We believe that the incorporation of potassium into the structure of catalyst particles would affect catalytic activity and stability in direct NO decomposition. The main objective of our study was the enhancement of the NO catalytic decomposition rate through the optimization of the preparation methods. Results obtained in this study have confirmed that the precipitation of Co, Mn, and Al nitrates with a K_2CO_3/KOH solution is a promising preparation method that provides a uniform distribution of the promoter (potassium) in the catalyst particles, appropriate catalyst basicity, and reducibility and slightly higher catalytic activity in direct NO decomposition.

2. Results

In this chapter, the authors describe the experimental results obtained with the Co–Mn–Al mixed oxide catalysts containing variable concentrations of potassium prepared by the gradual washing of precipitates of Co, Mn, and Al nitrate solutions with a K_2CO_3/KOH solution, and they compare them with the catalyst prepared by the K-impregnated wet cake Co–Mn–Al mixed oxide catalyst. The precursors and the catalysts were characterized by chemical analysis, differential thermogravimetry (DTG), XRD, N_2 physisorption, FTIR, H_2-TPR, CO_2-TPD, XPS, and the thermal alkali desorption method (SR-TAD), and their catalytic activity was evaluated in direct decomposition of NO.

2.1. Properties of the Dried Samples

Samples of dried Co–Mn–Al catalyst precursors modified with various amount of potassium were prepared by washing the precipitates. The process of precursor washing and filtration is described in detail in Supplementary Table S1. The wet cake of the catalyst precursor obtained by precipitation was gradually washed with demineralized water to obtain five samples with decreasing concentrations of potassium. One fifth of the original cake was removed and left to dry, and the remaining part of the cake was washed with 200 ml of water and filtered again. The procedure was repeated four times to obtain samples "1" to "5" with a gradual decrease of K concentration. In addition to potassium, small amounts of Na was found in the filtrates, most likely originating from the initial compounds used for catalyst preparation.

2.1.1. Phase Composition

X-ray diffraction is the main technique for the investigation of bulk structure, which is very important, as many of the catalyst characteristics depend on it. XRD measurements were used to characterize the dried catalysts precursors. The diffraction patterns of the precipitated samples dried at 60 °C confirmed the presence of a hydrotalcite-like compound (H) with peak maxima at 13.3, 27.0, 40.2, 41.3, 45.6, 70.9, and 72.9° (ICDD PDF-2 card No. 01-070-2151) and manganese carbonate (rhodochrosite R) with peaks maxima at 28.1, 36.6, 43.8, 48.4, 53.0, 58.4, 60.7, and 80.6° (ICDD PDF-2 card No. 00-007-0268) (Supplementary data, Figure S1). The gradual washing of the precursors and therefore, the decreasing K concentration, led to the continuous crystallization of the hydrotalcite-like compounds, manifesting itself in higher peaks of hydrotalcite. A slightly lower intensity of the basal (003) and (006) diffraction patterns (detected in the range of 2Theta from 10 to 30°) in comparison with non-basal patterns can indicate a lower concentration of preferably orientated layered double hydroxide (LDH) crystals. Apart from hydrotalcite, a K compound (KNO_3) at 2Theta 21.95, 34.07, 37.7, 39.55, 48.12, and 54.5 (°) was detected in two samples ("1" and "2") with the highest K concentration. Though K_2CO_3/KOH was used as the precipitant, KNO_3 was formed during precipitation in the solution by the recombination of ions (metal nitrates were used for catalyst preparation). A decreasing concentration of K in the samples led to the gradual diminution of peaks characteristic for K components in the diffractogram. Nevertheless, a small part of the most intensive peak at 2Theta = 54.5 (°) remained preserved, except for the "5" sample, in which a very low concentration of K was determined.

2.1.2. TG and DTG Measurements

Themogravimetry (TG) and differential thermogravimetry (DTG) measurements of the dried "1"–"5" samples showed (Table 1) three areas of maximum weight decrease manifesting in the ranges of 113–136, 193–213, and 448–558 °C. With decreasing K concentration in the mixed transition metal precursors, the position of the peaks shifted slightly to lower temperatures.

The weight change between 193–213 °C was the most significant (ca 13.5–18%). A somewhat smaller weight decrease (9.5–13.9%) was observed in the stage appearing at the lowest temperatures (113–136 °C). The weight loss in the high temperature region (438–558 °C) was the smallest and varied from 1.5 to 3.7%.

Table 1. Results of thermogravimetry (TG) and differential thermogravimetry (DTG) measurements of the dried samples.

Sample	DTG_1 T_{max}, °C	DTG_2 T_{max}, °C	DTG_3 T_{max}, °C	ΔTG_1 %	ΔTG_2 %	ΔTG_3 %
"1"	113	201	558	−9.5	−17.4	−3.7
"2"	131	206	Nd [a]	−11.9	−14.8	−2.7
"3"	132	201	438	−11.6	−14.0	−3.2
"4"	136	213	445	−13.9	−18.4	−1.5
"5"	136	193	448	−13.7	−13.5	−3.6

[a] Not detected.

To identify the processes taking place in various temperature regions during the temperature treatment of catalyst precursors, an analysis of effluents by mass spectroscopy was carried out during calcination. Three selected precursors differing in K concentration were calcined in helium (0.1 g, temperature ramp 20 K min^{-1}), and the results are shown in Figure 1. Desorption of H_2O was finished at ca 500 °C, with temperature maxima at ca 187 and 250 °C.

The decomposition of nitrates in the catalysts proceeds in two steps, manifesting as a presence of NO and NO_2 in the gas phase. A higher concentration of NO than NO_2 was observed for the catalysts with a K concentration higher than 1 wt.%. Carbonates present in the precursors were decomposed in three stages, the first one (at ca 140 °C) being of lowest intensity. The course of the two following regions (at ca 250 and 400 °C) changed in dependence on K concentration: With a lower K concentration, the temperature maxima shifted slightly to higher temperatures. It is very likely that oxyhydroxides decompose first, and K_2CO_3, which remains in the catalysts as a residue of the precipitating agent, decomposes at about 400 °C. The positions of all temperature peak maxima are summarized in Table 2.

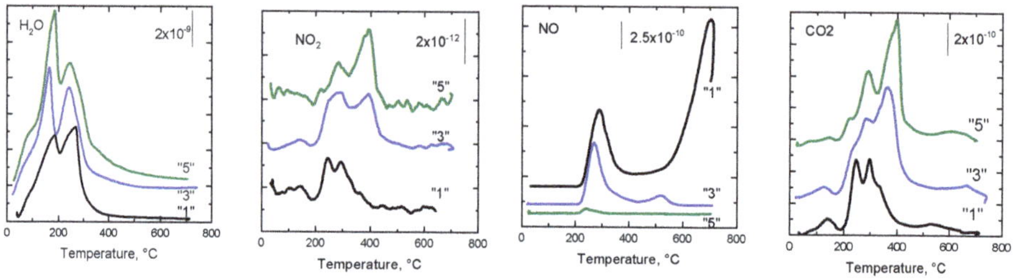

Figure 1. Composition of effluents in A observed by mass spectroscopy during calcination of the "1", "3", and "5" samples in helium (0.2 g of sample, 20 K min^{-1}).

Table 2. Positions of the temperature peak maxima (°C) observed during calcination of the selected precursors in helium.

Sample	H_2O	NO_2	NO	CO_2
"1"	184; 269	126; 242; 293	287; 702	144; 249; 297; 535
"3"	164; 245	139; 274; 400	272; 520	136; 289; 366; 668
"5"	95; 187; 245	106; 281; 400	242	148; 229; 290; 400; 628

It can be concluded that the first peak in the DTG measurements can be ascribed to the removal of water, the second one to the decomposition of carbonates and nitrates present in the metal oxyhydroxides, and the third one to the decomposition of carbonates and nitrates present in the precipitates as compounds.

2.2. *Properties of the Calcined Precursors*

2.2.1. Chemical Composition

The chemical compositions of the catalysts calcined for 4 h at 500 °C are shown in Table 3. Potassium concentration gradually decreased due to washing from 18.9 to 0.6 wt.%. All catalysts, except the one with the highest K concentration, showed molar ratios closely approaching that of the average formula Co:Mn:Al = 4:1.4:1.3. However, the molar ratio of K to Co decreased from the value of 4 for sample "1" to 0.1 for sample "5". The molar ratio of the K-impregnated wet cake catalyst (BP) approached that of the sample "3".

Table 3. Composition of the samples calcined 4 h at 500 °C determined by chemical analysis and molar ratios of the metals.

Sample	Wt.%				Molar Ratio
	K	Co	Mn	Al	Co:Mn:Al:K
"1"	18.90	28.2	6.0	3.0	4:0.9:0.9:4.0
"2"	8.20	40.0	8.8	4.0	4:1.3:1.2:1.7
"3"	2.39	45.4	10.2	4.7	4:1.5:1.5:0.5
"4"	0.92	46.8	10.2	4.7	4:1.6:1.5:0.2
"5"	0.60	45.6	9.9	4.4	4:0.9:1.4:0.1
BP	2.10	48.9	11.2	5.1	4:0.9:1.5:0.4

2.2.2. Phase Composition

The XRD patterns (Figure 2) of the precursors calcined for 4 h at 500 °C exhibited a Co_3O_4 spinel-like structure (labelled as S), which was dominant in all examined samples. Diffraction positions at 21.9, 36.3, 42.7, 52.3, 65.4, 69.8, and 77.2° and the intensity of the peaks are in agreement with the data published for this system (ICDD PDF-2 card No 01-074-1656). Traces of cryptomelane KMn_8O_{16} (ICDD PDF-2 card No 00-006-0547), labelled as C in the diffractograms overlay, seemed to be present in all calcined samples, especially in those having higher K concentration (the "1", "2", and "3" samples). Qiuhua Zhang et al. [17] identified cryptomelane in the calcined precursor formed by a reaction of $KMnO_4$ and maleic acid at moderate temperatures (300–600 °C). At temperatures above 600 °C, cryptomelane usually transforms into bixbyite (Mn_2O_3), but in the prepared samples, such compound was not detected. Except for the above mentioned phases, the peaks at 27.26, 39.17, 48.0, and 54.65 appeared in diffractograms that were ascribed to potassium nitrate (KNO_3) in trigonal crystallographic modification (ICDD PDF-2 card No 01-078-7937). A somewhat lower concentration of the orthorhombic modification of KNO_3 was registered at 24.3, 31.55, 34.29, 38.0, 45.79, 50.37, 61.25, 63.91, 79.37, and 81.71° (ICDD PDF-2 card No 01-074-2055) in samples "1" and "2", which were washed only partly. It is also obvious from diffractograms that the examined samples with a higher stage of washing showed better crystallinity than those only partly washed.

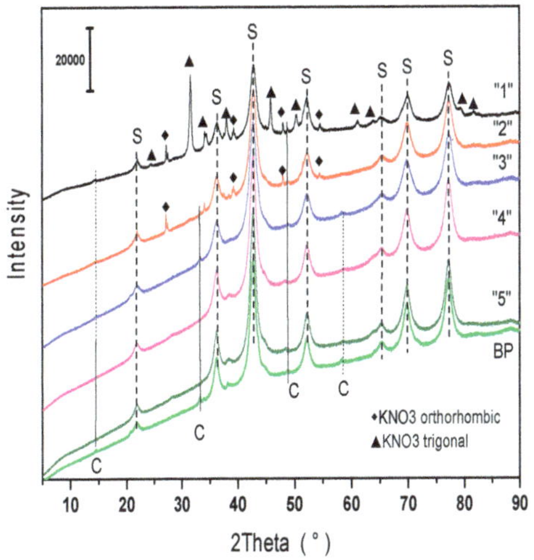

Figure 2. Diffraction patterns of the calcined K-precipitated Co–Mn–Al precursors containing various amounts of potassium. S—spinel, ▲—KNO_3 trigonal, ♦—KNO_3 orthorhombic, C—KMn_8O_{16} cryptomelane.

The effect of potassium concentration in the Co–Mn–Al oxides on spinel lattice parameter *a* and spinel coherent domains L_d can be seen in Table 4 and Figure 3, respectively. The crystallite size of the spinel was calculated using Scherrer's formula from the following diffraction lines (220), (311), (400), (511), and (440). The calculated value was corrected to the instrumental broadening (NIST SRM 660b). Recently, Kovanda et al. [18] described the behavior of potassium non-modified Co–Mn–Al mixed oxides during calcination. They found out that the decomposition of Co–Mn–Al layered double hydroxides above 200–260 °C proceeds with the formation of nanocrystalline spinels, and spinels are the only phases present up to 900 °C. At a temperature of about 500 °C, the segregation of the Co-rich spinel was observed, while the incorporation of manganese into the spinel lattice proceeded during its further recrystallization, and this process was accompanied by a lattice parameter decrease. The presence of potassium in Co–Mn–Al oxides could also affect the lattice parameter *a* of the spinel: One can imagine that in case of no potassium or just a small amount of it in the oxides, manganese enters into the spinel lattice during calcination and decreases the lattice parameter due to the partial distortion of the Co_3O_4 lattice. With an increasing amount of potassium in the sample, more manganese is bonded to potassium, and, therefore, manganese cannot enter the Co spinel lattice, which is why the spinel lattice parameter *a* could return to the values characteristic of the Co_3O_4 phase [19]. However, it is obvious from Table 4 that the presence of the increasing amount of potassium in the K-precipitated Co–Mn–Al oxides caused only small changes (±0.001 nm, less than 0.2 rel.%) in the lattice parameter of the spinel.

The dimension of the spinel coherent domain L_d appears to decrease from 9 to 6 nm when the K concentration increases to 2.4 wt.%. With a further increase of K concentration in the solids, the spinel coherent domain started increasing and reached the value of 9 nm at a K concentration equal to 18.9 wt.% (Figure 3). The course of the dependence seems to correspond to the changes in spinel lattice parameter, but the differences among L_d parameters were very small—the observation of the minimum in the dependence could be coincidental.

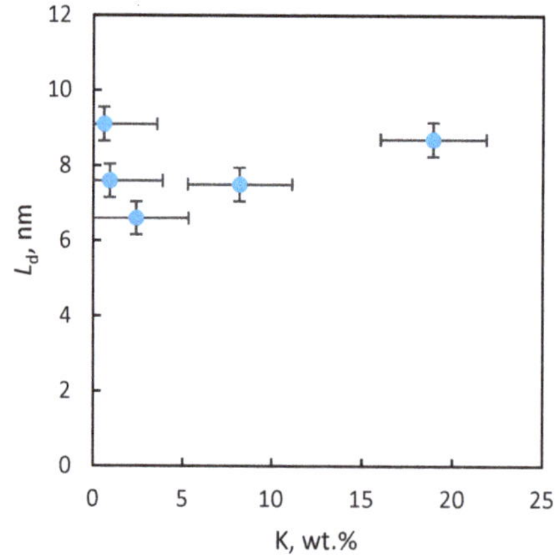

Figure 3. Dependence of spinel coherent domain L_d on concentration of K in the K-precipitated Co–Mn–Al mixed oxides catalysts.

2.2.3. Textural Properties

The porous structure of the K-precipitated catalysts was characterized by the adsorption/desorption of nitrogen at −196 °C, and the results are summarized in Table 4. The surface area of the catalysts S_{BET} with a concentration of potassium lower than 2.4 wt.% was about 100 m^2 g^{-1}. A higher concentration of potassium in the samples caused a diminution of surface area; the smallest surface was exhibited by the catalyst with 18.9 wt.% K ($S_{BET} < 1$ m^2 g^{-1}). Accordingly, total pore volume increased with the decreasing K concentration in the solids from 266 to 435 $mm^3{}_{liq}/g$. The volume of micropores was small in all cases (less than 5% of total pore volume). The experimental data document that the concentration of potassium in the catalyst higher than 8 wt.% had a negative effect on the porous structure of metal oxides. The phenomenon is likely connected with the collapsing of smaller pores and coalescence of larger pores.

Table 4. Characteristic physical–chemical data of the samples calcined 4 h at 500 °C.

Sample	S_{BET} (m^2/g)	S_{meso} (m^2/g)	V_{tot} ($mm^3{}_{liq}/g$)	V_{micro} ($mm^3{}_{liq}/g$)	L_d (nm)	a (nm)
"1"	<1	–	–	–	9	$0.81056 \pm 7 \times 10^{-5}$
"2"	58	32	266	13	8	$0.81187 \pm 17 \times 10^{-5}$
"3"	110	69	401	22	7	$0.81109 \pm 5 \times 10^{-5}$
"4"	105	66	435	21	8	$0.81209 \pm 4 \times 10^{-5}$
"5"	102	66	435	21	9	$0.81153 \pm 4 \times 10^{-5}$
BP	94	Nd. *	460	64	9	$0.81015 \pm 4 \times 10^{-5}$

* Not determined.

The adsorption–desorption isotherms of nitrogen (Figure S2 in Supplementary data) obtained over calcined catalysts confirm that there was not a large difference in pore size distribution among the catalysts with a potassium concentration lower than 2.4 wt.%.

2.2.4. Reducibility

The reducibility of the catalysts with different contents of potassium was studied by H_2-TPR in the temperature range of 25–900 °C. The H_2-TPR patterns of the K-precipitated catalysts are depicted in Figure 4. The catalysts were reduced in two main temperature regions, 200–500 °C and >500 °C. The low-temperature reduction peak represents the reduction of $Co^{III} \rightarrow Co^{II} \rightarrow Co^0$ in the Co_3O_4-like phase and the reduction of Mn^{IV} to Mn^{III} oxides. The reduction of $Mn^{III} \rightarrow Mn^{II}$ can take place in both temperature regions [19]. The reduction of the K containing compounds, like cryptomelane KMn_8O_{16} or others, also cannot be excluded in the low temperature region since the reduction of those species proceeds at temperatures between 200–450 °C [20–23]. In literature, the high temperature peak was ascribed to the reduction of Co and Mn ions surrounded by Al ions in the spinel-like phase [24] and to the reduction of Mn^{III}. The effect of K amount on reducibility in the fresh samples, calcined at 500 °C, is shown in Figure 4. The modification of the Co–Mn–Al mixed oxide by potassium caused significant changes in the obtained reduction profiles. With an increasing K content, a shift to lower temperatures accompanied by the broadening of the peak was observed. A new low temperature peak appeared as a shoulder at around 300 °C. The catalyst with the highest concentration of K (sample "1") was completely reduced in the temperature region 200–500 °C, while all the other catalysts showed a high-temperature reduction peak with a maximum between 614 and 755 °C. The temperature 755 °C was determined for the catalyst with the lowest K concentration. The position of the temperature peak maximum shifted to lower temperatures with increasing K concentration, reaching the value of 614 °C for sample "2". Thus, the positions of T_{max} of the reduction peaks in H_2-TPR (Table 5) reflected the concentration of K in the catalysts: Increasing the concentration of K (except the highest concentration 18.9 wt.%) led to the easier reduction of the metal mixed oxides. The reduction profile of the catalyst prepared by modified by impregnation of wet cake (BP) was similar as that of the sample "5".

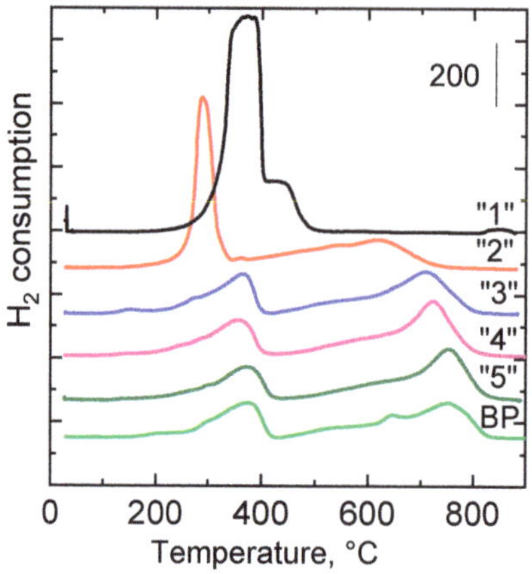

Figure 4. H_2-TPR profiles of the samples with different amount of K, calcined for 4 h at 500 °C.

Table 5. Positions of the peak maxima in TPR curves and consumptions of hydrogen and CO_2 in H_2-TPR and CO_2-TPD measurements, respectively.

Sample	H_2-TPR T_{max}, °C	H_2-TPR [a] mmol g^{-1}	CO_2-TPD T_{max}, °C	CO_2-TPD [a] mmol g^{-1}
"1"	376; 438 [b]	18.24	-	0.03
"2"	290; 614	12.62	120; 275; 444	0.18
"3"	362; 709	6.61	101; 203; 350	0.16
"4"	368; 726	6.29	101; 202	0.09
"5"	371; 756	5.67	101; 202	0.10
BP	375; 755	6.82	105; 277; 630	0.16

[a] in the range of 25–650 °C, [b] shoulder.

2.2.5. Basicity

Catalyst basicity is an important factor influencing the chemical reactivity of NO because of the acidic nature of the NO molecule [25]. In the CO_2-TPD profiles of mixed oxides catalysts, several types of basic sites can be recognized. Weak basic sites are attributed to –OH groups occurring on the surface of the catalyst, medium sites consist of oxygen bonded to metal as Me^{2+}-O^{2-} or Me^{3+}-O^{2-} pairs, and strong basic sites are assigned to isolated O^{2-} anions [26]. Individual peaks below 140 °C can be ascribed to weak basic sites, the peaks appearing in the range of 140–220 °C can be assigned to medium basic sites, and the peaks above 270 °C can be assigned to strong basic sites.

TPD-CO_2 measurements were performed with the aim to find differences in basicity of the fresh catalysts calcined at 500 °C. The effect of potassium content on the amount of basic sites is shown in Figure 5, where the course of CO_2 desorption was recorded on a mass spectrometer (mass contributions m/z = 16 were collected). To avoid damaging the mass spectrometer detector, the temperature rise was finished at about 650 °C. The CO_2-TPD profiles of the examined samples represent all types of basic sites—weak, medium, and strong—but a strict separation was not possible—even more than three types of sites can be noticeable in some cases. The catalyst with the lowest concentration of K (sample "5") contained weak and medium basic sites with some amount of very strong basic sites with a temperature maximum at 600 °C. The "4" catalyst, having a slightly higher concentration of K (0.9 wt.%) than sample "5", exhibited very similar profiles of CO_2 desorption. A further increase in K concentration (2.4 wt.%) led to a higher amount of medium basic sites with a maximum desorption at 220 °C (Table 5), but stronger basic sites corresponding to CO_2 desorption at about 350 °C also appeared in the catalyst. An even higher concentration of potassium in the catalyst (8.2 wt.%) resulted in a substantial increase of the amount of medium and stronger basic sites desorbing at 275 and 444 °C, respectively. The catalyst with very a high concentration of potassium (18.9 wt.%) exhibited quite different CO_2 desorption profiles: Weak and medium basic sites completely disappeared, and only very strong basic sites remained. The BP catalyst prepared by the K-impregnation of wet cake method (2.1 wt.% K) exhibited a similar amount of basic sites in the range of 25–650 °C as sample "3", having 2.4 wt.% K (Table 5). However, the CO_2 desorption course of the BP catalyst differed slightly from those of the other catalysts (Figure 5), as it showed a higher amount of stronger and very strong basic sites characterized by CO_2 desorption at T_{max} of about 277 and 630 °C, respectively.

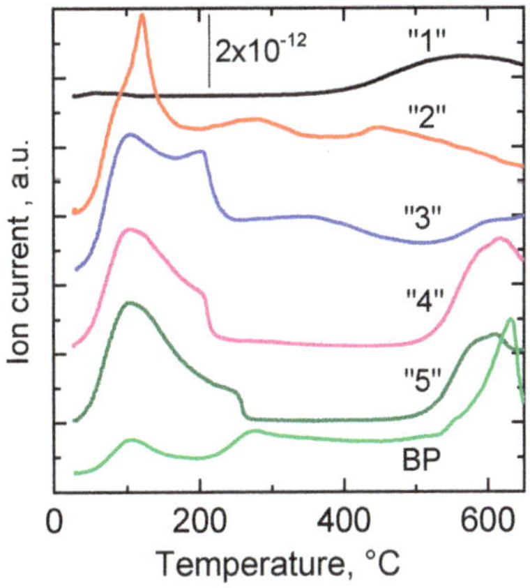

Figure 5. CO_2-TPD desorption profiles (in A) of the catalysts (calcined for 4 h at 500 °C) containing various amounts of K.

2.2.6. FTIR

Fourier transform infrared spectroscopy (FTIR) can identify various kinds of molecular bonds present in solid samples. The FTIR data of both the examined dried Co–Mn–Al–K hydrotalcite-like precursors and their calcined analogs are shown in Figure 6. In the FTIR spectra of the dried samples, a band at 1355 cm^{-1} dominated, which is ascribed, according to literature [27], to characteristic CO_3 antisymmetric stretching mode vibration. Its presence in the spectra, together with the presence of the OH stretching mode band at 3435 cm^{-1} (not shown) confirms the formation of hydrotalcite-like compounds during precipitation. In all dried samples, there was also the vibration at 860 cm^{-1}, which corresponds, according to literature, also to the K_2CO_3 out-of-plane bending mode [27]. In the case of sample "1", a band at 822 cm^{-1} corresponded to KNO_3 and/or $Al(NO_3)_3$ out-of-plane bending mode. A very wide band appeared at 738 cm^{-1}, which can indicate a metal-to-oxygen bond of very small particles. The band can be ascribed to Al–O vibrations in AlOOH, which could be also formed during metal nitrates precipitation. In the region below 600 cm^{-1}, other metal–oxygen bonds also occurred that cannot be ascribed with certainty.

The FTIR spectra of the calcined samples show unique bands at 657 cm^{-1} and approximately 552 cm^{-1}. Both bands can indicate the presence of Co_3O_4, MnO_2, Mn_3O_4 [28], or cryptomelane [29]. Potter et al. [29] attributed a band at 550 cm^{-1} unambiguously to cryptomelane. Though Co_3O_4 shows the closest IR spectrum to those of analyzed samples, the presence of cryptomelane cannot be excluded. Samples "1" and "2", with the highest concentration of K, also exhibited KNO_3 vibrations. By comparing the spectra with references, the presence of other substances in the catalysts was excluded. KNO_3 bands split into pairs 824 and 1372 cm^{-1} and pairs 833 and 1347 cm^{-1}, respectively, the first pair being more tightly bound, i.e., worse washed out. This splitting of the bands could be explained by the different structure of KNO_3 (calcite—trigonal and aragonite—orthorhombic). The aragonite structure arises from the violation of symmetry (for example by adsorption, change of environment, or recrystallization), thereby increasing the intensity of the originally inactive vibrations [30]. The calcined BP catalyst prepared by bulk promotion exhibits, in principle, the same IR spectrum as the K-precipitated catalyst having nearly the same K concentration (sample "3"). However, it showed slightly higher amount of nitrates (bands at about 1350 and 830 cm^{-1}).

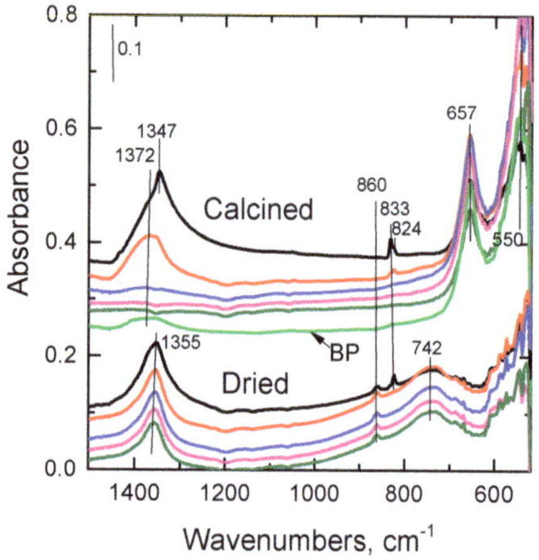

Figure 6. FTIR data of the dried and calcined K-precipitated Co–Mn–Al–K mixed oxides and K-impregnated wet cake catalyst (BP) catalyst. Curves from top to bottom: Samples "1", "2", ... , "5".

2.2.7. Surface Composition

The surface compositions in the near-surface region and chemical state of the elements of the fresh K-precipitated catalysts were determined by XPS. As the catalytic measurements proceeded at reaction temperatures 650–700 °C, the catalysts were calcined at 700 °C, and, therefore, the properties of the catalysts' surfaces calcined at 700 °C were examined. The surface concentrations of the elements were determined from the intensities (peak areas) divided by the corresponding response factor [31]. Carbon tape used for fixing of the samples to the holder manifested itself in a relatively high concentration of C (28–35 at. %) (Table 6). Nevertheless, the calibration of the spectra was carried out according to carbon (284.8 eV). In the fresh K-precipitated samples calcined at 700 °C, 44 at. % of O, 8.7 at. % of Co, 3.3 at. % of Mn, 7.2 at. % of Al, and 3.3 at. % of K—in average—were determined (Table 7). The gradual washing of the precursor with distilled water resulted in a progressive decrease of K concentration and, at the same time, in the enhancement of other metal elements in the catalysts. In all catalysts, manganese occurred in a higher oxidation state, very likely as Mn^{+4}, most likely MnO_2. Aluminum looked very similar in all samples and was probably Al_2O_3, though the position of the peak is shifted slightly in comparison with published data. In all samples, potassium occurred in an identical form, very likely bounded to metal oxide ($KMnO_4$, KMn_8O_{16}, $KCoO_2$, or a similar compound).

The deconvolution of oxygen spectra revealed two peaks (Table 6) with binding energies of about 529.8 and 531.5 eV. The first one at 529.8 eV can be ascribed to metal oxide (lattice oxygen O^{2-}), and the second at 531.5 eV can be ascribed to the adsorbed surface oxygen bound to metal oxides as O_2^-, O^-, or OH^- species.

In contrast to the K-precipitated catalysts, the K-impregnated wet cake (BP) catalyst calcined at 700 °C exhibited 54 at. % of O, 13.7 at. % of Co, 6.1 at. % of Mn, 8.1 at. % of Al, 5.2 at. % of K, and 12.5 at. % of C (Table 6). It is obvious that the surface Co concentration of the BP catalyst was the highest of all examined catalysts, and the K concentration was the second highest. The distinctions are certainly a result of different preparation procedures of the examined catalysts.

Table 6. Concentration of elements [at. %] in both the BP catalyst and the K-precipitated catalysts calcined at 700 °C and the relative quantum of two forms of oxygen in the decomposed of O 1s peak.

Sample	Co 2p At. %	Mn 2p At. %	Al 2p At. %	C 1s At. %	K 2p At. %	O 1s At. %	O 1s eV (rel. %)	O 1s eV (rel. %)
"1"	6.8	2.4	5.1	28.7	10.1	46.9	529.7 (45.8)	531.5 (54.2)
"2"	9.5	2.9	6.2	34.8	3.6	43.1	529.7 (63.5)	531.5 (36.5)
"3"	9.1	4.0	7.0	34.1	2.4	43.5	529.7 (67.8)	531.5 (32.2)
"4"	9.0	3.5	8.6	35.9	0.5	42.7	529.8 (56.7)	531.4 (43.3)
"5"	9.1	3.7	9.2	32.5	0.1	45.3	529.9 (59.1)	531.5 (40.8)
BP	13.7	6.1	8.1	12.5	5.2	54.4	530.0 (84.5)	531.7 (15.5)

The surface and bulk concentrations of the main catalyst components (in wt.%) are summarized in Table 7. A comparison of both groups of values indicates that the surface of the catalyst prepared by calcination of the non-washed K-precipitated catalyst precursor was substantially enriched by aluminum (2.5 times) and only slightly enriched by manganese and potassium (in both cases 1.2 times) at the expense of cobalt, which was reduced 0.8 times. The gradual washing of the precipitates by water led to the further enrichment of the catalyst surface by aluminum (3.3 times) at the expense of potassium. Manganese and cobalt surface concentrations practically did not change with the washing of the precipitates. Potassium concentration on the surface and in the bulk also practically did not differ with washing, which indicates the relatively high stability of potassium in the K-precipitated catalyst at calcination temperatures around 700 °C.

Somewhat different relation of surface and bulk composition can be observed for the BP catalyst. The surface concentration of K was more than four times higher than the bulk, while the surface concentration of cobalt was about two thirds of the bulk. Additionally in this sample, surface is enriched by aluminum. The data indicates that the addition of KNO_3 to the wet cake of Co–Mn–Al hydrotalcite-like led to the preferential adsorption of K to Co.

Table 7. Surface concentrations of active components in the calcined catalysts obtained by X-ray photoelectron spectrometry (XPS) in wt.% (in parenthesis there are concentrations of metals in wt.% determined by chemical analysis).

Sample	K	Co	Mn	Al	O
"1"	22.1 (18.9)	21.8 (28.2)	7.2 (6.0)	7.6 (3.0)	41.3 (43.9)
"2"	8.2 (8.2)	32.6 (40.0)	9.2 (8.8)	9.7 (4.0)	40.2 (39.0)
"3"	5.5 (2.4)	30.8 (45.4)	12.8 (10.2)	10.9 (4.7)	40.1 (37.3)
"4"	1.1 (0.9)	32.1 (46.8)	11.6 (10.2)	14.0 (4.7)	41.3 (37.4)
"5"	0.3 (0.6)	31.2 (45.6)	11.9 (9.9)	14.4 (4.4)	42.1 (39.5)
BP	8.3 (2.1)	33.1 (48.9)	13.9 (11.2)	9.0 (5.1)	35.7 (32.7)

2.2.8. Species-Resolved Thermal Alkali Desorption (SR-TAD)

The stability of alkali metals on the catalyst surface was studied by species resolved thermal alkali desorption (SR-TAD). The results of two selected samples prepared by two different methods of K promotion and calcined at 700 °C were compared (Figure 7). Since desorption of K atoms was more significant than the potassium desorption in the form of ions, only desorption of alkali metal atoms (atom fluxes) during heating was observed.

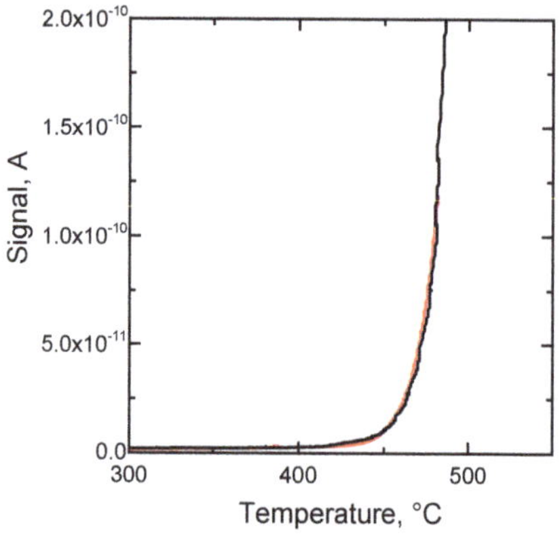

Figure 7. Atomic K desorption flux as a function of temperature during heating. Black line: K-precipitated sample "2", red line: BP catalyst, both calcined prior to measurements at 700 °C/4 h.

It is obvious from the temperature-dependent changes of atomic fluxes (Figure 9) that potassium desorption already occurs at temperatures higher than 450 °C, what means that the desorption of alkali metals is possible at the temperatures used for calcinations and during NO direct decomposition testing. The same non-monotonic curves in the desorption signal with rising temperature were observed for both samples, which indicates that the potassium promoter leaves the surface through a single energy barrier, and the distribution of the promoter on the catalysts surface here was homogenous [32]. Since the dependence was the same for both differently prepared samples, the resulting potassium surface state should be the same in both samples.

Interestingly, the K atomic desorption flux profiles were different for heating and cooling periods (not shown) for both samples. A slightly higher desorption flux during cooling was observed. A higher signal during cooling than during heating means that some potassium was segregated on the catalyst surface during heating. Different flux profiles obtained during heating and cooling point to potassium migration processes which took place during the measurement—diffusion inside the bulk, agglomeration on the surface, and thermal desorption [33]. These processes probably also take place during the catalytic reaction.

2.2.9. Activity of the Catalysts in Decomposition of NO

The effect of potassium amount on the catalytic activity of K promoted Co–Mn–Al mixed oxides in direct NO decomposition was studied over the catalysts prepared by coprecipitation of metal nitrates with a K_2CO_3/KOH solution and compared with the activity of the catalyst prepared by the K-impregnation of wet cake. Catalytic activity was determined over the catalysts calcined for 4 h at 700 °C. We found that activity of all samples slightly increased with the increasing time-on-stream. After 13 h, the activity reached stable values. In this initial period of catalyst testing, potassium was likely migrating from the interior of the catalyst grains to their surface. The steady state NO conversions obtained over the catalysts with various K concentrations at 650 and 700 °C are depicted in Figure 8 in comparison with that of the catalyst BP prepared by the K-impregnation of wet cake [19]. The highest conversion of NO (ca 61%) was observed at 700 °C in the presence of the catalyst containing about 8 wt.% K prepared by the coprecipitation of metal nitrates with a K_2CO_3/KOH solution. Lower and higher K content led to the decrease of NO conversion, and, for K content lower than 0.6 wt.%, NO catalytic decomposition was not observed at all. The promotional effect of K has already been well described in the case of N_2O catalytic decomposition over K-promoted Co–Mn–Al mixed oxides [34]. Potassium, due to its low ionization potential, enables the charge transfer to the transition metal cations, inducing an electric field gradient at the surface generated by the resulting dipole and the modification of the density of states characteristics. When K content was higher than the optimal amount, depolarization occurred, leading to the decrease of NO conversion.

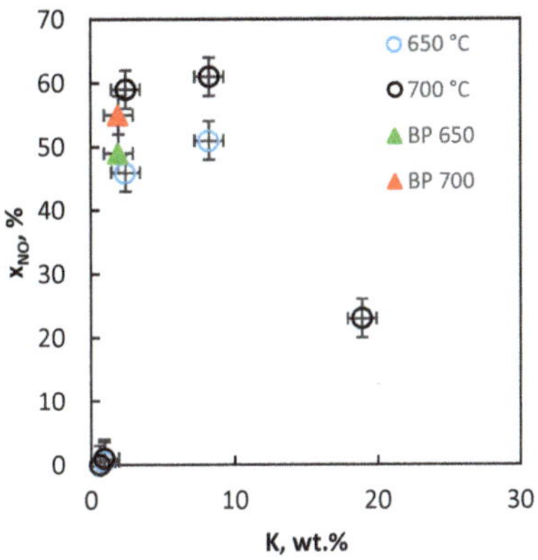

Figure 8. Dependence of nitric oxide (NO) conversion in direct NO decomposition over K-precipitated Co–Mn–Al oxide (circles) and K-impregnated wet cake (BP—triangles) Co–Mn–Al oxide catalysts on the concentration of potassium for two reaction temperatures. Reaction conditions: 1000 ppm NO balanced by N_2, T = 650 or 700 °C, WHSV = 6000 mL g^{-1} h^{-1}.

2.2.10. Effect of O_2 and CO_2 in the Reactant on Catalytic Activity

As flue gases from power plants contain some amount of other constituents, not only NO, the effect of oxygen or CO_2 in the reactant on catalytic activity was studied with the most active sample "2" at reaction temperature 700 °C. The concentration of both components varied gradually from 1 to 10 molar %. The obtained results are depicted in Figure 9. The effect of both added reaction components was similar: The highest decrease in activity was observed after the addition of their lowest concentration (1 mol %)—however, the effect on initial catalytic activity decrease differed: The presence of O_2 caused a slower decrease in activity than CO_2. However, higher concentrations of O_2 or CO_2 (about 4 mol %) showed a nearly identical decrease in catalytic activity. Very likely, the adsorption of both components either blocked adsorption of NO or slowed down desorption of N_2 and O_2.

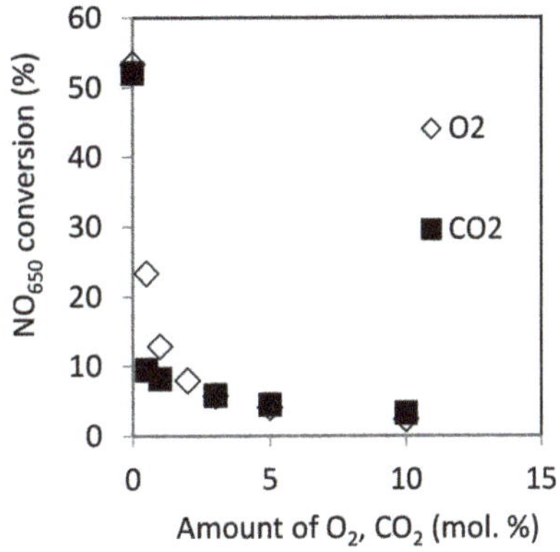

Figure 9. Dependence of NO conversion in direct decomposition at 650 °C on the amount of O_2 or CO_2 in the reaction mixture for sample "2" having 8.2 wt.% of K. Reaction conditions: 1000 ppm NO balanced by N_2, T = 700 °C, WHSV = 6 000 mL g^{-1} h^{-1}.

3. Discussion

In this paper, a method of the preparation of the potassium promoted Co–Mn–Al mixed oxide catalyst (K-precipitated) was examined. An alternative method of the preparation of potassium promoted Co–Mn–Al mixed oxide consists of the precipitation of metal nitrates with potassium salts (K_2CO_3/KOH) and of the consecutive gradual washing of the precipitate to obtain a specific concentration of K in the calcined catalyst. The application of the preparation procedure led to the catalysts with various concentration of potassium whose physical–chemical properties were studied. Their properties were compared with those of the catalyst prepared by the K-impregnation wet cake method.

As shown in the XRD profiles of the calcined catalysts, both kinds of catalysts with similar K content, the K-precipitated sample "3" and BP, contained identical crystallographic phases of the compounds (Figure 2). However, the dimension of coherent domain of spinel (L_d) was somewhat higher for the BP catalyst (9.3 nm) than for the K-precipitated catalyst with a corresponding K amount (6.6 nm)—see Table 4. This finding reflects a slightly lower surface area and a higher total volume of pores and of micropores of the BP catalyst.

FTIR measurements proved that the BP catalyst comprises more nitrates than the K-precipitated catalyst of corresponding composition (Figure 6). It could not cohere with a slightly lower concentration of potassium in the BP catalyst (2.1 wt.%) than the compared catalyst (2.4 wt.%). The phenomenon had to correspond with the way of KNO_3 addition: In the case of the BP catalyst, KNO_3 was likely deposited on the surface of the primary precipitated particles, while, in the case of the washed K-precipitated catalysts, KNO_3 was included in the structure of the hydrotalcite-like compound formed during the precipitation of metal nitrates with a solution of K_2CO_3/KOH.

The presence of nitrates was reflected in the basicity of the BP catalyst (Figure 5, Table 5), which exhibited a lower amount of stronger basic sites (T_{max} of CO_2 desorption > 220 °C) than the K-precipitated Co–Mn–Al oxide catalyst, with 2.4 wt.% K. Nitrates in the BP catalyst contributed to a higher degree of acidity of the catalyst surface and, thus, to lower basicity.

H_2-TPR did not show any significant difference in the reducibility of both kinds of the compared catalysts. The catalyst with ca 8 wt.% of K exhibited the lowest T_{max} temperatures in the TPR curves of all samples, thus indicating the easiest reducibility of the transition metals mixed oxides. The reduction profile of the catalyst prepared by K-impregnation of wet cake (BP) was similar to that of sample "5", which had a three times lower concentration of potassium and therefore documented the lower redox properties of the BP catalyst.

A comparison of the XPS and chemical analysis data for the K-precipitated and BP catalysts led to the following statement: The K-precipitation method of Co–Mn–Al catalysts made uniform distribution of potassium in the catalyst particles possible, while the K-impregnated wet cake method caused the significant enrichment of the catalyst surface by potassium at the expense of cobalt. That is why the optimum concentration of potassium (determined by chemical analysis), which was necessary to obtain the highest NO conversion, was lower [19] in the case of BP preparation method (about 2 wt.%) than in the case of the K-precipitation method (about 8 wt.%).

The catalytic activity of the K-precipitated Co–Mn–Al oxide catalysts in direct decomposition of NO varied with the concentration of K in the solids. It went through maximum at about 8 wt.% K in the solid and decreased when lower and higher K concentrations were present in the catalysts than this optimum value (Figure 8). The BP catalyst exhibited a slightly lower catalytic activity, caused, very likely, by a slightly lower surface area of the catalyst or lower amount of basic sites on the catalyst surface, which substantially affected the sorption of NO on the catalyst surface and/or a slightly lower reducibility of the catalyst.

In Figure 10, the dependence of NO conversions on number of K atoms per nm^2 of the catalyst is depicted. The optimum number of K atoms on the surface is around 10 atoms per nm^2. With a decrease of this number, the conversion of NO sharply decreased, and, at the value of 1 atom per nm^2 conversion of NO approached zero. On the other hand, the decrease in NO conversion with an increasing number of atoms per nm^2 was not so sharp, and, for that reason, a slightly higher number of K atoms per nm^2 in the case of K concentration fluctuation is better than vice versa.

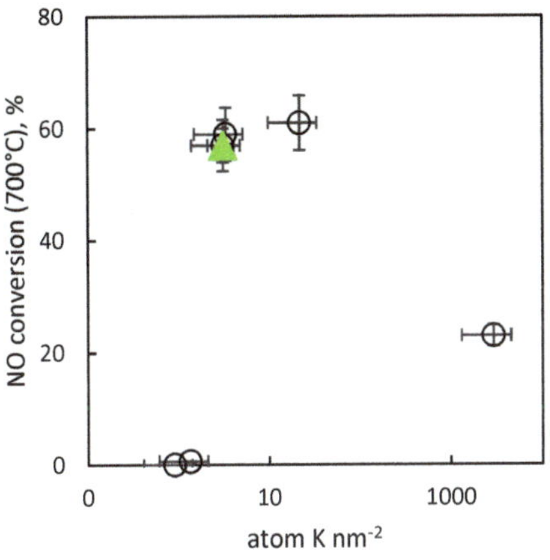

Figure 10. Dependence of NO conversion at 700 °C on the number of K atoms per nm^2 in Co–Mn–Al oxide catalysts. Circles—K-precipitated Co–Mn–Al oxide catalysts, triangles—K-impregnated wet cake Co–Mn–Al oxide (BP) catalyst.

4. Materials and Methods

4.1. Preparation of Catalysts

Co–Mn–Al mixed oxides modified by K were prepared by the co-precipitation of the 210 mL solution of 40 g $Co(NO_3)_2{\cdot}6H_2O$, 8.7 g $Mn(NO_3)_2{\cdot}H_2O$, and 12.9 g $Al(NO_3)_3{\cdot}9H_2O$ with the 225 mL aqueous solution of 14.5 g K_2CO_3 and 29.7 g KOH. The solutions were dosed to a vessel equipped with a magnetic Triga stirrer (500 revolutions min^{-1}), in which 100 mL of water were mixed with such an amount of KOH that the solution maintained a pH = 10. The rate of dosing of each solution was ca 1.9 mL/min, the pH of precipitation 10 ± 0.1, and the temperature was 22 °C. The resulting suspension was mixed for 60 min at laboratory temperature.

Then, the suspension was filtered off, and the precipitate was washed with ca 200 mL of water after removing ca one fifth of the precipitate to obtain different levels of K concentration in the samples. The washed samples were dried for 12 h at 60 °C and calcined at 500 °C. Numbers "1" to "5" labeled the prepared catalysts, with sample "1" having the highest concentration of K and sample "5" having the lowest.

For comparison, a catalyst prepared by the co-precipitation of transition metal nitrates by Na_2CO_3/NaOH solution and the subsequent impregnation of the resulting wet cake by a solution of KNO_3 (K-impregnated wet cake method) was labeled, after drying and calcination, as BP.

4.2. Characterization of the Samples

The content of metals in the prepared catalysts was determined by atomic emission spectroscopy with microwave plasma using an Avio 500 MP-AES (Perkin-Elmer, Chichester, UK) after the dissolution of the samples in diluted (2%) hydrochloric acid.

Phase composition and microstructural properties were determined using an X-ray powder diffraction (XRD) technique. XRD patterns were obtained using a Rigaku SmartLab diffractometer (Rigaku, Tokyo, Japan) equipped with a D/teX Ultra 250 detector. The source of X-ray irradiation was a Co tube (CoKα, λ_1 = 0.178892 nm, λ_2 = 0.179278 nm) operated at 40 kV and 40 mA. Incident and diffracted beam optics were equipped with 5° Soller slits; incident slits were set up to irradiate a 10 × 10 mm area of the sample (automatic divergence slits) constantly. The slits on the diffracted beam were set up to a fixed value of 8 and 14 mm. The powder samples were gently grinded using an agate mortar before analysis, pressed using microscope glass in a rotational sample holder, and measured in reflection mode (Bragg–Brentano geometry). The samples were rotated (30 rpm) during the measurement to eliminate the preferred orientation effect. The XRD patterns were collected in a 2θ range 5°–90° with a step size of 0.01° and speed of 0.5 deg.min^{-1}. The measured XRD patterns were evaluated using the PDXL 2 software (version 2.4.2.0) and compared with the PDF-2 database issued by ICDD, released in 2015. Nitrogen physisorption on catalyst powders (grain size 0.16–0.32 mm) was performed using an ASAP 2020 Micromeritics instrument (Norcross, Atlanta, GA, USA) after degassing at 105 °C for 24 h at 1 Pa vacuum. The adsorption–desorption isotherms of nitrogen at 77 K were evaluated by the standard Brunauer–Emmett–Teller (BET) procedure [35] for the p/p_0 range = 0.05–0.25 to calculate the specific surface area S_{BET}. Mesopore surface areas, S_{meso}, and micropore volume, V_{micro}, were determined by the *t*-plot method [36]. The total pore volume, V_{total}, was determined from the nitrogen adsorption isotherm at maximum p/p_0 (~0.995). The pore-size distribution (pore radius 10^0–10^2 nm) was evaluated from the adsorption branch of the nitrogen adsorption–desorption isotherm by the Barrett–Joyner–Halenda (BJH) method [37], assuming a cylindrical pore geometry. The Lecloux–Pirard standard isotherm [38] was used for the *t*–plot and for the pore–size distribution evaluation.

Temperature programmed reduction (H_2-TPR) measurements were performed with a H_2/N_2 mixture (10 mol % H_2), flow rate 50 mL min^{-1} and a linear temperature increase of 20 °C min^{-1} up to 900 °C. Changes in H_2 concentration were detected with a catharometer. A reduction of grained CuO (0.160–0.315 mm) was performed to calculate the absolute values of the hydrogen consumed during reduction of the samples.

The temperature-programmed desorption of CO_2 (CO_2-TPD) was carried out to examine basic properties of the catalysts surface. The measurements were accomplished with a 0.050 g sample in the temperature range of 20–900 °C, with helium as a carrier gas and CO_2 as the adsorbing gas. Prior to the CO_2–TPD measurement, the sample was heated in helium from 25 to 500 °C with a temperature ramp of 20 °C min^{-1}; then, the sample was cooled in helium to 25 °C. Ten doses of CO_2, 840 μL each, were applied to the catalyst sample at 30 °C before flushing with He for 1 h and heating at a rate of 20 °C min^{-1}. The composition of the gases evolved during the experiments was determined by a mass spectrometer (Balzers, Pfeiffer Vacuum, Asslar, Germany). The following mass contributions *m/z* were collected: 2–H_2 and 44-CO_2. The spectrometer was calibrated by dosing the known amount of CO_2 into the carrier gas (He) in every experiment. The H_2-TPR and CO_2–TPD experiments were evaluated using OriginPro 8.0 software with an accuracy of ±5%.

The thermal decomposition of dried materials was conducted by heating the 10 mg of samples to 700 °C at a rate of 2 °C/min, using a NETZSCH TG 209 F1 Libra (Selb, Germany) recording microbalances in a stream of inert gas (Ar) at a flow rate 50 mL/min. The sample was heated in the oven, the output of which was regulated by a programmable digital temperature controller. The flow of carrier gas was controlled by electronic mass flowmeters.

A Nicolet Model 360 Avatar FT-IR spectrometer (Analytical Instruments Brokers LLC, Golden Valley, MN, U.S.A.) was used in an attenuated total reflection (ATR) mode to obtain the spectra from catalysts between 360 and 4000 cm^{-1} (resolution 1.93 cm^{-1}, 300 scans, 1 s per scan) when the powder was pressed against ZnSe crystal (working range between 508 and 4000 cm^{-1}). All spectra were obtained at laboratory temperature (~22 °C) and atmospheric pressure.

The thermal stability of the alkali metal promoters was investigated by the species resolved thermal alkali desorption (SR-TAD) method. The experiments were carried out in a vacuum apparatus with a background pressure of 10^{-7} kPa. The samples in the form of wafers (13 mm diameter, 200 mg weight) were heated at the rate of 5 °C min^{-1} from room temperature up to 650 °C in a stepwise voltage-controlled mode. Then, they were cooled down at the same rate. The desorption flux of potassium atoms was determined by means of a surface ionization detector, consisting of a platinum wire heated by a current of 2.2 A, to approximately 1000 °C, with a positive potential of +120 V causing the ionization of the desorbed atoms and their acceleration towards the collector. During the measurements, the samples were biased with a positive potential +5 V to quench the thermal emission of electrons. In this way, the possibility of the reneutralization of ions by thermal electrons outside the surface was effectively eliminated. The resulting positive current was directly measured with a Keithley 6512 digital electrometer.

Superficial elemental analyses were performed by XPS (X-ray photoelectron spectrometry ESCA 3400, Kratos, Manchester, UK) at a base pressure better than 5×10^{-7} Pa, using the polychromatic Mg X-ray source (Mg Kα, 1253.4 eV). The composition of the elements was determined without any annealing. For the spectra, the Shirley background was subtracted, and the elemental compositions of layers were calculated from the corresponding areas.

The catalytic decomposition of NO to N_2 and O_2 was performed in an integral fixed bed stainless steel reactor of 5 mm internal diameter (650 °C or 700 °C, 0.5 g, 49 mL min^{-1}). To some catalytic runs, oxygen (0–10 mol%) was added to an inlet gas composed of 1000 ppm NO in N_2. The catalysts (fraction 0.16–0.316 mm) pre-calcined at 700 °C were activated in 50 mL of N_2 + O_2 min^{-1} (101325 Pa, 20 °C) for 1 h at 650 °C. Then, the NO catalytic decomposition at 650 °C was measured at least for 15 h. After this period, when a stable performance was observed, the temperature dependence of conversion was launched with cooling rate of 5 °C min^{-1}, and the catalysts activity was measured for 3 h at each temperature (640 °C, 620 °C, 600 °C, 580 °C, and 560 °C). After the conversion curve measurement, the reactor was heated back to 650 °C in order to check the stability of the catalyst. In case the performance was stable, the catalyst was heated to 700 °C, and the steady state conversion at 700 °C was measured. When the oxygen was added to the inlet mixture, it took 1 h at minimum to achieve steady state performance. Infrared analyzers for the online analysis of NO (Ultramat 6, Siemens, Karlsruhe, Germany) and N_2O (Sick) were used. During all measurements, no N_2O and no NO_2 were detected, as proven by the low-temperature NO_2/NO catalytic convertor (TESO Ltd.). The activity of the catalysts was determined as conversions of NO from the relation $X_{NO} = (c_{NO}{}^0 - c_{NO})/c_{NO}{}^0$, where X_{NO} is NO conversions, $c_{NO}{}^0$ is the initial NO concentration, and c_{NO} is the NO concentration at reactor outlet.

5. Conclusions

The investigated process of the K promoted Co–Mn–Al catalyst preparation consisting of the precipitation of transition metal nitrates with a solution of K_2CO_3/KOH is simpler than the other possible ways of preparation, e.g., the K-impregnation of wet cake method. When using the examined process of preparation, it is possible to obtain a catalyst with a desired K concentration in the solids without any additional preparation steps. The concentration of potassium affected catalytic activity of the K-precipitated catalysts in direct NO decomposition, the highest being at ca 8 wt.% of K, while that of the K-impregnated wet cake samples was about 2 wt.%. We believe that the incorporation of potassium into the catalyst particle structure positively influences catalyst stability, as the potassium inside the catalyst particles serves as its reservoir when it desorbs from the surface during catalytic reaction.

Supplementary Materials: The following are available online at http://www.mdpi.com/2073-4344/9/7/592/s1, Figure S1: Diffraction patterns of the dried K-precipitated Co-Mn-Al compounds containing various amounts of potassium. H – hydrotalcite-like compounds, R – rhodochrosite, K – KNO_3, Figure S2: Adsorption-desorption isotherms of nitrogen for K-precipitated Co-Mn-Al mixed oxides containing various amounts of potassium in wt. %: curve 2 – 8.20, curve 3 – 2.39, curve 4 – 0.92, curve 5 – 0.60, Table S1: Details of the preparation method and concentration of K and Na (mg/L) in filtrate and washing waters.

Author Contributions: Conceptualization, K.J.; methodology, K.J. and K.P.; validation, T.B., L.O., A.M., A.K. (Andrzej Kotarba), and K.J.; investigation, J.B., T.B., K.P., K.K., A.K. (Anna Klegová), A.M., V.J., and M.K.; data curation, T.B., K.P., K.K., J.B., A.K. (Anna Klegová), and M.K.; writing—original draft preparation, K.J.; writing—review and editing, K.J. and K.P.; supervision, L.O.; project administration, K.P., L.O., and K.J.

Funding: The authors thank Czech Science Foundation (project No. 18-19519S) for financial support. The work was also supported from ERDF "Institute of Environmental Technology – Excellent Research" (No. CZ.02.1.01/0.0/0.0/16_019/0000853). Experimental results were accomplished by using Large Research Infrastructures ENREGAT and CATPRO supported by the Ministry of Education, Youth and Sports of the Czech Republic under projects No. LM2018098 and No. LM2015039.

Acknowledgments: The authors thank L. Soukupová for the chemical analysis of the samples and H. Šnajdaufová for the measurement of porous structure of the catalysts.

Conflicts of Interest: The authors declare no conflict of interest.

References

1. Bu, Y.F.; Ding, D.; Gan, L.; Xiong, X.H.; Cai, W.; Tan, W.T.; Zhong, Q. New insights into intermediate-temperature solid oxide fuel cells with oxygen-ion conducting electrolyte act as a catalyst for NO decomposition. *Appl. Catal. B* **2014**, *418*, 158–159. [CrossRef]
2. Janssen, J. *Environmental Catalysis—Stationary Sources, Environmental Catalysis*; Wiley-VCH Verlag GmbH: Weinheim, Germany, 2008; pp. 119–179.
3. Li, J.; Chang, H.; Ma, L.; Hao, J.; Yang, R.T. Low-temperature selective catalytic reduction of NOx with NH_3 over metal oxide and zeolite catalysts—A review. *Catal. Today* **2011**, *175*, 147–156. [CrossRef]
4. Javed, M.T.; Irfan, N.; Gibbs, B.M. Control of combustion-generated nitrogen oxides by selective non-catalytic reduction. *J. Environ. Manage.* **2007**, *83*, 251–289.
5. Wu, Y.; Dujardin, C.; Lancelot, C.; Dacquin, J.P.; Parvulescu, C.M.; Henry, C.R.; Neisius, T. Catalytic abatement of NO and N_2O from nitric acid plants: A novel approach using noble metal-modified perovskites. *J. Catal.* **2015**, *328*, 236–247. [CrossRef]
6. Haneda, M.; Hamada, H. Recent progress in catalytic NO decomposition. *C. R. Chim.* **2016**, *19*, 1254–1265. [CrossRef]
7. Falsig, H.; Bligaard, T. Trends in catalytic NO decomposition over transition metal surfaces. *Top. Catal.* **2007**, *45*, 117–120. [CrossRef]
8. Brown, W.A.; King, D.A. NO Chemisorption and Reactions on Metal Surfaces: A New Perspective. *J. Phys. Chem. B* **2000**, *104*, 2578–2595. [CrossRef]
9. Falsig, H.; Bligaard, T.; Christensen, C.H.; Nørskov, J.K. Direct NO decomposition over stepped transition-metal surfaces. *Pure Appl. Chem.* **2007**, *79*, 1895–1903. [CrossRef]
10. Wu, Z.; Xu, L.; Zhang, W.; Ma, Y.; Yuan, Q.; Jin, Y.; Yang, J.; Huang, W. Structure sensitivity of low-temperature NO decomposition on Au surfaces. *J. Catal.* **2013**, *304*, 112–122. [CrossRef]
11. Haneda, M.; Kintaichi, Y.; Hamada, H. Catalytic active site for NO decomposition elucidated by surface science and real catalyst. *Appl. Catal. B* **2005**, *55*, 169–175. [CrossRef]
12. Bonzel, H.P.; Bradshaw, A.M.; Ertl, G. *Physics and Chemistry of Alkali Metal Adsorption*; Elsevier: Amsterdam, The Netherlands, 1989.
13. Mross, W.D. Alkali doping in heterogeneous catalysis. *Catal. Rev.* **1983**, *25*, 591–637.
14. Niemantsverdriet, J.W. *Spectroscopy in Catalysis*; Wiley-VCH Verlag GmbH & Co.: Weinheim, Germany, 2007; p. 271.
15. Pacultová, K.; Draštíková, V.; Chromčáková, Ž.; Bílková, T.; Mamulová-Kutláková, K.; Kotarba, A.; Obalová, L. On the stability of alkali metal promoters in Co mixed oxides during direct NO catalytic decomposition. *Mol. Catal.* **2017**, *428*, 33–40. [CrossRef]
16. Wang, Y.; Hu, X.; Zheng, K.; Zhang, H. Effect of precipitants on the catalytic activity of Co–Ce composite oxide for N2O catalytic decomposition. *Reac. Kinet. Mech. Cat.* **2018**, *123*, 707–721. [CrossRef]
17. Zhang, Q.; Luo, J.; Vileno, E.; Steven, L.; Suib, S.L. Synthesis of cryptomelane-type manganese oxides by microwave heating. *Chem. Mater.* **1997**, *9*, 2090–2095. [CrossRef]
18. Kovanda, K.; Rojka, T.; Dobešová, J.; Machovič, V.; Bezdička, P.; Obalová, L.; Jirátová, K.; Grygar, T. Mixed oxides obtained from Co and Mn containing layered double hydroxides: Preparation, characterization, and catalytic properties. *J. Solid State Chem.* **2006**, *179*, 812–823. [CrossRef]
19. Pacultová, K.; Bílková, T.; Klegová, A.; Karásková, K.; Fridrichová, D.; Martaus, A.; Jirátová, K.; Kiška, T.; Balabánová, J.; Koštejn, M.; et al. Direct NO decomposition over K-promoted Co-Mn-Al mixed oxides. *Catalysts* **2019**, *9*. (submitted).
20. Klyushina, A.; Pacultová, K.; Karásková, K.; Jirátová, K.; Ritz, M.; Fridrichová, D.; Volodarskaja, A.; Obalová, L. Effect of preparation method on catalytic properties of Co-Mn-Al mixed oxides for N2O decomposition. *J. Mol. Catal. A Chemical.* **2016**, *425*, 237–247. [CrossRef]

21. Santos, V.P.; Soares, O.S.G.P.; Bakker, J.J.W.; Pereira, M.F.R.; Órfão, J.J.M.; Gascon, J.; Kapteijn, F.; Figueiredo, J.L. Structural and chemical disorder of cryptomelane promoted by alkali doping: Influence on catalytic properties. *J. Catal.* **2012**, *293*, 165–174. [CrossRef]
22. Becerra, M.E.; Arias, N.P.; Giraldo, O.H.; López Suárez, F.E.; Illán Gómez, M.J.; Bueno López, A. Soot combustion manganese catalysts prepared by thermal decomposition of $KMnO_4$. *App. Catal. B Environ.* **2011**, *102*, 260–266. [CrossRef]
23. Da Costa-Serra, J.F.; Chica, A. Catalysts based on Co-Birnessite and Co-Todorokite for the efficient production of hydrogen by ethanol steam reforming. *Intern. J. Hydrog. Energy* **2018**, *43*, 16859–16865. [CrossRef]
24. Obalová, L.; Karásková, K.; Wach, A.; Kustrowski, P.; Mamulová-Kutláková, K.; Michalik, S.; Jirátová, K. Alkali metals as promoters in Co–Mn–Al mixed oxide for N_2O decomposition. *Appl. Catal. A Gen.* **2013**, *462–463*, 227–235. [CrossRef]
25. Imanaka, N.; Masui, T. Advances in direct NOx decomposition catalysts. *Appl. Catal. A Gen.* **2012**, *431–432*, 1–8. [CrossRef]
26. Smoláková, L.; Frolich, K.; Troppová, I.; Kutálek, P.; Kroft, E.; Čapek, L. Determination of basic sites in Mg–Al mixed oxides by combination of TPD-CO_2 and CO_2 adsorption calorimetry. *J. Therm. Anal. Calorim.* **2017**, *127*, 1921–1929. [CrossRef]
27. Nyquist, R.A.; Nagel, R.O. *Infrared Spectra of Inorganic Compounds*; Academic Press: New York, NY, USA, 1971.
28. Bentley, F.F.; Smithson, L.D.; Rozek, A.L. *Infrared Spectra and Characteristic Frequencies 700–300 cm-1: A Collection of Spectra, Interpretation, and Bibliography*; Interscience: New York, NY, USA, 1986.
29. Potter, R.M.; Rossman, G.R. The tetravalent manganese oxides: Identification, hydration, and structural relationships by infrared spectroscopy. *Am. Mineral.* **1979**, *64*, 1199–1218.
30. Horák, M.; Papoušek, D. *Infračervená spektra a struktura molekul: použití vibrační spektrokopie při určování struktury molekul*, 1st ed.; Academia: Praha, Czech Republic, 1976; 836p.
31. Scofield, J.H. Hartree-Slater subshell photoionization cross-sections at 1254 and 1487 eV. *J. Electron Spectrosc. Relat. Phenom.* **1976**, *8*, 129–137. [CrossRef]
32. Borowiecki, T.; Denis, A.; Rawski, M.; Gołębiowski, A.; Stołecki, K.; Dmytrzyk, J.; Kotarba, A. Studies of potassium-promoted nickel catalysts for methane steam reforming: Effect of surface potassium location. *Appl. Surf. Sci.* **2014**, *300*, 191–200. [CrossRef]
33. Kaspera, W.; Wojas, J.; Molenda, M.; Kotarba, A. Parallel migration of potassium and oxygen ions in hexagonal tungsten bronze—Bulk diffusion, surface segregation and desorption. *Solid State Ionics* **2016**, *297*, 1–6. [CrossRef]
34. Obalová, L.; Maniak, G.; Karásková, K.; Kovanda, F.; Kotarba, A. Electronic nature of potassium promotion effect in Co-Mn-Al layered double hydroxide on the catalytic decomposition of N2O. *Catal. Commun.* **2011**, *12*, 1055–1058. [CrossRef]
35. Brunauer, S.; Emmett, P.H.; Teller, E. Adsorption of gases in multimolecular layers. *J. Am. Chem. Soc.* **1938**, *60*, 309–319. [CrossRef]
36. Deboer, J.H.; Lippens, B.C.; Linsen, B.G.; Broekhof, J.C.; Vandenhe, A.; Osinga, T.J. The t-curve of multimolecular N_2-adsorption. *J. Colloid Interface Sci.* **1966**, *21*, 405–414. [CrossRef]
37. Barrett, E.P.; Joyner, L.G.; Halenda, P.P. The Determination of Pore Volume and Area Distributions in Porous Substances. I. Computations from Nitrogen Isotherms. *J. Am. Chem. Soc.* **1951**, *73*, 373–380.
38. Lecloux, A.; Pirard, J.P. The importance of standard isotherms in the analysis of adsorption isotherms for determining the porous texture of solids. *J. Colloid Interface Sci.* **1979**, *70*, 265–281. [CrossRef]

catalysts

MDPI

Article

Co-Mn-Al Mixed Oxides Promoted by K for Direct NO Decomposition: Effect of Preparation Parameters

Kateřina Pacultová [1,*], Tereza Bílková [1,2], Anna Klegova [1], Kateřina Karásková [1], Dagmar Fridrichová [1,3], Květa Jirátová [4], Tomáš Kiška [1,2], Jana Balabánová [4], Martin Koštejn [4], Andrzej Kotarba [5], Wojciech Kaspera [5], Paweł Stelmachowski [5], Grzegorz Słowik [6] and Lucie Obalová [1]

1 Institute of Environmental Technology, VŠB-Technical University of Ostrava, 17. listopadu 15/2172, CZ-70800 Ostrava-Poruba, Czech Republic

2 Faculty of Material Science and Technology, VŠB-Technical University of Ostrava, 17. listopadu 15/2172, CZ 70800 Ostrava-Poruba, Czech Republic

3 Centre ENET, VŠB-Technical University of Ostrava, 17. listopadu 15/2172, CZ-70800 Ostrava-Poruba, Czech Republic

4 Institute of Chemical Process Fundamentals of the CAS, v.v.i., Rozvojová 2/135, CZ-16501 Praha 6-Suchdol, Czech Republic

5 Faculty of Chemistry, Jagiellonian University, Gronostajowa 2, PL-30387 Krakow, Poland

6 Faculty of Chemistry, Department of Chemical Technology, Maria Curie—Skłodowska University, Plac Marii Curie—Skłodowskiej 3, PL-20-031 Lublin, Poland

* Correspondence: katerina.pacultova@vsb.cz; Tel.: +420-597-327-327

Received: 27 May 2019; Accepted: 3 July 2019; Published: 9 July 2019

Abstract: Fundamental research on direct NO decomposition is still needed for the design of a sufficiently active, stable and selective catalyst. Co-based mixed oxides promoted by alkali metals are promising catalysts for direct NO decomposition, but which parameters play the key role in NO decomposition over mixed oxide catalysts? How do applied preparation conditions affect the obtained catalyst's properties? Co_4MnAlO_x mixed oxides promoted by potassium calcined at various conditions were tested for direct NO decomposition with the aim to determine their activity, stability and selectivity. The catalysts were prepared by co-precipitation of the corresponding nitrates and subsequently promoted by KNO_3. The catalysts were characterized by atomic absorption spectrometry (AAS)/inductive coupled plasma (ICP), X-ray photoelectron spectrometry (XPS), XRD, N_2 physisorption, temperature programmed desorption of CO_2 (TPD-CO_2), temperature programmed reduction by hydrogen (TPR-H_2), species-resolved thermal alkali desorption (SR-TAD), work function measurement and STEM. The preparation procedure affects physico-chemical properties of the catalysts, especially those that are associated with the potassium promoter presence. The addition of K is essential for catalytic activity, as it substantially affects the catalyst reducibility and basicity—key properties of a deNO catalyst. However, SR-TAD revealed that potassium migration, redistribution and volatilization are strongly dependent on the catalyst calcination temperature—higher calcination temperature leads to potassium stabilization. It also caused the formation of new phases and thus affected the main properties—S_{BET}, crystallinity and residual potassium amount.

Keywords: nitric oxide; catalytic decomposition; potassium promoter; cobalt-based mixed oxide

---

1. Introduction

Nitrogen oxides (NO_x) are produced by anthropogenic activities, mainly by combustion of fossil fuels in automotive engines and power plants. NO accounts for more than 95% of these nitrogen oxides emissions [1]. Nowadays, there are two types of technologies that help to reduce NO emissions—selective catalytic reduction (SCR) and selective non-catalytic reduction (SNCR). However, in both these processes, a reducing agent (ammonia, urea) has to be used. NO can also be decomposed directly to N_2 and O_2. The disadvantage of this simple method is that the reaction takes place at high temperature (>1100 °C) due to the high value of activation energy (~335 kJ/mol) [2]. High activation energy of NO decomposition can be decreased by the presence of a suitable catalyst.

Many catalysts for direct NO decomposition have been tested; however, none of them have been sufficiently stable, active and selective at economically feasible temperatures yet. The studied catalysts are mainly based on precious metals (Pt, Pd, Rh) [3–5], zeolites [6–8] or single [9,10] and mixed oxide catalysts [11–13]. The results suggest that further increase in catalytic activity would be difficult to achieve with metals or alloys alone or simple metal oxides. A more active system requires the presence of different kind of active sites necessary for each individual step of the NO decomposition reaction, such as NO adsorption, surface reaction of intermediates and oxygen and nitrogen desorption [14,15]. These can be obtained for example in mixed oxides systems.

Co-based mixed oxides promoted by alkali metals were used as catalysts for various reactions and they showed interesting results in NO decomposition [9,11]. From previous studies, it is known that alkali metals volatilize from transition metal oxides at temperatures higher than 500 °C [16–18], which could affect long-term stability of the catalyst [11]. The above mentioned results imply that the stability of alkali metal promoters has to be improved for viable application of these catalysts.

The literature showed that the alkali promoter stability could be improved by modifications of the preparation procedure. One possibility is changing the way of introducing an alkali metal into the catalyst (changing the catalyst bulk structure) during synthesis. The calcination procedure of catalyst precursors is another very important optimizing parameter, since calcination conditions (e.g., temperature, time) strongly influence the stability of alkali metal-containing catalysts. The advantage of high temperature calcination of K/Co-Al mixed oxide regarding N_2O decomposition was published by Cheng [19]. The rearrangement of the surface alkali metal species depending on the calcination temperature leading to different stability of the obtained catalytic system was also reported elsewhere [18,20].

For a tailored synthesis of an active catalyst, understanding the reaction mechanism is also important. Although several works have been devoted to the study of the NO_x decomposition mechanism [2,21–23], details of the mechanisms and the exact nature of active sites, especially for NO catalytic decomposition on mixed oxides, are still a matter of debate, as well as the relationship between the method of preparation, physico-chemical properties and catalytic performance. The NO molecule can adsorb either dissociatively or molecularly depending on the type of active metal and on the conditions of adsorption; moreover, chemisorption of the reactants can cause adsorbate-induced reconstruction of the surface [2]. Dissociation of NO on the surface often depends on surface temperature, surface coverage, crystal plane and the concentration of surface defects [3]. Tsujimoto [24] along with Hong et al. [25] have recently demonstrated that control of the surface basicity of the catalyst is also important in terms of enhancing NO decomposition activity. An active catalyst should allow easy desorption of oxygen atoms from the catalyst surface, which is a very important reaction step. Moreover, catalysts with appreciable activity for direct NO decomposition should provide stronger adsorption of nitrogen compared to that of oxygen [26].

The above studies motivated us to prepare the Co–Mn–Al mixed oxide catalyst promoted by potassium to study its deNO catalytic properties more systematically. The bulk promotion method was used for modifying the catalyst by potassium, since it is supposed that incorporation of the promoter into the structure improves the promoter activity and stability [27]. The main objective was the enhancement of NO catalytic decomposition activity and stability through optimizing the conditions (calcination temperature and time) of the preparation method and the content of the alkali promoter.

The addition of potassium was proven to be important for catalyst activity. There is an optimal amount of K responsible for the formation of surface basic sites. The sufficient amount and right type of basic sites play a key role in direct NO decomposition, as well as optimal catalyst reducibility.

The SR-TAD measurements clearly showed that potassium migration, redistribution and finally volatilization are directly linked to the catalyst calcination temperature.

2. Results

2.1. Characterization of Catalysts

2.1.1. Chemical Composition

The results of chemical analysis of selected fresh catalysts promoted by different amount of potassium, calcined at four different temperatures (500, 600, 700 and 800 °C) and for two different time periods (4 and 12 h) are summarized in Table 1. The calculated molar ratio is close to the values set during preparation Co:Mn:Al = 4:1:1.

Table 1. Physicochemical properties of selected prepared catalysts.

Sample *	Co (wt.%)	Mn (wt.%)	Al (wt.%)	Co:Mn:Al (Molar Ratio)
K(II)/500/4	52.8	11.2	5.1	4.0:0.9:0.8
K(II)/600/4	49.7	11.9	5.4	4.0:1.0:1.0
K(II)/700/4	51.5	11.4	5.4	4.0:1.0:0.9
K(II)/800/4	45.1	12.0	5.6	4.0:1.1:1.1
K(I)/500/12	48.1	11.7	5.3	4.0:1.0:1.0
K(II)/500/12	44.2	11.3	5.1	4.0:1.1:1.0
K(III)/500/12	46.9	11.0	5.5	4.0:1.0:1.0
K(IV)/500/12	51.0	10.5	5.4	4.0:0.9:0.9
K(0)/600/4	51.0	16.0	6.4	4.0:1.3:1.1

* Sample denotation: K(*x*)/*y*/*z*, where *x* means K content in wt.%, *y* means calcination temperature (°C) and *z* means calcination time (h). For details, see Section 4.1.

Chemical analysis was also focused on the content of potassium, since its loss due to thermal vaporization was expected [28] (Table 2). A decrease of K concentration was observed with increasing calcination temperature and calcination time. While the increase in temperature from 500 to 600 °C caused almost no decrease in K amount, 20% of K weight was lost during calcination at 700 and 800 °C (4 h). The prolonged calcination time also caused a small drop in K amount (see samples K(II)/500/4 and K(II)/500/12). It is important to notice that simply comparing the potassium content in wt.% is not possible because of the associated oxygen release from the catalysts during calcination at high temperatures.

Table 2. Selected experimental parameters (K content, specific surface area, particle size and unit cell parameter) for the investigated fresh and used catalysts.

Sample	K_{fresh} (wt.%)	K_{used} (wt.%)	$S_{BET\ fresh}$ (m^2/g)	$S_{BET\ used}$ (m^2/g)	$L_{c\ fresh}$ (nm)	a_{fresh} (nm)
K(II)/500/4	2.1	1.8	94	33	9	0.81163 ± 0.00004
K(II)/600/4	2.0	2.0	72	23	12	0.81075 ± 0.00007
K(II)/700/4	2.0	1.8	50	21	16	0.81485 ± 0.00004
K(II)/800/4	1.6	1.4	23	16	28	0.81512 ± 0.00004
K(I)/500/12	0.7	0.6	73	42	10	0.81079 ± 0.00010
K(II)/500/12	1.9	1.8	87	32	13	0.81085 ± 0.00011
K(III)/500/12	2.4	2.5	73	44	12	0.81051 ± 0.00005
K(IV)/500/12	3.1	2.7	69	40	13	0.81051 ± 0.00005
K(0)/600/4	0	0	49	12	13	0.81210 ± 0.00008

A small loss of K (12.5%) was also observed during the catalytic tests, which corresponds to the results published in our previous work [11]. However, this trend was surprisingly not confirmed for all sets of the catalysts (not shown), which can be caused by a non-homogenous distribution of potassium in the samples (caused by its bulk and surface migration and/or vaporization) and the complexity of potassium content determination.

2.1.2. Textural Properties

The surface area of all catalysts was examined using N_2 physisorption (one point Brunauer–Emmett–Teller (BET) method) and selected results are summarized in Table 2. The specific surface areas varied from 16 to 94 m^2/g and from 12 to 44 m^2/g for fresh and used catalysts, respectively. Specific surface area decreased with rising calcination temperature and calcination time as well. A decrease in surface area of about 20 m^2/g was observed for each 100 °C jump in the temperature used for calcination. A sharp decrease of surface area was also observed after catalytic measurements.

Pore size distributions of the selected samples reflecting the effect of K amount, calcination time and calcination temperature are shown in Figure 1. The catalysts calcined at 500 °C showed unimodal pore size distribution with a maximum around 30 nm. Longer calcination time did not affect pore size distribution of the catalysts (Figure 1a). Lower potassium amount caused a slight shift of pore sizes to higher values (Figure 1b). The same trend is visible in Figure 1c: After calcination at higher temperatures accompanied by a decrease in K amount, the pore size shifted to higher values along with the creation of new pores with diameters higher than 60 nm. The trend corresponds to the decrease in surface area after calcination at higher temperatures and is probably connected with the collapse of smaller pores and coalescence of the larger ones.

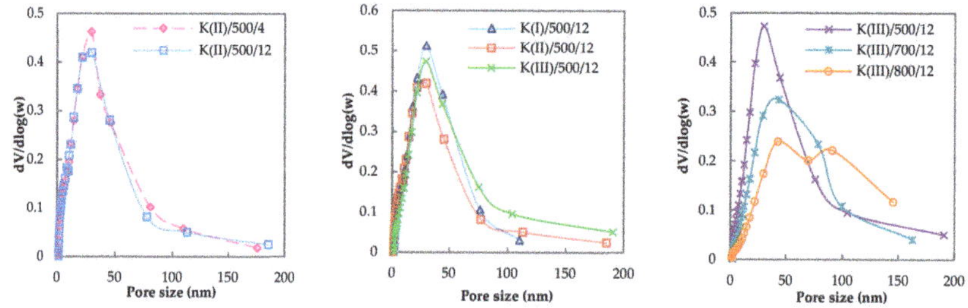

(a) (b) (c)

Figure 1. Pore size distribution of selected catalysts: (**a**) Effect of calcination time for samples K(II)/500/*z*, (**b**) effect of K amount for samples K(*x*)/500/12 and (**c**) effect of calcination temperature for samples K(III)/*y*/12.

2.1.3. Phase Composition

Spinel-type mixed oxide phase was detected in all samples by X-ray powder diffraction analysis (Figures S1–S3 in Supplementary Materials). Very small diffraction line at about 38° present in the samples calcined at lower temperatures can be ascribed to Mn_2O_3 (PDF-2 card No. 01-071-0636; see Supplementary Materials, Figures S1–S3) or can simply be an artifact of the measurement, K-beta line of the main spinel diffraction. Intensities of diffraction lines depend on the position of atoms in the spinel unit cell, while the position depends on the shape and size of the unit cell [29]. The ratio between spinel diffraction lines (111)/(220) and (220)/(444) is a measure of the occupancy of the tetrahedral cation sites [30,31]. In our case lattice parameter *a* (Figure 2a), mean coherent domain size corresponding to crystallinity (Figure 2b), and (220)/(440) intensities ratio *I* (Figure 2c) increased with rising calcination temperature. All these parameters reflect the continuous process of gradual thermal crystallization of the spinel, i.e., continuous conversion of disordered material into well X-ray diffracting oxides.

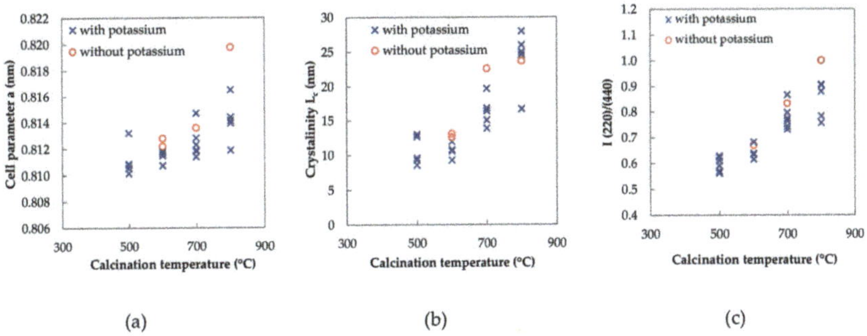

Figure 2. Effect of calcination temperature on (**a**) spinel lattice parameter *a*, (**b**) coherent domain size L_c and (**c**) *I* (220)/(440) intensities ratio.

The thermal behavior of a Co–Mn–Al mixed oxide system unmodified by potassium was previously described by Kovanda et al. [30]. They found that after decomposition of layered double hydroxides above 200–260 °C, nanocrystalline spinels are formed and spinels are the only phases present up to 900 °C. However, at temperatures of about 500 °C the segregation of Co rich spinel was observed, while the incorporation of manganese into the spinel lattice proceeded during its further recrystallization with increasing temperature accompanied by a lattice parameter increase, approaching the value typical for Co_3O_4 (samples calcined at 500 °C) and the value typical for Co_2MnO_4 (samples calcined at 800 °C).

After careful inspection of the cell parameter changes, it was found out that L_c and *a* did not depend on the K amount for samples calcined at 500 and 600 °C, while for samples calcined at 700 and especially at 800 °C, the obvious dependence on potassium concentration could be observed (Figure 3a,b).

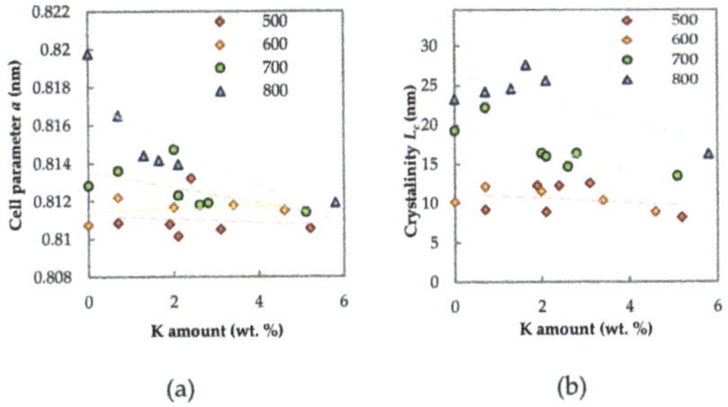

Figure 3. Dependence of (**a**) spinel lattice parameter *a* on K amount and (**b**) crystallinity L_c on K amount. The numbers in the legend mean the calcination temperature.

With increasing calcination temperature and/or higher amount of potassium, in addition to the spinel phase, a new phase containing potassium was also detected (Figures S1–S3 in Supplementary Materials). The nature and stoichiometry of the new phase depended on the temperature as well as on the potassium amount. Two different phases containing potassium were observed (Table 3)—cryptomelane KMn_8O_{16} (molar ratio K/Mn = 0.13; PDF-2, card No. 00-006-0547), which was clearly distinguishable (marked as A) and another potassium manganese oxide phase (marked as B), which could be identified as $K_2Mn_4O_8$ (PDF-2, card No. 00-016-0205) or $K_{1.39}Mn_3O_6$ (PDF-2, card No. 01-080-7317), both having higher potassium/manganese molar ratio than cryptomelane (K/Mn = 0.5). The results of X-ray diffraction (XRD) K-phase determination are shown in Table 3. Cryptomelane (A) was present in all samples calcined at 500 °C and decomposed during calcination at above 600 °C, forming the potassium manganese oxide phase (B). For the samples calcined at 600 °C, the type of present phase depended on the K content: For lower potassium amount cryptomelane was formed while for potassium amount higher that 1 wt.%, the potassium manganese oxide phase was formed. The phase transformation depending on temperature for samples containing approximately 2 wt.% of potassium is illustrated in Figure S1 in the Supplementary Materials; the phase changes depending on potassium amount for samples calcined at 600 and 800 °C are illustrated in Figures S2 and S3, respectively. Longer calcination time did not affect the type of K-containing phase, only its degree of crystallinity (not shown). The changes observed for samples calcined below the catalytic reaction temperature (650 and/or 700 °C) are in accordance with previous findings (Figures S1 and S2), since after the catalytic reaction, the potassium phase in A form was not present and only type B phases were observed in all tested samples (not shown).

Table 3. K-containing phases detected by XRD in Co–Mn–Al mixed oxide with different K-content and calcined at different temperatures (fresh samples).

Calcination Temperature y (°C)	Sample			
	K(I)/y/4	K(II)/y/4	K(III)/y/4	K(IV)/y/4
500	A	A	A	A
600	A	A	B	n.d.
700	B	B	B	B
800	B	B	B	B

A—KMn_8O_{16} (cryptomelane), B—$K_2Mn_4O_8$ or $K_{1,39}Mn_3O_6$, n.d.—not determined.

Based on the XRD findings described above, it can be concluded that potassium is bonded mainly to manganese. When there is not any potassium present in the sample or only a small amount, with increasing calcination temperature manganese enters the spinel lattice as can be inferred from the changes in lattice parameter, similarly as in a K-free Co_4MnAlO_x spinel [28]. On the other hand, when a higher amount of potassium is present, more manganese is bonded to potassium and cannot enter the spinel lattice at higher calcination temperatures and for that reason, the lattice parameters returned to the values characteristic of the Co_3O_4 phase.

2.1.4. Surface Composition

The surface composition in the near-surface region and the chemical state of the elements over three selected samples was examined using X-ray photoelectron spectrometry (XPS). The following samples were chosen in order to elucidate the effect of potassium presence as well as the effect of calcination temperature:

(i) Sample without K,
(ii) Fresh sample modified by potassium calcined at 500 °C,
(iii) Used sample modified by potassium, which means that its structure corresponds to the sample calcined at 700 °C and can be changed somehow by catalytic reaction intermediates.

The surface concentrations of elements were determined (Table 4) from the intensities (peak areas) divided by the corresponding response factor [32]. Carbon correction was done for all samples. Apart from the main elements—Co, Mn, Al, K and Na–the elements O and C were also determined. The K(0)/670/4 catalyst contains 4.18 at. % of Na on the surface. It remained in the catalyst after incomplete washing of the precipitate by distilled water. The catalyst K(II)/500/12 prepared by bulk promotion of the precipitate with K salt contains no Na and the finding indicates that the rest of sodium in the precipitate was interchanged with potassium during the bulk promotion procedure. The concentrations of oxygen and carbon on the surface were disregarded, and the atomic concentrations of metals were converted into weight percentages to compare them with the values obtained from inductive coupled plasma (ICP) analysis (Table 5).

Table 4. Surface concentrations determined using X-ray photoelectron spectrometry (XPS) data evaluation in at. %.

Sample	Co	Mn	Al	O	C	K	Na
	(at. %)	(at. %)	(at. %)	(at. %)	(at. %)	(at. %)	(at. %)
K(0)/670/4	9.08	4.55	7.66	44.41	30.12	0	4.18
K(II)/500/12	9.37	4.07	8.25	44.26	30.99	3.07	0
K(II)/500/12_used	8.81	4.17	8.54	45.59	27.1	5.79	0

Table 5. Comparison of bulk (from inductive coupled plasma (ICP)) and surface (from XPS) catalysts composition in wt.%.

Sample	Co (wt.%)		Mn (wt.%)		Al (wt.%)		K (wt.%)		Na (wt.%)
	Bulk	Surface	Bulk	Surface	Bulk	Surface	Bulk	Surface	Surface
K0/670/4	52.0	49.2	11.7	23.0	5.4	19.0	0.0	0	n.d.
KII/500/12	45.6	49.4	11.0	0.0	4.4	19.9	2.5	10.7	0
KII/500/12_used	44.0	43.1	10.1	19.0	5.0	19.1	1.9	18.8	0

As one can see from Table 5, the values of bulk and surface concentrations are not the same. The following relations between surface and bulk concentrations can be seen (Table 6): Compared to bulk content, surface of the catalysts shows identical concentration of cobalt, nearly twice as high concentration of manganese and substantially higher concentrations of aluminum and potassium (when it is present). The findings indicate that the catalyst particles contain equal cobalt concentration throughout the whole volume, which is not changed during catalyst use. The surface of the non-promoted catalyst is enriched by Al and Mn. The fresh K-promoted catalyst shows four times higher K concentration on the surface than in the bulk and it means that K introduced by the bulk promotion method is located preferentially on the catalyst surface. The K, Mn and Al surface enrichment was also observed in our previous works [33,34].

Table 6. Ratio of weight percentages of surface and bulk concentrations of the metals.

Sample	Surface/Bulk (Weight Ratio)			
	Co	Mn	Al	K
K(0)/670/4	0.9	2.0	3.5	0
K(II)/500/12	1.1	1.8	4.5	4.3
K(II)/500/12_used	1.0	1.9	3.8	9.9

After the catalytic tests, further enrichment of the catalyst surface by K proceeds. The finding confirms gradual diffusion of K from the bulk of the catalyst due to its movement from lower parts of the catalyst particles to the surface during the tests at high reaction temperatures (>650 °C) and also subsequent volatilization since chemical analysis showed reduced concentration of K in the catalyst after catalytic tests.

Co 2p, Mn 2p, Al 2p, O 1s, K 2p and C 1s photoelectron spectra are presented in Figure S4 in Supplementary Materials for the K(0)/670/4, K(II)/500/12 and K(II)/500/12_used catalysts. In all samples, the deconvolution of the Co $2p_{3/2}$ region showed two peaks, main and satellite, with maxima at binding energy (BE) 780.5 and 786.4 eV.

Distinction between Co^{3+} and Co^{2+} is very difficult, as the difference between corresponding peaks is very small (Co^{3+}: 779.19-779.81 eV, Co^{2+}: 781.13 – 781.49 eV [35–40]). For identifying the cobalt chemical state, the energy separation of 2p3/2 to 2p1/2 and satellite structure of Co 2p spectra were used leading to the conclusion that both Co^{2+} and Co^{3+} chemical states are present. The shift of binding energies of Co 2p3/2 electrons towards higher values compared to Co_3O_4 (BE = 779.5 eV) could be related to the presence of other components (Al, Mn) [33].

Relative percentage of the Co $2p_{3/2}$ satellite peak (Table 7) correlates with Co^{2+}. Therefore, K(II)/500/12_used catalyst spectrum indicated the highest and K(0)/670/4 the lowest amount of Co^{2+}. Manganese Mn $2p_{3/2}$ spectra showed one maximum at 641.9 eV indicating higher oxidation states. As the peak was quite broad, the catalysts consist of a mixture of at least two manganese states with BE corresponding to Mn_3O_4 and MnO_2 [41]. Based on our previous research, Mn 2p3/2 was fitted by two peaks corresponding to Mn^{3+} (641.4–641.7 eV) and Mn^{4+} (642.6–642.9 eV) [33]. The data showed that K(II)/500/12_used catalyst contained higher amount of Mn^{4+} than the others. Aluminum in all samples occurred in the form of Al^{3+}.

Table 7. Binding energies and relative percentage (in parentheses) of fitted areas.

Sample	Co $2p_{3/2}$	Co $2p_{3/2}$ sat.	Mn $2p_{3/2}$	Mn $2p_{3/2}$	O 1s	O 1s	O 1s
	eV (%)	eV (%)	eV (%)	eV (%)	eV (%)	eV (%)	eV (%)
K(0)/670/4	780.5 (82)	786.4 (18)	641.4 (57)	642.6 (43)	529.9 (66)	531.4 (27)	532.7 (7)
K(II)/500/12	780.5 (70)	786.3 (24)	641.7 (58)	642.9 (42)	529.8 (57)	531.3 (33)	532.7 (10)
K(II)/500/12_used	780.5 (76)	786.4 (30)	641.4 (47)	642.6 (53)	529.7 (53)	531.2 (28)	532.8 (19)

Oxygen spectra were composed of three peaks at 529.91–529.61 (metal oxides), 531.42–531.13 (C–O single bond, –OH group) and 532.76–532.67 (C=O double bond) eV. Approximately 50–60 relelative % of oxygen signal is formed by oxygen bonded to metals, but particular metal oxides were not possible to distinguish from the spectra. The rest of oxygen is bonded to carbon. For the samples containing potassium (both fresh and after catalytic tests) a change in the shape of the peaks can be seen. The changes indicate a slightly different oxidation state of metals.

Analysis of potassium in the fresh sample shows one state (BE 292.7 eV for K $2p_{3/2}$); for the used sample, potassium shows two states (BE 292.7 and 294.8 eV for K $2p_{3/2}$). The common state for both fresh and used samples is a K–Mn–O phase, while additional state of potassium in the used sample (BE 294.8 eV) can be ascribed to K oxidized slightly during the catalytic reaction.

Obtained results indicate that K is affecting the oxidation state of Mn and Co in spite of the finding that K is associated only with Mn in the potassium manganese oxide phase (XRD). The more K on the surface, the more Co^{2+} was found on the surface at the expense of Co^{3+}, indicating that the presence of K causes the reduction of surface Co^{3+} to Co^{2+}. Conversely, in the case of Mn, the proportion of Mn^{4+} at the expense of Mn^{3+} increases with increasing K content on the surface, i.e., the presence of K causes oxidation of surface Mn^{3+} to Mn^{4+}. This can be explained as follows: Since the amount of potassium located in the catalyst in the potassium manganese oxide phase was not determined, there can still be some amorphous potassium present in the vicinity of cobalt, which can influence the Co oxidation state. No potassium phases were found out via high-resolution transmission electron microscopy (HRTEM) imaging, which suggests that some potassium was present as amorphous species. Moreover, the catalyst consists of a mixed spinel oxide where both manganese and cobalt can be in different oxidation states. In case the state of manganese is changed e.g., by interaction with potassium, it will also cause a change in the cobalt oxidation state in the mixed oxide to maintain the neutrality and/or phase parameters. It is known [28] that potassium adsorption on metal surfaces results in substantial lowering of the surface potential on sites adjacent to a potassium atom, and a small, but still significant, lowering of the potential on sites located further away. The long-range effect is caused by a cumulative effect of all potassium atoms on the surface. The promotional effect of K has been already well described in the case of N_2O catalytic decomposition over K-promoted Co–Mn–Al mixed oxides [42]. Potassium, due to its low ionization potential, enables the charge transfer to the transition metal cations inducing an electric field gradient at the surface generated by the resulting dipole and modification of the density of states characteristics.

2.1.5. Reducibility

Reducibility of prepared catalysts was studied by temperature programmed reduction by hydrogen (TPR-H_2). The TPR patterns of the selected fresh and used non-promoted and K-promoted catalysts are depicted in Figure 4. The non-promoted catalyst was reduced in two main temperature regions, 200–600 °C and >600 °C. The maximum temperature used during TPR-H_2 was 600 °C in order to avoid alkali vaporization and damage of the temperature conductivity detector (TCD)—for that reason only low temperature peaks can be visible and discussed. Both reduction maxima consist of overlapping peaks corresponding to the co-effect of several reducible species. The low-temperature peak represents the reduction of $Co^{III} \rightarrow Co^{II} \rightarrow Co^0$ in the Co_3O_4-like phase and the reduction of Mn^{IV} to Mn^{III} oxides. $Mn^{III} \rightarrow Mn^{II}$ reduction can takes place in both temperature regions [43]. In the K-promoted samples (Figure 4b,c), the reduction of K containing phases in the low temperature region also cannot be excluded since the reduction of those species proceeds at temperatures between 200–450 °C [44–46]. The high temperature peak was attributed in the literature to the reduction of Co and Mn ions surrounded by Al ions in a spinel-like phase [47] and to the reduction of Mn^{III}.

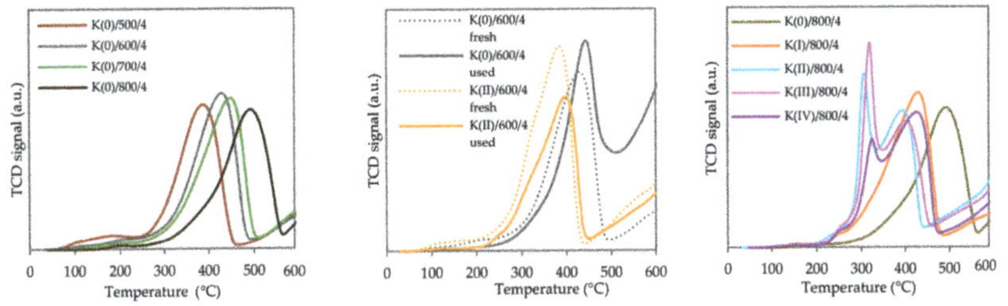

(a) (b) (c)

Figure 4. TPR-H_2 results of (**a**) non-promoted samples calcined at 500–800 °C, (**b**) fresh (dashed lines) and used potassium non-promoted and promoted samples calcined at 600 °C and (**c**) fresh samples calcined at 800 °C containing different amount of K.

For non-promoted samples, the different calcination temperature affected the position of the main low-temperature peak and showed different behavior compared to the promoted samples. While in both cases the low temperature peak shifted a little to higher temperatures for all samples due to the catalytic reaction, the peak increased after the catalytic reaction for non-promoted samples, in contrast to K-promoted samples where the peak area and especially the area of the low-temperature peak decreased after catalytic testing (Figure 4b).

The effect of the K amount for fresh samples calcined at 800 °C is shown in Figure 4c and for used samples calcined at 500 °C it is shown in Figure S5 in Supplementary Materials. The term "used samples" means that they underwent NO catalytic decomposition tests, i.e., in a state corresponding to calcination at 700 °C and influenced by the course of NO decomposition reaction as well. The modification of the Co–Mn–Al mixed oxide by potassium caused significant changes in the obtained reduction profiles. With increasing K content, a shift to lower temperatures accompanied by the broadening and subsequent splitting of the peak into two was observed (Figure 4c). New low temperature maximum appeared at around 300 °C. However, the trend of the new low temperature maximum differs depending on the calcination temperature. For used samples (corresponding to calcination temperature 700 °C), the peak shifted to lower temperatures with increasing amount of potassium. In the case of samples calcined at 800 °C, the peak shifted to lower temperatures at first and then moved back to higher temperatures. The maxima of the main peak shifted to lower temperatures with increasing K amount (Figure S6 in Supplementary Materials) regardless of the used calcination temperature.

The effect of calcination temperature on catalysts reducibility while maintaining the same K amount was also observed. With increasing calcination temperature, the low-temperature peak shifted to higher temperatures, and started dividing into two with low temperature maximum at around 300 °C (Figure S7). A different situation was observed for the samples containing the lowest amount of K (K(I)series), where no apparent changes related to calcination temperature were observed (Figure S8).

2.1.6. Basicity

Catalyst basicity is an important factor influencing chemical reactivity towards NO because of the acidic nature of the NO molecule [48]. In the temperature programmed desorption of CO_2 (TPD-CO_2) profile, several types of basic sites in mixed oxides were reported. Weak basic sites represent –OH groups on the surface of the catalyst, medium sites consist of oxygen in Me^{2+}–O^{2-}, Me^{3+}–O^{2-} pairs and strong basic sites correspond to the isolated O^{2-} anions [49]. However, strict separation of individual peaks above 140 °C reflecting the presence of medium (140–220 °C) or strong (>270 °C) basic sites is not common in the literature.

TPD-CO_2 measurements were performed with the aim to find differences in basicity of the prepared catalysts. Since the catalytic activity achieved at a steady state is closer to the state of the used sample (sample after catalytic reaction) than to the sample in the fresh state, the amount and type of basic sites were mainly studied on the used samples. The comparison of the TPD-CO_2 profile of fresh and used catalysts is shown in Figure 5a. The effect of calcination temperature and potassium content on the amount of basic sites is shown in Figure 5b,c.

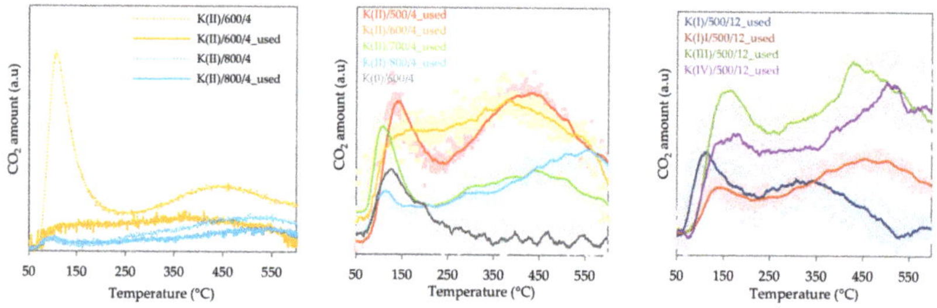

(a) (b) (c)

Figure 5. Temperature programmed reduction (TPD) CO_2 results: (**a**) Fresh (dashed line) and used (full line) samples calcined at 600 or 800 °C for 4 h, (**b**) effect of calcination temperature over K(II)/*y*/4_used samples calcined at different temperatures and (**c**) effect of K amount over used samples (K(*x*)/500/12_used) calcined at 500 °C for 12 h.

In the TPD profiles of our samples, the presence of all types of basic sites was clear or observable but in line with literature results, making a strict distinction was also not possible in our case; even more than three types of sites could be visible in some cases. The sample without the K promoter contains only weak basic sites (Figure 5b), and their amount is much lower than on fresh K-promoted samples. The amount of basic sites changed after the catalytic reaction. The used samples also had a lower amount of medium and strong basic sites in comparison with fresh samples and the type of sites has also changed—the temperature maxima of medium and strong basic sites are lower for the used samples. The most noticeable difference can be seen for the K(II)/600/4 fresh sample containing significant amount of weak basic sites, which disappeared after calcination at a higher temperature and also after the catalytic reaction. This can be explained by the fact that samples calcined at temperatures lower than the reaction temperature were re-calcined during the catalytic tests, so the changes reflect not only reaction-induced changes (the presence of intermediates and reaction products on the surface) but also temperature-induced structural changes observed by XRD and TPR-H_2.

It was found out for all measured samples that the amount of medium and strong basic sites expressed as area under TPD-CO_2 curve from the given temperature up to the end of desorption (600 °C) was linearly dependent on the total amount of basic sites (Figure 6a). The CO_2 consumption decreased with increasing calcination temperature (Figure 6b) and increased with increasing amount of potassium (Figure 6c). From Figure 6c it can be deduced that potassium (0–3.1 wt.%) linearly influenced the amount of basic sites determined in the temperature range higher than 250–400 °C (the intercept of the regression line is almost zero), which is in agreement with our previous work [34].

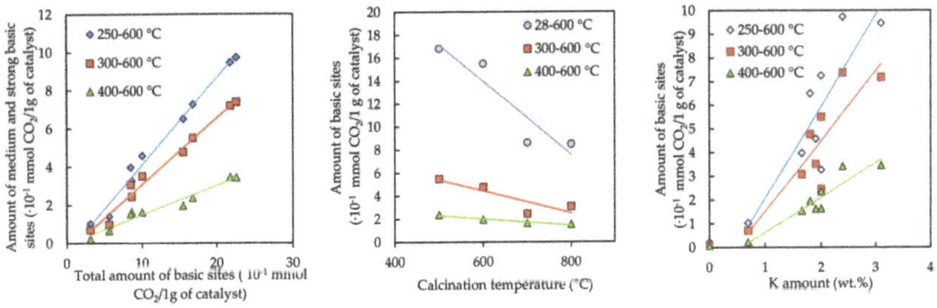

(a) (b) (c)

Figure 6. TPD CO_2 results: (**a**) Dependence of the amount of medium and strong basic sites on the total amount of basic sites in temperature range of 0–600 °C, (**b**) dependence of the amount of basic sites on the calcination temperature and (**c**) dependence of the amount of basic sites on the potassium amount. Numbers in legend mean temperature region from which the amount of sites was calculated.

2.1.7. Phase Composition and Surface Elementary Mapping—TEM and STEM-EDS Analyses

Places with diffraction patterns typical for Co_3O_4 and Mn_3O_4 mixed oxides together with places with amorphous structure (no orderly lattice planes) were identified in the HRTEM images with FFT (fast Fourier transform; S9). This may suggest that some of the constituent elements form an amorphous phase. The results are in agreement with Co and Mn chemical states on the catalysts' surface determined by the XPS measurements and the presence of spinel phase found via XRD.

Scanning transmission microscopy using energy-dispersive X-Ray (EDS) analysis was used for 2D atomic resolution chemical mapping of selected samples. Figure 7 shows results of scanning-transmission electron microscopy (STEM) combined with energy-dispersive X-Ray (EDS) analysis for the K(II)600/4 fresh (a) and K(II)/600/4_used (b) samples. It is obvious that independently of the catalyst state, K is distributed non-homogenously and correlates with the presence of Mn, while the dispersion of cobalt and aluminum is different and more uniform. This indicates that potassium is preferentially bonded to manganese. However, the lack of identified potassium phases in TEM images may suggest that potassium is mainly present in amorphous form (potassium oxide) and/or is non-homogeneously distributed over the surface. This finding correlates with XRD results, where the only potassium containing phase was the K–Mn–O phase (not K–Co–O phase) of the amorphous character, present only in low amounts and thus difficult to detect by XRD.

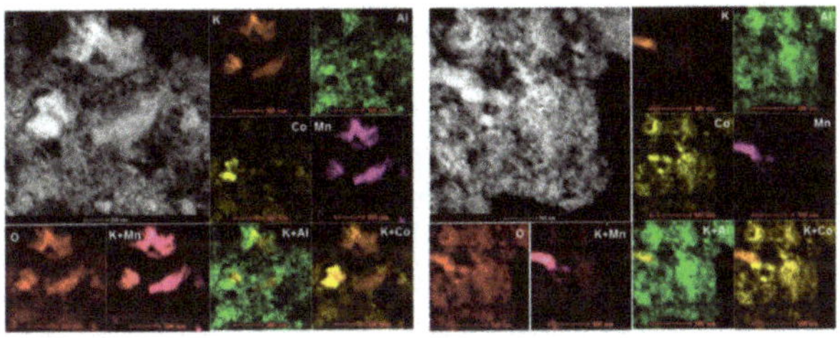

(a) (b)

Figure 7. STEM-energy-dispersive X-Ray (EDS) analysis of K(II)/600/4 fresh (**a**) and used (**b**).

No significant difference was observed between the fresh and used sample.

Particle size measurements of fresh and used samples (Figure 8) showed that after the catalytic reaction, the particles get larger, which is consistent with the results of XRD measurements.

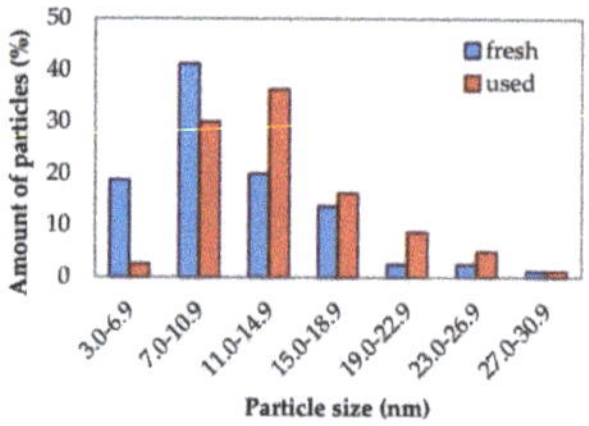

Figure 8. Particle size distribution in fresh and used K(II)/600/4 sample.

2.1.8. Work Function

Since it was previously found that catalytic activity correlates with electronic properties of the catalyst surface for oxidation–reduction reactions [50], the influence of potassium doping on work function (WF) was examined for K(IV) series. Potassium, due to its low ionization potential, transfers charge to the catalyst and by formation of the $K^{\delta+1}$–$O_{surf}{}^{\delta-1}$ surface dipoles modifies the catalyst work function. According to previous studies [42], the lower the work function of the catalyst, the easier is the oxygen release and thus a correlation of electronic properties, reactivity and reducibility can be expected.

The measurement of WF was done at different conditions (in air or vacuum at room temperature or at 150 °C). Since the WF value is very sensitive to the actual surface state, the measurement is strongly influenced by external conditions, so it is more suitable to compare the magnitude of changes in the WF value for the same conditions. In all cases only very small changes of WF were observed for our samples (Table S10). The WF values in vacuum at room temperature are indicative of partial cleaning of the surface. The decrease of WF from 500 to 800 °C shows a decrease in stability of the weakly adsorbed surface water. Since the K desorption starts before the surface is completely cleaned, it may be inferred that at these temperatures, a surface that is partially covered with water is taking part in the reaction.

Relatively low values of work function indicate the appropriate redox properties of tested catalysts, however a direct correlation of WF values and catalytic activity was not found in contrast to N_2O decomposition [42]. This could mean that redox properties are important but are not the only factor influencing catalytic activity.

2.1.9. Species-Resolved Thermal Alkali Desorption

The stability of potassium on the catalyst surface was studied by species-resolved thermal alkali desorption (SR-TAD). Samples calcined at different temperatures (600–800 °C) containing two different amounts of potassium were tested. The maximum SR-TAD measurement temperatures are different for each sample, since the used experimental set up did not allow fixing the temperature to a constant value. The desorption flux of potassium during heating up to 650 °C was dominated by atoms in comparison to ions. This is in line with the obtained relatively low work function values, which means that the electron exchange process is easy. During the thermal desorption experiment the K-surface bond is broken and the obtained desorption parameters (flux intensity, activation energy, Arrhenius pre-exponential factor) describe not only the potassium surface stability but also contain information about its surface state, including dispersion [51]. From the temperature-dependent changes of atomic fluxes, it is obvious that potassium desorption from fresh samples can be measured in vacuum already at temperatures of 400 °C and higher. For that reason, desorption of potassium from the catalyst is possible at the temperatures used for calcination and during NO direct decomposition testing.

In the case of samples with lower amount of K (K(II) samples), non-monotonic curves of the desorption signal dependent on temperature were observed for used samples, while fresh samples showed a monotonic increase. The exponential components of the signal were always dominant. This indicates that the potassium promoter predominantly leaves the surface through a single energy barrier. The local maxima, which appeared at lower temperatures, represent loosely-bonded potassium, and indicate its non-homogenous distribution on the catalysts surface [52]. Redistribution of potassium species upon thermal treatment can thus be identified. There are three different K species in the K(II)/800/4 used sample, which were not present in the fresh samples. Very low measured desorption flux intensities indicate that only small amounts of these species are present. Formation of different K species can be connected with a chemical reaction, e.g., formation of nitrates or due to other reason, e.g., formation of K_2O or clusters of K. First peak could belong to the K–K bond e.g., from K_2O, third peak could be KNO_3 due to a reaction of K with NO. The origin of the second peak was unknown. The second peak was missing for K(II)/600/4 used. However, since no deactivation was observed in long term catalytic experiments (Section 2.2. Direct NO decomposition), the new K species were not significant for the catalytic reaction.

For K(IV) samples calcined at 500 and 600 °C only the third peak of loosely bonded potassium species was present, probably representing KNO_3 species (Figure 9b). The samples calcined at higher temperatures were characterized only by a monotonic increase of signal, which means that loosely bonded potassium species were already removed from the surface. However, the intensities were an order of magnitude higher and were shifted to somewhat higher temperatures than for K(II) samples (Figure 9a). In this case the KNO_3 species were probably residuals from the catalyst's synthesis, which did not decompose during calcination below 600 °C [53].

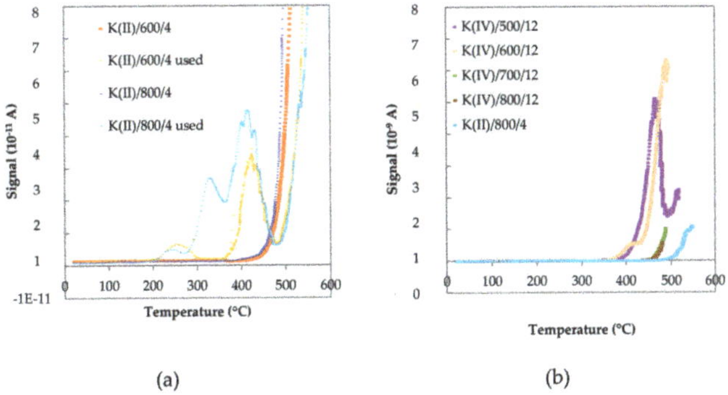

(a) (b)

Figure 9. Atomic K desorption flux as a function of temperature for (**a**) K(II)/600/4 and K(II)/800/4 fresh and used samples and (**b**) used K(II)/800/4 and K(IV)/*y*/12 samples.

It is interesting that the variance between K atomic desorption flux as a function of the temperature obtained during heating and during cooling was different for different samples (not shown). Higher desorption flux during heating than during cooling was observed for K(IV)/500/4 and K(IV)/600/4; higher desorption flux during cooling was observed for K(IV)/700/4, K(IV)/800/4 and K(II)/700/4; almost the same flux during cooling and heating was found for K(II)/600/4 and crossing of the lines during heating and cooling was visible for K(II)/800/4 and for used samples. If the signal during heating is higher than during cooling it means that some potassium was already lost during the heating period while when the dependence was the opposite it means that potassium was being accumulated and segregated on the surface during heating. The K(IV)/500/4 and K(IV)/600/4 had the highest amount of potassium and were calcined at the lowest temperatures. For that reason, the observed process of potassium loss was more pronounced for these samples than for samples with lower amounts of potassium or samples already used in the reaction. Based on the different flux profiles obtained during the heating and cooling processes it could be concluded that potassium migration takes place during SR-TAD measurement via three processes: Diffusion inside the bulk, agglomeration on the surface and thermal desorption [54]. These processes also probably take place during the stabilization period of the catalytic reaction, which for this type of catalyst requires 10–20 h (see Section 2.2.1 Long term stability).

The desorption activation energies for fresh and used samples determined from the linear parts of the corresponding Arrhenius plots during cooling, assuming first order kinetics, are given in Table 8. The activation energies of desorption correspond to the strength of a surface chemical bond, which breaks during the desorption process and can be used as a parameter for the evaluation of K surface stability [52]. The obtained values of E_a clearly showed that the activation energy was decreasing with rising calcination temperature and only a very small difference was detected for fresh and used samples (Table 8). Different activation energies revealed changes in the potassium surface state. The attribution of the observed particular activation energies to specific potassium states could be done based on the reference data published in [18], where E_a ~2–2.5 eV was found to be characteristic for potassium located on cobalt spinel oxide, while E_a ~1.4–1.8 was attributed to potassium placed on aluminum oxide support. In our case, the decrease of activation energies caused by the increase of catalyst calcination temperature could be connected to potassium migration.

Table 8. Activation energies of K desorption for selected samples.

Sample	Activation Energy (eV)
K(IV)/500/12	2.7
K(IV)/600/12	2.6
K(IV)/700/12	2.3
K(IV)/800/12	2.2
K(II)/600/4	2.3
K(II)/700/4	2.3
K(II)/800/4	1.9
K(II)/600/4 used	2.1
K(II)/800/4 used	1.8

2.2. *Direct NO Decomposition*

Direct NO decomposition was studied over all samples described in Section 4.1 Catalyst preration. Two types of measurements, given in Section 4.3 Catalyst measurements were performed.

2.2.1. Long Term Stability

The effect of calcination temperature, calcination time and potassium amount on the long term stability of K-promoted Co–Mn–Al mixed oxides in direct NO decomposition was studied. Time dependences of NO conversion over K-promoted Co–Mn–Al mixed oxides calcined at 500, 600, 700 or 800 °C are depicted in Figure 10a. Time dependences of NO conversion over K-promoted Co–Mn–Al mixed oxides with different amount of potassium promoter are depicted in Figure 10b. Time dependences of NO conversion over K-promoted Co–Mn–Al mixed oxides calcined at 500 °C for 4 h or 12 h are depicted in Figure 10c. All promoted samples showed similar trends. A stable performance was achieved after a 20 h period and was maintained until the end of the catalytic test—for 80 h.

During the catalytic tests, the specific surface area decreased and potassium volatilization took place. For that reason, it was supposed that longer calcination time could shorten the stabilization period of NO decomposition. However, the sample calcined for 12 h showed similar stabilization period (time as well as trend). It means that processes that take place during calcination differ from processes taking place during the stabilization period of NO catalytic decomposition and the latter processes are connected with reaction itself—in other words with the adsorption/desorption processes, the surface coverage by reactants or reaction intermediates or the reaction-induced surface reconstruction, rather than with strictly physical processes.

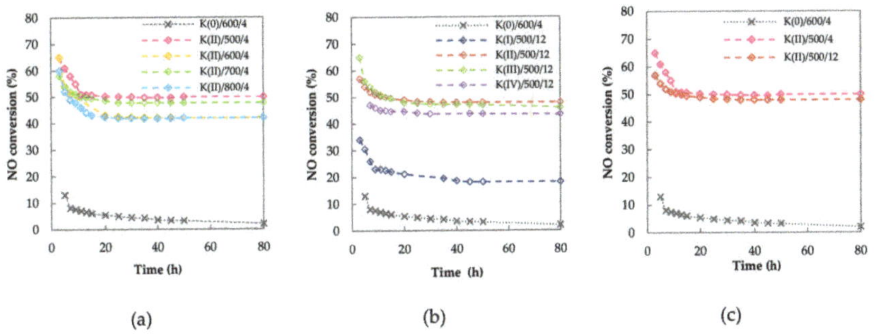

Figure 10. Long-term stability of (**a**) K(II)/*y*/4, (**b**) K(*x*)/500/12 and (**c**) K(II)/500/*z* and K(0)/600/4 in direct NO catalytic decomposition. Conditions: 1000 ppm NO balanced by N_2, T = 650 °C, Gas Hourly Space Velocity (GHSV) = $6\,l\,g^{-1}\,h^{-1}$.

In previous studies it was proposed that the stabilization period is connected with alkali metal vaporization [11], which surely takes place within the time on stream. However, it is interesting that the stabilization process in all cases consists of two parts. The first one is characterized by a fast drop in NO conversion and takes place up to 10 h. Then the conversion drop tends to slow down and lasts for the next 10 h. Since the samples were calcined and some potassium already vaporized during calcination it seems peculiar that the rate of potassium loss followed the fast trend in activity drop (first region) and it seemed that besides alkali vaporization there could be another reason, especially for the first part of the stabilization process of NO catalytic decomposition.

2.2.2. Catalytic Activity

The effect of calcination temperature, calcination time and potassium amount on the catalytic activity of K-promoted Co–Mn–Al mixed oxide in direct NO decomposition was studied in order to define the most suitable conditions at which the most active and stable catalyst can be prepared.

Temperature dependences of NO conversion over K(II) samples calcined for different time periods and at different temperatures are depicted in Figure 11a. The effect of calcination time on catalytic activity was not observed. In Figure 11a, there is moreover a comparison of obtained steady state conversions for K(II)/500/4 sample when different procedures of catalytic measurement were applied. Surprisingly, similar conversions were obtained in case when the sample was heated as well as in case when the sample was stabilized first at 650 °C and then was cooled down. For comparison, the NO conversions over the samples from the same set K(II) calcined at different temperatures are also shown in Figure 11a. No significant differences in conversions were found out for these catalysts. Since a change in the K–Mn–O phase of used samples was observed using XRD, it can be concluded that the effect of K on NO conversion is independent on the form of the K–Mn–O phase in which potassium was incorporated. Gradual crystallization of the spinel up to 800 °C (visible from characterizations) did not affect the catalytic activity.

A different situation was observed for the samples containing a lower amount of potassium (K(I) series, around 0.7 wt.% K, Figure 11 b). The activity measured via a heating process was lower than activity measured via a standard cooling process. Moreover, the samples calcined at higher temperatures always exhibited higher activity than the samples calcined at lower temperatures.

The temperature dependence of NO conversion over K-promoted samples containing various amount of K promoter is shown on Figure 11c. The obtained conversions were different; however, the dependence on K content was not linear. Deeper insight into the observed phenomenon is given in Section 3.

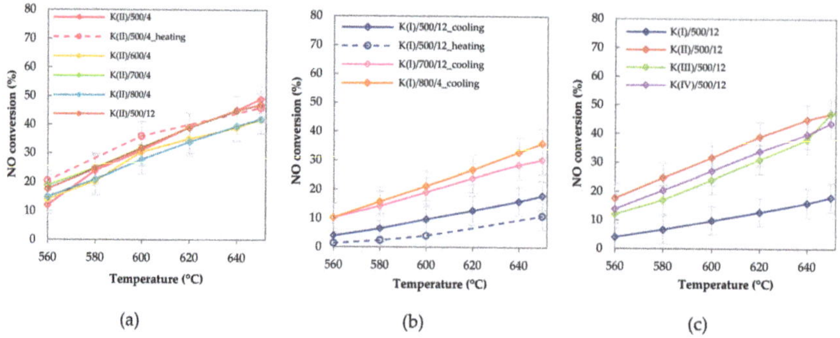

Figure 11. Temperature dependence of the NO conversion (**a**) the effect of calcination temperature and time, K(II)/*y*/*z* catalysts, (**b**) effect of different procedures of catalytic measurement, K(I)/*y*/*z* catalysts and the (**c**) effect of K content, K(*x*)/500/12 catalysts. Conditions: 1000 ppm NO balanced by N_2, $GHSV = 6\,l\,g^{-1}\,h^{-1}$.

3. Discussion

The aim of the work was to correlate physico-chemical properties of catalyst samples with their catalytic activity and stability and to define appropriate conditions for catalyst preparation and parameters influencing de-NO activity. The determination of appropriate preparation conditions is necessary for further optimization of the catalyst. The defined properties can help avoiding phase transformation processes during the catalytic reaction and allow making presumptions about the alkali metal amount in the calcined sample and determining the appropriate way of catalyst testing to compare samples, which are truly comparable, and not only apparently so (which are actually in different states).

Catalytic activity is often affected by the specific surface area of the catalysts. In our case, no dependence of the catalytic activity on surface area was found out. The samples containing a similar amount of K had similar activity regardless of the specific surface area determined both for the fresh and the used catalysts (Figure 12a).

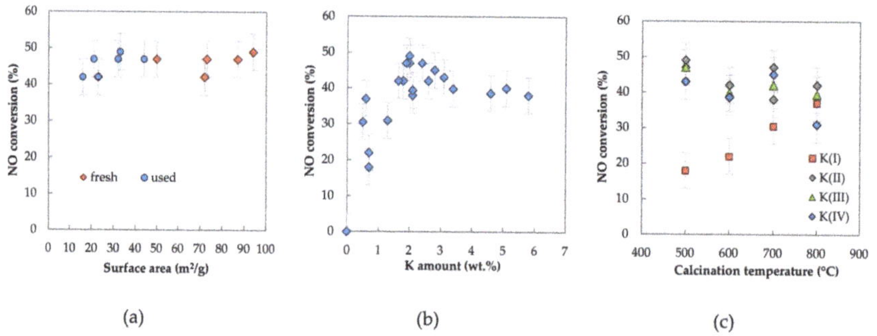

Figure 12. Dependence of NO conversion at 650 °C on (**a**) the surface area of fresh and used K promoted Co–Mn–Al mixed oxides containing similar amount of K (2 wt.%), (**b**) K amount and (**c**) calcination temperature. Conditions: 1000 ppm NO balanced by N_2, GHSV = $6\,l\,g^{-1}\,h^{-1}$.

The most evident parameter influencing catalytic activity was the K amount; however, the effect of the K amount was not so straightforward. While NO conversions over catalysts with K amount higher than 1 wt.% had similar values, the sample K(I)/500/12 (0.7 wt.% K) was significantly less active. For this reason, the dependence of NO conversion on the K amount determined via atomic absorption spectrometry (AAS) in fresh samples was depicted for all tested samples regardless of time and temperature used for calcination to elucidate the effect of K amount on the resulting catalytic activity (Figure 12b). It is obvious that NO conversion reached a maximum at around 2–3 wt.% K and with further K increase, a gradual decrease of NO was observed. Haneda et al. [22] reported that NO decomposition takes place at the interface between the alkali metal and mixed oxide and that an optimal K/Co ratio for the reaction exists. Their results showed a similar trend as our results—initial increase of activity with increasing potassium amount followed by a plateau.

A very interesting feature was observed when the conversion of NO was plotted versus the calcination temperature (Figure 12c). For samples containing more than 1 wt.% K, the conversion of NO did not depend on the calcination temperature up to 700 °C and a small decrease was observed for samples calcined at 800 °C. This was especially noticeable for the samples containing higher amounts of potassium. The fully opposite trend was observed for samples containing below 1 wt.% K—the conversion increased monotonously with rising temperature and samples calcined at 800 °C reached the same conversion as samples containing a higher amount of potassium. The redistribution of potassium on the catalyst surface due to thermal treatment probably took place and the process was sensitive to potassium amount. This is in accordance with the TAD results, where thermally induced processes of potassium diffusion, migration and desorption were confirmed.

NO conversion is not dependent only on the weight percentage of potassium in the sample, but also depends on the normalized potassium loading expressed as number of potassium atoms per m^2 surface area. NO conversion sharply increased at first and after reaching a critical amount of potassium loading (2 K atoms/nm^2), NO conversion slowly decreased with increasing normalized potassium loading (Figure 13).

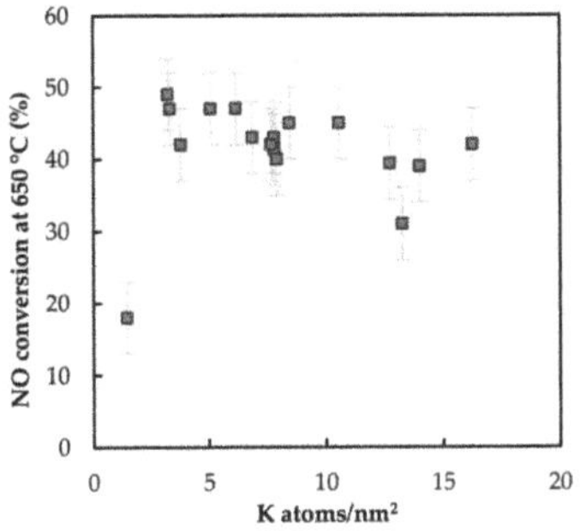

Figure 13. The dependence of NO conversion at 650 °C on normalized potassium loading.

The potassium promoter enhances additional catalyst basicity, thus the effect of K amount on activity should also be related to basic properties of the catalysts. A close correlation of the amount of desorbed CO_2 and the NO decomposition activity on Co_3O_4 mixed oxides was published by Haneda [55], where the activity of supported alkaline earth metal oxide catalysts was not related only to the number of basic sites, but also to the strength of basic sites, irrespective of the type of alkaline earth elements. Hong [56] claimed that the addition of alkaline earth components increased the number of basic sites together with their basicity and also correlated NO decomposition activity with the catalyst basicity. Palomares [57] studied multifunctional catalysts/storage materials based on mixed oxides derived from modified layered double hydroxides and concluded that the role of cobalt sites is to oxidize NO to NO_2. The formed NO_2 was quickly adsorbed on the basic sites of the catalyst to form nitrites and nitrates on the catalyst surface [57]; the addition of sodium increased the alkalinity of the system resulting in a higher storage capacity of NO_x as nitrates. For that reason, alkali metals are often present in NO_x storage/reduction catalysts.

In Figure 6c the results of TPD CO_2 are given also for the samples without potassium, which are not active in NO decomposition ("zero point"). It is evident that when the area of peaks above 250 and 300 °C were taken into account, the "zero point" suited the trend, while in other cases (below and above these temperatures) the zero point did not match the shown dependences. Therefore, it was assumed that basic sites having temperature maxima between 250 and 400 °C could play an important role in NO catalytic decomposition. Simultaneously, the most active samples were those with basic sites characterized by the maximum in TPD spectra at about 430 °C (Figure 14a). The type of sites seems to be more important than their amount if a minimal critical amount of potassium (and thus basic sites) was ensured. A minimal critical amount of potassium, >1 wt.% K, could be estimated from Figure 14b where NO conversion remained constant after a certain amount of basic sites.

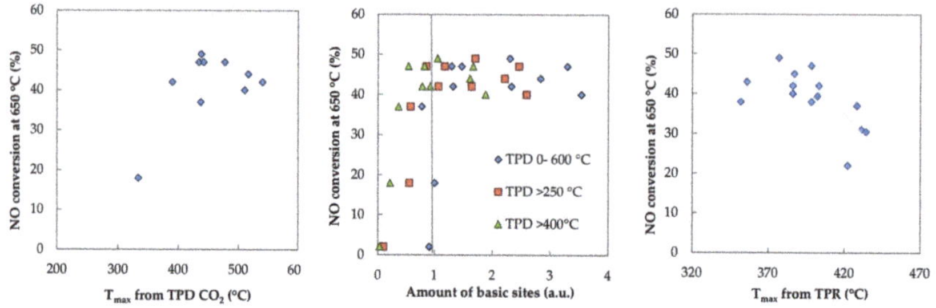

(a) (b) (c)

Figure 14. (**a**) The dependence of NO conversion on the temperature maximum of the main peak acquired from TPD CO_2 results, (**b**) the dependence of NO conversion on the temperature maximum of the main peak acquired from TPR-H_2 and (**c**) the dependence of the NO conversion on the amount of basic sites.

The catalyst's activity was also connected with its reducibility—the optimal temperature (370–400 °C) of the main reduction peak was necessary for achieving the best catalytic performance (Figure 14c). It means that the optimal strength of the transition metal–oxygen bond is necessary for achieving a high activity for NO decomposition. Since the temperature of the maxima of the main peak shifted to lower temperatures with increasing K amount (Figure S6 in Supplementary Materials), the optimum K amount ensuring desired reducibility is essential. Nevertheless, the reducibility is not the only factor influencing catalyst activity, since the work function did not correlate directly with catalyst activity.

In general, NO direct catalytic decomposition proceeds via the following steps [58]: NO adsorption, reactions of adsorbed NO on the surface leading to formation of NO_x species, decomposition of surface NO_x species and desorption of N_2 and O_2. The surface ionic NO_x species are commonly mentioned reaction intermediates of NO decomposition. Haneda [22,58] proposed that over alkali metal doped Co_3O_4 catalysts, NO_2^- species probably react with another NO adspecies resulting in the formation of N_2O_x intermediates, which decompose very fast into gaseous N_2 and adsorbed oxygen species. In the literature, it was most often considered that oxygen desorption is the rate determining step; however, some other steps like the attack of second NO molecule on the oxygen vacancy, NO_2 formation, N_2O formation and decomposition were also considered by different authors [59,60]. Anyway, the suggestion that the formation of N_2 is directly caused from the collision of two activated nitrogen atoms and that formation of O_2 is combined by the two adsorbed oxygen atoms is valid across a lot of different studies [60].

The active catalytic material has to ensure different functionalities according to the NO decomposition mechanism. The role of transition metals is to oxidize NO to NO_2 [61], while the formed NO_2 is adsorbed on the basic sites of the catalyst to form nitrites and nitrates. Here the charge neutrality for the formation of NO_2^- or NO_3^- species can be retained by oxidation of Me^{x+} cation to form Me^{x+1}, where Me is Co or Mn. In order to close the catalytic cycle via oxygen desorption, the presence of transition metal with appropriate redox properties is again important.

Our results show that potassium presence in Co–Mn–Al mixed oxide influences not only basic properties connected to NO adsorption and NO_x storage, but also redox properties, which are necessary for the activation of NO molecule and oxygen desorption.

4. Materials and Methods

4.1. Catalyst Preparation

The Co–Mn–Al layered double hydroxide precursor with the Co:Mn:Al molar ratio of 4:1:1 was prepared by coprecipitation of corresponding nitrates in an alkaline Na_2CO_3/NaOH solution. The precipitate was filtered off, washed with water, dried and calcined at 500, 600, 670, 700 or 800 °C for 4 or 12 h in static air.

Samples modified with potassium promoter were prepared by the bulk promotion method. The washed precipitate of Co–Mn–Al layered double hydroxide precursor was dispersed in an aqueous KNO_3 solution of different concentrations and the dried filtration cake was calcined in air at 500–800 °C for 4 or 12 h and denoted as K(x)/y/z, where x means modification by different K amounts (the same sample group means that sample was modified by the same amount of K before calcination), y means calcination temperature (°C) and z means calcination time (h).

The catalytic activity was measured for 24 samples (K(II)/500/4, K(II)/600/4, K(II)/700/4, K(II)/800/4, K(I)/500/12, K(II)/500/12, K(III)/500/12, K(IV)/500/12, K(I)/600/4, K(I)/800/4, K(II)/700/12, K(III)/700/12, K(IV)/700/12, K(III)/800/12, K(IV)/800/12, K(III)/600/12, K(IV)/600/12, K(I)/700/12, K(0)/800/4, K(0)/700/4, K(0)/600/4, K(0)/670/4, K(V)/700/4 and K(V)/800/4), the characterization was done only for selected samples.

4.2. Catalyst Characterization

Chemical composition of the laboratory prepared catalysts was determined by atomic absorption spectrometry (AAS) or inductive coupled plasma (ICP) spectrometry. K, Co, Mn and Al were analyzed using AAS on a ContrAA 700 atomic absorption spectrometer (Analytik Jena AG, Jena, Germany) using flame atomization technique. Approximately 50 mg of the sample in powder form was weighed in teflon vessels of a SK-15 rotor of a microwave digestion system (Milestone Ethos Up, Sorisole, Italy). Nitric (2 mL) and hydrochloric (6 mL) acids were added (both p.a. (proanalysis) purity, Penta chemicals, Prague, Czech Republic) to form aqua regia and vessels were closed. The digestion program has three steps with power output set to maximum (1800 W, not applied full time): (1) 10-min ramp from laboratory temperature to 220 °C; (2) 15 min hold at 220 °C and (3) 30 min cooling. After cooling, vessels were opened, the sample solutions transferred into 50 mL volumetric flasks, washings were added to the sample and the flasks were filled up to 50 mL with deionized water. With each batch of digestions one blank digestion was performed. For each element different wavelength and flame type was applied: 766.49 nm and C_2H_2/Air for K, 240.73 nm and C_2H_2/Air for Co, 279.48 nm and C_2H_2/Air for Mn and 396.15 nm and C_2H_2/N_2O for Al.

For ICP measurements the Agilent 725 ICP OES Spectrometer (Agilent Technologies, Mulgrave, Australia) was used. 0.1 g of sample in powder form was dissolved in aqua regia and transferred to a 250 mL volumetric flask. Metals (K, Co, Mn, Al) were analyzed on an emission spectrometer with inductively coupled plasma using the calibration curve method with standard solutions by the Merck company.

Phase composition and microstructural properties were determined using the X-ray powder diffraction (XRD) technique. XRD patterns were obtained using Rigaku SmartLab powder diffractometer (Rigaku, Tokyo, Japan) with a D/teX Ultra 250detector. The source of X-ray irradiation was Co tube (CoKα, $\lambda1$ = 0.178892 nm, $\lambda2$ = 0.179278 nm) operated at 40 kV and 40 mA. Incident and diffracted beam optics were equipped with 5° Soller slits; incident slits were set up to irradiate a 10 × 10 mm area of the sample (automatic divergence slits) constantly. Slits on the diffracted beam were set up to a fixed value of 8 and 14 mm. The powder samples were gently grinded using agate mortar before analysis and pressed using microscope glass in a rotational sample holder and measured in the reflection mode (Bragg–Brentano geometry). The samples rotated (30 rpm) during the measurement to eliminate the preferred orientation effect. The XRD patterns were collected in a 2θ range 5–90° with a step size of 0.01° and speed of 0.5° min^{-1}. Measured XRD patterns were evaluated using the PDXL 2 software (version 2.4.2.0) and compared with the PDF-2database, release 2015.XRD patterns were analyzed using the LeBail method (software PDXL2) to refine lattice parameters of the spinel like phase. Background of the patterns was determined using B-Spline function, peak shapes were modeled with a pseudo-Voigt function accounting for a peak asymmetry due to axial divergence. Mean coherent domain corresponding to spinel crystallite size was calculated using the Scherrer's formula with correction to the instrumental broadening. Mean coherent domain corresponding to spinel crystallite size was calculated using Scherrer's formula from spinel hkl reflections (220), (211), (400), (511) and (440). The calculated value was corrected to the instrumental broadening using silicon standard (NIST SRM 640d).

Superficial elemental analyses were performed by XPS (X-ray Photoelectron Spectrometer ESCA 3400, Kratos, Manchester, England; base pressure better than $5{\cdot}10^{-7}$ Pa), using a polychromatic Mg X-ray source (Mg Kα, 1253.4 eV). Powder for the XPS measurement was impressed into a carbon tape (double sided adhesive carbon tape, SPI Supplies). The carbon tape was completely covered by the powder; therefore, only a very low signal originated from the carbon tape itself. Composition of the powder was determined without any annealing (Ar^+ ion gun, 0.5 kV, 10 mA, 90 s). For all spectra, the Shirley background was subtracted and elemental compositions of layers were calculated from the areas of Co 2p, Mn 2p, Al 2p, O 1s, C 1s, Na and K 2p. The carbon correction was done by setting the position of C 1s peak to 284.8 eV.

The electron microscope Tecnai G2 20 X-TWIN FEI Company, equipped with: A LaB6 gun, HAADF detector and an EDS spectrometer (energy dispersive X-Ray spectroscopy) (FEI company, Eindhoven, The Netherlands) was used to display the prepared catalysts. Microscopic studies of the catalysts were carried out at an accelerating voltage of the electron beam equal to 200 kV. The elements mapping was carried out in the STEM mode by collecting point by point the EDS spectrum of each of the corresponding pixels in the map. The collected maps were presented in the form of a matrix of pixels with the color mapped significant element and the intensity corresponding to the percentage of the element. Before STEM measurement, the catalysts were grinded in an agate mortar to a fine powder. The resulting powders were mixed with 99.8% ethanol (POCH) to form a slurry, which was subsequently inserted into an ultrasonic homogenizer for 20 s. Then, the slurry containing the catalyst was pipetted and supported on a 200 mesh copper grid covered with Lacey Formvar and stabilized with carbon (Ted Pella Company, Redding, USA) and left on the filter paper until the ethanol has evaporated. Subsequently, the samples deposited on the grid were inserted into a holder and placed into the electron microscope.

N_2 physisorption at −196 °C was performed on AutoChem II (Micromeritics, Atlanta, GA, USA) and the one point BET method was used for specific surface area evaluation. Before analysis, each sample was degassed for one hour at 450 °C in He flow of 50 mL/min. After degassing, the mixture of 30 mol. % N_2 and 70 mol. % He was applied to the sample that was immersed in liquid nitrogen. The amount of N_2 adsorbed at liquid nitrogen temperature was used to calculate the surface area.

Nitrogen physisorption measurements were also carried out on the 3Flex adsorption apparatus (Micromeritics, Atlanta, GA, USA). The nitrogen adsorption–desorption isotherms were measured at 77 K after 120 h degassing of the samples at 350 °C. The specific surface area, S_{BET}, was calculated according to the classical Brunauer-Emmett-Teller (BET) theory for the p/p_0 range = 0.05–0.30. Pore-size distribution was calculated from the adsorption branch of the nitrogen adsorption–desorption isotherm using the Barret–Joyner–Halenda (BJH) method, the empirical Broekhoff–De Boer standard isotherm and assuming cylindrical pore geometry.

The temperature programmed reduction by hydrogen (TPR-H_2) was carried out on Autochem II 2920 (Micromeritics, Atlanta, GA, USA) with a sample amount 0.07 g. Prior to the TPR experiments, the samples were outgassed in the flow of pure argon (50 mL/min) at 600 °C for 60 min, cooled down to 30 °C in the same atmosphere and afterwards the TPR runs were performed in the flow of 10 mol.% H_2/Ar (50 mL/min) with a heating rate 20 °C/min in the temperature range of 30–600 °C. After reaching 600 °C, the temperature was kept constant for 20 min. A cold trap (−78 °C) was applied to eliminate the water evolved during experiments.

Temperature programmed desorption of CO_2 (TPD-CO_2) was carried out on Autochem II 2920 (Micromeritics, Atlanta, GA, USA) connected on-line to a mass spectrometer (Prevac). A sample amount of 0.07 g was used. The procedure involved the activation of the samples in flowing He (50 mL/min) at 600 °C (1 h), cooling to 28 °C (adsorption temperature), and adsorbing CO_2 from a stream of He-CO_2 (50 %; 1 h). The flow rate of the gas mixture was 50 mL/min. To remove physically adsorbed CO_2, the samples were purified for 110 min in the helium stream (50 mL/min) at 28 °C and subsequent desorption (TPD-CO_2) was carried out in the temperature range of 28–600°C (20 °C/min) under helium flow (50 mL/min).

Thermal stability of alkali metal promoters was investigated by the species resolved thermal alkali desorption (SR-TAD) method. The experiments were carried out in a vacuum apparatus with a background pressure of 10^{-7} kPa. The samples in the form of wafers (10 mm diameter, 100 mg weight) were heated up from room temperature to 650 °C in a stepwise mode at the rate of 5 °C min^{-1}. The desorption flux of potassium atoms was determined by means of a surface ionization detector, whereas the flux of ions was determined with an ion collector. During the measurements, the samples were biased with a positive potential (+10 and +100 V for atoms and ions, respectively) to quench the thermal emission of electrons and additionally, in the case of ions, to accelerate them towards the collector. In this way, the possibility of reneutralization of ions by thermal electrons outside of the surface is effectively eliminated. In all measurements, the resulting positive current was directly measured with a Keithley 6512 digital electrometer (Cleveland, USA).

The redox properties of the catalysts were investigated by work function measurement using the Kelvin Probe (KP) method. The experiments were carried out in a vacuum apparatus with a background pressure of 10^{-7} kPa. The samples in the form of wafers (10 mm diameter, 100 mg weight) were measured first at room temperature and pressure, then at room temperature and vacuum, then heated to 400 °C at the rate of 20 °C min^{-1} and then after 30 min left to cool down to 150 °C and measured at this temperature, and after cooling down to room temperature measured again. For the measurements a KP 6500 (McAllister Technical Services, Coeur d'Alene, Idaho, USA) reference electrode with a work function ϕ_{ref} = 4.1 eV was used. The work function of a sample was evaluated based on 20 independent values of contact potential difference between the sample and the reference electrode.

Characterization of the catalysts was done for fresh samples but in some cases also for samples used in catalytic tests, since some changes of the physico-chemical properties induced by the NO decomposition reaction as well as resulting from long term high temperature exposition were expected.

4.3. Catalytic Measurements

Catalytic NO decomposition was performed in an integral fixed bed stainless steel reactor of 6 mm internal diameter. The catalyst bed contained 0.5 g of the sample with a particle size of 0.16–0.315 mm. Total flow was kept at 49 mL min^{-1} (20 °C, atmospheric pressure). An infrared spectrometer Ultramat 6 (Siemens, Karlsruhe, Germany) was used for NO online analysis. The low-temperature NO_2/NO converter (TESO Ltd., Prague, Czech Republic) was connected at the bypassed line before the NO analyzer and the gas for analysis was periodically switched through the convertor in order to analyze the sum of NO_x and thus control the amount of NO_2. The presence of N_2O species was controlled on an FTIR spectrometer Antaris IGS (Nicolet, Prague, Czech Republic). The measurement error of NO conversion was determined by repeated measurements as ±5% (absolute error).

The typical procedure was as follows: After catalyst activation (heating in N_2, then 1 h in N_2 at 650 °C), the NO catalytic decomposition at 650 °C was measured for 50 h to reach steady state. After this period, when stable performance was achieved, the temperature dependence of NO conversion was determined with cooling rate of 5 °C min^{-1} and the catalysts activity was measured for 3 h at each temperature (640 °C, 620 °C, 600 °C, 580 °C and 560 °C). After the conversion curve measurement, the reactor was heated back to 650 °C and catalysts stability after 80 h was evaluated. In case the performance was stable, the catalyst was heated to 700 °C and steady state NO conversion at 700 °C was measured. In order to elucidate phases evolution in the studied K/Co–Mn–Al system and to choose the best calcination temperature related to catalytic activity as well as to catalyst stability, the calcination temperatures of 500 and 600 °C were also applied.

For two selected samples (K(II)/500/4, K(II)/600/4), the catalytic activity measurement was performed not only by the procedure given above but it was also adjusted as follows: After catalyst activation (heating in N_2, then 1 h in N_2 at 650 °C), the NO catalytic decomposition at 500 °C was measured until reaching stable performance. After that the temperature was raised to 560 °C and the measurement was performed until reaching stable performance. The same step was repeated for 600 and 650 °C. For the rest of the samples calcined at 500 or 600 °C measured by the described typical procedure, it is necessary to keep in mind that they were actually re-calcined to 650 °C during the activation and measurement of NO catalytic decomposition at 650 °C.

5. Conclusions

From the results obtained in this work, we can conclude that Co-Mn-Al oxides promoted by potassium are active in direct NO decomposition. The activity is given by the diversity of active surface sites. The presence of potassium improves catalysts' basicity and reducibility, both factors influencing the catalytic deNO activity.

The calcination temperature affects the specific surface area, pore size, reducibility, basicity, the alkali metal volatilization process, the amount of residual potassium (after volatilization) and phase composition of the catalyst. XRD measurements showed that the presence of alkali metal influenced the position of atoms in the spinel unit cell as well as the shape and size of the unit cell. The calcination time affects the specific surface area, amount of residual potassium and catalyst crystallinity. The type of the potassium-containing phase is temperature dependent; however, if a sufficient amount of potassium promoter is present at the surface, the surface state of K does not influence NO conversion. The best catalytic performance was achieved for catalysts containing more than 1 wt.% of K. The calcination temperatures of 700 and 800 °C were found to be the best for direct NO decomposition regarding achieved activity as well as potassium promoter stability.

The schematic diagram of the catalyst changes taking place during preparation and NO catalytic decomposition is given in Figure 15.

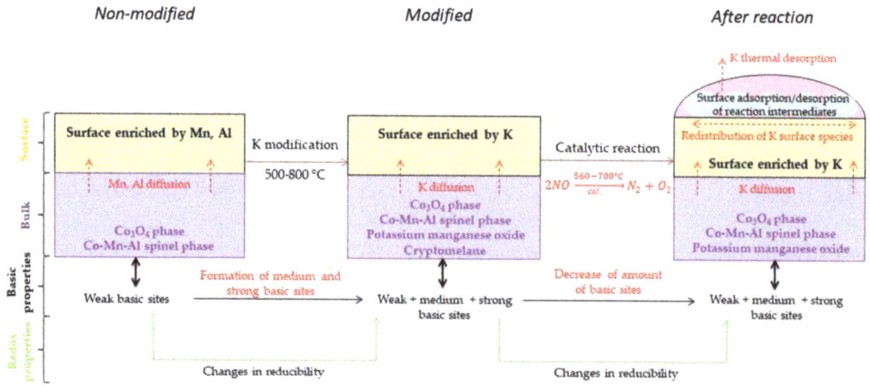

Figure 15. The schematic diagram of the potassium modified Co–Mn–Al mixed oxide catalyst changes during preparation and NO catalytic decomposition.

The resulting potassium phases differ according to the applied potassium amount and the temperature used for calcination, which are subject to subsequent changes (diffusion, migration, redistribution, volatilization) during the catalytic process. However, if the minimal critical amount of potassium is reached (>1 wt.% K), a stable and active catalyst is obtained.

Supplementary Materials: The following are available online at http://www.mdpi.com/2073-4344/9/7/593/s1, Figure S1: XRD results for samples K(II)/*y*/4., Figure S2: XRD results for samples K(*x*)/600/*z*., Figure S3: XRD results for samples K(*x*)/800/*z*., Figure S4: XPS results., Figure S5: TPR H_2—effect of potassium amount., Figure S6: T_{max} (from TPR H_2) dependence on K amount, Figure S7: TPR H_2—effect of calcination temperature for samples K(II)/*y*/4., Figure S8: TPR H_2—effect of calcination temperature for samples K(I)/*y*/*z*., Table S9: Work function of selected samples.

Author Contributions: Conceptualization, K.P. and T.B.; methodology, K.P.; validation, T.B., L.O., A.K. and K.J.; investigation, T.B., K.P., K.K., A.K., W.K., T.K., D.F., M.K., P.S. and J.B.; data curation, T.B., D.F., K.P., K.K., A.K., W.K., P.S. and T.K.; writing—original draft preparation, K.P..; writing—review and editing, K.P. and T.B.; supervision, L.O.; project administration, K.P., L.O. and K.J.

Funding: This research was funded by ERDF "Institute of Environmental Technology—Excellent Research" (No. CZ.02.1.01/0.0/0.0/16_019/0000853), by the Ministry of Education, Youth and Sports of the Czech Republic- project No. LM2012039 and LM2015039, by Czech Science Foundation—project No. 18-19519S and by VSB-TU internal student project No. SP2019-91.

Acknowledgments: The authors are grateful to Lenka Matějová for nitrogen physisorption and Alexandr Martaus for XRD analyses.

Conflicts of Interest: The authors declare no conflict of interest. The funders had no role in the design of the study; in the collection, analyses, or interpretation of data; in the writing of the manuscript, or in the decision to publish the results.

References

1. Falsig, H.; Bligaard, T.; Rass-Hansen, J.; Kustov, A.; Christensen, C.; Nørskov, J. Trends in catalytic NO decomposition over transition metal surfaces. *Top. Catal.* **2007**, *45*, 117–120. [CrossRef]
2. Garin, F. Mechanism of NO_x decomposition. *Appl. Catal. A* **2001**, *222*, 183–219. [CrossRef]
3. Brown, W.A.; King, D.A. NO Chemisorption and Reactions on Metal Surfaces: A New Perspective. *J. Phys. Chem. B* **2000**, *104*, 2578–2595. [CrossRef]
4. Inderwildi, O.R.; Jenkins, S.J.; King, D.A. When adding an unreactive metal enhances catalytic activity: NO_x decomposition over silver–rhodium bimetallic surfaces. *Surf. Sci.* **2007**, *601*, L103–L108. [CrossRef]
5. Kumar, A.; Medhekar, V.; Harold, M.P.; Balakotaiah, V. NO decomposition and reduction on Pt/Al_2O_3 powder and monolith catalysts using the TAP reactor. *Appl. Catal. B* **2009**, *90*, 642–651. [CrossRef]
6. Modén, B.; Da Costa, P.; Fonfé, B.; Lee, D.K.; Iglesia, E. Kinetics and Mechanism of Steady-State Catalytic NO Decomposition Reactions on Cu–ZSM5. *J. Catal.* **2002**, *209*, 75–86. [CrossRef]
7. Ganemi, B.; Björnbom, E.; Paul, J. Conversion and in situ FTIR studies of direct NO decomposition over Cu-ZSM5. *Appl. Catal. B* **1998**, *17*, 293–311. [CrossRef]
8. Shi, Y.; Pan, H.; Li, Z.; Zhang, Y.; Li, W. Low-temperature decomposition of NO_x over Fe–Mn/H-beta catalysts in the presence of oxygen. *Catal. Commun.* **2008**, *9*, 1356–1359. [CrossRef]
9. Haneda, M.; Kintaichi, Y.; Hamada, H. Surface reactivity of prereduced rare earth oxides with nitric oxide: New approach for NO decomposition. *Phys. Chem. Chem. Phys.* **2002**, *4*, 3146–3151. [CrossRef]
10. Hamada, H.; Kintaichi, Y.; Sasaki, M.; Ito, T. Silver-promoted cobalt oxide catalysts for direct decomposition of nitrogen oxides. *Chem. Lett.* **1990**, *19*, 1069–1070. [CrossRef]
11. Pacultová, K.; Draštíková, V.; Chromčáková, Ž.; Bílková, T.; Kutláková, K.M.; Kotarba, A.; Obalová, L. On the stability of alkali metal promoters in Co mixed oxides during direct NO catalytic decomposition. *Mol. Catal.* **2017**, *428*, 33–40. [CrossRef]
12. Haneda, M.; Kintaichi, Y.; Bion, N.; Hamada, H. Alkali metal-doped cobalt oxide catalysts for NO decomposition. *Appl. Catal. B Environ.* **2003**, *46*, 473–482. [CrossRef]
13. Cheng, J.; Wang, X.; Ma, C.; Hao, Z. Novel Co-Mg-Al-Ti-O catalyst derived from hydrotalcite-like compound for NO storage/decomposition. *J. Environ. Sci.* **2012**, *24*, 488–493. [CrossRef]
14. Park, P.W.; Kil, J.K.; Kung, H.H.; Kung, M.C. NO decomposition over sodium-promoted cobalt oxide. *Catal. Today* **1998**, *42*, 51–60. [CrossRef]

15. Neurock, M. Theory-Aided Catalyst Design. In *Design of Heterogeneous Catalysts*; Wiley-VCH Verlag GmbH & Co. KGaA: Weinheim, Germany, 2009; pp. 231–258. [CrossRef]
16. Kotarba, A.; Kruk, I.; Sojka, Z. Energetics of Potassium Loss from Styrene Catalyst Model Components: Reassignment of K Storage and Release Phases. *J. Catal.* **2002**, *211*, 265–272. [CrossRef]
17. Bieniasz, W.; Trębala, M.; Sojka, Z.; Kotarba, A. Irreversible deactivation of styrene catalyst due to potassium loss—Development of antidote via mechanism pinning. *Catal. Today* **2010**, *154*, 224–228. [CrossRef]
18. Grzybek, G.; Wójcik, S.; Legutko, P.; Gryboś, J.; Indyka, P.; Leszczyński, B.; Kotarba, A.; Sojka, Z. Thermal stability and repartition of potassium promoter between the support and active phase in the K-$Co_{2.6}Zn_{0.4}O_4$|α-Al_2O_3 catalyst for N_2O decomposition: Crucial role of activation temperature on catalytic performance. *Appl. Catal. B* **2017**, *205*, 597–604. [CrossRef]
19. Cheng, H.; Huang, Y.; Wang, A.; Li, L.; Wang, X.; Zhang, T. N_2O decomposition over K-promoted Co-Al catalysts prepared from hydrotalcite-like precursors. *Appl. Catal. B* **2009**, *89*, 391–397. [CrossRef]
20. Kotarba, A.S.G.; Ciura, K.; Indyka, P.; Grybos, J.; Grzybek, G.; Legutko, P.; Stelmachowski, P.; Sojka, Z. Alkali Promotion for Enhancing the Activity of Bulk and Supported Cobalt Spinel Catalysts for Low-Temperature deN_2O, KRAcat de N_2O. In Proceedings of the International Thematic Workshop KRACat-deN_2O, Krakow, Poland, 4–6 May 2016..
21. Cheng, J.; Wang, X.; Yu, J.; Hao, Z.; Xu, Z.P. Sulfur-Resistant NO Decomposition Catalysts Derived from Co–Ca/Ti–Al Hydrotalcite-like Compounds. *J. Phys. Chem. C* **2011**, *115*, 6651–6660. [CrossRef]
22. Haneda, M.; Nakamura, I.; Fujitani, T.; Hamada, H. Catalytic Active Site for NO Decomposition Elucidated by Surface Science and Real Catalyst. *Catal. Surv. Asia* **2005**, *9*, 207–215. [CrossRef]
23. Winter, E.R.S. The catalytic decomposition of nitric oxide by metallic oxides. *J. Catal.* **1971**, *22*, 158–170. [CrossRef]
24. Tsujimoto, S.; Masui, T.; Imanaka, N. Fundamental Aspects of Rare Earth Oxides Affecting Direct NO Decomposition Catalysis. *Eur. J. Inorg. Chem.* **2015**, *2015*, 1524–1528. [CrossRef]
25. Hong, W.-J.; Iwamoto, S.; Hosokawa, S.; Wada, K.; Kanai, H.; Inoue, M. Effect of Mn content on physical properties of CeO_x–MnOy support and BaO–CeO_x–MnOy catalysts for direct NO decomposition. *J. Catal.* **2011**, *277*, 208–216. [CrossRef]
26. Falsig, H.; Bligaard, T.; Christensen, C.H.; Nørskov, J.K. Direct NO decomposition over stepped transition-metal surfaces. *Pure Appl. Chem.* **2007**, *79*, 1895–1903. [CrossRef]
27. Jakubek, T.; Kaspera, W.; Legutko, P.; Stelmachowski, P.; Kotarba, A. Surface versus bulk alkali promotion of cobalt-oxide catalyst in soot oxidation. *Catal. Commun.* **2015**, *71*, 37–41. [CrossRef]
28. Niemantsverdriet, J.W. *Spectroscopy in Catalysis: An Introduction*; Willey: Hoboken, NJ, USA, 2007. [CrossRef]
29. Raghuvanshi, S.; Mazaleyrat, F.; Kane, S.N. $Mg_{1-x}Zn_xFe_2O_4$ nanoparticles: Interplay between cation distribution and magnetic properties. *AIP Adv.* **2017**, *8*, 047804. [CrossRef]
30. Kovanda, F.; Rojka, T.; Dobešová, J.; Machovič, V.; Bezdička, P.; Obalová, L.; Jirátová, K.; Grygar, T. Mixed oxides obtained from Co and Mn containing layered double hydroxides: Preparation, characterization, and catalytic properties. *J. Solid State Chem.* **2006**, *179*, 812–823. [CrossRef]
31. Deraz, N.M. Formation and Characterization of Cobalt Aluminate Nano-Particles. *Int. J. Electrochem. Sci.* **2013**, *8*, 4036–4046.
32. Scofield, J.H. Hartree-Slater subshell photoionization cross-sections at 1254 and 1487 eV. *J. Electron. Spectrosc. Relat. Phenom.* **1976**, *8*, 129–137. [CrossRef]
33. Obalová, L.; Pacultová, K.; Balabánová, J.; Jirátová, K.; Bastl, Z.; Valášková, M.; Lacný, Z.; Kovanda, F. Effect of Mn/Al ratio in Co–Mn–Al mixed oxide catalysts prepared from hydrotalcite-like precursors on catalytic decomposition of N_2O. *Catal. Today* **2007**, *119*, 233–238. [CrossRef]
34. Obalová, L.; Karásková, K.; Jirátová, K.; Kovanda, F. Effect of potassium in calcined Co–Mn–Al layered double hydroxide on the catalytic decomposition of N_2O. *Appl. Catal. B* **2009**, *90*, 132–140. [CrossRef]
35. Biesinger, M.C.; Payne, B.P.; Grosvenor, A.P.; Lau, L.W.M.; Gerson, A.R.; Smart, R.S.C. Resolving surface chemical states in XPS analysis of first row transition metals, oxides and hydroxides: Cr, Mn, Fe, Co and Ni. *Appl. Surf. Sci.* **2011**, *257*, 2717–2730. [CrossRef]
36. Gautier, J.L.; Rios, E.; Gracia, M.; Marco, J.F.; Gancedo, J.R. Characterisation by X-ray photoelectron spectroscopy of thin $Mn_xCo_{3-x}O_4(1 \geq x \geq 0)$ spinel films prepared by low-temperature spray pyrolysis. *Thin Solid Films* **1997**, *311*, 51–57. [CrossRef]

37. Todorova, S.; Kolev, H.; Holgado, J.P.; Kadinov, G.; Bonev, C.; Pereñíguez, R.; Caballero, A. Complete n-hexane oxidation over supported Mn–Co catalysts. *Appl. Catal. B* **2010**, *94*, 46–54. [CrossRef]
38. Zhou, M.; Cai, L.; Bajdich, M.; García-Melchor, M.; Li, H.; He, J.; Wilcox, J.; Wu, W.; Vojvodic, A.; Zheng, X. Enhancing Catalytic CO Oxidation over Co_3O_4 Nanowires by Substituting Co^{2+} with Cu^{2+}. *ACS Catal.* **2015**, *5*, 4485–4491. [CrossRef]
39. Chen, Z.; Kronawitter, C.X.; Koel, B.E. Facet-dependent activity and stability of Co_3O_4 nanocrystals towards the oxygen evolution reaction. *PCCP* **2015**, *17*, 29387–29393. [CrossRef]
40. Bhatnagar, A.; Jain, A.K. A comparative adsorption study with different industrial wastes as adsorbents for the removal of cationic dyes from water. *J. Colloid Interface Sci.* **2005**, *281*, 49–55. [CrossRef]
41. Kim, S.C.; Shim, W.G. Catalytic combustion of VOCs over a series of manganese oxide catalysts. *Appl. Catal. B* **2010**, *98*, 180–185. [CrossRef]
42. Obalová, L.; Maniak, G.; Karásková, K.; Kovanda, F.; Kotarba, A. Electronic nature of potassium promotion effect in Co–Mn–Al mixed oxide on the catalytic decomposition of N_2O. *Catal. Commun.* **2011**, *12*, 1055–1058. [CrossRef]
43. Klyushina, A.; Pacultová, K.; Karásková, K.; Jirátová, K.; Ritz, M.; Fridrichová, D.; Volodarskaja, A.; Obalová, L. Effect of preparation method on catalytic properties of Co-Mn-Al mixed oxides for N_2O decomposition. *J. Mol. Catal. A Chem.* **2016**, *425*, 237–247. [CrossRef]
44. Santos, V.P.; Soares, O.S.G.P.; Bakker, J.J.W.; Pereira, M.F.R.; Órfão, J.J.M.; Gascon, J.; Kapteijn, F.; Figueiredo, J.L. Structural and chemical disorder of cryptomelane promoted by alkali doping: Influence on catalytic properties. *J. Catal.* **2012**, *293*, 165–174. [CrossRef]
45. Becerra, M.E.; Arias, N.P.; Giraldo, O.H.; López Suárez, F.E.; Illán Gómez, M.J.; Bueno López, A. Soot combustion manganese catalysts prepared by thermal decomposition of $KMnO_4$. *Appl. Catal. B* **2011**, *102*, 260–266. [CrossRef]
46. Da Costa-Serra, J.F.; Chica, A. Catalysts based on Co-Birnessite and Co-Todorokite for the efficient production of hydrogen by ethanol steam reforming. *Int. J. Hydrog. Energy* **2018**, *43*, 16859–16865. [CrossRef]
47. Obalová, L.; Karásková, K.; Wach, A.; Kustrowski, P.; Mamulová-Kutláková, K.; Michalik, S.; Jirátová, K. Alkali metals as promoters in Co–Mn–Al mixed oxide for N_2O decomposition. *Appl. Catal. A* **2013**, *462–463*, 227–235. [CrossRef]
48. Imanaka, N.; Masui, T. Advances in direct NO_x decomposition catalysts. *Appl. Catal. A* **2012**, *431–432*, 1–8. [CrossRef]
49. Smoláková, L.; Frolich, K.; Troppová, I.; Kutálek, P.; Kroft, E.; Čapek, L. Determination of basic sites in Mg–Al mixed oxides by combination of TPD-CO_2 and CO_2 adsorption calorimetry. *J. Therm. Anal. Calorim.* **2017**, *127*, 1921–1929. [CrossRef]
50. Maniak, G.; Stelmachowski, P.; Kotarba, A.; Sojka, Z.; Rico-Pérez, V.; Bueno-López, A. Rationales for the selection of the best precursor for potassium doping of cobalt spinel based deN_2O catalyst. *Appl. Catal. B* **2013**, *136–137*, 302–307. [CrossRef]
51. Gálvez, M.E.; Ascaso, S.; Stelmachowski, P.; Legutko, P.; Kotarba, A.; Moliner, R.; Lázaro, M.J. Influence of the surface potassium species in Fe–K/Al_2O_3 catalysts on the soot oxidation activity in the presence of NO_x. *Appl. Catal. B* **2014**, *152–153*, 88–98. [CrossRef]
52. Borowiecki, T.; Denis, A.; Rawski, M.; Gołębiowski, A.; Stołecki, K.; Dmytrzyk, J.; Kotarba, A. Studies of potassium-promoted nickel catalysts for methane steam reforming: Effect of surface potassium location. *Appl. Surf. Sci.* **2014**, *300*, 191–200. [CrossRef]
53. Gimenez, P.; Fereres, S. Effect of Heating Rates and Composition on the Thermal Decomposition of Nitrate Based Molten Salts. *Energy Procedia* **2015**, *69*, 654–662. [CrossRef]
54. Kaspera, W.; Wojas, J.; Molenda, M.; Kotarba, A. Parallel migration of potassium and oxygen ions in hexagonal tungsten bronze—Bulk diffusion, surface segregation and desorption. *Solid State Ion.* **2016**, *297*, 1–6. [CrossRef]
55. Haneda, M.; Tsuboi, G.; Nagao, Y.; Kintaichi, Y.; Hamada, H. Direct Decomposition of NO Over Alkaline Earth Metal Oxide Catalysts Supported on Cobalt Oxide. *Catal. Lett.* **2004**, *97*, 145–150. [CrossRef]
56. Hong, W.-J.; Iwamoto, S.; Inoue, M. Direct NO decomposition over a Ce–Mn mixed oxide modified with alkali and alkaline earth species and CO_2-TPD behavior of the catalysts. *Catal. Today* **2011**, *164*, 489–494. [CrossRef]

57. Palomares, E.; Uzcátegui, A.; Franch, C.; Corma, A. Multifunctional catalyst for maximizing NO_x oxidation/storage/reduction: The role of the different active sites. *Appl. Catal. B* **2013**, *142–143*, 795–800. [CrossRef]
58. Haneda, M.; Kintaichi, Y.; Hamada, H. Reaction mechanism of NO decomposition over alkali metal-doped cobalt oxide catalysts. *Appl. Catal. B* **2005**, *55*, 169–175. [CrossRef]
59. Haneda, M.; Hamada, H. Recent progress in catalytic NO decomposition. *Comptes Rendus Chim.* **2016**, *19*, 1254–1265. [CrossRef]
60. Zhu, J.; Thomas, A. Perovskite-type mixed oxides as catalytic material for NO removal. *Appl. Catal. B* **2009**, *92*, 225–233. [CrossRef]
61. Jabłońska, M.; Palomares, A.E.; Chmielarz, L. NOx storage/reduction catalysts based on Mg/Zn/Al/Fe hydrotalcite-like materials. *Chem. Eng. J.* **2013**, *231*, 273–280. [CrossRef]

catalysts

Article

The Effect of CeO_2 Preparation Method on the Carbon Pathways in the Dry Reforming of Methane on Ni/CeO_2 Studied by Transient Techniques

Constantinos M. Damaskinos [1], Michalis A. Vasiliades [1], Vassilis N. Stathopoulos [2] and Angelos M. Efstathiou [1,*]

1 Heterogeneous Catalysis Laboratory, Chemistry Department, University of Cyprus, Nicosia 2109, Cyprus
2 Laboratory of Chemistry and Materials Technology, General Department, School of Sciences, National and Kapodistrian University of Athens, GR-34400 Athens, Greece
* Correspondence: efstath@ucy.ac.cy; Tel.: +357-22-892776

Received: 30 May 2019; Accepted: 18 July 2019; Published: 21 July 2019

Abstract: The present work discusses the effect of CeO_2 synthesis method (thermal decomposition (TD), precipitation (PT), hydrothermal (HT), and sol-gel (SG)) on the carbon pathways of dry reforming of methane with carbon dioxide (DRM) applied at 750 °C over 5 wt% Ni/CeO_2. In particular, specific transient and isotopic experiments (use of ^{13}CO, $^{13}CO_2$, and $^{18}O_2$) were designed and conducted in an attempt at providing insights about the effect of support's preparation method on the concentration (mg g_{cat}^{-1}), reactivity towards oxygen, and transient evolution rates (μmol g_{cat}^{-1} s^{-1}) of the *inactive* carbon formed under (i) CH_4/He (methane decomposition), (ii) CO/He (reverse Boudouard reaction), and (iii) the copresence of the two (CH_4/CO/He, use of ^{13}CO). Moreover, important information regarding the relative contribution of CH_4 and CO_2 activation routes towards carbon formation under DRM reaction conditions was derived by using isotopically labelled $^{13}CO_2$ in the feed gas stream. Of interest was also the amount, and the transient rate, of carbon removal via the participation of support's labile active oxygen species.

Keywords: DRM; nickel; cerium dioxide; transient experiments; lattice oxygen; isotopes

1. Introduction

Nowadays, a great academic and research interest is seen aimed at exploring the gradual replacement of conventional fossil fuels towards energy production through the utilization of alternative and renewable energy sources such as Natural Gas (NG) and Bio-Gas (BG). The driving force behind it are the findings of NG reservoirs rich in CO_2 (>40 vol%) [1–3] and renewable bio-gas [4,5], which can be used in the development of technologies, such as the dry reforming of methane (DRM: $CH_4 + CO_2 \rightarrow 2CO + 2H_2$, $\Delta H^0 = +261$ kJ mol^{-1}), as more environmentally friendly processes in many aspects [6]. The latter is enforced as it uses two major greenhouse gases (CH_4 and CO_2), while at the same time produces a favorable H_2/CO gas ratio (~1) for the Fischer–Tropsch [7,8] synthesis towards liquid fuels, but also for other processes in the production of chemicals (DME, MeOH, ammonia) [9,10]. In addition, the low operational cost of DRM in comparison with the already used steam methane reforming (SMR) and partial oxidation of methane (POM) technologies, makes its use very attractive [11–14]. However, the main obstacle for the development of an industrial DRM technology is the catalyst's deactivation due to carbon accumulation, especially over Ni-supported [15] solids, which are mainly used due to their low cost and wide availability. The formation of inactive carbon, in the form of filaments, graphite, and whiskers, mainly is derived from the CH_4 decomposition ($CH_4 \rightarrow C\text{-s} + 2H_2$, $\Delta H^0 = +75$ kJ mol^{-1}) and Boudouard reaction ($2CO \rightarrow C\text{-s} + CO_2$, $\Delta H^0 = -172$ kJ mol^{-1}). Thus, the design of a suitable Ni-based catalyst supported on reducible metal oxides emerged (e.g., use of CeO_2, Zr^{4+}-, Pr^{3+}-, Ti^{4+}-doped CeO_2, La_2O_3, Nb_2O_5) [16–25] since the latter supports possess oxygen storage capacity (OSC), oxygen vacancies, and high oxygen mobility, leading to carbon gasification rates that significantly reduce carbon accumulation rates, but also provide high thermal stability for the supported Ni catalysts [26,27]. The Ce-based materials owe their advantages against non-reducible metal oxides to the undergoing of fast change in the $Ce^{4+} \leftrightarrow Ce^{3+}$ oxidation state (redox behavior), leading to an oxygen release, and vice versa to an oxygen storage, in the ceria-based stable crystal structure [28,29].

Several studies reveal the effect of preparation method of CeO_2 nanoparticles for use in a wide variety of applications, and they argue that such solids form different surface defects by exhibiting more surface atoms than their bulk counterparts [30–36]. Such nanoparticles could have various morphological and structural differences (nanorods, nanowires, nano-cubes, etc.) with different surface area, pore volume, and mean pore diameter, thus the synthesis method seems to play an important role [37–41].

In spite of recent efforts to develop suitable CeO_2-supported Ni catalysts exhibiting high DRM catalytic activity and carbon resistance [42], fundamental understanding of the effect of support synthesis method on the contribution of the carbon deposition and removal routes has not been reported yet, to the best of our knowledge. The synthesis of CeO_2 via different methods [43,44] could lead to several variations in its physicochemical properties, but also to the metal surface when ceria is used as support. The latter is well demonstrated as due to the existence of strong metal support interactions (SMSI) between Ni particles and CeO_2 [45,46].

Transient methods (step-gas switches, use of isotopes, and temperature programmed oxidation or hydrogenation) performed over supported metal catalysts provided important information about the carbon paths in the DRM reaction, and relationships between the catalytic activity and coke formation. Furthermore, rival reaction mechanisms and rate determined steps (RDS) under DRM reaction conditions (working catalyst surface) can be elucidated. For example, Schuurman and Mirodatos [47] suggested that on Ni/SiO_2 catalyst the RDS is the recombination of atomic C (derived from CH_4 dissociation) and atomic O (derived from CO_2 dissociation) over the Ni surface. On the other hand, Slagtern et al. [48] observed that on Ni/La_2O_3 catalyst, CH_4 is activated on Ni as opposed to CO_2, which is activated on La_2O_3 support (or metal-support interface) towards carbonate-like species formation. Advanced kinetic and mechanistic studies to elucidate the carbon paths in DRM with the use of isotopes ($C^{18}O_2$, $^{13}CH_4$, $^{13}CO_2$) were recently performed to a large extent by our group [17–20,46], but also in some other works [47,49–51]. In these works, the significant participation of lattice oxygen of reducible metal oxide supports (e.g., doped ceria-based materials) towards removal of carbon to form CO was proved experimentally by ^{18}O transient isotopic experiments followed by DRM reaction. Also, the quantification of origin of carbon (CH_4 vs. CO_2 activation route) was probed as a function of reaction T and catalyst composition.

The present work aims to address the effect of CeO_2 support synthesis method on the carbon pathways in the dry reforming of methane over 5 wt% Ni/CeO_2 catalysts, where this is reported for the first time to our knowledge. Of particular interest was to investigate differences on (i) the concentration of inactive carbon and its reactivity towards oxygen, (ii) the relative contribution of CH_4 and CO_2 activation routes to the total carbon formation on the catalytic surface via methane decomposition and Boudouard reactions, and (iii) the participation of labile support's lattice oxygen towards carbon removal, and to what extent. For this purpose, various transient and isotopic experiments followed by temperature programmed oxidation (TPO) were performed.

2. Results

2.1. Catalysts Surface Texture and Structural Properties

The BET specific surface area (SSA, $m^2\ g^{-1}$), mean pore diameter (d_p, nm), and the specific pore volume (V_p, $cm^3\ g^{-1}$) of the four CeO_2 solid supports prepared by different methods, namely: Thermal decomposition (TD), precipitation (PT), hydrothermal (HT), and sol-gel (SG), are given in Table S1 (Electronic Supplementary Information, ESI). The SSA was found to be in the 5.6–50 $m^2\ g^{-1}$ range, the d_p in the 6.7–22.5 nm range, and the V_p in the 0.029–0.203 $cm^3\ g^{-1}$ range. The CeO_2-HT solid exhibited the largest value of SSA and V_P (50 $m^2\ g^{-1}$, 0.203 $cm^3\ g^{-1}$), and a pore size of d_p = 15.8 nm, as opposed to CeO_2-PT (5.6 $m^2\ g^{-1}$, 0.032 $cm^3\ g^{-1}$) with the largest pore size (22.5 nm). The powder XRD diffractograms of 5 wt% Ni supported on the various CeO_2 solid supports are given in Figure S1. The CeO_2 support exhibits the cubic structure [46], and after using the Scherrer equation and the CeO_2 (111) diffraction line, the mean primary crystal size (d_C, nm) of support was estimated. Similarly, after using the NiO (111) diffraction line, the particle size (d_{NiO}, nm) of NiO was also estimated. The latter value was then used to estimate the particle size (d_{Ni}, nm) of Ni^0 via Equation (3), and the obtained results are reported in Table S1. There was not any shift of the NiO (111) 2θ diffraction peak (see Figure S1B) among the different samples, however, variations of the mean Ni particle size (8.4–20.8 nm) and the mean primary crystal size of ceria support (11.5–43.1 nm) were observed (Table S1). The latter results find good support by the literature as will be discussed in Section 3.1.

2.2. H_2 Temperature-Programmed Desorption (H_2-TPD)

Figure 1 presents H_2-TPD traces of the 5 wt% Ni supported on the various CeO_2 solids. It is clearly seen that the CeO_2 preparation method resulted in drastic changes of the H_2 desorption kinetic features in terms of strength of hydrogen binding states (T_M, peak maximum temperature) and their corresponding surface coverage (area under a given desorption peak). Thus, the different preparation method of CeO_2 support followed by the same Ni deposition method (wet impregnation) led to differences in the heterogeneity of the Ni surface (e.g., distribution of strength of hydrogen chemisorption sites on surface Ni). In the low-temperature range of 50–200 °C, the amount of H_2 desorbed (μmol g^{-1}) was significantly larger for CeO_2-HT (23.8 μmol g^{-1}) compared to the other three ceria-supported Ni catalysts (CeO_2-PT, -TD, and -SG), where similar amounts were found (8.8, 8.0, and 6.2 μmol g^{-1}, respectively). In the high-temperature range of 200–500 °C, the amount of H_2 desorbed follows a different order: Ni/CeO_2-TD (25.3 μmol g^{-1}) > Ni/CeO_2-PT (17.3 μmol g^{-1}) > Ni/CeO_2-HT (17 μmol g^{-1}) > Ni/CeO_2-SG (8.5 μmol g^{-1}). Of interest is the fact that all ceria-supported Ni catalysts present three main desorption peaks. However, shoulders to these main desorption peaks appear at different temperatures. For example, the Ni/CeO_2-TD presents three main desorption peaks centered at 57, 222, and 354 °C with shoulder at the falling part of the 3rd peak (Figure 1a), Ni/CeO_2-PT at 57, 303, and 393 °C with clear shoulders at the falling part of the 1st peak, the rising part of 2nd peak, and the falling part of 3rd peak (Figure 1b). Ni/CeO_2-HT presents the three main desorption peaks centered at 95, 258, and 362 °C with shoulders at the falling part of 3nd peak (Figure 1c), whereas Ni/CeO_2-SG at 79, 157, and 383 °C with shoulders at the low-T side of 3nd peak (Figure 1d).

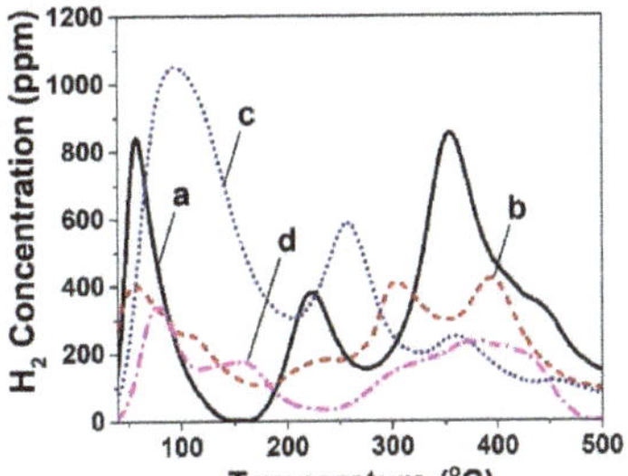

Figure 1. H_2 temperature-programmed desorption (H_2-TPD) traces obtained over 5 wt% Ni/$CeO_{2-\delta}$ catalysts prepared by (**a**) Thermal Decomposition (TD), (**b**) Precipitation (PT), (**c**) Hydrothermal (HT), and (**d**) Sol Gel (SG) method; F_{He} = 50 NmL min^{-1}; β = 30 °C min^{-1}; W = 0.3 g.

The Ni dispersion (D_{Ni}, %) of the given solids was estimated based on the total amount of H_2 desorbed (Figure 1), and results are presented in Table S1 (ESI). The lowest dispersion of Ni was found when CeO_2-SG was used as support (3.4%), followed by the CeO_2-PT (6.1%), CeO_2-TD (7.8%), and the CeO_2-HT (9.6%, highest dispersion). Thus, the Ni particle size (d_{Ni}, nm) estimated via Equation (4) was found to be: 10.1, 12.4, 15.9, and 28.5 nm for the CeO_2-HT, -TD, -PT, and -SG, respectively. The latter results were supported by those obtained from the powder XRD analyses (Section 2.1, Figure S1 and Table S1) and the HR-TEM (Section 2.3), but also with those reported previously [20].

2.3. Transmission Electron Microscopy (TEM) Studies

HR-TEM images obtained over the fresh 5 wt% Ni/CeO_2-HT (the support was prepared by the hydrothermal method, HT) calcined in air for 4 h at 750 °C are given in Figure S2 (ESI). It is seen that dispersed Ni nanoparticles of ~8–12 nm in size were observed, in good agreement with the H_2-TPD and powder XRD results.

2.4. Scanning Electron Microscopy (SEM) Studies

SEM images obtained over the fresh CeO_2-supported Ni solids are presented in Figure S3 (ESI). The secondary particle size (agglomerates) of the catalyst's support was in the range of 10–50 nm, where different porous structures were derived after using the four different CeO_2 synthesis methods.

2.5. Catalytic Performance Studies in DRM

Figure 2A presents catalytic performance results in terms of specific integral rate (mol g_{cat}^{-1} min^{-1}) of CH_4 conversion and H_2/CO gas product ratio obtained after 30 min of DRM at 750 °C over the 5 wt% Ni supported on differently prepared CeO_2 solids. The Ni/CeO_2-HT presented the highest catalytic activity (5.5 mmol g_{cat}^{-1} min^{-1}), while Ni/CeO_2-SG the lowest activity (2.0 mmol g_{cat}^{-1} min^{-1}), thus a significant difference by a factor of ~2.9 existed between these two catalysts. On the other hand, the H_2/CO gas product ratio did not follow the activity differences shown in Figure 2A (compare Figure 2A,B) since the most active Ni/CeO_2-HT exhibited a value of ~1.2 (similar to -TD and -PT) and the least active Ni/CeO_2-SG presented a value of ~1.1. These results clearly demonstrated that the series of four catalysts presented different orders in terms of H_2 and CO reaction selectivity. It should be noted that all four catalytic systems presented X_{CH4} (%) and X_{CO2} (%) larger than 80% (81–94%), with H_2-yields larger than 45% (48–59%) and H_2/CO gas product ratio close to the desired value of ~1, tested at the same GHSV (30,000 h^{-1}) (see Table S2).

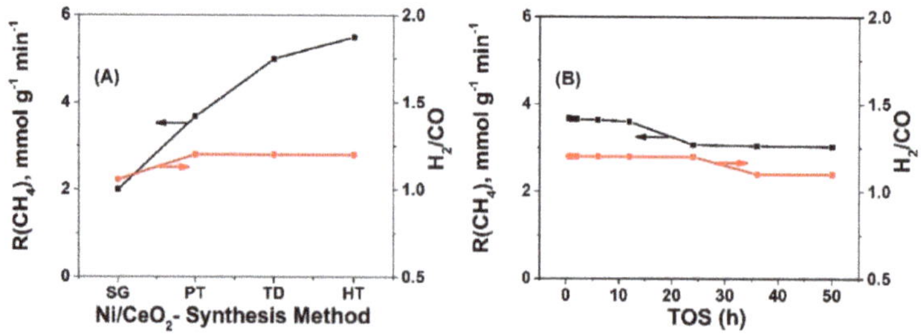

Figure 2. (**A**) Specific integral rates of CH_4 conversion (mmol g_{cat}^{-1} min^{-1}) and H_2/CO gas product ratio obtained after 30 min of DRM at 750 °C (GHSV ~30,000 h^{-1}) over the four catalysts; (**B**) Stability test in terms of integral rate of CH_4 conversion conducted over 50 h of TOS on 5 wt% Ni/CeO_2-PT catalyst; GHSV ~30,000 h^{-1}.

Figure 2B presents the stability test (up to 50 h on TOS) for the 5 wt% Ni/CeO_2-PT catalyst, which exhibited the least amount of accumulated carbon (mg g_{cat}^{-1}) after 12 h on TOS among the series of catalysts. It was clearly seen that after up to ~12 h on TOS, the catalyst's activity remained practically constant, while a drop by ~17.5% in the integral rate of methane conversion occurred after 50 h on TOS (see also Table S3). Similar results were also observed for the other three catalyst compositions (not reported). The comparative activity behavior based on 30-min on TOS shown in Figure 2A is thus very representative for the true effect of ceria support synthesis method.

It's worth mentioning that regarding the CO_2 conversion (%), this was found to be lower than that of CH_4 for all catalytic systems except in the case of Ni/CeO_2-SG. The latter result was similar to that obtained over other CeO_2-supported Ni catalysts [17,50], and is mainly attributed to the effect of reverse water-gas shift (RWGS) side reaction. It will be shown in the following Section 2.6, that the four catalysts, for their activity performance depicted in Figure 2A, also exhibited significantly different amounts of carbon accumulation due to their different CeO_2 support preparation method.

2.6. Characterization of Carbon Formed under Different Reaction Conditions

2.6.1. Dry Reforming of Methane ($^{12}CO_2$/$^{12}CH_4$) at Steady-State Reaction Conditions

Transient response curves of CO_2 obtained during temperature-programmed oxidation (TPO) of carbon deposited over the four Ni/CeO_2 catalysts after 12 h in DRM (20 vol% CH_4/20 vol% CO_2/60 vol% He) at 750 °C are presented in Figure 3A. The Ni/CeO_2-PT led to a lower carbon accumulation, ca. ~3.8 times (30.7 vs. 116.1 mg C g_{cat}^{-1}) compared to the Ni/CeO_2-HT catalyst, with the other two catalysts, Ni/CeO_2-TD and Ni/CeO_2-SG showing a decrease by 1.8 and 1.4 times (66.2 and 80.4 mg C g_{cat}^{-1}), respectively. In the case of Ni/CeO_2-TD and Ni/CeO_2-PT catalysts, a main peak starting at 450 °C and ending at 750 °C with peak maximum at ~630 °C was observed, whereas in the case of Ni/CeO_2-HT, a wider main peak was observed, which was centered at ~670 °C. As opposed to the latter behavior, the Ni/CeO_2-SG (Figure 3Ad) presented likely several types of carbon, since it started reacting with oxygen at ~500 °C with a shoulder at 600 °C and a main peak at 700 °C, but a clear sharp peak at 750 °C was also observed; the latter might have also been the result of a hot spot in the catalytic bed formed at these high temperatures given the large exotherm of carbon oxidation to CO_2.

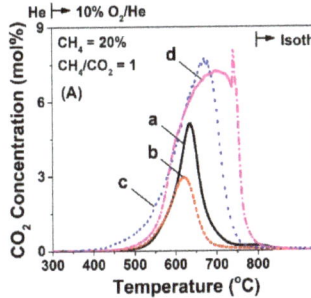

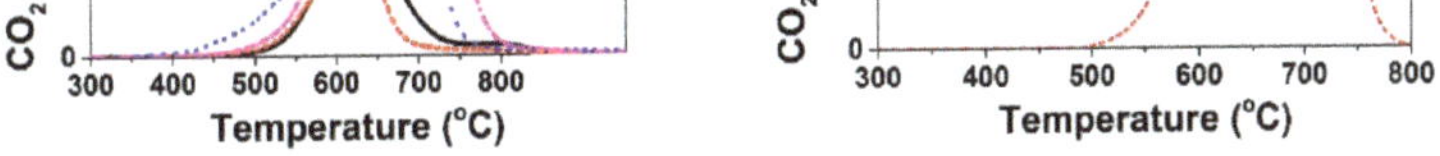

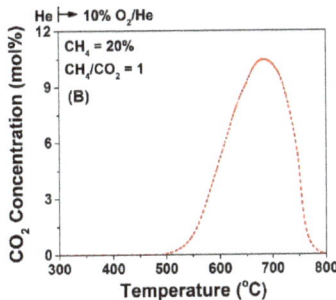

Figure 3. Transient response curves of CO_2 concentration obtained during TPO of carbon formed after (**A**) 12 h of DRM (20% CH_4/20% CO_2/He; 50 NmL min^{-1}; GHSV ~30,000 h^{-1}) at 750 °C over 5 wt% Ni/CeO_2 prepared by (a) Thermal decomposition (TD), (b) Precipitation (PT), (c) Hydrothermal (HT), and (d) Sol-Gel (SG) method; (**B**) TPO trace of carbon formed after 50 h in DRM over the 5 wt% Ni/CeO_2-PT catalyst.

The Ni/CeO_2-PT catalyst, which led to the lowest amount of carbon deposition, was also tested for longer time-on-stream (ca. 50 h, see Figure 2B, Table S3), and the TPO trace recorded is presented in Figure 3B. The amount of carbon deposition was increased when the TOS increased from 12 h to 50 h, ca. 147.1 vs. 30.7 mg C $g^{-1}{}_{cat}$ (see also Table S3). These results will be discussed below in relation to a synergy observed for carbon accumulation between CH_4 and CO presence in the same gas mixture compared to the CH_4 decomposition and Boudouard reaction contribution alone.

2.6.2. Isotopically Labelled Dry Reforming of Methane ($^{13}CO_2/^{12}CH_4$)

Figure 4 presents $^{13}CO_2$ and $^{12}CO_2$ transient response curves recorded during TPO of the carbon formed after 30 min in isotopically labelled DRM (5 vol% $^{13}CO_2$/5 vol% $^{12}CH_4$/45 vol% Ar/45 vol% He) at 750 °C over the Ni/CeO$_2$-TD, Ni/CeO$_2$-HT, and Ni/CeO$_2$-SG catalysts. It's worth mentioning that the Ni/CeO$_2$-PT, where the support was prepared by the precipitation method (CeO$_2$-PT), exhibited non-measurable amounts of carbon, and neither ^{12}CO nor ^{13}CO signals were recorded in the MS. The TPO traces of $^{13}CO_2$ and $^{12}CO_2$ were different in shape among the three catalytic systems, and this was largely attributed to the different carbon oxidation kinetics influenced by the type of carbon deposited, and its reactivity towards oxygen. The $^{13}CO_2$-TPO trace originated from the $^{13}CO_2$ activation route during DRM, while that of $^{12}CO_2$-TPO from the $^{12}CH_4$ activation route. Furthermore, the three catalysts presented different amounts of carbon formed via the two activation routes but also a different total amount of carbon, which was estimated by integrating the respective TPO traces. The contribution of each reactant (CH_4 vs. CO_2) to the carbon formation under DRM reaction conditions was estimated based on the ratio of $^{12}CO_2/^{13}CO_2$ (TPO traces). It was shown that in all three catalytic systems, CH_4 decomposition is the dominant route, but to a different extent. More precisely, the Ni/CeO$_2$-TD (Figure 4A) and Ni/CeO$_2$-HT (Figure 4B) presented $^{12}C/^{13}C = 1.6$ and 1.8, respectively, as opposed to the Ni/CeO$_2$-SG catalyst (Figure 4C), where CH_4 decomposition contributed in a significantly higher extent ($^{12}C/^{13}C = 4.7$). In addition, the total amount of carbon was found to be larger in the case of Ni/CeO$_2$-HT (29.5 μmol g^{-1}) compared to Ni/CeO$_2$-SG and Ni/CeO$_2$-TD (28.1 and 11.1 μmol g^{-1}, respectively). The latter results agree with those presented in Section 2.6.1, where the feed gas stream (5 vs. 20 vol% of reactants) and the TOS (30 min vs. 12 h) were much different.

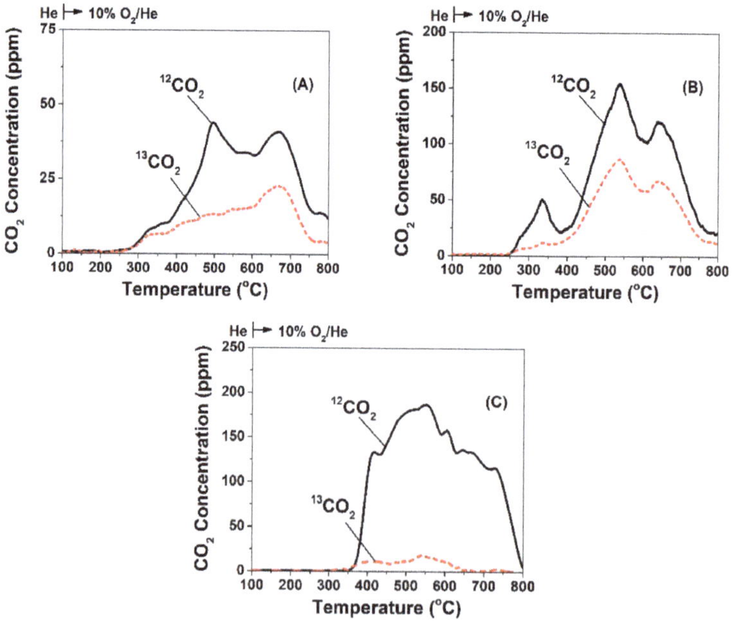

Figure 4. Temperature-programmed oxidation (TPO) of carbon to $^{12}CO_2$ and $^{13}CO_2$ formed after 30 min in 5 vol% $^{13}CO_2$/5 vol% $^{12}CH_4$/45 vol% Ar/45 vol% He (50 NmL min^{-1}; GHSV ~30,000 h^{-1}) at 750 °C over (**A**) 5 wt% Ni/CeO$_2$-TD, (**B**) 5 wt% Ni/CeO$_2$-HT, and (**C**) 5 wt% Ni/CeO$_2$-SG.

2.6.3. Transient Methane Decomposition (CH_4/He) Reaction

Figure 5 shows transient evolution rates of CH_4 consumption and H_2 and CO gas formation (the only gaseous reaction products observed), obtained during the step gas switch He → 20 vol% CH_4/1% Ar/He (30 min) made at 750 °C over the four 5 wt% Ni supported on CeO_2 carriers prepared by different synthesis methods. The differences in the initial transient rate values, but also their shapes, are apparent. It should be mentioned at this point that the latter rates appeared very small when the reaction was performed over the supports alone. The different kinetics of CH_4 decomposition, over each of the four catalytic surfaces presented in Figure 5A, led also to different H_2 transient formation rates (Figure 5B), similar in shape with those of CH_4 consumption (Figure 5A). On the other hand, the rate of CO formation was the result of carbon removal by the support's lattice oxygen, which followed largely different kinetics (compare Figure 5B,C). In particular, the H_2 transient rates in the case of Ni/CeO_2-TD and Ni/CeO_2-PT passed through a maximum a short time after the switch (<10 s), as opposed to Ni/CeO_2-HT and Ni/CeO_2-SG, which passed through a maximum after 25 s in CH_4/Ar/He feed gas stream. Also, the latter catalyst presented only a slight decrease in the reaction rates after maximum rate was achieved (practically a plateau in the rate is obtained (Figure 5Ad,Bd,Cd). It has been discussed that these transient features reflect the Ni metal surface's ability to decompose methane over the remaining empty sites with time on stream, leading to carbon structure dependent deposition with different kinetics [18,20].

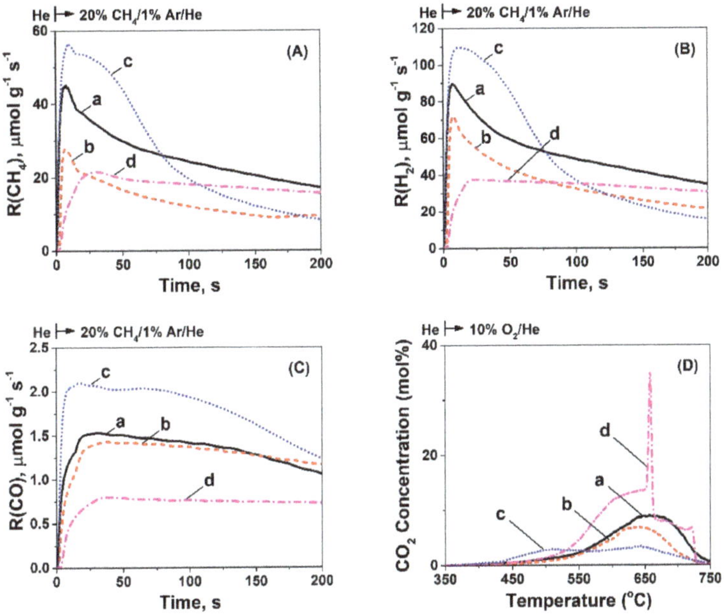

Figure 5. Transient rates (μmol $g^{-1}s^{-1}$) of CH_4 consumption (**A**), H_2 (**B**), and CO (**C**) formation, as a function of time after the gas switch He → 20 vol% CH_4/1% Ar/He (50 NmL min^{-1}; GHSV ~30,000 h^{-1}) at 750 °C. (**D**) Transient response curves of CO_2 concentration obtained during TPO of carbon formed after 30 min of methane decomposition (20% CH_4/1% Ar/He) at 750 °C over 5 wt% Ni/CeO_2 prepared by (a) Thermal decomposition (TD), (b) Precipitation (PT), (c) Hydrothermal (HT), and (d) Sol-Gel (SG) method.

The carbon accumulated over the ceria-supported Ni catalytic surface can diffuse towards the Ni-support interface, where it was gasified to form CO(g) by the support's labile lattice oxygen, and this chemical process was likely responsible for the delay in the peak maximum, as shown in Figure 5C.

However, lattice oxygen diffusion towards carbon formed on Ni and/or Ni-ceria support interface can also be considered, as discussed in Section 3.2. The H- and C-material balances close within less than 5% (Table 1). In particular, the amount of CH_4 decomposed was found to be the same (7.8 mmol g^{-1}) for Ni/CeO_2-PT and Ni/CeO_2-HT, an amount which increases by about 1.3 and 1.6 times for Ni/CeO_2-SG and Ni/CeO_2-TD, ca. 10 and 12.2 mmol g^{-1}, respectively. On the other hand, the amount of CO formed was found to be 0.9 mmol g_{cat}^{-1} in the cases of Ni/CeO_2-PT and Ni/CeO_2-HT, but slightly lower in the case of Ni/CeO_2-TD and Ni/CeO_2-SG, ca. 0.8 and 0.6 mmol g_{cat}^{-1}, respectively. The amount of H_2 produced was lower in the case of Ni/CeO_2 -HT (14.9 mmol g_{cat}^{-1}) compared to Ni/CeO_2-PT (16.4 mmol g_{cat}^{-1}), Ni/CeO_2-SG (22 mmol g_{cat}^{-1}), and Ni/CeO_2-TD (26.4 mmol g_{cat}^{-1}). Of interest is the amount of labile oxygen of the ceria support contributing to the gasification of carbon towards CO(g), which could be quantified by estimating the ratio between the CO production and CH_4 consumption, as shown in Table 1. This ratio was found to be the same (0.12) for the Ni/CeO_2-PT and Ni/CeO_2-HT catalysts, but significantly lower in the case of Ni/CeO_2-TD (0.07) and Ni/CeO_2-SG (0.06), showing clearly the lower contribution of O_L (active labile oxygen) towards CO(g).

Table 1. Quantity of CH_4 consumed, H_2 and CO formed (mmol g^{-1}), and molar ratio of CO/CH_4 obtained after 30 min of methane decomposition (20% CH_4/He) conducted at 750 °C. Also shown is the amount of carbon deposited (mmol g^{-1}), which was obtained after TPO following 30 min of methane decomposition.

Catalyst (5 wt% Ni)	CH_4 Consumption (mmol g^{-1})	H_2 Production (mmol g^{-1})	CO Production (mmol g^{-1})	CO/CH_4	Carbon Deposition (mmol g^{-1})
CeO_2 -TD	12.2	26.4	0.8	0.07	12.1
CeO_2 -PT	7.8	16.4	0.9	0.12	8.3
CeO_2 -HT	7.8	14.9	0.9	0.12	6.8
CeO_2 -SG	10	22	0.6	0.06	10.7

Following the 30 min CH_4 decomposition performed at 750 °C over the Ni/CeO_2 catalysts, temperature-programmed oxidation was performed to estimate the amount of carbon and its reactivity towards oxygen. The TPO traces in terms of CO_2 concentration (mol%) are depicted in Figure 5D, and the amount of carbon deposited (mmol g^{-1}) is reported in Table 1. The latter results were in harmony with the amount of CH_4 consumed and H_2 produced, as reported above. However, it should be mentioned at this point that these values did not agree with the results regarding the amount of carbon deposited during DRM, where Ni/CeO_2-HT was found to accumulate more carbon. As it will be discussed in the following sections, we argue that the carbon formation rate, and that of carbon removal towards CO formation, are largely influenced when both CH_4 and CO_2 (or CO) are present over the ceria-supported Ni catalyst surface to be compared to the case when CH_4, CO_2, or CO is only present.

2.6.4. Transient Carbon Monoxide Dissociation (CO/He) Reaction

The transient rates of CO(g) consumption obtained during the step-gas switch He → 20 vol% CO/1 vol% Ar/He (750 °C, 30 min) over the four catalysts are presented in Figure 6A. It was clearly shown that during the reverse Boudouard reaction, two peak maxima were present, as opposed to the case of CH_4 decomposition reaction (Figure 5A). The first very sharp peak was formed immediately (t_{max} ~5 s) after the switch from inert He to CO/Ar/He, followed by a fast decay, while the second peak appeared at t_{max} ~20 s and was followed by a slower rate of CO consumption. Thus, the kinetics involved in both the initial very sharp and the slower transient rates of carbon monoxide dissociation are strongly affected by differences in the four catalytic surfaces.

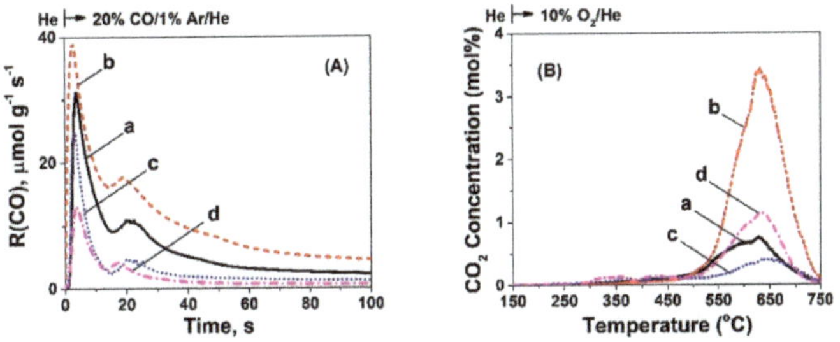

Figure 6. Transient rates (μmol $g^{-1}s^{-1}$) of (**A**) CO consumption as a function of time after the gas switch He → 20 vol% CO/1 vol% Ar/He (50 NmL min^{-1}; GHSV ~30,000 h^{-1}) at 750 °C. (**B**) Transient response curves of CO_2 concentration obtained during TPO of carbon formed after 30 min of CO dissociation at 750 °C over 5 wt% Ni/CeO_2 prepared by (a) Thermal decomposition (TD), (b) Precipitation (PT), (c) Hydrothermal (HT), and (d) Sol-Gel (SG) method.

Figure 6B presents the TPO traces obtained after the reaction with 20 vol% CO/1 vol% Ar/He (30 min) at 750 °C. The four catalytic surfaces showed one main peak centered at ~600 °C with shoulders, revealing likely the existence of different carbon structures, oxidized with different kinetics. Table 2 presents the amount of carbon deposited during the 30 min reaction with CO/Ar/He. The largest amount of carbon deposited was found to be on the Ni/CeO_2-PT (2.7 mmol g^{-1}) followed by the Ni/CeO_2-TD (1.1 mmol g^{-1}), Ni/CeO_2-HT (0.8 mmol g^{-1}), and Ni/CeO_2-SG (0.6 mmol g^{-1}) catalysts. The C-material balance closes within less than 5% in all cases (Table 2).

Table 2. Quantity of CO consumption and CO_2 formation (mmol g^{-1}) obtained after 30 min 20% CO/He at 750 °C. Also shown is the amount of carbon deposition (mmol g^{-1}) obtained after TPO following 30 min of CO disproportionation.

Catalyst (5 wt% Ni)	CO Consumption (mmol g^{-1})	CO_2 Production (mmol g^{-1})	Carbon Deposition (mmol g^{-1})
CeO_2 - TD	1.9	1.0	1.2
CeO_2 - PT	4.9	2.2	2.7
CeO_2 - HT	1.7	0.8	0.8
CeO_2 - SG	1.1	-	0.6

2.6.5. Isotopically Labelled Competitive ($^{13}CO/^{12}CH_4$) Reaction towards Carbon Formation

Figure 7 presents CO (C,D) and CO_2 (A,B) concentration (mol%) profiles obtained after temperature-programmed oxidation (TPO) following a 20-min treatment of the Ni/CeO_2-TD (a), Ni/CeO_2-PT (b), Ni/CeO_2-HT (c), and Ni/CeO_2-SG (d) catalysts with 2.5 vol% $^{12}CH_4$/2.5 vol% ^{13}CO/2 vol% Kr/Ar/He gas mixture at 750 °C. The amount of deposited carbon, but also its reactivity towards oxygen (shape and position of TPO trace), were clearly different among the four catalysts. It was shown that all catalytic systems exhibited differences in the shape of $^{12}CO_2/^{12}CO$ and $^{13}CO_2/^{13}CO$ response curves as the result of the oxidation of carbon originated from the $^{12}CH_4$ decomposition and ^{13}CO dissociation routes, respectively. Moreover, the amount of carbon derived from each route is different (area under the TPO trace), and the contribution of each route to the formation of carbon is estimated by considering the $^{12}C/^{13}C$ ratio. It is illustrated that in all cases, except Ni/CeO_2-TD (0.97), CH_4 decomposition was dominant but to a different extent. In particular, Ni/CeO_2-PT showed a ratio of $^{12}C/^{13}C$ = 1.06, Ni/CeO_2-HT $^{12}C/^{13}C$ = 1.61, and the Ni/CeO_2-SG a ratio of $^{12}C/^{13}C$ = 1.5 (Table 3).

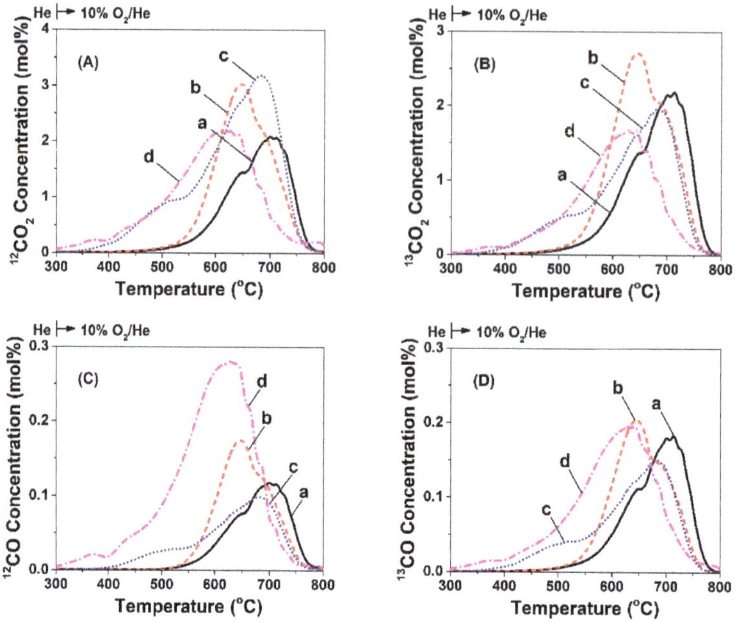

Figure 7. Temperature-programmed oxidation (TPO) of carbon to ^{12}C- and ^{13}C-containing CO_2 (**A**,**B**) and CO (**C**,**D**) formed after 20 min in 2.5 vol% ^{13}CO/2.5 vol% $^{12}CH_4$/2 vol% Kr/Ar/He (50 NmL min^{-1}; GHSV ~30,000 h^{-1}) at 750 °C over 5 wt% Ni/CeO_2 prepared by (a) Thermal decomposition (TD), (b) Precipitation (PT), (c) Hydrothermal (HT), and (d) Sol-gel (SG) method.

Table 3. Quantity of ^{12}CO, ^{13}CO, $^{12}CO_2$, and $^{13}CO_2$ (mmol g^{-1}) formed during TPO following 20 min of reaction with 2.5 vol% ^{13}CO/2.5 vol% $^{12}CH_4$/2 vol% Kr/Ar/He at 750 °C over all the Ni/CeO_2 catalysts. Also shown is the total amount of "carbon" (mmol g^{-1}), and the ratio ^{12}C to ^{13}C in the products.

Catalyst 5 wt% Ni/CeO_2	^{12}CO Production (mmol g^{-1})	^{13}CO Production (mmol g^{-1})	$^{12}CO_2$ Production (mmol g^{-1})	$^{13}CO_2$ Production (mmol g^{-1})	$^{12}C/^{13}C$	Carbon Deposition (mmol g^{-1})
CeO_2-TD	0.14	0.23	2.6	2.6	0.97	5.6
CeO_2-PT	0.14	0.18	2.6	2.4	1.06	5.3
CeO_2-HT	0.16	0.23	5.2	3.1	1.61	8.7
CeO_2-SG	0.39	0.25	3.5	2.4	1.50	6.5

In addition, the largest amount of carbon was deposited over the Ni/CeO_2-HT catalyst, ca. 1.6, 1.5, and 1.3 times higher compared to Ni/CeO_2-PT, Ni/CeO_2-TD, and Ni/CeO_2-SG (8.7 vs. 5.3, 5.6, and 6.5 mmol g^{-1}, respectively). Furthermore, multiple peaks/shoulders appeared in the TPO traces, showing that at least two kinds of carbon were formed after CH_4/CO gas treatment of the four catalysts, but to a different extent. More precisely, the Ni/CeO_2-HT (c) and Ni/CeO_2-SG (d), which revealed the highest amount of carbon deposition during DRM and CH_4/CO gas treatments, reveal reaction of carbon with oxygen in the range 300–800 °C as opposed to Ni/CeO_2-TD (a) and Ni/CeO_2-PT (b), where carbon oxidation occurs in the 500–800 °C range (more strongly bound carbon species but of lower amount). The latter results are in harmony with the results obtained under DRM reaction conditions (see Sections 2.6.1 and 2.6.2), and these will be discussed next in relation to the competitive contribution of CH_4 decomposition and CO dissociation towards carbon formation and removal rates.

2.7. Participation of Support's Lattice Oxygen under DRM Conditions

Figure 8A shows the transient evolution rate of $^{18}O_2$ consumption estimated upon the 10 min isotopic exchange of the ^{16}O ceria lattice oxygen (surface and bulk) with gaseous $^{18}O_2$, and that due mainly to the oxidation of Ni to $Ni^{18}O$ (less to the exchange of ^{16}O with $^{18}O_2$ in $Ni^{16}O_x$, see Section 4.5) at the step-gas switch Ar → 2 vol% $^{18}O_2$/2 vol% Kr/Ar at 750 °C. It is seen that the ceria-supported Ni catalysts show similar $^{18}O_2$ consumption rates during the 10 min exchange, except the Ni/CeO_2-PT, but all four catalysts showed a similar exchangeable amount of ^{16}O which was found to be between 10.2–12.4 mmol O g^{-1} (within less than 20%), as shown in Table S6. It should be noted that the maximum amount of ^{18}O consumed, and which is related to Ni oxidation, was 0.85 mmol g^{-1}. The latter illustrates that both the initial rates of $^{16}O/^{18}O$ exchanged, but also the surface and bulk mobility of ^{16}O species that were exchanged with $^{18}O_2$ were influenced by the CeO_2 synthesis method and Ni particles size only, to a small extent.

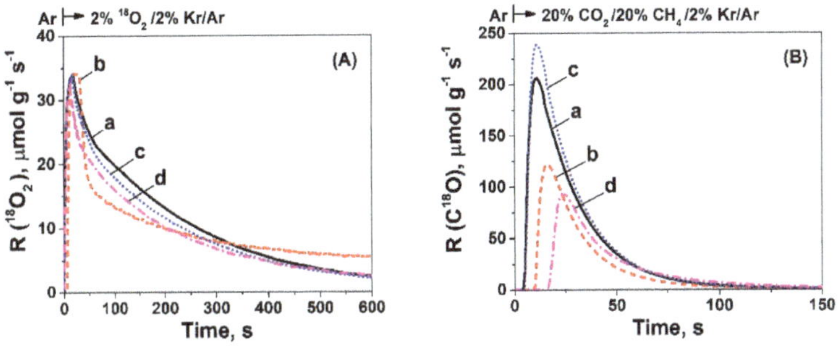

Figure 8. Transient rates (μmol $g^{-1}s^{-1}$) of (**A**) $^{18}O_2$ consumption during $^{16}O/^{18}O$ exchange at 750 °C after the gas switch: Ar → 2% $^{18}O_2$/2% Kr/Ar, (**B**) $C^{18}O$ formation obtained during the switch from Ar → 20% CH_4/20% CO_2/2% Kr/Ar/He (t) over 5 wt% Ni/CeO_2 prepared by (a) Thermal decomposition (TD), (b) Precipitation (PT), (c) Hydrothermal (HT), and (d) Sol-Gel (SG) method. W_{cat} = 0.02 g.

Figure 8B presents the transient rates of $C^{18}O$(g) formation over the four ceria-supported Ni catalysts obtained during the step-gas switch Ar → 20 vol% CH_4/20 vol% CO_2/2 vol% Kr/Ar at 750 °C, following the 10 min $^{16}O/^{18}O$ exchange (Figure 8A). It was observed that all four catalysts present similar shapes of the transient rate of $C^{18}O$(g) formation, however, they differ on their time delays and quantity (area under the transient curve). In particular, the Ni/CeO_2-TD and Ni/CeO_2-HT exhibit similar time delays (~5 s), followed by the Ni/CeO_2-PT (~10 s) and Ni/CeO_2-SG (~15 s). In addition, the amount of $C^{18}O$(g) produced over the four catalysts, after subtracting the equivalent amount of ^{18}O stored in the Ni during the $^{18}O_2$ gas treatment (0.85 mmol g^{-1}), was found to be 1.05 times larger for Ni/CeO_2-HT (6.5 mmol g^{-1}) compared to Ni/CeO_2-TD (5.9 mmol g^{-1}), 1.85 times larger compared to Ni/CeO_2-PT (3.5 mmol g^{-1}), and 1.96 times larger compared to Ni/CeO_2-SG (3.3 mmol g^{-1}). Considering the ratio of equivalent ^{18}O in $C^{18}O$ (Figure 8B) to ^{18}O exchanged (Figure 8A) as the contribution of the available amount of ^{18}O to the carbon gasification, the highest value was found to result from the Ni/CeO_2-HT (0.57), followed by Ni/CeO_2-TD (0.48), Ni/CeO_2-SG (0.32), and Ni/CeO_2-PT (0.31). The latter results (shown also in Table S6) will be discussed in the next section regarding the importance of participation of support's lattice oxygen to the carbon gasification rate.

3. Discussion

3.1. Structural Properties and Catalytic Performance of Ni Supported on CeO_2 Solids

The 5 wt% Ni supported on CeO_2 carriers prepared by different preparation methods exhibited largely different catalytic activity under DRM (20 vol% CO_2/20 vol% CH_4/He, 750 °C) (Figure 2) and structural and morphological differences were apparent among them. The high conversions achieved in our work, are close to the calculated equilibrium values for the used feed gas composition at 1 atm total pressure [52]. The 5 wt% Ni/CeO_2-HT catalyst with the largest surface area (~50 m^2 g^{-1}) consisted of smaller ceria mean primary crystallite size (~11.5 nm), smaller Ni mean particle size (~8.4 nm), and 11.5% Ni dispersion (Table S1, Figure 9). On the other hand, the 5 wt% Ni/CeO_2-SG with smaller surface area (~14.5 m^2 g^{-1}), consisted of larger ceria primary crystallite size (~43.1 nm) and Ni mean particle size (~20.8 nm; Ni dispersion 4.7%). The latter results are in good agreement with the literature [17,19,20,44,46,53–56]. The structural heterogeneity of the CeO_2 surface had strong effect on the deposition of Ni species. In fact, it was reported [45] that NiO (10 wt%) deposited on ceria nanoparticles of cubic shape was homogeneously dispersed. Yahi et al. [43] used three different preparation methods (microemulsion, sol-gel, and autocombustion) to synthesize CeO_2, on which 15 wt% of Ni was deposited. They clearly showed, via XRD and TPR studies, that NiO could be present with good crystallization in different phases (i.e., monoclinic and cubic phase for the auto-combustion and sol-gel, and cubic only phase for the microemulsion), which depended on the different preparation method of the ceria support. The authors [43] also reported different pore volume, surface area, and particle size by changing the preparation method, results of which are in good agreement with the present work. Xu et al. [55] prepared three Ni/Al_2O_3 catalytic systems of the same nominal composition by varying the preparation method, namely: Impregnation, water-in-oil-microemulsion, and sol-gel. By using XRD, TEM, and TPR, they showed crystalline structural differences, both for the support and Ni, which led to similar catalytic performance, but differences in the coking resistant, and thus in catalyst stability, as seen also in the present work. They argued that the latter differences might be due to strong metal–support interactions leading to differences on Ni particles size and dispersion. The latter results are in good agreement with those reported by other research groups [44,45,57], where a non-conventional synthesis method, namely the precipitation ionic exchange, led to cubic phases of CeO_2 and NiO (verified via FEG-SEM and XRD). In a recent study, Lykaki et al. [28] showed that the hydrothermal method (among other research works which are well reported there-in) led to well defined ceria nanorods of high specific surface area and with improved redox properties.

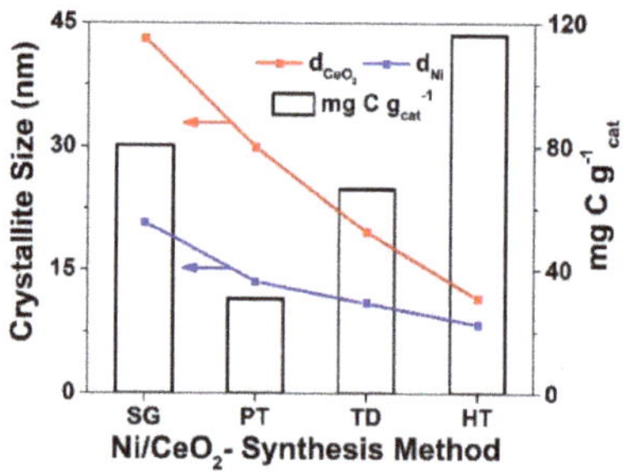

Figure 9. Comparative graph of the crystallite size (d_{CeO2} and d_{Ni}, nm) and amount of carbon deposition (mg C g^{-1}) after 12 h in DRM as a function of ceria support synthesis method.

The H_2-TPD traces (Figure 1) obtained over the four catalytic systems suggest large differences in the electronic structure of the Ni supported metal surfaces (distribution of binding strength between H and surface Ni atoms (E_{Ni-H}, kcal mol^{-1})). The electronic structure of Ni surface (different faces and defects), as well shown in the literature, influenced the rates of carbon formation and diffusion on the Ni surface and towards the support during both CH_4 decomposition and Boudouard reaction [58–65]. The morphological differences presented in the CeO_2 carrier through SEM images (Figure S3) seem to play a role for the induced differences in the electronic structure of the Ni surface (Figure 1), as reported also in previous publications [27,46]. HR-TEM (Figure S2) also suggested different morphologies for the Ni nanoparticles for the given Ni/CeO_2-HT catalyst, where ceria preparation method also influenced the mean Ni particle size as illustrated in Figure 9. The amount of carbon measured after

12 h of DRM (Figure 9, Table S5), and the transient results of CH_4 and CO decomposition reactions (Figures 5–8), tend therefore to suggest that the different morphology of ceria, as the result of different preparation method applied, induced Ni morphological/surface structural differences, thus surface nickel electronic modifications. These in turn govern the DRM activity behavior and carbon deposition rate of the various ceria-supported Ni catalysts [58].

Based on the CH_4-activity results reported in Figure 2 and the Ni dispersion values over the same catalysts (Table S1), it appears that Ni/CeO_2-HT contained a smaller number of active sites than Ni/CeO_2-SG catalyst. On the other hand, the former catalyst produced more carbon (larger carbon accumulation rates) under DRM reaction conditions (Figures 3 and 9), whereas the opposite was seen under transient methane decomposition reaction conditions (Figure 5); the initial rates of CH_4 dissociation to H_2 and the deposited amount of carbon (Figure 5D) were higher on Ni/CeO_2-HT than Ni/CeO_2-SG catalysts (Table 1). As illustrated in Figure 9, there seemed to be no clear relationship between d_{CeO2} and d_{Ni} and the amount of carbon deposition for at least 12 h after DRM. As is discussed below, the carbon formation and removal rates during DRM cannot be influenced only by the Ni metal and ceria support particle size in a clear monotonic way. The Ni-C strength, diffusion of carbon species on the Ni surface, and oxygen diffusion from the ceria support that the Ni phase and Ni-ceria interface (where carbon is formed) should all influence carbon deposition rate [46].

A possible explanation on the above results regarding carbon deposition on Ni/CeO_2 as influenced differently by the DRM, CO/He, and CH_4/He reaction conditions seems to be the competition of CH_4 and CO for the same Ni catalytic active sites, even though under DRM higher energy barriers are needed during the first and second steps of CH_4 dissociation ($CH_4 \rightarrow CH_3 + H$ and $CH_3 \rightarrow CH_2 + H$), and carbon dimer formation (C_2, carbon dimer is the first crucial step for inactive carbon formation). It was reported that both CH_4 and CO preferentially dissociate on Ni(111) surface, with the former to also favorably dissociate on Ni(100) and Ni(110) surfaces to a similar extent [66–74].

The main focus of this work was the effect of preparation method of CeO_2 used as support of Ni (5 wt% Ni/CeO_2) on the carbon deposition rates of the main routes (CH_4 decomposition and reverse Boudouard reaction), and on carbon removal rate (participation of support's O_L). It was shown that the preparation method influenced the Ni particle size and its morphology, and in turn its surface electronic structure, thus its catalytic performance for the DRM reaction at 750 °C. The various temperature-programmed and step-gas concentration transient experiments (including the use of isotopes) provided important information for the better understanding of the carbon pathways during DRM, to be discussed next.

3.2. Rates of Carbon Deposition and Removal under DRM Reaction Conditions

As mentioned in the Introduction section, supported Ni catalysts suffer from large amounts of carbon deposition under DRM reaction conditions. The amount of carbon deposited over the catalytic surface should be considered as the net rate between carbon formation (CH_4 decomposition and Boudouard reaction) and the carbon removal (e.g., participation of support's lattice oxygen). The carbon removal rate via the participation of lattice oxygen was probed by the transient response curves depicted in Figure 8B after partially exchanging active ceria support lattice $^{16}O_L$ with $^{18}O_L$. The carbon removal rate by this chemical step can be written to a first approximation as shown by the following Equation (1):

$$R_{C18O}\ (t) = k\ \theta_{OL}(t)\ \theta_C(t) \tag{1}$$

where k is an effective rate constant for the reaction step (2), θ_{OL} is the surface coverage of support lattice oxygen able to participate in reaction step (2), and θ_C is the surface coverage of carbon formed during DRM. In Equation (1), k might be considered as an average reactivity of more than one kind of carbon, whilst θ_{OL} (t) is also determined by the rate of surface O_L diffusion towards carbon. These two important kinetic parameters describing the rate of carbon gasification via Equation (1) are likely to depend on the Ni particle size/morphology as well as CeO_2 primary particle size.

$$C\text{-}s + {}^{18}O\text{-}_L \rightarrow C^{18}O(g) + s + Vo \quad (2)$$

For reaction step (2), s is a catalytic site at the metal-support interface, the support or both, and V_O is a surface oxygen vacancy of ceria support.

Initial carbon formation rates (recorded over a clean catalyst surface) and total amount of carbon accumulated during 30 min treatment of the catalysts were measured by performing transient experiments at 750 °C with 20 vol% of CH_4 reactant in the feed (Figure 5), similar to DRM conditions, and by the reverse Boudouard reaction or the CO dissociation alone (Figure 6), using 20 vol% CO (similar composition obtained in the DRM depicted in Figure 2). In addition, the individual amount of carbon derived from each route (CH_4 vs. CO) when both gases were present in the feed stream was also estimated for probing any synergy effects on the accumulation of carbon (Figure 7).

It is clearly shown that both the initial rate of carbon formation (Figures 5A and 6A) and the total amount of carbon formed (Table S5) over the four catalytic surfaces was at least 10 times larger in the case of CH_4 decomposition compared to the reverse Boudouard reaction. This result is in very good agreement with the TPO results obtained following the isotopic DRM reaction ($^{13}CO_2/^{12}CH_4$/He, Figure 4) and the isotopic $^{13}CO/^{12}CH_4$/He experiment (Figure 7), which both quantified the origin of carbon accumulation. Thus, the first conclusion is that CH_4 activation route was dominant and the one controlling the rate of carbon formation, however, the competition of CH_4 and CO activation for same catalytic sites, as clearly demonstrated, should be highly considered. In particular, the Ni/CeO_2-HT catalyst (CeO_2 prepared by the hydrothermal method) led to a smaller (~1.8 times) initial rate of carbon formation via CH_4 decomposition (Figure 5A) and CO dissociation (~3.5 times, Figure 6A) compared to the Ni/CeO_2-TD and Ni/CeO_2-PT catalysts, respectively. At this point it would be of interest to mention the effect of DRM reaction temperature on the origin of carbon deposition (CH_4 vs. CO_2 activation route). Vasiliades et al. [17,20] reported similar $^{12}CH_4/^{13}CO_2$/He isotopic DRM experiments as those reported in Section 2.6.2 (Figure 4) at 550 and 750 °C over 5 wt% Ni/$Ce_{1-x}M_xO_2$ (M = Zr^{4+}, Pr^{3+}) catalysts, the support of which (including pure CeO_2) was prepared by the citrate sol-gel method. It was illustrated that at the low-T of 550 °C, a higher contribution to carbon deposition was obtained via the CO_2 activation route (reverse Boudouard reaction: 2 CO → CO_2 + C) as opposed to the reaction T of 750 °C.

A careful comparison could be also made on the transient rates of CO formation during the CH_4/He treatment (Figure 5C), where the Ni/CeO_2-HT catalyst revealed significantly larger initial rate (~1.5 times) of its labile oxygen towards carbon gasification to CO(g) compared to the Ni/CeO_2-PT and Ni/CeO_2-TD catalysts, and even larger (~3 times) in the case of Ni/CeO_2-SG catalyst. The latter results are in a good agreement with the experimental findings shown in Figure 8B, where gasification of the formed carbon towards $C^{18}O$(g) formation under DRM reaction conditions takes place by the participation of support's $^{18}O_L$. The amount of available labile oxygen for $^{16}O/^{18}O$ isotopic exchange was found to be similar for the four supported Ni catalysts, a fact that suggests that morphological differences in their metal and support do not influence this specific process at 750 °C.

Considering the transient rates of $C^{18}O$(g) obtained over the four catalytic systems (Figure 8B), it was apparent that Ni/CeO_2-HT catalyst had activated a higher amount of lattice ^{18}O (11.4 mmol g^{-1}) by a factor of ~1.1 compared to Ni/CeO_2-SG (10.2 mmol g^{-1}, Table S6). Given the fact that the amount of carbon accumulated during DRM after 12 h was ~1.5 times larger in the case of Ni/CeO_2-HT compared to Ni/CeO_2-SG (see Section 2.6.1), it might be suggested that the effective rate constant k (Equation (1)) must be considered larger in the former than the latter catalyst. This result is important since it can prove that during DRM, the rate of carbon deposition on Ni/CeO_2-HT must be considered larger than on Ni/CeO_2-SG, a result in harmony with the transient CH_4 decomposition studies described in Section 2.6.3 (Figure 5A). Moreover, considering the integral rates of CH_4 conversion reported in Figure 2A, the carbon deposited by CH_4 during the 2 min transient shown in Figure 8B (end of rate of carbon removal by ^{18}O lattice oxygen) could be estimated. Then, the ratio of the amount of carbon removed by ^{18}O lattice oxygen as $C^{18}O$ (see Figure 8B, Table S6) to the amount of carbon deposited via

CH_4 decomposition could be estimated. This ratio was found to follow the order: CeO_2-HT > CeO_2-TD > CeO_2-SG > CeO_2-PT. The implication of this is that the reason that Ni/CeO_2-HT experienced the largest amount of carbon accumulation after 12 h in DRM (see Table S5), ~3.8 times larger than that of Ni/CeO_2-PT, should not be considered to be due to its inferior ability compared to the other ceria supports to provide mobile lattice oxygen for carbon gasification, at least for the first 30 min of TOS. It was suggested that carbon deposition and removal rates could change with longer time-on-stream as Ni surface and ceria support start to accommodate carbon deposits. Thus, deep understanding of the carbon accumulation with TOS and the intrinsic reasons for this is required for the given DRM ceria-supported Ni catalytic system.

It is noteworthy to be mentioned at this point that the differences in the delays of $C^{18}O$(g) that appeared during the switch from the inert gas to the DRM feed gas among the different catalysts (Figure 8B) were due to the different transient kinetics of reduction of the initially oxidized Ni surface (after $^{16}O/^{18}O$ exchange), as previously reported [46,75].

The temperature-programmed oxidation profiles of the carbon accumulation over a reduced metal surface after CH_4 decomposition or CO disproportionation alone or in the presence of both carbon sources illustrated that the co-presence of CH_4 and CO largely enhances the rate of carbon deposition.

4. Materials and Methods

4.1. Catalysts Synthesis

4.1.1. Cerium Dioxide (CeO_2) Supports

Sol-Gel Method

The CeO_2-SG metal oxide support was prepared using the modified citrate sol-gel method. The Ce metal precursor of $Ce(NO_3)_3.6H_2O$ (Sigma Aldrich, > 99% purity) was diluted in a beaker containing 100 mL solution of 1:1 (*v*/*v*) ratio of deionized H_2O and propanol-1. Citric acid (CA) was added for the creation of 1:1.5 M_{tot}:CA, where M_{tot} refers to the total molar concentration of metal ions in the solution, and similarly, the CA (molar concentration of citric acid). The pH of the solution was continuously adjusted (pH ~2.0) by adding HNO_3 (5M), with the solution to be under stirring at 70 °C. The resulting gel-like yellowish material was dried at 120 °C for 12 h, prior to its thermal heating with 1 °C min^{-1} under static air from room T to 500 °C. The sample was then kept at 500 °C for 6 h and its temperature was further increased to 750 °C (β = 5 °C min^{-1}) and kept for additional 4 h before cooled down to room T.

Thermal Decomposition Method

The CeO_2-TD metal oxide support was prepared using the thermal decomposition method. An appropriate amount of $Ce(NO_3)_3.6H_2O$ was dried in static air at 120 °C for 12 h, and after being cooled down to room T, its temperature was increased with 1 °C min^{-1} to 500 °C, where it was kept for 6 h. The temperature of the resulting material was then further increased to 750 °C (β = 5 °C min^{-1}), where it was kept for additional 4 h before cooled down to room T.

Hydrothermal Method

The CeO_2-HT metal oxide support was prepared using the hydrothermal method. During this method, 40 M NaOH (pH ~12.5) and 0.13 M $Ce(NO_3)_3{\cdot}6H_2O$ aqueous solutions were mixed (75 mL:175 mL), under vigorous stirring until a purplish milky slurry was formed. The milky slurry with total volume of 250 mL was kept under continuous stirring for 1 h and then transferred in a 1 L Teflon bottle and heated for 48 h at 90 °C. The reaction product was then cooled down to room temperature and the solid product was collected by filtration. The collected solid was rinsed with deionized water until pH neutralization to remove any co-precipitate salts. Drying and calcination procedures were performed as described in the thermal decomposition method.

Precipitation Method

The CeO_2-PT metal oxide support was prepared using the precipitation method. In the latter method, ammonia solution (25% *v*/*v*) as precipitation agent was added dropwise at room temperature and under continuous stirring in a 0.5 M aqueous solution of $Ce(NO_3)_3 \cdot 6H_2O$ until pH reached the value of 10, conditions that were controlled for 3 h. The resulting solution was then filtered, and the precipitate material was dried and calcined as described in the thermal decomposition method.

4.1.2. Wetness Impregnation of CeO_2 Supports with Ni (5 wt% Ni/CeO_2)

The resulting CeO_2 supports from the various synthesis procedures were grinded prior to Ni metal deposition. A given amount of each of the oxidic ceria support was diluted in an aqueous solution of $Ni(NO_3)_2 \cdot 6H_2O$ (Sigma-Aldrich, >99% purity) so as to be impregnated with 5 wt% Ni nominal loading. The resulting slurry was dried overnight at 120 °C, followed by cooling to room T. The temperature of the solid was then increased under static air to 750 °C, where it was kept for 4 h. The resulting material was named "fresh catalyst", and prior to any catalytic measurements it was in situ reduced in pure H_2 gas (1 bar, 50 NmL min^{-1}) at 700 °C for 2 h.

4.2. Catalysts Characterization

4.2.1. Powder X-ray Diffraction (XRD)

Powder X-ray diffractograms of the calcined CeO_2-supported Ni catalysts were recorded by using a Shimadzu 6000 Series diffractometer (CuKa radiation, $\lambda = 0.15418$ nm, Kyoto, Japan) in the 20–80° 2θ range (2° min^{-1}, 0.02° increment). By using the Scherrer equation [17], the lattice parameter (α, Å), the mean primary crystallite size (d_c, nm) of the ceria pseudo-cubic structure, and the mean crystal size of NiO were estimated. The latter was used to estimate the Ni mean particle size (d_{Ni}, nm) as of Equation (3), after the assumption that Ni and NiO preserve the same particle geometrical shape:

$$d\,(Ni, nm) = d\,(NiO, nm) \times 0.847. \quad (3)$$

4.2.2. Surface Texture (BET/BJH)

The BET specific surface area (SSA, $m^2\,g^{-1}$), the total pore volume (V_p, $cm^3\,g^{-1}$), and the mean pore size (d_p, nm) of the CeO_2-supported nickel catalysts and their supports alone were determined based on N_2 adsorption/desorption isotherms measured at 77 K with a Micromeritics Gemini 2360 surface area and pore size analyzer (Norcross, Georgia, United States). Prior to any measurements, the sample was degassed in N_2 gas flow at 300 °C for 4 h.

4.2.3. H_2 Temperature-Programmed Desorption (TPD)

The effect of Ni supported on different CeO_2 supports on the hydrogen chemisorption and desorption behavior was investigated by using the H_2-TPD technique. A 0.3 g sample was first reduced in situ in hydrogen gas at 700 °C for 2 h prior to He-purge and increase of its temperature to 750 °C, until H_2 signal reached its background value (desorption of any spilled-over hydrogen on the support). The reactor was then cooled down to room temperature and a switch from He to 0.5 vol% H_2/He (30 min) gas mixture was performed. The catalyst was then purged for 10 min in He flow, and the temperature was subsequently increased to 750 °C (TPD, $\beta = 30$ °C min^{-1}, 50 NmL min^{-1}). The H_2 (m/z = 2) signal was continuously monitored with online mass spectrometer (MS, Balzers, Omnistar 1–200 amu, Pfeiffer Vacuum, Asslar, Germany), and the MS signal was converted into concentration (ppm) by using a certified standard gas mixture (0.95 vol% H_2/He). The Ni dispersion (D_{Ni}, %) was estimated after assuming an H_2 chemisorption stoichiometry of $H/Ni_s = 1$, where the Ni mean primary particle size (d_{Ni}, nm) was estimated by using the following Equation (4) [76]:

$$d_{Ni}\,(nm) = 0.97/D_{Ni}. \quad (4)$$

4.2.4. Transmission Electron Microscopy (TEM)

The fresh 5 wt% Ni/CeO_2-HT supported Ni catalyst was characterized with a JEOL (JEM-2100) high-resolution transmission electron microscopy system (HR-TEM) (Tokyo, Japan), operated at 200 kV (resolution point 0.23 nm, lattice 0.14 nm). Selected specimens were prepared by dispersion of the powdered catalyst in water, and spread onto a carbon-coated copper grid (200 mesh), while images were recorded by means of films (Kodak SO-163).

4.2.5. Scanning Electron Microscopy (SEM)

The morphology of the fresh CeO_2-supported Ni solids was characterized by using scanning electron microscope (SEM, JEOL JSM-6610 LV, Tokyo, Japan), equipped with a BRUKER type QUANTAX 200 energy dispersive spectrometer (EDS). The effect of different method of synthesis of the CeO_2 was studied using secondary electron images (SEI). EDS analysis was performed for determining the chemical composition of the solids.

4.3. *Catalytic Performance of CeO_2-Supported Ni in DRM*

Catalytic measurements were performed using a Micro-activity reactor system (MA-REF from PID Eng & Tech, Madrid, Spain) equipped with a tubular quartz reactor (i.d. = 6 mm), and the experimental apparatus used was described elsewhere [20]. The catalytic bed was prepared by grinding (grain powder size less than 106 μm) and mixing an appropriate amount of Ni/CeO_2 catalyst with SiC (1 cat:1 SiC (*w*/*w*)) in order to achieve a gas hourly space velocity (GHSV) of ~30,000 h^{-1}. Due to the differences in the ceria solid powder prepared by the four different methods, the amount of catalyst for each Ni/CeO_2 was varied (W_{cat} = 0.072–0.167), while the total gas flow rate was kept the same (50 NmL min^{-1}). The catalytic performance of the solids was examined at 750 °C for 30 min with a DRM feed gas composition of 20 vol% CH_4/20 vol% CO_2/60 vol% He. The conversions of CH_4 and CO_2 (X_{CH4} and X_{CO2}, %) were calculated by using Equation (5). The effluent gas stream from the micro-reactor was analyzed through online MS and infrared gas analyzers (Horiba, VA-3000, Kyoto, Japan) for H_2 (m/z = 2), CH_4 (m/z = 15) and CO, CO_2, respectively. Calibration of the signals from the MS and IR gas analyzers was made by using certified calibration gas mixtures (1.06 vol% CO/1.02 vol% CH_4/0.95 vol% H_2/He and 2.55 vol% CO_2/He). The product yields (Y_{H2} and Y_{CO}, %) were estimated via Equations (6) and (7):

$$X_Y(\%) = \frac{F_Y^{in} - F_Y^{out}}{F_Y^{in}} \times 100 \tag{5}$$

$$Y_{H_2}(\%) = \frac{F_{H_2}^{out}}{2F_{CH_4}^{in}} \times 100 \tag{6}$$

$$Y_{CO}(\%) = \frac{F_{CO}^{out}}{F_{CH_4}^{in} - F_{CO}^{in}} \times 100 \tag{7}$$

where, F^{in} and F^{out} are the molar flow rates (mol s^{-1}) of reactant Y (CH_4 or CO_2) and product (H_2 or CO) at the inlet and outlet of the reactor, respectively. The F^{out} was estimated based on the total volume flow rate at the outlet of the reactor (measured at 1 bar and room T), and the mole fraction of the component measured by the above-mentioned gas analysis system.

4.4. *Characterization of Carbon Formed under Different Reaction Conditions*

4.4.1. Dry Reforming of Methane ($^{12}CO_2$/$^{12}CH_4$) Reaction

The reactivity of carbon towards oxygen and its amount (mg C g_{cat}^{-1} or wt%) accumulated after 12 h of DRM at 750 °C over the catalysts investigated in this work were estimated by performing temperature-programmed oxidation (TPO) experiments following DRM. A purge with He (20 min) was applied after the 12 h DRM reaction with the reactor's temperature to increase to 800 °C until no MS signals was identified for CH_4, CO_2, H_2, and CO. The catalyst's temperature was then decreased to 100 °C followed by a feed gas switched from He to 10 vol% O_2/He (50 NmL min^{-1}). The catalyst's temperature was subsequently increased to 800 °C with a heating ramp of 30 °C min^{-1} (TPO). During the latter switch, the signals of CO (m/z = 28) and CO_2 (m/z = 44) were continuously monitored with the MS and CO/CO_2 infrared gas analyzer, and then converted into mol% based on certified calibration gas mixtures (1.06 vol% CO/He and 2.55 vol% CO_2/He).

4.4.2. Isotopically Labelled Dry Reforming of Methane ($^{13}CO_2$/$^{12}CH_4$) Reaction

Isotopically labelled DRM mixture (5 vol% $^{13}CO_2$/5 vol% $^{12}CH_4$/45 vol% Ar/45 vol% He; 50 NmL min^{-1}; GHSV ~30,000 h^{-1}) was used for 30 min at 750 °C, followed by TPO, in order to investigate the relative contribution of CH_4 and CO_2 activation routes towards carbon accumulation (μmol g^{-1} and mg g_{cat}^{-1}) over the examined ceria-supported Ni catalytic systems. The ^{12}C-containing TPO traces referred to the CH_4 activation route contribution on the amount of carbon, whereas the ^{13}C-containing TPO traces referred to the CO_2 activation route. More precisely, after 30 min in DRM, a He purge was performed for 10 min prior to the temperature increase to 800 °C (until no MS signals for CO and CO_2 were observed). The reactor was then cooled down to 200 °C, and the feed gas was switched from He to 10 vol% O_2/He (50 NmL min^{-1}), followed by TPO to 800 °C (β = 30 °C min^{-1}). The signals for ^{12}CO, ^{13}CO, $^{12}CO_2$, and $^{13}CO_2$ (m/z = 28, 29, 44, and 45, respectively) were continuously monitored by MS and their quantification (mol%) was made by using certified gas mixtures (1.06 vol% ^{12}CO/He, 10 vol% ^{13}CO/Ar, 2.55 vol% $^{12}CO_2$/He, and 10 vol% $^{13}CO_2$/Ar).

4.4.3. Methane Decomposition (CH_4/He) Reaction

The reduced CeO_2-supported Ni catalysts were exposed to 20% CH_4/He for 30 min in order to measure the initial rate of CH_4 decomposition and its subsequent rate evolution, which is one of the main routes of inactive carbon formation under DRM conditions. The transient responses of H_2 (m/z = 2), CH_4 (m/z = 15), and CO (m/z = 28) were followed during the step-gas switch He → 20% CH_4/He (750 °C, 30 min, 50 NmL min^{-1}; GHSV ~30,000 h^{-1}) by online MS. The latter switch was followed by a 10 min He purge, while the temperature was increased to 800 °C (until background values were reached for the CO and CO_2 MS signals). The catalyst was then cooled down to 200 °C and the feed was switched from He to 10 vol% O_2/He (50 NmL min^{-1}) to perform a TPO run (β = 30 °C min^{-1}). The transient evolution of CO (m/z = 28) and CO_2 (m/z = 44) was continuously monitored with MS, and their quantification was made using certified calibration gas mixtures (1.06 vol% CO/1.02 vol% CH_4/0.95 vol% H_2/He and 2.55 vol% CO_2/He).

4.4.4. Carbon Monoxide Dissociation (CO/He) Reaction

The second main route of inactive carbon formation during DRM, that of reverse Boudouard reaction, was investigated by performing over the 5 wt% Ni/CeO_2 catalysts the step-gas switch He → 20% CO/He (750 °C, 30 min, 50 NmL min^{-1}; GHSV ~30,000 h^{-1}), where the evolution of CO (m/z = 28) and CO_2 (m/z = 44) were continuously monitored with MS. The latter gas switch was followed by a He purge (10 min) and temperature increase to 800 °C, where the catalyst was kept at this temperature until the CO and CO_2 MS signals reached their respective background value. The reactor's temperature was then decreased to 200 °C, where a switch to 10 vol% O_2/He (50 NmL min^{-1}) gas mixture was made for a TPO run to 800 °C (β = 30 °C min^{-1}). During TPO, the mass numbers (m/z) of 28 and 44 were followed by MS, and quantification was made by considering certified calibration gas mixtures (1.06 vol% CO/He and 2.55 vol% CO_2/He).

4.4.5. Isotopically Labelled Competitive ($^{13}CO/^{12}CH_4$) Reaction towards Carbon Formation

The relative contribution of the two main routes towards inactive carbon accumulation under DRM reaction conditions (CH_4 decomposition and reverse Boudouard reaction) was investigated by exposing the catalysts over an isotopically labelled mixture consisting of 2.5 vol% ^{13}CO/2.5 vol% $^{12}CH_4$/2 vol% Kr/Ar/He (50 NmL min^{-1}; GHSV ~30,000 h^{-1}) at 750 °C for 20 min. The gas-flow was then switched to He for a 10 min purge and the temperature was increased to 800 °C until the ^{12}CO, ^{13}CO, $^{12}CO_2$, and $^{13}CO_2$ MS signals reached their respective background value. The catalyst was then cooled in He flow to 200 °C and a switch to 10 vol% O_2/He (50 NmL min^{-1}) gas mixture was made for a TPO run (increase T to 800 °C, β = 30 °C min^{-1}). The effluent gas stream was continuously monitored by MS for ^{12}CO, ^{13}CO, $^{12}CO_2$, and $^{13}CO_2$ (m/z = 28, 29, 44, and 45, respectively), and quantification of the MS signals was made by using the previously mentioned (Section 4.4.2) calibration gas mixtures. It's worth mentioning that the ^{12}C-containing TPO traces refer to the $^{12}CH_4$ contribution on the amount of carbon accumulation, whereas the ^{13}C-containing TPO traces refer to the ^{13}CO route.

4.5. *Participation of Support's Lattice Oxygen in DRM Reaction Conditions*

The partial $^{16}O/^{18}O$ isotopic exchange of ceria support's lattice ^{16}O was performed over pre-reduced Ni/CeO_2 catalysts (W_{cat} = 0.02 g) at 750 °C for 10 min prior to the dry reforming of methane reaction. This designed experiment probes for the extent of contribution of support's lattice oxygen in the carbon-path under DRM conditions [18,19]. More precisely, after 2 h reduction of catalyst with pure H_2 (1 bar) at 700 °C, the feed was switched to Ar for 10 min with subsequent increase of the temperature to 750 °C, until no H_2 (m/z = 2), $^{16}O_2$ (m/z = 32), and $^{16}O^{18}O$ (m/z = 34) MS signals were recorded. The exchange of support lattice oxygen and the oxidation of Ni/NiO_x to $Ni^{18}O$ with $^{18}O_2$(g) was then made by exposing the catalyst to 2 vol% $^{18}O_2$/2 vol% Kr/Ar/He (10 min, 50 NmL min^{-1}). During the exchange process, the signals of $^{16}O_2$, $^{16}O^{18}O$, $^{18}O_2$, and Kr (m/z = 32, 34, 36, and 84, respectively) were recorded continuously with online MS, which then converted into concentration (mol%) by using appropriate material balances [18] from which the amount of oxygen exchanged (mol ^{16}O g^{-1}) can be estimated. A 10 min He purge then followed, and the feed gas was then switched to 20 vol% CH_4/20 vol% CO_2/He (50 NmL min^{-1}). During the latter DRM reaction step, the MS signals of 30, 44, 46, 48, and 84 ($C^{18}O$, $C^{16}O_2$, $C^{16}O^{18}O$, $C^{18}O_2$, and Kr, respectively) were continuously monitored, and then converted into concentration (mol%) by using appropriate calibration gases. It was assumed same sensitivities for the $C^{16}O$ and $C^{18}O$ (m/z = 30) gases. The contribution of $C^{18}O_2$ (m/z = 48) and $C^{16}O^{18}O$ (m/z = 46) to the m/z=30 were carefully subtracted from the m/z = 30 ($C^{18}O$) signal recorded by using a standard $C^{16}O_2$/He gas mixture and considering the same contribution of m/z = 44 to m/z = 28 for the m/z = 48 and m/z = 46 to m/z = 30. The formation of $C^{18}O$(g) during DRM, following the oxygen $^{16}O/^{18}O$ isotopic exchange step, is clearly described in our previous publications [20,46], where *$^{18}O_L$ of support* can react with carbon formed on the catalyst surface.

5. Conclusions

The main conclusions derived from the present work are as follows:

(a) The 5 wt% Ni supported on CeO_2 carriers prepared by four different methods exhibited obvious structural and morphological differences, which led to large differences in catalytic activity under DRM reaction conditions at 750 °C.

(b) The Ni/CeO_2-PT (ceria was prepared by the precipitation method) exhibited the lowest amount of carbon formation among the four catalytic systems. A notable reduction of carbon deposition by ~3.8, 1.8, and 1.4 times was observed after 12 h in DRM (20 vol% CH_4, CO_2/CH_4 = 1) compared to Ni/CeO_2-HT, Ni/CeO_2-TD, and Ni/CeO_2-SG, respectively. This precipitation route might lead also to lower carbon deposits in DRM for other CeO_2-based supported Ni catalysts in attempting to develop low-carbon resistant DRM catalytic systems in the future.

(c) Based on various transient and other isotopic experiments, it was shown that a large pool of oxygen contributes to the gasification of carbon formed in DRM towards the formation of CO, thus offering an important path for carbon removal from the catalyst during DRM.

(d) The origin of carbon deposition was found to be largely determined by the CH_4 activation route in all four catalytic systems but to a different extent.

(e) The Ni/CeO_2-HT and Ni/CeO_2-PT catalysts presented similar amount of CH_4 decomposed (CH_4/He reaction), which was found to increase by ~1.6 and 1.3 times for Ni/CeO_2-TD and Ni/CeO_2-SG catalysts. On the other hand, the ratio between the CO formation and CH_4 consumption was found to be the same for Ni/CeO_2-HT and Ni/CeO_2-PT but significantly lower in the case of Ni/CeO_2-TD and Ni/CeO_2-SG catalysts, indicating the higher ability of the former solids to remove deposited carbon by the participation of their ceria support lattice oxygen.

(f) During the reverse Boudouard reaction (CO/He reaction), the largest amount of carbon deposited was found to be on Ni/CeO_2-PT followed by Ni/CeO_2-TD, Ni/CeO_2-HT, and Ni/CeO_2-SG solids.

(g) Despite the fact that on Ni/CeO_2-HT a lesser amount of carbon was deposited during CH_4 decomposition, during CO disproportionation compared to the other catalysts, the amount of carbon deposition observed after 12 h in DRM (20 vol% CH_4, CO_2/CH_4 = 1) was the largest. This behavior could be justified by the enhancement of carbon deposition in the co-presence of CH_4 and CO which occurs in a larger extent over Ni/CeO_2-HT as proved experimentally.

Supplementary Materials: The following are available online at http://www.mdpi.com/2073-4344/9/7/621/s1, Figure S1: Powder X-ray diffractograms of 5 wt% Ni supported on (a) CeO_2-TD, (b) CeO_2-PT, (c) CeO_2-HT and (d) CeO_2-SG carriers in the (A) 20–70° 2θ and (B) 35–45° 2θ region (diffraction peaks of NiO), Figure S2: Representative HR-TEM images of the calcined (air, 750 °C/4 h) 5 wt% Ni/CeO_2-HT catalyst. Left graph: magnification at 50 nm unit scale; Right graph: magnification at 10 nm unit scale, Figure S3: SEM images of the fresh Ni/CeO_2-SG (top left), Ni/CeO_2-PT (top right), Ni/CeO_2-HT (down left) and Ni/CeO_2-TD (down right), Table S1: Textural and structural characterization of 5 wt% Ni/CeO_2 (-TD, -PT, -HT and -SG) DRM fresh catalysts, Table S2: Catalytic activity in terms of CH_4, CO_2 conversion (X_{CH4}, X_{CO2}, %), H_2 Yield (%) and H_2/CO gas product ratio obtained after 30 min in DRM at 750 °C for the four ceria supports prepared by different methods, Table S3: Catalytic stability performance in terms of CH_4, CO_2 conversion (X_{CH4}, X_{CO2}, %), H_2 Yield (%), H_2/CO gas product ratio and carbon deposition (mg C $g^{-1}{}_{cat}$) obtained during DRM (20% CH_4/20% CO_2/He) at 750 °C over the 5 wt% Ni/CeO_2-PT solid, Table S4: Catalytic activity in terms of CH_4, CO_2 conversion (X_{CH4}, X_{CO2}, %), H_2 Yield (%) and H_2/CO gas product ratio obtained after 30 min in DRM (5 vol% $^{13}CO_2$/5 vol% $^{12}CH_4$/He) at 750 °C, Table S5: Carbon accumulation (mg C $g_{cat}{}^{-1}$) estimated via TPO followed individual reactions over all catalysts at 750 °C; 20 vol% CO_2/20 vol% CH_4/He (12 h), 5 vol% $^{13}CO_2$/5 vol% $^{12}CH_4$/He (30 min), 20 vol% CH_4/He (30 min), 20 vol% CO/He (30 min), 2.5 vol% ^{13}CO/2.5 vol% $^{12}CH_4$/He (20 min), Table S6: 18O consumption (mmol g^{-1}) during $^{16}O/^{18}O$ exchange, $C^{18}O$ formation (mmol g^{-1}) during DRM following $^{16}O/^{18}O$ oxygen exchange, and $C^{18}O/^{18}O$ ratio.

Author Contributions: Conceptualization, A.M.E., V.N.S., and M.A.V.; methodology, A.M.E., V.N.S., and M.A.V.; validation, M.A.V. and C.M.D.; formal analysis, C.M.D.; investigation, C.M.D.; resources, A.M.E.; data curation, C.M.D.; writing—original draft preparation, M.A.V.; writing—review and editing, A.M.E.; visualization, A.M.E. and M.A.V.; supervision, A.M.E. and M.A.V.; project administration, M.A.V. and C.M.D.; funding acquisition, A.M.E.

Funding: This research was funded by the Research Committee of the University of Cyprus.

Acknowledgments: The authors are grateful to Maria Kollia (Research Associate) of the University of Patras for performing the HR-TEM studies.

Conflicts of Interest: The authors declare no conflict of interest.

References

1. Barbosa, L.C.; Araújo, O.Q.F.; de Medeiros, J.L. Carbon capture and adjustment of water and hydrocarbon dew-points via absorption with ionic liquid [Bmim][NTf2] in offshore processing of CO_2-rich natural gas. *J. Nat. Gas Sci. Eng.* **2019**, *66*, 26–41. [CrossRef]
2. De Carvalho Reis, A.; de Medeiros, J.L.; Nunes, G.C.; Araújo, O.Q.F. Upgrading of natural gas ultra-rich in carbon dioxide: Optimal arrangement of membrane skids and polishing with chemical absorption. *J. Clean. Prod.* **2017**, *165*, 1013–1024. [CrossRef]

3. De Oliveira Arinelli, L.; Trotta, T.A.F.; Teixeira, A.M.; de Medeiros, J.L.; Araújo, O.Q.F. Offshore processing of CO_2 rich natural gas with supersonic separator versus conventional routes. *J. Nat. Gas Sci. Eng.* **2017**, *46*, 199–221. [CrossRef]
4. Lachén, J.; Durán, P.; Menéndez, M.; Peña, J.A.; Herguido, J. Biogas to high purity hydrogen by methane dry reforming in TZFBR + MB and exhaustion by Steam-Iron Process. Techno–economic assessment. *Int. J. Hydrogen Energy* **2018**, *43*, 11663–11675. [CrossRef]
5. Ugarte, P.; Durán, P.; Lasobras, J.; Soler, J.; Menéndez, M.; Herguido, J. Dry reforming of biogas in fluidized bed: Process intensification. *Int. J. Hydrogen Energy* **2017**, *42*, 13589–13597. [CrossRef]
6. Gao, J.; Hou, Z.; Lou, H.; Zheng, X. Dry (CO_2) Reforming. In *Fuel Cells: Technologies for Fuel Processing*; Elsevier: Amsterdam, The Netherlands, 2011; pp. 191–221. ISBN 9780444535634.
7. Navas-Anguita, Z.; Cruz, P.L.; Martín-Gamboa, M.; Iribarren, D.; Dufour, J. Simulation and life cycle assessment of synthetic fuels produced via biogas dry reforming and Fischer-Tropsch synthesis. *Fuel* **2019**, *235*, 1492–1500. [CrossRef]
8. Clarkson, J.; Ellis, P.R.; Humble, R.; Kelly, G.J.; McKenna, M.; West, J. Deactivation of alumina supported cobalt FT catalysts during testing in a Continuous-stirred tank reactor (CSTR). *Appl. Catal. A Gen.* **2018**, *550*, 28–37. [CrossRef]
9. Leonzio, G. State of art and perspectives about the production of methanol, dimethyl ether and syngas by carbon dioxide hydrogenation. *J. CO2 Util.* **2018**, *27*, 326–354. [CrossRef]
10. Venvik, H.J.; Yang, J. Catalysis in microstructured reactors: Short review on small-scale syngas production and further conversion into methanol, DME and Fischer-Tropsch products. *Catal. Today* **2017**, *285*, 135–146. [CrossRef]
11. Aramouni, N.A.K.; Touma, J.G.; Tarboush, B.A.; Zeaiter, J.; Ahmad, M.N. Catalyst design for dry reforming of methane: Analysis review. *Renew. Sustain. Energy Rev.* **2018**, *82*, 2570–2585. [CrossRef]
12. Jang, W.J.; Shim, J.O.; Kim, H.M.; Yoo, S.Y.; Roh, H.S. A review on dry reforming of methane in aspect of catalytic properties. *Catal. Today* **2018**, *324*, 15–26. [CrossRef]
13. Mah, A.X.Y.; Ho, W.S.; Bong, C.P.C.; Hassim, M.H.; Liew, P.Y.; Asli, U.A.; Kamaruddin, M.J.; Chemmangattuvalappil, N.G. Review of hydrogen economy in Malaysia and its way forward. *Int. J. Hydrogen Energy* **2019**, *44*, 5661–5675. [CrossRef]
14. Sovacool, B.K. Reviewing, Reforming, and Rethinking Global Energy Subsidies: Towards a Political Economy Research Agenda. *Ecol. Econ.* **2017**, *135*, 150–163. [CrossRef]
15. Damyanova, S.; Pawelec, B.; Arishtirova, K.; Fierro, J.L.G. Ni-based catalysts for reforming of methane with CO_2. *Int. J. Hydrogen Energy* **2012**, *37*, 15966–15975. [CrossRef]
16. Usman, M.; Wan Daud, W.M.A.; Abbas, H.F. Dry reforming of methane: Influence of process parameters—A review. *Renew. Sustain. Energy Rev.* **2015**, *45*, 710–744. [CrossRef]
17. Makri, M.M.; Vasiliades, M.A.; Petallidou, K.C.; Efstathiou, A.M. Effect of support composition on the origin and reactivity of carbon formed during dry reforming of methane over 5 wt% $Ni/Ce_{1-x}M_xO_{2-\delta}$ ($M = Zr^{4+}$, Pr^{3+}) catalysts. *Catal. Today* **2016**, *259*, 150–164. [CrossRef]
18. Vasiliades, M.A.; Djinović, P.; Pintar, A.; Kovač, J.; Efstathiou, A.M. The effect of CeO_2-ZrO_2 structural differences on the origin and reactivity of carbon formed during methane dry reforming over NiCo/CeO_2-ZrO_2 catalysts studied by transient techniques. *Catal. Sci. Technol.* **2017**, *7*, 5422–5434. [CrossRef]
19. Vasiliades, M.A.; Djinović, P.; Davlyatova, L.F.; Pintar, A.; Efstathiou, A.M. Origin and reactivity of active and inactive carbon formed during DRM over $Ni/Ce_{0.38}Zr_{0.62}O_{2-\delta}$ studied by transient isotopic techniques. *Catal. Today* **2018**, *299*, 201–211. [CrossRef]
20. Vasiliades, M.A.; Makri, M.M.; Djinović, P.; Erjavec, B.; Pintar, A.; Efstathiou, A.M. Dry reforming of methane over 5 wt% $Ni/Ce_{1-x}Pr_xO_{2-\delta}$ catalysts: Performance and characterisation of active and inactive carbon by transient isotopic techniques. *Appl. Catal. B Environ.* **2016**, *197*, 168–183. [CrossRef]
21. Simonov, M.N.; Rogov, V.A.; Smirnova, M.Y.; Sadykov, V.A. Pulse Microalorimetry Study of Methane Dry Reforming Reaction on Ni/Ceria-Zirconia Catalyst. *Catalysts* **2017**, *7*, 268. [CrossRef]
22. Pakhare, D.; Spivey, J. A review of dry (CO_2) reforming of methane over noble metal catalysts. *Chem. Soc. Rev.* **2014**, *43*, 7813–7837. [CrossRef]
23. Zhang, G.; Liu, J.; Xu, Y.; Sun, Y. A review of CH_4–CO_2 reforming to synthesis gas over Ni-based catalysts in recent years (2010–2017). *Int. J. Hydrogen Energy* **2018**, *43*, 15030–15054. [CrossRef]

24. Muraza, O.; Galadima, A. A review on coke management during dry reforming of methane. *Int. J. Energy Res.* **2015**, *39*, 1196–1216. [CrossRef]
25. Arora, S.; Prasad, R. An overview on dry reforming of methane: Strategies to reduce carbonaceous deactivation of catalysts. *RSC Adv.* **2016**, *6*, 108668–108688. [CrossRef]
26. Lykaki, M.; Pachatouridou, E.; Iliopoulou, E.; Carabineiro, S.A.C.; Konsolakis, M. Impact of the synthesis parameters on the solid state properties and the CO oxidation performance of ceria nanoparticles. *RSC Adv.* **2017**, *7*, 6160–6169. [CrossRef]
27. Cortés Corberán, V.; Rives, V.; Stathopoulos, V. Recent Applications of Nanometal Oxide Catalysts in Oxidation Reactions. In *Advanced Nanomaterials for Catalysis and Energy*; Sadykov, V.A., Ed.; Elsevier: Amsterdam, The Netherlands, 2018; pp. 227–293. ISBN 978-0-12-814807-5.
28. Lykaki, M.; Pachatouridou, E.; Carabineiro, S.A.C.; Iliopoulou, E.; Andriopoulou, C.; Kallithrakas-Kontos, N.; Boghosian, S.; Konsolakis, M. Ceria nanoparticles shape effects on the structural defects and surface chemistry: Implications in CO oxidation by Cu/CeO_2 catalysts. *Appl. Catal. B Environ.* **2018**, *230*, 18–28. [CrossRef]
29. Zhang, D.; Du, X.; Shi, L.; Gao, R. Shape-controlled synthesis and catalytic application of ceria nanomaterials. *Dalt. Trans.* **2012**, *41*, 14455–14475. [CrossRef]
30. Tok, A.I.Y.; Boey, F.Y.C.; Dong, Z.; Sun, X.L. Hydrothermal synthesis of CeO_2 nano-particles. *J. Mater. Process. Technol.* **2007**, *190*, 217–222. [CrossRef]
31. Ilgaz Soykal, I.; Sohn, H.; Miller, J.T.; Ozkan, U.S. Investigation of the reduction/oxidation behavior of cobalt supported on nano-ceria. *Top. Catal.* **2014**, *57*, 785–795. [CrossRef]
32. Yin, L.; Wang, Y.; Pang, G.; Koltypin, Y.; Gedanken, A. Sonochemical synthesis of cerium oxide nanoparticles-Effect of additives and quantum size effect. *J. Colloid Interface Sci.* **2002**, *246*, 78–84. [CrossRef]
33. Piumetti, M.; Andana, T.; Bensaid, S.; Russo, N.; Fino, D.; Pirone, R. Study on the CO Oxidation over Ceria-Based Nanocatalysts. *Nanoscale Res. Lett.* **2016**, *11*, 165. [CrossRef]
34. Lu, X.H.; Zheng, D.Z.; Gan, J.Y.; Liu, Z.Q.; Liang, C.L.; Liu, P.; Tong, Y.X. Porous CeO_2 nanowires/nanowire arrays: Electrochemical synthesis and application in water treatment. *J. Mater. Chem.* **2010**, *20*, 7118–7122. [CrossRef]
35. Tang, W.X.; Gao, P.X. Nanostructured cerium oxide: Preparation, characterization, and application in energy and environmental catalysis. *MRS Commun.* **2016**, *6*, 311–329. [CrossRef]
36. Kovacevic, M.; Mojet, B.L.; Van Ommen, J.G.; Lefferts, L. Effects of Morphology of Cerium Oxide Catalysts for Reverse Water Gas Shift Reaction. *Catal. Lett.* **2016**, *146*, 770–777. [CrossRef]
37. Wu, Z.; Li, M.; Howe, J.; Meyer, H.M.; Overbury, S.H. Probing defect sites on CeO_2 nanocrystals with well-defined surface planes by raman spectroscopy and O_2 adsorption. *Langmuir* **2010**, *26*, 16595–16606. [CrossRef]
38. He, H.; Yang, P.; Li, J.; Shi, R.; Chen, L.; Zhang, A.; Zhu, Y. Controllable synthesis, characterization, and CO oxidation activity of CeO_2 nanostructures with various morphologies. *Ceram. Int.* **2016**, *42*, 7810–7818. [CrossRef]
39. Liu, J.; Li, Y.; Zhang, J.; He, D. Glycerol carbonylation with CO_2 to glycerol carbonate over CeO_2 catalyst and the influence of CeO_2 preparation methods and reaction parameters. *Appl. Catal. A Gen.* **2016**, *513*, 9–18. [CrossRef]
40. Tan, J.P.Y.; Tan, H.R.; Boothroyd, C.; Foo, Y.L.; He, C.B.; Lin, M. Three-dimensional structure of CeO_2 nanocrystals. *J. Phys. Chem. C* **2011**, *115*, 3544–3551. [CrossRef]
41. Montini, T.; Melchionna, M.; Monai, M.; Fornasiero, P. Fundamentals and Catalytic Applications of CeO_2-Based Materials. *Chem. Rev.* **2016**, *116*, 5987–6041. [CrossRef]
42. Tana; Zhang, M.; Li, J.; Li, H.; Li, Y.; Shen, W. Morphology-dependent redox and catalytic properties of CeO_2 nanostructures: Nanowires, nanorods and nanoparticles. *Catal. Today* **2010**, *148*, 179–183. [CrossRef]
43. Yahi, N.; Menad, S.; Rodríguez-Ramos, I. Dry reforming of methane over Ni/CeO_2 catalysts prepared by three different methods. *Green Process. Synth.* **2015**, *4*, 479–486. [CrossRef]
44. Gonzalez-Delacruz, V.M.; Holgado, J.P.; Pereñíguez, R.; Caballero, A. Morphology changes induced by strong metal-support interaction on a Ni-ceria catalytic system. *J. Catal.* **2008**, *257*, 307–314. [CrossRef]
45. Gonzalez-Delacruz, V.M.; Ternero, F.; Pereñíguez, R.; Caballero, A.; Holgado, J.P. Study of nanostructured Ni/CeO_2 catalysts prepared by combustion synthesis in dry reforming of methane. *Appl. Catal. A Gen.* **2010**, *384*, 1–9. [CrossRef]

46. Damaskinos, C.M.; Vasiliades, M.A.; Efstathiou, A.M. The effect of Ti^{4+} dopant in the 5 wt% Ni/$Ce_{1-x}Ti_xO_{2-δ}$ catalyst on the carbon pathways of dry reforming of methane studied by various transient and isotopic techniques. *Appl. Catal. A Gen.* **2019**, *579*, 116–129. [CrossRef]
47. Schuurman, Y.; Mirodatos, C. Uses of transient kinetics for methane activation studies. *Appl Catal A* **1997**, *151*, 305–331. [CrossRef]
48. Slagtern, A.; Schuurman, Y.; Leclercq, C.; Verykios, X.; Mirodatos, C. Specific Features Concerning the Mechanism of Methane Reforming by Carbon Dioxide over Ni/La_2O_3 catalyst. *J. Catal.* **1997**, *172*, 118–126. [CrossRef]
49. Bobin, A.S.; Sadykov, V.A.; Rogov, V.A.; Mezentseva, N.V.; Alikina, G.M.; Sadovskaya, E.M.; Glazneva, T.S.; Sazonova, N.N.; Smirnova, M.Y.; Veniaminov, S.A.; et al. Mechanism of CH_4 Dry Reforming on Nanocrystalline Doped Ceria-Zirconia with Supported Pt, Ru, Ni, and Ni–Ru. *Top. Catal.* **2013**, *56*, 958–968. [CrossRef]
50. Sadykov, V.A.; Gubanova, E.L.; Sazonova, N.N.; Pokrovskaya, S.A.; Chumakova, N.A.; Mezentseva, N.V.; Bobin, A.S.; Gulyaev, R.V.; Ishchenko, A.V.; Krieger, T.A.; et al. Dry reforming of methane over Pt/PrCeZrO catalyst: Kinetic and mechanistic features by transient studies and their modeling. *Catal. Today* **2011**, *177*, 140–149. [CrossRef]
51. Ferreira-Aparicio, P.; Marquez-Alvarez, C.; Rodrıguez-Ramos, I.; Schuurman, Y.; Guerrero-Ruiz, A.; Mirodatos, C. A Transient Kinetic Study of the Carbon Dioxide Reforming of Methane over Supported Ru Catalysts. *J. Catal.* **1999**, *184*, 202–212. [CrossRef]
52. York, A.P.E.; Xiao, T.C.; Green, M.L.H.; Claridge, J.B. Methane Oxyforming for Synthesis Gas Production. *Catal. Rev.* **2007**, *49*, 511–560. [CrossRef]
53. Yuan, K.; Zhong, J.Q.; Zhou, X.; Xu, L.; Bergman, S.L.; Wu, K.; Xu, G.Q.; Bernasek, S.L.; Li, H.X.; Chen, W. Dynamic Oxygen on Surface: Catalytic Intermediate and Coking Barrier in the Modeled CO_2 Reforming of CH_4 on Ni (111). *ACS Catal.* **2016**, *6*, 4330–4339. [CrossRef]
54. Peymani, M.; Alavi, S.; Arandiyan, H.; Rezaei, M. Rational Design of High Surface Area Mesoporous Ni/CeO_2 for Partial Oxidation of Propane. *Catalysts* **2018**, *8*, 388. [CrossRef]
55. Xu, S.; Zhao, R.; Wang, X. Highly coking resistant and stable Ni/Al_2O_3 catalysts prepared by W/O microemulsion for partial oxidation of methane. *Fuel Process. Technol.* **2004**, *86*, 123–133. [CrossRef]
56. Yang, H.; Whitten, J.L. Dissociative chemisorption of CH_4 on Ni(111). *J. Chem. Phys.* **1992**, *96*, 5529–5537. [CrossRef]
57. Kratzer, P.; Hammer, B.; No/rskov, J.K. A theoretical study of CH_4 dissociation on pure and gold-alloyed Ni(111) surfaces. *J. Chem. Phys.* **2002**, *105*, 5595–5604. [CrossRef]
58. Papadopoulou, C.; Matralis, H.; Verykios, X.E. Utilization of Biogas as a Renewable Carbon Source: Dry Reforming of Methane. In *Catalysis for alternative energy generation*; Springer: New York, NY, USA, 2012; pp. 57–126.
59. Helveg, S.; Lopez-Cartes, C.; Sehested, J.; Hansen, P.L.; Clausen, B.S.; RostrupNielsen, J.R.; Abild-Pedersen, F.; Nørskov, J.K. Atomic-scale imaging of carbon nanofibre growth. *Nature* **2004**, *427*, 426–429. [CrossRef]
60. Toebes, M.L.; Bitter, J.H.; van Dillen, A.J.; de Jong, K.P. Impact of the structure and reactivity of nickel particles on the catalytic growth of carbon nanofibers. *Catal. Today* **2002**, *76*, 33–42. [CrossRef]
61. Liu, J.X.; Zhang, B.Y.; Chen, P.P.; Su, H.Y.; Li, W.X. CO Dissociation on Face-Centered Cubic and Hexagonal Close-Packed Nickel Catalysts: A First-Principles Study. *J. Phys. Chem. C* **2016**, *120*, 24895–24903. [CrossRef]
62. Choudhary, T.V.; Goodman, D.W. Methane activation on Ni and Ru model catalysts. *J. Mol. Catal. A Chem.* **2000**, *163*, 9–18. [CrossRef]
63. Abild-Pedersen, F.; Nørskov, J.K.; Rostrup-Nielsen, J.R.; Sehested, J.; Helveg, S. Mechanisms for catalytic carbon nanofiber growth studied by ab initio density functional theory calculations. *Phys. Rev. B* **2006**, *73*, 115419. [CrossRef]
64. Wang, X.; Yuan, Q.; Li, J.; Ding, F. The transition metal surface dependent methane decomposition in graphene chemical vapor deposition growth. *Nanoscale* **2017**, *9*, 11584–11589. [CrossRef]
65. Yang, K.; Zhang, M.; Yu, Y. Theoretical insights into the effect of terrace width and step edge coverage on CO adsorption and dissociation over stepped Ni surfaces. *Phys. Chem. Chem. Phys.* **2017**, *19*, 17918–17927. [CrossRef]
66. Beebe, T.P.; Goodman, D.W.; Kay, B.D.; Yates, J.T. Kinetics of the activated dissociative adsorption of methane on the low index planes of nickel single crystal surfaces. *J. Chem. Phys.* **1987**, *87*, 2305–2315. [CrossRef]

67. Wang, S.G.; Cao, D.B.; Li, Y.W.; Wang, J.; Jiao, H. Chemisorption of CO_2 on nickel surfaces. *J. Phys. Chem. B* **2005**, *109*, 18956–18963. [CrossRef]
68. Burghgraef, H.; Jansen, A.P.J.; van Santen, R.A. Methane activation and dehydrogenation on nickel and cobalt: A computational study. *Surf. Sci.* **1995**, *324*, 345–356. [CrossRef]
69. Wind, T.L.; Falsig, H.; Sehested, J.; Moses, P.G.; Nguyen, T.T.M. Comparison of mechanistic understanding and experiments for CO methanation over nickel. *J. Catal.* **2016**, *342*, 105–116. [CrossRef]
70. Engbæk, J.; Lytken, O.; Nielsen, J.H.; Chorkendorff, I. CO dissociation on Ni: The effect of steps and of nickel carbonyl. *Surf. Sci.* **2008**, *602*, 733–743. [CrossRef]
71. Arevalo, R.L.; Aspera, S.M.; Escaño, M.C.S.; Nakanishi, H.; Kasai, H. Tuning methane decomposition on stepped Ni surface: The role of subsurface atoms in catalyst design. *Sci. Rep.* **2017**, *7*, 13963. [CrossRef]
72. Schouten, F.C.; Gijzeman, O.L.J.; Bootsma, G.A. Reaction of methane with nickel single crystal surfaces and the stability of surface nickel carbides. *Bull. Sociétés Chim. Belges* **1979**, *88*, 541–547. [CrossRef]
73. Vasiliades, M.A.; Damaskinos, C.M.; Kyprianou, K.K.; Kollia, M.; Efstathiou, A.M. The effect of Pt on the carbon pathways in the dry reforming of methane over Ni-Pt/$Ce_{0.8}Pr_{0.2}O_{2-\delta}$ catalyst. *Catal. Today* **2019**. [CrossRef]
74. Li, B.; Zhang, S. Methane reforming with CO_2 using nickel catalysts supported on yttria-doped SBA-15 mesoporous materials via sol-gel process. *Int. J. Hydrogen Energy* **2013**, *38*, 14250–14260. [CrossRef]
75. Polychronopoulou, K.; Costa, C.N.; Efstathiou, A.M. The steam reforming of phenol reaction over supported-Rh catalysts. *Appl. Catal. A Gen.* **2004**, *272*, 37–52. [CrossRef]
76. Christou, S.Y.; Costa, C.N.; Efstathiou, A.M. A Two-Step Reaction Mechanism of Oxygen Release from Pd/CeO_2: Mathematical Modelling Based on Step Gas Concentration Experiments. *Top. Catal.* **2004**, *30*, 325–331. [CrossRef]

catalysts

MDPI

Article

Comparative Study of Strategies for Enhancing the Performance of Co_3O_4/Al_2O_3 Catalysts for Lean Methane Combustion

Andoni Choya, Beatriz de Rivas, Jose Ignacio Gutiérrez-Ortiz and Rubén López-Fonseca *

Chemical Technologies for Environmental Sustainability Group, Department of Chemical Engineering, Faculty of Science and Technology, University of the Basque Country UPV/EHU, P.O. Box 644, E-48080 Bilbao, Spain; andoni.choya@ehu.eus (A.C.); beatriz.derivas@ehu.eus (B.d.R.); joseignacio.gutierrez@ehu.eus (J.I.G.-O.)

* Correspondence: ruben.lopez@ehu.es; Tel.: +34-94-601-5985

Received: 19 June 2020; Accepted: 7 July 2020; Published: 8 July 2020

Abstract: Spinel-type cobalt oxide is a highly active catalyst for oxidation reactions owing to its remarkable redox properties, although it generally exhibits poor mechanical, textural and structural properties. Supporting this material on a porous alumina can significantly improve these characteristics. However, the strong cobalt–alumina interaction leads to the formation of inactive cobalt aluminate, which limits the activity of the resulting catalysts. In this work, three different strategies for enhancing the performance of alumina-supported catalysts are examined: (i) surface protection of the alumina with magnesia prior to the deposition of the cobalt precursor, with the objective of minimizing the cobalt–alumina interaction; (ii) coprecipitation of cobalt along with nickel, with the aim of improving the redox properties of the deposited cobalt and (iii) surface protection of alumina with ceria, to provide both a barrier effect, minimizing the cobalt–alumina interaction, and a redox promoting effect on the deposited cobalt. Among the examined strategies, the addition of ceria (20 wt % Ce) prior to the deposition of cobalt resulted in being highly efficient. This sample was characterized by a notable abundance of both Co^{3+} and oxygen lattice species, derived from the partial inhibition of cobalt aluminate formation and the insertion of Ce^{4+} cations into the spinel lattice.

Keywords: cobalt oxide; methane oxidation; modified alumina; magnesium oxide; nickel cobaltite; ceria; lattice distortion; oxygen mobility

1. Introduction

The commercialization of vehicles driven by natural gas engines is a widely accepted strategy to mitigate the emissions associated with transport, which is one of the largest emitting sectors of greenhouse effect gases, due to their reduced CO_2, NO_x, particles and hydrocarbons emissions [1,2]. However, the consolidation of this type of vehicles in the automotive fleet requires the control of the residual (around 1%) unburned methane (the main component of natural gas) emissions from the engine, since this pollutant possesses a powerful greenhouse effect potential (around 25 times that of CO_2 in a 100 years period). Owing to the high chemical stability of this compound and its low concentration in the flue gases (<1% vol.), the low-temperature catalytic oxidation appears as an attractive solution for this purpose. In order for this treatment not to significantly increase the cost of the engine, the selected catalyst must present a high activity with a reasonable economic investment.

Traditionally, the most commonly applied catalysts have been based on noble metals such as palladium or platinum [3,4]. However, this type of catalyst is generally expensive, due to the high cost of the noble metals and their relatively high metallic loading (2–4 wt %), and is prone to deactivation by sintering [5,6]. Spinel oxides based on transition metals can be a promising catalytic system for the oxidation of lean methane, due to their lower cost, higher availability and relatively good activity for the oxidation of CO and light hydrocarbons [7,8]. More specifically, bulk spinel-type cobalt oxide (Co_3O_4) is regarded as a highly interesting substitute to noble metals for catalytic oxidation of trace amounts of methane on the basis of its excellent redox properties [9–12]. However, when prepared by conventional synthesis methodologies [13,14], the textural and structural properties of this oxide tend to be poor. Although some routes can partially overcome this problem, they are normally too complex and difficult to scale up to the industrial operation [15,16].

One possible alternative could be to use a porous media as a support for cobalt oxide, in order to increase the amount of active surface area available for reaction. This generally translates into and enhancement of the physico-chemical properties of the resulting catalyst, but the probability of the occurrence a strong cobalt–support interaction, causing a negative effect on the redox properties of the deposited cobalt, is high. More specifically, when the selected support is high surface gamma alumina, this interaction provokes the formation of a cobalt aluminate phase ($CoAl_2O_4$), which is characterized by a poor reducibility and a consequent low specific activity for oxidation reactions [17,18].

A proposed solution would be the modification of the properties of the alumina with the aim of tuning its affinity for the Co_3O_4 deposited over it. This can be carried out by incorporating some chemical promoters to the alumina before the deposition of Co_3O_4, or by adding the promoters after the formation of the final Co/Al_2O_3 catalyst. In this sense, Liotta et al. [19] reported that adding Ba during the sol–gel synthesis of Al_2O_3 decreased diffusion of the Co^{2+} ions into the structure of the alumina after Co deposition. Similarly, Park et al. [20] and Park et al. [21], on different studies, observed that adding phosphorus to Al_2O_3 resulted in the formation of a mixed $AlPO_4$ phase that partially suppressed the formation of $CoAl_2O_4$. On the other hand, Cheng et al. [22] found out that the addition of Mn or Fe during the synthesis of a Cu-Co/Al_2O_3 catalyst improved the redox properties of both active metals. In addition, El-Shobaky et al. [23] reported an increased activity for CO oxidation of Co_3O_4/Al_2O_3 catalysts doped with small amounts of Mn and/or La.

Similarly, an attractive strategy to improve the performance of this type of catalyst could be using magnesium as a modifier. In this sense, Riad [24] observed that alumina coated with magnesium and prepared by coprecipitation presented better properties than bare alumina. Cobalt catalysts can also be supported over pure magnesium oxide, which results in systems with enhanced activity due to the magnesium–cobalt interaction, as reported by Ulla et al. [25] and Ji et al. [26].

Considering this background, in the present paper various enhancing strategies for γ-alumina-supported cobalt catalysts were examined in order to determine their efficiency for improving the activity of the resulting catalysts for the complete oxidation of methane under lean conditions with respect to an unmodified reference Co/Al_2O_3 catalyst (Co/Al sample). The three investigated approaches were the following:

(i) Superficial protection of alumina with magnesia prior to the deposition of cobalt (Co/Mg-Al sample). The cobalt content was fixed at 30 wt % and the Mg/Co molar was 0.25.

(ii) Co-deposition of nickel and cobalt over alumina in order to promote the redox properties of Co_3O_4 (Co-Ni/Al sample). Since nickel oxide can show a significant activity of methane oxidation [27,28], the total metallic loading was fixed at 30 wt % while using a Ni/Co molar ratio of 0.20.

(iii) Coating the alumina surface with ceria before the addition of cobalt in order to eventually serve as both a surface protector for alumina and redox promoter for cobalt oxide (Co/Ce-Al sample). The cobalt content was fixed at 30 wt % and the Ce/Co molar was 0.20.

2. Results

2.1. Physico-Chemical Characterization

Along with their chemical composition as determined by Wavelength dispersive X-ray fluorescence (WDXRF), the textural properties of the synthesized catalysts in terms of the BET (Brunauer–Emmett–Teller) surface and pore volume are listed in Table 1. Additionally, the properties of bare alumina and the modified supports with MgO and CeO_2 are included for comparative purposes. All supports and catalysts exhibited type IV adsorption–desorption isotherms (Figure 1) with H2 type hysteresis cycles, characteristic of mesoporous materials with a relatively broad size distribution. The addition of ceria provoked an appreciable decrease (16%) in the surface area whereas the effect of MgO deposition was virtually negligible. After incorporating the active metallic oxide, the surface area decreased by around 19–32%. Hence, the specific surface area of the catalysts varied in the 93–113 $m^2\ g^{-1}$ range while the pore volume was between 0.27 and 0.35 $cm^3\ g^{-1}$.

Table 1. Physico-chemical properties of the supported cobalt catalysts.

Sample	Co Loading, wt %	Me Loading, wt %	Me/Co Molar Ratio	BET Surface, $m^2\ g^{-1}$	Pore Volume, $cm^3\ g^{-1}$	Cophase Crystallite Size, nm
Al_2O_3	-	-	-	139	0.56	-
Mg-Al	-	6.7	-	145	0.50	-
Ce-Al	-	20.6	-	117	0.42	-
Co/Al	27.9	0.0	0	108	0.29	29
Co/Mg-Al	28.7	3.0	0.25	99	0.27	23
Co-Ni/Al	23.2	4.8	0.21	113	0.35	21
Co/Ce-Al	29.5	12.4	0.18	93	0.30	23

Me stands for Mg, Ni or Ce.

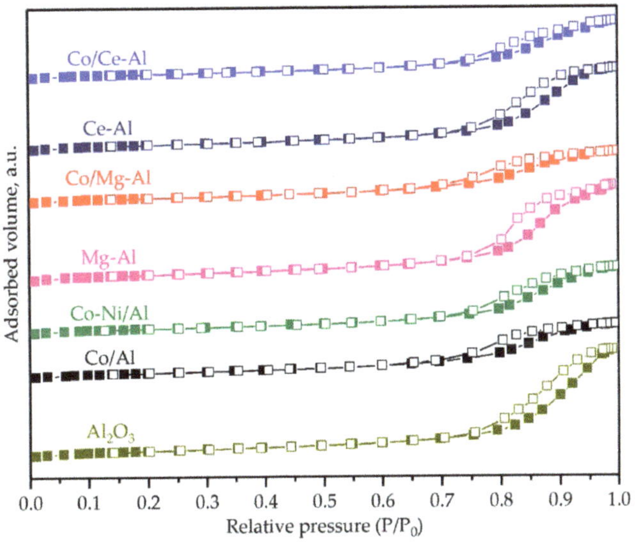

Figure 1. N_2 physisorption isotherms of the supported cobalt catalysts.

The pore size distributions of the supports and the cobalt catalysts are shown in Figure 2. The blank alumina exhibited a bimodal distribution with maxima at 110 and 150 Å, while the distributions were unimodal, centered at 110 Å, over the MgO- and CeO_2 modified supports, thereby showing that the promoter oxides were preferentially deposited on the larger pores of the alumina. On the other hand, all cobalt catalysts exhibited unimodal distributions centered at 90 Å, thus evidencing that cobalt was also massively located on the larger pores of the supports.

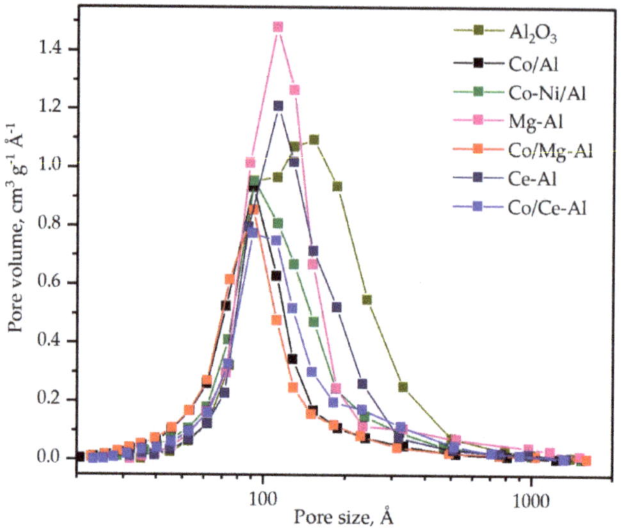

Figure 2. Pore size distributions of the supported cobalt catalysts.

The structural characterization of the catalysts was performed by XRD. The diffractograms in Figure 3 revealed signals of a cubic spinel phase (2θ = 19.1°, 31.3°, 37.0°, 45.1°, 59.4° and 65.3°) in the four cobalt catalysts that could be assigned to the presence of both Co_3O_4 (ICDD 00-042-1467) and $CoAl_2O_4$ (ICDD 00-044-0160) formed by the interaction of the deposited cobalt with the alumina [29,30]. The relative occurrence of these two phases could not be determined since both oxides essentially exhibit the same diffraction signals at similar positions.

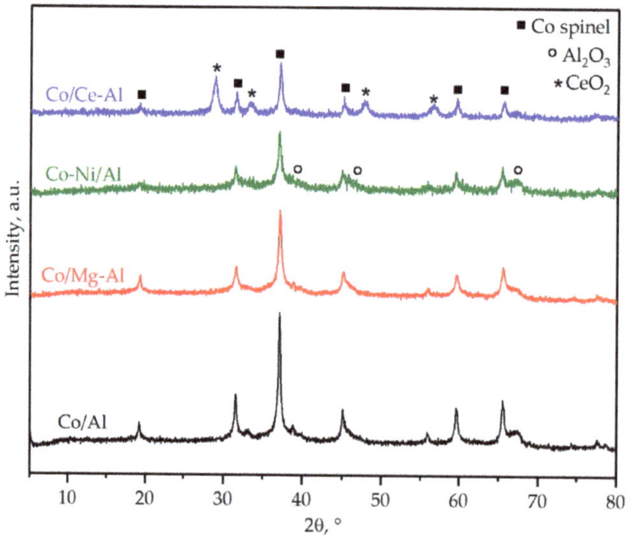

Figure 3. XRD patterns of the supported cobalt catalysts.

On the other hand, weak signals from a cubic phase with low crystallinity were detected at 2θ = 37.7, 45.9 and 66.9° that could be attributed to the gamma-alumina (ICDD 01-074-2206). Additionally, various signals from a fluorite-like phase of CeO_2 were detected at 2θ = 28.5°, 33.3°, 47.5° and 56.4° for the Co/Ce-Al catalyst. Due to the lower amount of Mg and Ni (<5 wt %), no signals attributable to MgO and NiO were noticed over the Co/Mg-Al and the Co-Ni/Al samples, respectively.

The average crystallite size of the cobalt spinel phase was estimated from the full width half maximum (FWHM) of the characteristic peak located at 37.1°, assignable to the (311) crystalline plane, by applying the Scherrer equation (Table 1). The crystallite size for the catalyst supported over bare alumina was 29 nm, while it was slightly smaller (21–23 nm) over the modified catalysts. It must be noted that sizes calculated from the refinement were not substantially different from those obtained from single peak data. Furthermore, a close-up view of this signal (Figure 4) revealed a significant 2θ shift towards lower angle values, thus denoting an enlargement of the cell size of the spinel due to the insertion of cations of the promoters into its lattice. In this sense, the Co-Ni/Al catalyst showed the largest growth (a_0 = 8.122 Å) with respect to the catalyst supported over bare alumina (a_0 = 8.096 Å). This remarkable growth may indicate the presence of the nickel cobaltite mixed spinel ($NiCo_2O_4$) in this sample [28,31].

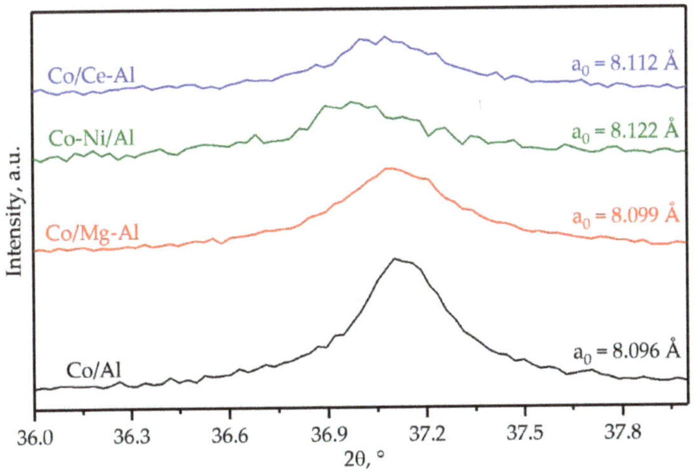

Figure 4. Close up view of the XRD patterns of the supported cobalt catalysts in the 2θ = 36–38° range.

A closer inspection of the position and width of the A_{1g} mode of the modified catalysts could be helpful in determining a possible distortion of the Co_3O_4 lattice due to the insertion of cations of the promoters. This influence was analyzed in terms of the shift and the FWHM value of this signal (Figure 5). The shift of the A_{1g} signal of the modified catalysts with respect to the reference Co/Al sample ranged between 4 and 17 cm^{-1}, with the Co-Ni/Al catalyst exhibiting the largest shift. In addition, the FWHM for the modified catalysts was between 26 and 30 cm^{-1}, while it was 13 cm^{-1} for the unmodified catalyst. These results revealed that the most marked lattice distortion occurred over the Co-Ni/Al catalyst, in line with the results from XRD. However, the Raman spectra of this sample did not show any detectable signals related to the presence of $NiCo_2O_4$ in contrast to what the observed the enlargement of the cell size of the crystalline spinel phase suggested. Therefore, the amount of nickel cobaltite present in this catalyst was probably low.

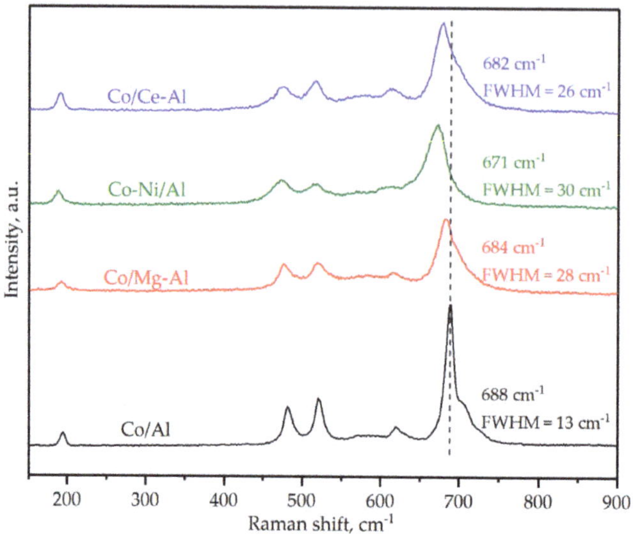

Figure 5. Raman spectra of the supported cobalt catalysts.

Figure 6 shows the $Co2p_{3/2}$ XPS spectra of the samples. The spectra of all catalysts were composed of five different signals. As for the Co/Al sample the two features with the lowest binding energies, located at 779.5 and 781.0 eV, were assigned to Co^{3+} and Co^{2+} ions, respectively, while the signal centered at 783.2 eV was attributed to the presence of CoO in the surface of the samples [32,33]. The two signals with the highest binding energies (786.2 and 790.1 eV) were identified as the satellite signals from Co^{3+} and Co^{2+} ions, respectively. The position of the main signal of the $Co2p_{3/2}$ spectra can be used as an indication of the predominant cobalt species on the surface of each sample. Thus, higher binding energy values suggested a higher abundance of Co^{2+} cations, while lower binding energies pointed out a favored presence of Co^{3+} cations. In this sense, the position of the main signal of the unmodified catalyst was 781.1 eV, thereby evidencing a significant presence of Co^{2+} ions, in line with its higher $CoAl_2O_4$ content as deduced by Raman spectroscopy. Conversely, the main signal for the Co/Mg-Al sample was centered at 780.8 eV assignable to a slightly lower Co^{2+} content. This observation was not coherent with the results from Raman spectroscopy that evidenced that the presence of $CoAl_2O_4$ in this sample was appreciably limited. For this reason, the increased Co^{2+} presence was not attributed to the presence of cobalt aluminate, but rather to the probable formation of a Co-Mg solid solution, as already reported elsewhere [34,35].

On the other hand, the main signals of the Co-Ni/Al and Co-Ce/Al catalysts exhibited a significant shift towards lower binding energies with respect to the Co/Al sample. Thus, both catalysts displayed their main signal centered at about 780.1 eV, thus evidencing their higher content in Co^{3+} cations. Additionally, the $Ni2p_{3/2}$ spectra of the Co-Ni/Al catalyst (not shown) were deconvoluted into five signals. Hence, three main contributions at 853.7, 855.3 and 857.3 eV were attributed to the presence of Ni^{2+} as NiO, Ni^{2+} belonging to a spinel lattice and Ni^{3+} cations, respectively, while the signals at 861.2 and 866.5 eV were assigned to the satellites of Ni^{2+} and Ni^{3+} cations, respectively. The relatively low binding energy of the signal associated with the presence of Ni^{2+} cations in a spinel lattice (855.3 eV) also denoted that these species were mainly a part of the $NiCo_2O_4$ spinel instead of the $NiAl_2O_4$, for which the binding energy of that signal would be expectedly higher (ca. 856.0 eV), as shown by other studies [36,37].

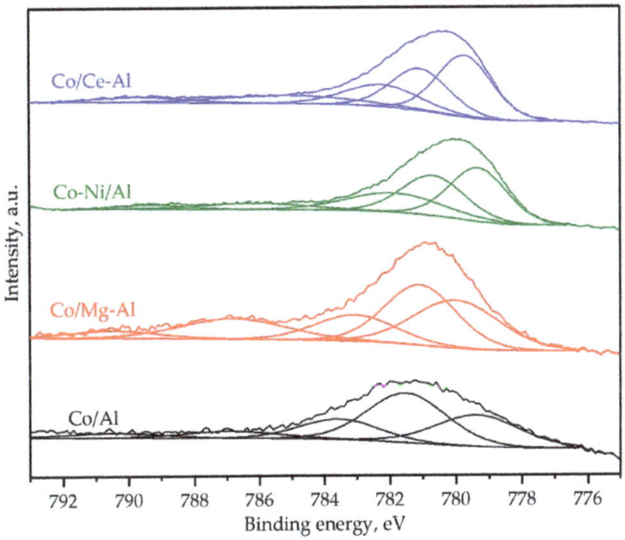

Figure 6. Co2p$_{3/2}$ XPS spectra of the supported cobalt catalysts.

The deconvolution and integration of the XPS spectra allowed for a quantitative analysis of the composition of the surface, as shown in Table 2. Firstly, it must be noticed that the surface cobalt loading of the Co/Al sample and particularly Co/Mg-Al sample was significantly lower than their bulk content, which, on one hand, suggested that a notable fraction of Co species was as $CoAl_2O_4$ and, on the other hand, evidenced a marked formation of the aforementioned Co-Mg solid solution. These results would be in agreement with the relatively low Co^{3+}/Co^{2+} molar ratios of these catalysts (0.67 and 0.94, respectively).

Table 2. Surface composition of the supported cobalt catalysts.

Catalyst	Co, wt %	Me, wt %	Co^{3+}/Co^{2+} Molar Ratio	O_{ads}/O_{latt} Molar Ratio
Co/Al	22.6 (27.9)	-	0.67	1.41
Co/Mg-Al	13.3 (28.7)	4.4 (3.0)	0.94	1.02
Co-Ni/Al	25.4 (23.2)	8.2 (4.8)	1.21	0.94
Co/Ce-Al	32.9 (29.5)	3.2 (12.4)	1.38	0.77

The values in brackets correspond to the bulk composition. Me stands for Mg, Ni or Ce.

On the other hand, a substantial surface cobalt enrichment was noticed for the Co-Ni/Al and Co/Ce-Al catalysts, thus pointing out that deposited cobalt was preferentially located on the external surface and it did not tend to strongly interact with the support. This in turn was in good agreement with the higher Co^{3+}/Co^{2+} molar ratios of these samples (1.21 and 1.38, respectively). It is worth pointing out that this higher abundance of Co^{3+} species at the surface was accompanied by a more notable presence of lattice oxygen species in the Co-Ni/Al and Co/Ce-Al catalysts, as shown in Table 2. This type of oxygen species is usually involved in the oxidation of methane by a Mars-van Krevelen mechanism [38,39]. As revealed by our results, its abundance was optimized over the Co/Ce-Al catalyst. It is quite reasonable to expect that the Ce^{3+}/Ce^{4+} relative abundance can play a role in controlling the activity of the Co/Ce-Al. Unfortunately, since the amount of Ce at the surface of the catalyst is very low, a proper deconvolution of the Ce3d XPS spectrum devoted to the evaluation of the Ce^{3+}/Ce^{4+} molar ratio was not possible. In this sense, we have shown elsewhere [40] that an inverse relationship between the Ce^{3+}/Ce^{4+} and the Co^{3+}/Co^{2+} molar ratios, as determined by XPS, for a series of Ce-doped bulk Co_3O_4 catalysts with a Ce/Co molar ratio between 0.03 and 0.14 (corresponding to 5–20 wt % Ce).

These results could be explained in terms of the equilibrium $Ce^{3+} + Co^{3+} \leftrightarrow Ce^{4+} + Co^{2+}$, established by the requirement of charge balance within the cations of the spinel lattice. Since our XRD and Raman measurements suggested the insertion of Ce cations into the Co_3O_4 lattice in the Co/Ce-Al sample, the aforementioned effect could also be occurring in this catalyst. This would explain the higher surface Co^{3+}/Co^{2+} molar ratio among all the samples.

The redox properties of the cobalt catalysts were investigated by temperature-programmed reduction with hydrogen (H_2-TPR) and temperature-programmed reaction with methane in the absence of oxygen (CH_4-TPRe). As widely accepted, the H_2-TPR profile of the base alumina-supported Co_3O_4 catalyst (Figure 7) evidenced a two-step reduction process [41,42]. A first H_2 uptake at low temperatures (<550 °C) was associated with the reduction of Co_3O_4, which in turn could be subdivided into two contributions located at around 270–310 °C, assignable to the reduction of Co^{3+} ions into Co^{2+}, and at 350–400 °C, attributable to the subsequent reduction of Co^{2+} into metallic cobalt. This second H_2 uptake did not take the shape of a single peak, but instead was formed of at least two different contributions, which suggested the presence of Co^{2+} species with varying reducibilities. The presence of this type of species was much more noticeable for the Co/Mg-Al catalysts, where the temperature window for this second reduction event extended from 300 to 550 °C. On the other hand, the marked H_2 consumption located at high temperatures (>600 °C) over the four catalysts was attributed to the reduction of cobalt aluminate [43].

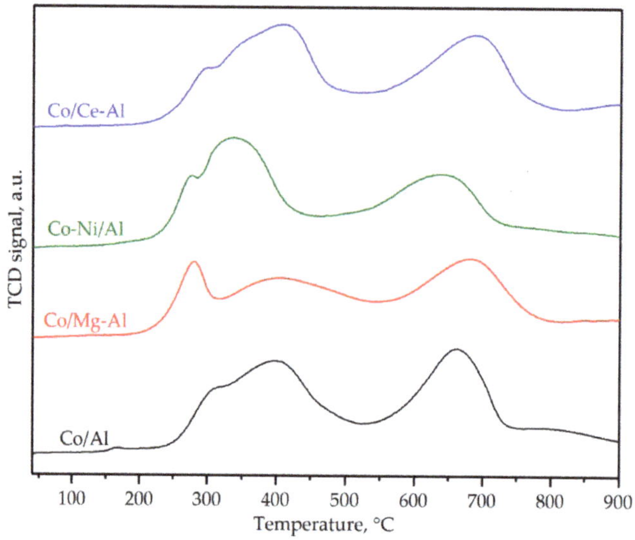

Figure 7. Temperature-programmed reduction with hydrogen (H_2-TPR) profiles of the supported cobalt catalysts.

The relative amount of each type of Co species present in the various catalysts was estimated by deconvolution of the experimental TPR profiles. The threshold temperature of 550 °C was taken as a criterion to distinguish between easily reducible cobalt species (low-temperature uptake), namely free Co_3O_4 (200–450 °C) and some mixed cobalt-metal species (450–550 °C) such as cobalt-magnesium species, and hardly reducible cobalt species (high-temperature uptake) in the form of cobalt aluminate (>550 °C). The results are summarized in Table 3.

Table 3. Characterization by H_2-TPR and temperature-programmed reaction with methane in the absence of oxygen (CH_4-TPRe) of the supported cobalt catalysts.

	H_2-TPR			CH_4-TPRe
Catalyst	Low-Temperature H_2 Uptake, mmol g_{Co}^{-1}	High-Temperature H_2 Uptake, mmol g_{Co}^{-1}	Onset Reduction Temperature, °C	Low-Temperature O_2 Consumption, mmol g_{Co}^{-1}
Co/Al	9.8	10.2	225	0.28
Co/Mg-Al	10.3	10.4	195	0.61
Co-Ni/Al	12.5	12.9	175	0.83
Co/Ce-Al	13.6	8.5	220	0.88

Firstly, it must be noticed that the H_2 uptake at low temperatures, associated with species with a high reducibility, mainly free of Co_3O_4, was larger for the three modified catalysts with respect to the Co/Al sample. The increase in H_2 uptake was 5% for the Co/Mg-Al, 28% for the Co-Ni/Al and 39% for the Co/Ce-Al, thus showing that nickel and cerium were efficient redox promoters for cobalt. On one hand, for the Co-Ni/Al catalyst, the increase in H_2 uptake at a low temperature could be due to the presence of the $NiCo_2O_4$ spinel, which possesses a higher specific H_2 uptake than Co_3O_4. On the other hand, for the Co/Ce-Al sample, the increase could be related to the reduction of the surface of the ceria, which is known to occur at around 450–500 °C [44], and also the improved reducibility of cobalt oxide due to the insertion of cerium cations into its lattice. Additionally, the onset reduction temperature of the three modified catalysts was clearly lowered. The most noticeable temperature shift occurred for the catalyst modified with nickel (50 °C), which could be due to the presence of nickel cobaltite, while it was much less significant for the Co/Ce-Al catalyst (5 °C).

On the other hand, the H_2 uptake assignable to the presence of $CoAl_2O_4$ was only lower for the Co/Ce-Al catalyst (8.5 mmol g_{Co}^{-1}) with respect to the Co/Al sample (10.2 mmol g_{Co}^{-1}); while it was notably higher (12.9 mmol g_{Co}^{-1}) over the Co-Ni/Al catalyst. This could be due to ceria acting as a physical barrier between cobalt and alumina, whereas both cobalt and nickel were directly deposited on the alumina in the case of the Co-Ni/Al sample. Therefore, the cobalt–alumina interaction was not apparently limited over this sample. Alternatively, given that the Co/Mg-Al catalyst showed a comparable H_2 uptake at high temperatures (10.4 mmol g_{Co}^{-1}), it could be thought that such a barrier effect could be ruled out in this catalyst as well. However, it must be noticed that a certain fraction of this H_2 uptake would probably be due to the partial reduction of a Co-MgO solid solution. Therefore, the total H_2 uptake associated with the presence of cobalt aluminate was probably lower than that of the Co/Al sample.

As a complement of the H_2-TPR analysis, the reactivity of oxygen species present in the examined catalysts was also analyzed by temperature-programmed reaction with methane in the absence of oxygen (CH_4-TPRe). The evolution of evolved CO_2, CO and H_2 was monitored by mass spectrometry. In general, the CO_2 profiles evidenced a two-step reaction process, as depicted in Figure S1, Supplementary Material. The low-temperature reaction step was attributed to the complete oxidation of methane by lattice oxygen species associated with Co^{3+} ions. No CO or H_2 formation was detected at this temperature interval. The amount of CO_2 evolved from the complete oxidation reaction was barely perceptible over the Co/Al sample. Above 525–550 °C the methane partial oxidation occurred, where methane reacted with low-mobility oxygen species associated with Co^{2+} ions, yielding significant amounts of CO and H_2 along with CO_2 [14]. It must be noted that the occurrence of this second process was not observed over the Co/Al catalyst, thus suggesting that it could only take place at temperatures higher than 600 °C.

The comparatively larger formation of CO_2 above 500 °C made the proper analysis of the obtained results in the low-temperature range rather difficult. For this reason, Figure 8 only focuses on the evolution of the CO_2 yield between 100 and 550 °C. It was thus clearly evidenced that the three investigated enhancing strategies were efficient for increasing the amount of reactive oxygen in the resulting cobalt catalysts. More specifically, the O_2 consumption at low temperatures (Table 3) increased from 0.28 mmol g_{Co}^{-1} for the Co/Al sample up to 0.61–0.88 mmol g_{Co}^{-1} for the modified catalysts, with the Co/Ce-Al achieving the highest value. Note that the amount of reacted O_2 was relatively limited over the Co/Mg-Al sample in spite of its higher Me/Co molar ration (0.25). Additionally, this low-temperature step peaked at lower temperatures with respect to the reference cobalt catalyst. The temperature shift was as high as 67 °C for the Co-Ni/Al catalyst. Such remarkable improvement could be due to the presence of $NiCo_2O_4$ in this sample, which has been already reported to be highly reducible and active for methane oxidation elsewhere [27,45].

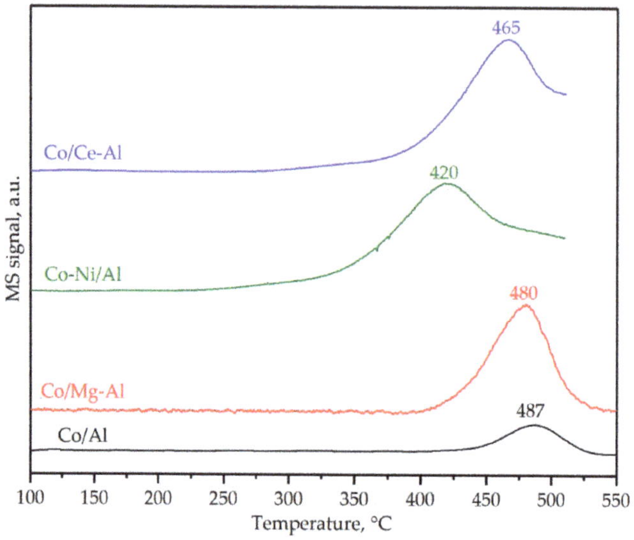

Figure 8. Close up view of the CH_4-TPRe profiles of the supported cobalt catalysts in the 100–550 °C temperature range.

Given the notable redox behavior of the Co-Ni/Al and Co/Ce-Al catalysts, these were examined by high angle annular dark field—scanning transmission electronic microscopy (HAADF–STEM) coupled with electron energy loss spectroscopy (EELS) or energy dispersive X-ray spectrometry (EDX), respectively. Elemental maps (Figure 9 and Figure S2, Supplementary Material) were generated for certain areas in each sample to allow studying the spatial distribution of cobalt and nickel or cerium in the bimetallic catalysts. These results evidenced that both Ni and Ce were homogenously present over the supported nanoparticles. In the case of nickel, this probably means that it was forming nickel cobaltite. The Co/Ce-Al sample also contained small nanoparticles of cerium species (as ceria) of around 10–20 nm.

2.2. Catalytic Performance

Three consecutive light-off tests were carried out for each catalyst. The second and third cycles were characterized by an identical light-off curve. Hence, the light-off curves corresponding to the third cycle of each catalyst are shown in Figure 10. It must be noticed that all four samples presented 100% selectivity towards CO_2. It was clear that the three modified catalysts resulted were remarkably more efficient than the base Co/Al catalyst. In this sense, the corresponding T_{50} value (Table 4) was lowered by 15 (Co/Mg-Al), 50 (Co-Ni/Al) and 70 °C (Co/Ce-Al). Note that the intrinsic activity of the Co-free alumina supports (pure Al_2O_3, Mg-Al and Ce-Al) was negligible at 600 °C.

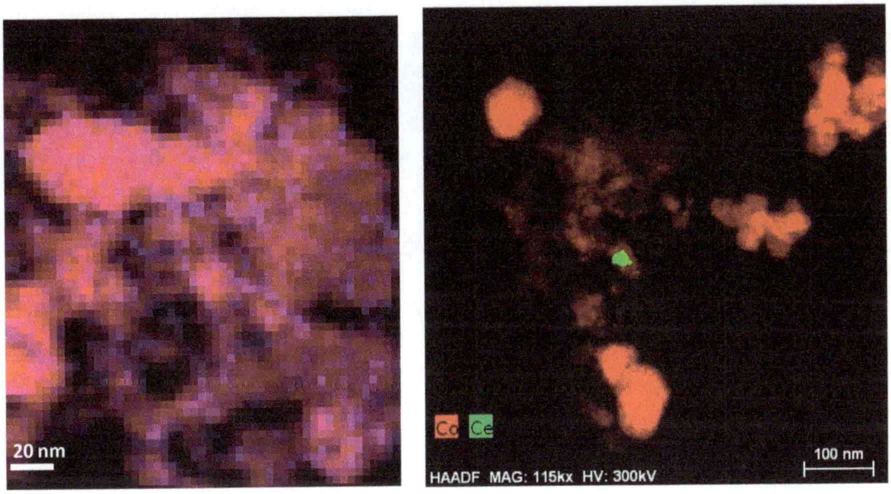

Figure 9. High angle annular dark field–scanning transmission electron microscopy (HAADF–STEM) images of the Co-Ni/Al (**left**) and Co/Ce-Al (**right**) catalysts coupled to electron energy loss spectroscopy (EELS; Co (red) and Ni (blue)) and energy dispersive X-ray spectrometry (EDX; Co (red) and Ce (green)) elemental distribution.

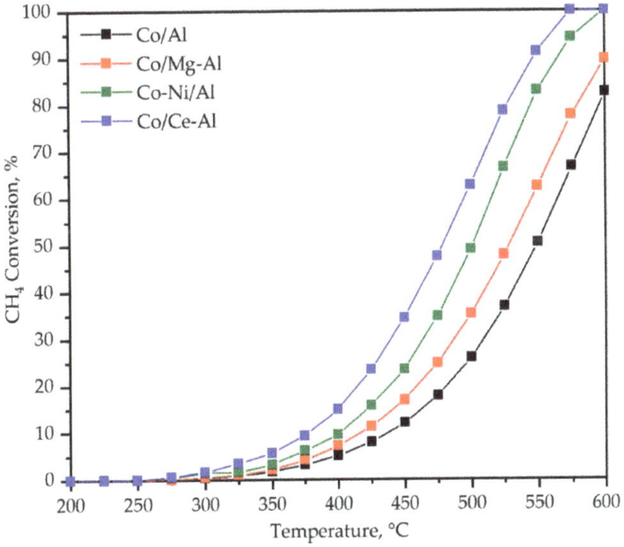

Figure 10. Light off curves of the supported cobalt catalysts.

Table 4. Kinetic results of the supported cobalt catalysts.

Catalyst	T_{50}, °C	Specific Reaction Rate at 400 °C, mmol g_{Co}^{-1} h^{-1}	E_a, kJ mol^{-1}
Co/Al	550	1.2	82 ± 2
Co/Mg-Al	535	1.5	83 ± 2
Co-Ni/Al	500	2.5	84 ± 2
Co/Ce-Al	480	3.2	80 ± 2

The apparent activation energy of the four evaluated catalysts (Table 4) were estimated by the integral method, assuming that the reaction kinetics followed a pseudo-order one for methane and a zeroth order for oxygen, as it is usual for the Mars–van Krevelen mechanism in the presence of excess oxygen [46,47]. The corresponding linearized plots are depicted in Figure S2. The activation energies were around 80–84 kJ mol^{-1}, a value comparable to those reported in other studies for methane oxidation over Co_3O_4-based catalysts [48,49]. Thus, it could be concluded that the main active species in all four catalysts was free of Co_3O_4.

On the other hand, the specific reaction rates of the various catalysts were estimated by the differential method (for methane conversions lower than 20%). In this analysis, for the sake of proper comparison, the specific activity of the various catalysts were calculated at the same temperature (400 °C) and normalized per gram of cobalt. The results are shown in Table 4. All three modified samples achieved higher reaction rates (1.5–3.2 mmol CH_4 g_{Co}^{-1} h^{-1}) with respect to the unmodified counterpart (1.2 mmol CH_4 g_{Co}^{-1} h^{-1}). The catalyst promoted with ceria (Co/Ce-Al), in particular, presented the highest reaction rate, which was almost three times higher than that of the catalyst supported over bare alumina.

In view of these results, modifying the alumina support with ceria prior to the deposition of cobalt would be a quite promising strategy to be developed for improving the activity of cobalt catalysts. Interestingly, when cobalt was deposited over the Ce-modified support, a dual beneficial effect was evidenced. On one hand, ceria was found to act as a physical barrier between alumina and deposited cobalt, limiting the cobalt–alumina interaction and the subsequent cobalt aluminate formation. On the other hand, the interaction between cobalt and ceria led to a partial incorporation of cerium ions into the lattice of Co_3O_4. Both phenomena in turn led to a larger abundance of Co^{3+} and therefore to a promoted mobility of the lattice oxygen species at low temperatures with ceria loading as shown in Figure 11.

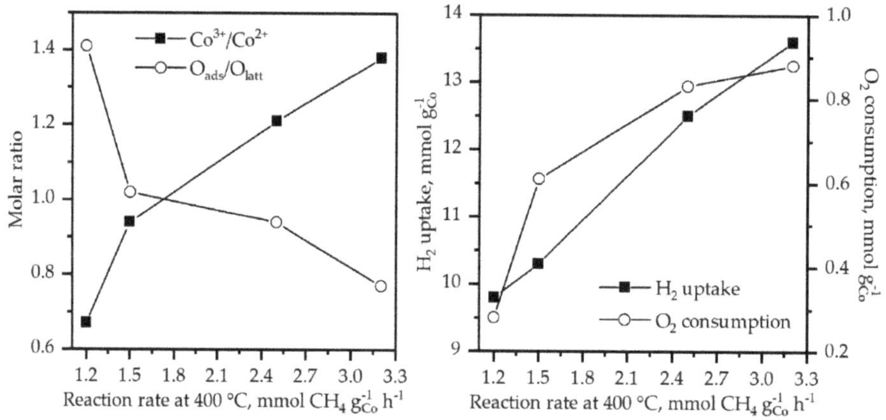

Figure 11. Relationship among specific reaction rate and: surface Co^{3+}/Co^{2+} and O_{ads}/O_{latt} molar ratios (**left**); low-temperature H_2 uptake from H_2-TPR and low-temperature O_2 consumption from CH_4-TPRe (**right**).

3. Materials and Methods

3.1. Catalyst Preparation

A commercial γ-alumina kindly provided by Saint Gobain was used as the base support. Alternatively, the alumina was modified with MgO (Mg-Al) or CeO_2 (Ce-Al), which were incorporated by precipitation, starting with magnesium nitrate and cerium nitrate, respectively, followed by a calcination step (600 °C for 4 h).

Four alumina supported cobalt oxide catalysts, namely Co/Al, Co-Ni/Al, Co/Mg-Al and Co/Ce-Al, were prepared following a simple precipitation route as detailed elsewhere [50]. This methodology consisted of the precipitation of aqueous solutions of cobalt nitrate hexahydrate and eventually nickel nitrate hexahydrate with adjusted concentrations by the drop-by-drop addition of a sodium carbonate 1.2 M solution at a constant temperature of 80 °C until pH 8.5 was achieved. Afterwards, the precipitates were thoroughly washed with deionized water in order to remove the residual sodium ions that are known to be detrimental for the activity of the resulting catalysts [40]. The washed precipitates were dried at 110 °C for 16 h and then subjected to calcination at 600 °C for 4 h in static air. The Me/Co molar ratio (where M = Mg, Ni or Ce) of the modified catalysts was 0.20 for the Co-Ni/Al and Co/Ce-Al samples and 0.25 for the Co/Mg-Al catalyst, as dictated from a previous optimization for this type of cobalt catalysts [51,52]. The total amount of active metal (Co and/or Ni) was fixed at 30 wt %.

3.2. *Characterisation Techniques*

Wavelength dispersive X-ray fluorescence (WDXRF) determined the composition of the synthesized catalysts. Previously, a boron pearl glass was obtained by mixing the corresponding catalyst with a commercial flux agent (Spectromelt A12) with a 1:20 mass ratio. This mixture was then melted at 1200 °C. Measurements were performed under vacuum with PANalytical AXIOS sequential WDXRF spectrometer equipped with a Rh tube and three different detectors (gas flow, scintillation and Xe sealed).

Specific surface area (BET method), pore volume (BJH method) and pore size distribution (BJH method) were estimated by low-temperature (−196 °C) N_2 physisorption with a Micromeritics TriStar II instrument. The samples were submitted to degassing before analysis with flowing N_2 in a Micromeritics SmartPrep sample preparation system at 300 °C for 10 h.

Powder X-ray diffraction (XRD) and Raman spectroscopy were used for the structural characterization of the cobalt catalysts. XRD data were collected on an X'PERT-PRO X-ray diffractometer using Cu Kα radiation (λ = 1.5406 Å) and a Ni filter at 40 kV and 40 mA. The patterns were collected with the 2θ range from 10 to 80° with a step size of 0.026° and a counting time of 2.0 s. The cell size of the Co-spinel phase was calculated by profile matching of the whole diffractogram using FullProf.2k software. Raman measurements were performed on a Renishaw InVia Raman spectrometer using a 514 nm laser source (ion-argon laser, Modu-Laser), scanning from 150 to 1200 cm^{-1}. For each analysis, 20 s were employed and 5 scans were accumulated

X-ray photoelectron spectroscopy (XPS) was employed for characterizing the structure and chemical composition and the electronic structure at the surface level. The analysis was carried out with a SPECS system coupled to a Phoibos 150 1D analyzer and a DLD (Delay−Line Detector)-monochromatic radiation source.

The redox behavior of the cobalt catalysts was examined by temperature-programmed reduction with hydrogen (H_2-TPR) with a Micromeritics Autochem 2920 equipment with a 5%H_2/Ar stream. The samples were first conditioned at 300 °C for 30 min with a flowing 5% O_2/He mixture. Next, they were cooled down to 50 °C in an inert stream. The reduction process was conducted up to 600 °C with a heating rate of 10 °C min^{-1}. In addition, the reducibility of the samples was also examined by temperature-programmed reaction with methane (CH_4-TPRe) with a 5% CH_4/He mixture.

Elemental maps of Co, Ni and Ce were obtained by electron energy loss spectroscopy (EELS) or energy dispersive X-ray spectrometry (EDX), both in the scanning transmission electron microscopy—high angle annular dark field (STEM–HAADF) mode (FEI Titan Cubed operating at 300 kV and Philips CM200 operating at 200 kV), with a Tridiem Energy Filter (Gatan) and Super-X detector (ChemiSTEM) as detectors, respectively. The Co-Ni/Al and Co/Ce-Al samples were characterized by EELS-STEM and EDS-STEM, respectively.

3.3. *Catalytic Activity Determination*

The catalytic performance was examined in a fixed bed reactor (Microactivity by PID Eng & Tech S.L., Alcobendas, Madrid). A multipoint K type thermocouple was fixed to the middle of the catalyst bed in order to control the reaction temperature. Of the catalyst 1 g (sieve fraction of 0.25–0.3 mm) diluted with 1 g of inert quartz (sieve fraction of 0.5–0.8 mm) was used. A gaseous mixture (500 cm^3 min^{-1}) of CH_4 (1 vol.%), O_2 (10%) and N_2 (89%) was continuously supplied at a space velocity of around 30,000 h^{-1} (300 mL CH_4 g^{-1} h^{-1}) under atmospheric pressure.

Catalytic conversion was evaluated in the 200–600 °C range each 25 °C. The products were analyzed with an on-line gas chromatography (Agilent Technologies 7890N) equipped with thermal conductivity detector (TCD), using a PLOT 5A molecular sieve column (analysis of CH_4, O_2, N_2 and CO) and a PLOT U column (CO_2 analysis). The methane conversion is referred to the yield of CO_2. Kinetic results were checked not to be controlled by both mass and heat transfer limitations, following the criteria proposed by Eurokin [53,54] (see Table S1, Supplementary Material).

4. Conclusions

Three strategies for enhancing the behavior of alumina-supported Co_3O_4 catalysts for oxidation of lean methane were compared. These approaches focused on two main objectives, namely minimizing the formation of inactive cobalt aluminate and promoting the intrinsic activity of the deposited cobalt oxide. Thus, our attention was focused on the surface protection of alumina with magnesia, redox promotion of Co_3O_4 with nickel oxide and surface protection of alumina with ceria, which eventually may also act as a redox promoter for Co_3O_4. These samples were extensively characterized by WDXRF, BET measurements, XRD, Raman spectroscopy, XPS, H_2-TPR, CH_4-TPRe and STEM-EELS/EDX.

Firstly, as for the evaluation of the influence of MgO on the catalytic behavior, magnesia was loaded onto the alumina support prior to Co_3O_4 addition. The incorporation of magnesia hardly affected the textural properties of the blank alumina support, probably due to notable surface area of this promoter. After incorporating cobalt, deposited MgO prevented Co_3O_4 from reacting with the alumina, thereby limiting the generation of inactive cobalt aluminate. On the other hand, a cobalt–magnesium interaction was favored, thereby resulting in better redox properties of the cobalt oxide with a marked shift of the reduction onset temperature by around 30 °C.

Secondly, a bimetallic cobalt-nickel catalyst supported over alumina was synthesized in order to examine the effect of coprecipitating small amounts of nickel (5 wt %) along with the cobalt precursor. The resulting Ni-Co catalyst exhibited good textural properties, with only a slight loss of specific surface with respect to the bare alumina. Combined results from XRD, XPS, Raman spectroscopy and STEM-EELS evidenced that nickel was homogeneously present on the surface and induced the formation of trace amounts of $NiCo_2O_4$, because of the partial insertion of Ni^{2+} cation into the lattice of Co_3O_4. The strong cobalt-nickel interaction promoted the redox properties of the resulting Ni-Co samples. Thus, when compared with the unmodified cobalt catalyst, the reduction onset temperature was noticeably shifted (around 50 °C) to lower temperatures and the specific H_2 uptake in the low temperature range increased to a considerable extent. Furthermore, the higher mobility of active oxygen species over this sample was also evidenced by the temperature-programmed reaction with methane in the absence of gaseous oxygen.

Finally, the cobalt addition over a cerium-coated alumina was examined. Ceria was efficiently dispersed on the support in view of the reduced impact on the textural properties. A dual effect of ceria on the properties of deposited cobalt was evidenced. On one hand, ceria, like magnesia, partially inhibit the formation of undesired cobalt aluminate. More interestingly, a strong interaction between cobalt oxide and ceria was found that ultimately resulted in the insertion of cerium atoms into the spinelic lattice of Co_3O_4. Consequently, a higher abundance of Co^{3+} species at the cost of Co^{2+} was evidenced, thereby promoting the mobility of active lattice oxygen species.

The comparison of the catalytic behavior of the modified catalysts revealed that the most suitable strategy was the addition of cerium to the alumina, prior to the deposition of the cobalt precursor. The resulting optimal catalyst reduced its T_{50} value by 70 °C with respect to the reference catalyst supported over bare alumina, and exhibited a specific reaction rate around three times higher in comparison with the reference Co/Al catalyst.

Supplementary Materials: The following are available online at http://www.mdpi.com/2073-4344/10/7/757/s1, Figure S1: CH_4-TPRe profiles of the supported cobalt catalysts, Figure S2: Additional HAADF-STEM images of the Co-Ni/Al (left) and Co/Ce-Al (right) catalysts coupled to EELS (Co (red) and Ni (blue)) and EDX (Co (red) and Ce (green)) elemental distribution, Figure S3: Pseudo-first order fit for the experimental data over the supported cobalt catalysts, Table S1: Series of recommendations and criteria for accurate analysis of intrinsic reaction rates (as evaluated for the Co/Ce-Al catalyst at 400 °C).

Author Contributions: Conceptualization, A.C. and R.L.-F.; Methodology, A.C., B.d.R. and J.I.G.-O.; Formal Analysis, A.C., B.d.R. and R.L.-F.; Investigation, A.C.; Writing—Original Draft Preparation, A.C., B.d.R. and R.L.-F.; Writing—Review and Editing, A.C., J.I.G.-O. and R.L.-F.; Supervision, J.I.G.-O. and R.L.-F.; Funding Acquisition, B.d.R., J.I.G.-O. and R.L.-F. All authors have read and agreed to the published version of the manuscript.

Funding: This research was funded by the Ministry of Economy and Competitiveness (CTQ2016-80253-R AEI/FEDER, UE), Basque Government (IT1297-19) and the University of The Basque Country UPV/EHU (PIF15/335).

Acknowledgments: The author wish to thank the technical and human support provided by SGIker (UPV/EHU) and the Advanced Microscopy Laboratory of the University of Zaragoza.

Conflicts of Interest: The authors declare no conflict of interest.

References

1. Raj, A. Methane emission control. *Johns. Matthey Technol. Rev.* **2016**, *60*, 228–235. [CrossRef]
2. Khan, M.I.; Yasmin, T.; Shakoor, A. Technical overview of compressed natural gas (CNG) as a transportation fuel. *Renew. Sust. Energy Rev.* **2015**, *51*, 785–797. [CrossRef]
3. Chen, J.; Arandiyan, H.; Gao, X.; Li, J. Recent advances in catalysts for methane combustion. *Catal. Surv. Asia* **2015**, *19*, 140–171. [CrossRef]
4. Choudhary, T.V.; Banerjee, S.; Choudhary, V.R. Catalysts for combustion of methane and lower alkanes. *Appl. Catal. A Gen.* **2002**, *234*, 1–23. [CrossRef]
5. Velin, P.; Ek, M.; Skoglundh, M.; Schaefer, A.; Raj, A.; Thompsett, D.; Smedler, G.; Carlsson, P. Water inhibition in methane oxidation over alumina supported palladium catalysts. *J. Phys. Chem. C* **2019**, *123*, 25724–25737. [CrossRef]
6. Persson, K.; Pfefferle, L.D.; Schwartz, W.; Ersson, A.; Järås, S.G. Stability of palladium-based catalysts during catalytic combustion of methane: The influence of water. *Appl. Catal. B Environ.* **2007**, *74*, 242–250. [CrossRef]
7. Hu, J.; Zhao, W.; Hu, R.; Chang, G.; Li, C.; Wang, L. Catalytic activity of spinel oxides $MgCr_2O_4$ and $CoCr_2O_4$ for methane combustion. *Mater. Res. Bull.* **2014**, *57*, 268–273. [CrossRef]
8. Ma, Z. Cobalt oxide catalysts for environmental remediation. *Curr. Catal.* **2014**, *3*, 15–26. [CrossRef]
9. Pu, Z.; Zhou, H.; Zheng, Y.; Huang, W.; Li, X. Enhanced methane combustion over Co_3O_4 catalysts prepared by a facile precipitation method: Effect of aging time. *Appl. Surf. Sci.* **2017**, *410*, 14–21. [CrossRef]
10. Zasada, F.; Piskorz, W.; Janas, J.; Grybos, J.; Indyka, P.; Sojka, Z. Reactive oxygen species on the (100) facet of cobalt spinel nanocatalyst and their relevance in $^{16}O_2/^{18}O_2$ isotopic exchange, *de*N_2O, and *de*CH_4 processes—A theoretical and experimental account. *ACS Catal.* **2015**, *5*, 6879–6892. [CrossRef]
11. Liotta, L.F.; Wu, H.; Pantaleo, G.; Venezia, A.M. Co_3O_4 nanocrystals and Co_3O_4-MO_x binary oxides for CO, CH_4 and VOC oxidation at low temperatures: A review. *Catal. Sci. Technol.* **2013**, *3*, 3085–3102. [CrossRef]
12. Liotta, L.F.; Di Carlo, G.; Pantaleo, G.; Venezia, A.M.; Deganello, G. Co_3O_4/CeO_2 composite oxides for methane emissions abatement: Relationship between Co_3O_4-CeO_2 interaction and catalytic activity. *Appl. Catal. B Environ.* **2006**, *66*, 217–227. [CrossRef]
13. Videla, A.H.M.; Stelmachowski, P.; Ercolino, G.; Specchia, S. Benchmark comparison of Co_3O_4 spinel-structured oxides with different morphologies for oxygen evolution reaction under alkaline conditions. *J. Appl. Electrochem.* **2017**, *47*, 295–304. [CrossRef]

14. Chen, Z.; Wang, S.; Liu, W.; Gao, X.; Gao, D.; Wang, M.; Wang, S. Morphology-dependent performance of Co_3O_4 via facile and controllable synthesis for methane combustion. *Appl. Catal. A Gen.* **2016**, *525*, 94–102. [CrossRef]
15. Fei, Z.; He, S.; Li, L.; Ji, W.; Au, C. Morphology-directed synthesis of Co_3O_4 nanotubes based on modified Kirkendall effect and its application in CH_4 combustion. *Chem. Commun.* **2012**, *48*, 853–855. [CrossRef] [PubMed]
16. Hu, L.; Peng, Q.; Li, Y. Selective synthesis of Co_3O_4 nanocrystal with different shape and crystal plane effect on catalytic property for methane combustion. *J. Am. Chem. Soc.* **2008**, *130*, 16136–16137. [CrossRef] [PubMed]
17. Wang, Q.; Peng, Y.; Fu, J.; Kyzas, G.Z.; Billah, S.M.R.; An, S. Synthesis, characterization, and catalytic evaluation of $Co_3O_4/\gamma\text{-}Al_2O_3$ as methane combustion catalysts: Significance of Co species and the redox cycle. *Appl. Catal. B Environ.* **2015**, *168-169*, 42–50. [CrossRef]
18. Solsona, B.; Davies, T.E.; Garcia, T.; Vázquez, I.; Dejoz, A.; Taylor, S.H. Total oxidation of propane using nanocrystalline cobalt oxide and supported cobalt oxide catalysts. *Appl. Catal. B Environ.* **2008**, *84*, 176–184. [CrossRef]
19. Liotta, L.F.; Pantaleo, G.; Macaluso, A.; Di Carlo, G.; Deganello, G. CoO_x catalysts supported on alumina and alumina-baria: Influence of the support on the cobalt species and their activity in NO reduction by C_3H_6 in lean conditions. *Appl. Catal. A Gen.* **2003**, *245*, 167–177. [CrossRef]
20. Park, S.; Kwak, G.; Lee, Y.; Jun, K.; Kim, Y.T. Effect of H_2O on slurry-phase Fischer–Tropsch synthesis over alumina-supported cobalt catalysts. *Bull. Korean Chem. Soc.* **2018**, *39*, 540–547. [CrossRef]
21. Park, J.; Yeo, S.; Kang, T.; Heo, I.; Lee, K.; Chang, T. Enhanced stability of Co catalysts supported on phosphorus-modified Al_2O_3 for dry reforming of CH_4. *Fuel* **2018**, *212*, 77–87. [CrossRef]
22. Cheng, J.; Yu, J.; Wang, X.; Li, L.; Li, J.; Hao, Z. Novel CH_4 combustion catalysts derived from Cu-Co/X-Al (X = Fe, Mn, La, Ce) hydrotalcite-like compounds. *Energy Fuels* **2008**, *22*, 2131–2137. [CrossRef]
23. El-Shobaky, H.G.; Shouman, M.A.; Attia, A.A. Effect of La_2O_3 and Mn_2O_3-doping of Co_3O_4/Al_2O_3 system on its surface and catalytic properties. *Colloids Surf. A Physicochem. Eng. Asp.* **2006**, *274*, 62–70. [CrossRef]
24. Riad, M. Influence of magnesium and chromium oxides on the physicochemical properties of γ-alumina. *Appl. Catal. A Gen.* **2007**, *327*, 13–21. [CrossRef]
25. Ulla, M.A.; Spretz, R.; Lombardo, E.; Daniell, W.; Knözinger, H. Catalytic combustion of methane on Co/MgO: Characterisation of active cobalt sites. *Appl. Catal. B Environ.* **2001**, *29*, 217–229. [CrossRef]
26. Ji, S.; Ji, S.; Wang, H.; Flahaut, E.; Coleman, K.S.; Green, M.L.H. Catalytic combustion of methane over cobalt-magnesium oxide solid solution catalysts. *Catal. Lett.* **2001**, *75*, 65–71. [CrossRef]
27. Trivedi, S.; Prasad, R. Selection of cobaltite and effect of preparation method of $NiCo_2O_4$ for catalytic oxidation of CO–CH_4 mixture. *Asia-Pac. J. Chem. Eng.* **2017**, *12*, 440–453. [CrossRef]
28. Lim, T.H.; Cho, S.J.; Yang, H.S.; Engelhard, M.H.; Kim, D.H. Effect of Co/Ni ratios in cobalt nickel mixed oxide catalysts on methane combustion. *Appl. Catal. A Gen.* **2015**, *505*, 62–69. [CrossRef]
29. Dumond, F.; Marceau, E.; Che, M. A study of cobalt speciation in Co/Al_2O_3 catalysts prepared from solutions of cobalt-ethylenediamine complexes. *J. Phys. Chem. C* **2007**, *111*, 4780–4789. [CrossRef]
30. Pérez-Ramírez, J.; Mul, G.; Kapteijn, F.; Moulijn, J.A. In situ investigation of the thermal decomposition of Co-Al hydrotalcite in different atmospheres. *J. Mater. Chem.* **2001**, *11*, 821–830. [CrossRef]
31. Marco, J.F.; Gancedo, J.R.; Gracia, M.; Gautier, J.L.; Ríos, E.; Berry, F.J. Characterization of the nickel cobaltite, $NiCo_2O_4$, prepared by several methods: An XRD, XANES, EXAFS, and XPS study. *J. Solid State Chem.* **2000**, *153*, 74–81. [CrossRef]
32. Lukashuk, L.; Yigit, N.; Rameshan, R.; Kolar, E.; Teschner, D.; Hävecker, M.; Knop-Gericke, A.; Schlögl, R.; Föttinger, K.; Rupprechter, G. Operando insights into CO oxidation on cobalt oxide catalysts by NAP-XPS, FTIR, and XRD. *ACS Catal.* **2018**, *8*, 8630–8641. [CrossRef] [PubMed]
33. Dupin, J.; Gonbeau, D.; Vinatier, P.; Levasseur, A. Systematic XPS studies of metal oxides, hydroxides and peroxides. *Phys. Chem. Chem. Phys.* **2000**, *2*, 1319–1324. [CrossRef]
34. Wu, L.; Jiao, D.; Wang, J.; Chen, L.; Cao, F. The role of MgO in the formation of surface active phases of $CoMo/Al_2O_3$-MgO catalysts for hydrodesulfurization of dibenzothiophene. *Catal. Commun.* **2009**, *11*, 302–305. [CrossRef]
35. Cazzanelli, E.; Kuzmin, A.; Mariotto, G.; Mironova-Ulmane, N. Study of vibrational and magnetic excitations in $Ni_cMg_{1-c}O$ solid solutions by Raman spectroscopy. *J. Phys. Condens. Matter* **2003**, *15*, 2045–2052. [CrossRef]

36. Trivedi, S.; Prasad, R.; Gautam, S.K. Design of active $NiCo_2O_{4-\delta}$ spinel catalyst for abatement of CO-CH_4 emissions from CNG fueled vehicles. *AIChE J.* **2018**, *64*, 2632–2646. [CrossRef]
37. Pettiti, I.; Gazzoli, D.; Benito, P.; Fornasari, G.; Vaccari, A. The reducibility of highly stable Ni-containing species in catalysts derived from hydrotalcite-type precursors. *RSC Adv.* **2015**, *5*, 82282–82291. [CrossRef]
38. Zasada, F.; Grybos, J.; Budiyanto, E.; Janas, J.; Sojka, Z. Oxygen species stabilized on the cobalt spinel nano-octahedra at various reaction conditions and their role in catalytic CO and CH_4 oxidation, N_2O decomposition and oxygen isotopic exchange. *J. Catal.* **2019**, *371*, 224–235. [CrossRef]
39. Zasada, F.; Janas, J.; Piskorz, W.; Gorczynska, M.; Sojka, Z. Total oxidation of lean methane over cobalt spinel nanocubes controlled by the self-adjusted redox state of the catalyst: Experimental and theoretical account for interplay between the Langmuir-Hinshelwood and Mars-Van Krevelen mechanisms. *ACS Catal.* **2017**, *7*, 2853–2867. [CrossRef]
40. Choya, A.; de Rivas, B.; González-Velasco, J.R.; Gutiérrez-Ortiz, J.I.; López-Fonseca, R. Oxidation of residual methane from VNG vehicles over Co_3O_4-based catalysts: Comparison among bulk, Al_2O_3-supported and Ce-doped catalysts. *Appl. Catal. B Environ.* **2018**, *237*, 844–854. [CrossRef]
41. Zheng, Y.; Yu, Y.; Zhou, H.; Huang, W.; Pu, Z. Combustion of lean methane over Co_3O_4 catalysts prepared with different cobalt precursors. *RSC Adv.* **2020**, *10*, 4490–4498. [CrossRef]
42. Liotta, L.F.; Di Carlo, G.; Pantaleo, G.; Deganello, G. Catalytic performance of Co_3O_4/CeO_2 and Co_3O_4/CeO_2–ZrO_2 composite oxides for methane combustion: Influence of catalyst pretreatment temperature and oxygen concentration in the reaction mixture. *Appl. Catal. B Environ.* **2007**, *70*, 314–322. [CrossRef]
43. Ji, Y.; Zhao, Z.; Duan, A.; Jiang, G.; Liu, J. Comparative study on the formation and reduction of bulk and Al_2O_3-supported cobalt oxides by H_2-TPR technique. *J. Phys. Chem. C* **2009**, *113*, 7186–7199. [CrossRef]
44. de Rivas, B.; Sampedro, C.; Ramos-Fernández, E.V.; López-Fonseca, R.; Gascon, J.; Makkee, M.; Gutiérrez-Ortiz, J.I. Influence of the synthesis route on the catalytic oxidation of 1,2-dichloroethane over CeO_2/H-ZSM5 catalysts. *Appl. Catal. A Gen.* **2013**, *456*, 96–104. [CrossRef]
45. Huang, Y.; Fan, W.; Long, B.; Li, H.; Qiu, W.; Zhao, F.; Tong, Y.; Ji, H. Alkali-modified non-precious metal 3D-$NiCo_2O_4$ nanosheets for efficient formaldehyde oxidation at low temperature. *J. Mater. Chem. A* **2016**, *4*, 3648–3654. [CrossRef]
46. Genty, E.; Siffert, S.; Cousin, R. Investigation of reaction mechanism and kinetic modelling for the toluene total oxidation in presence of CoAlCe catalyst. *Catal. Today* **2019**, *333*, 28–35. [CrossRef]
47. Bahlawane, N. Kinetics of methane combustion over CVD-made cobalt oxide catalysts. *Appl. Catal. B Environ.* **2006**, *67*, 168–176. [CrossRef]
48. Stefanov, P.; Todorova, S.; Naydenov, A.; Tzaneva, B.; Kolev, H.; Atanasova, G.; Stoyanova, D.; Karakirova, Y.; Aleksieva, K. On the development of active and stable Pd-Co/γ-Al_2O_3 catalyst for complete oxidation of methane. *Chem. Eng. J.* **2015**, *266*, 329–338. [CrossRef]
49. Paredes, J.R.; Díaz, E.; Díez, F.V.; Ordónez, S. Combustion of methane in lean mixtures over bulk transition-metal oxides: Evaluation of the activity and self-deactivation. *Energy Fuels.* **2009**, *23*, 86–93. [CrossRef]
50. Choya, A.; de Rivas, B.; Gutiérrez-Ortiz, J.I.; López-Fonseca, R. Effect of residual Na^+ on the combustion of methane over Co_3O_4 bulk catalysts prepared by precipitation. *Catalysts* **2018**, *8*, 427. [CrossRef]
51. Choya, A.; de Rivas, B.; González-Velasco, J.R.; Gutiérrez-Ortiz, J.I.; López-Fonseca, R. Oxidation of lean methane over cobalt catalysts supported on ceria/alumina. *Appl. Catal. A Gen.* **2020**, *591*, 117381. [CrossRef]
52. Choya, A.; de Rivas, B.; González-Velasco, J.R.; Gutiérrez-Ortiz, J.I.; López-Fonseca, R. On the beneficial effect of MgO promoter on the performance of Co_3O_4/Al_2O_3 catalysts for combustion of dilute methane. *Appl. Catal. A Gen.* **2019**, *582*, 117099. [CrossRef]
53. EUROKIN Spreadsheet on Requirements for Measurement of Intrinsic Kinetics in the Gas-Solid Fixed-Bed Reactor. Available online: http://eurokin.org/ (accessed on 29 April 2020).
54. Aranzabal, A.; González-Marcos, J.A.; Ayastuy, J.L.; González-Velasco, J.R. Kinetics of Pd/alumina catalysed 1,2-dichloroethane gas-phase oxidation. *Chem. Eng. Sci.* **2006**, *61*, 3564–3576. [CrossRef]

catalysts

MDPI

Article

The Support Effects on the Direct Conversion of Syngas to Higher Alcohol Synthesis over Copper-Based Catalysts

Xiaoli Li [1,2], Junfeng Zhang [1], Min Zhang [3], Wei Zhang [3], Meng Zhang [1,2], Hongjuan Xie [1], Yingquan Wu [1] and Yisheng Tan [1,4,*]

1 State Key Laboratory of Coal Conversion, Institute of Coal Chemistry, Chinese Academy of Sciences, Taiyuan 030001, China; lixiaoli@sxicc.ac.cn (X.L.); zhangjf@sxicc.ac.cn (J.Z.); zhangmeng@sxicc.ac.cn (M.Z.); xiehj@sxicc.ac.cn (H.X.); wuyq@sxicc.ac.cn (Y.W.)
2 School of Chemical Science, University of Chinese Academy of Sciences, Beijing 100049, China
3 Technology Center, Shanxi Lu'an Mining (Group) Co. Ltd., Changzhi 046204, China; zmin925@163.com (M.Z.); 15934381920@163.com (W.Z.)
4 National Engineering Research Center for Coal-Based Synthesis, Institute of Coal Chemistry, Chinese Academy of Sciences, Taiyuan 030001, China
* Correspondence: tan@sxicc.ac.cn; Tel./Fax: +86-351-404-4287; +86-351-404-4287

Received: 21 January 2019; Accepted: 9 February 2019; Published: 21 February 2019

Abstract: The types of supports employed profoundly influence the physicochemical properties and performances of as-prepared catalysts in almost all catalytic systems. Herein, Cu catalysts, with different supports (SiO_2, Al_2O_3), were prepared by a facile impregnation method and used for the direct synthesis of higher alcohols from CO hydrogenation. The prepared catalysts were characterized using multiple techniques, such as X-ray diffraction (XRD), N_2 sorption, H_2-temperature-programmed reduction (H_2-TPR), temperature-programmed desorption of ammonia (NH_3-TPD), X-ray photoelectron spectroscopy (XPS) and in situ Fourier-transform infrared spectroscopy (FTIR), etc. Compared to the Cu/Al_2O_3 catalyst, the Cu/SiO_2 catalyst easily promoted the formation of a higher amount of C1 oxygenate species on the surface, which is closely related to the formation of higher alcohols. Simultaneously, the Cu/Al_2O_3 and Cu/SiO_2 catalysts showed obvious differences in the CO conversion, alcohol distribution, and CO_2 selectivity, which were probably originated from differences in the structural and physicochemical properties, such as the types of copper species, the reduction behaviors, acidity, and electronic properties. Besides, it was also found that the gap in performances in two kinds of catalysts with the different supports could be narrowed by the addition of potassium because of its neutralization to surface acidy of Al_2O_3 and the creation of new basic sites, as well as the alteration of electronic properties.

Keywords: CO hydrogenation; higher alcohols; support effects; Cu-based catalysts

1. Introduction

Higher alcohols are attracting considerable attentions owing to their broad applications, such as fuels, fuel additives, and feedstock for the production of various chemicals and polymers [1–3]. With increasing concerns for environmental pollution and depletion of non-renewable petroleum resources, there is a growing interest in the direct synthesis of oxygenates, especially higher alcohols synthesis via syngas derived from coal, natural gas, or biomass [4]. Generally, the catalysts suitable for higher alcohols synthesis can be divided into the following classes: (I) Rh-based catalysts [5,6], (II) the modified Fischer–Tropsch catalysts [7,8], (III) Mo-based catalysts [9], and (IV) the modified Cu-based catalysts for methanol synthesis [10–14]. Non-noble Cu-based catalysts, due to their comparable high activity, are regarded as one kind of the most promising candidates for higher alcohols synthesis [4,13,15].

With respect to Cu-based catalysts, remarkably important advances have been made and reported owing to the simple preparation method and full utilization of active components [2,4,10–18]. It has been well documented that interaction between metal oxide and the support significantly improved the dispersion of the active species [19,20], stabilized active species [12], and promoted the generation of new inter-phases [16,17,21,22], thereby strongly influencing the catalytic performance [19–29]. Lemonidou et al. [19] compared the catalytic activities of the three Ni-Mo catalysts supported by activated carbon (AC), Al_2O_3, and ZrO_2, respectively. They revealed that the activity was closely related to the dispersion of the active phase on support surface, and AC support with a higher surface area was helpful for the exposure of more active Ni-O-Mo sites. Y. Khodakov et al. [20] found that Cu-Co supported on Al_2O_3, due to relatively high metal dispersion and formation of copper cobalt bimetallic species, exhibited much higher alcohol selectivity than that supported on other materials. Wang et al. [21] studied the Al_2O_3-supported Cu-Co bimetallic catalysts for CO hydrogenation and revealed that the employment of Al_2O_3 can significantly increase the interaction between cobalt and copper particles compared with unsupported catalysts, thereby improving the selectivity of the catalysts to higher alcohols. Lee et al. [23] investigated the effect of supports (ZnO, MgO, and Al_2O_3) on the activity of Cu-Co catalysts for the hydrogenation reaction of CO and suggested that the high surface area and strong interaction between active centers and support played a vital role in improving alcohol formation. By comparing the Cu-Zn catalysts with and without γ-Al_2O_3, Choi [30] et al. pointed out that the selectivity of higher alcohols and CO conversion over a Cu-Zn catalyst supported on γ-Al_2O_3 were higher than 1.8 and 2.7 times that of a Cu-Zn catalyst without γ-Al_2O_3, respectively. They further found that a refractory $CuAl_2O_4$, formed via the thermal reaction of CuO and Al^{3+}, was able to enhance the long-term stability by increasing the resistance to sintering of the catalyst. Sun et al. [31] studied methanol synthesis from CO_2 hydrogenation over micro-spherical SiO_2 support Cu/ZnO catalysts and found that the catalytic activity was enhanced as a result of the small Cu particle size and uniform metal dispersion. Co-Cu bimetallic catalysts with SiO_2 support have been thoroughly investigated for higher alcohols synthesis from syngas by Han et al [32]. It suggested that CoCu bimetallic particles covered by Cu atoms were responsible for alcohols synthesis. Ma et al. [22,33,34] reported that the improvement of Cu dispersion was mainly ascribed to the generation of copper phyllosilicate ($Cu_2SiO_5(OH)_2$) caused by enhanced metal-support interactions, which was quite vital for the high activity and stability in the ethanol synthesis. The above literature clearly showed that Al_2O_3 and SiO_2 are good support candidates to prepare the catalyst with good performance in the synthesis of higher alcohols.

Our group has also spent considerable effort to study the Al_2O_3 and SiO_2 supported Cu-based catalysts for the higher alcohols synthesis from syngas [12,16,17,35]. In our latest work, we found that the interaction between Cu and Al_2O_3 support on K-Cu/Al_2O_3 catalysts could be effectively tuned by changing the calcination temperature, which led to the different distribution of CuO, $CuAl_2O_4$, and $CuAlO_2$ on the catalysts and strongly affected the reaction behaviors in the direct synthesis of ethanol from syngas [16,17]. For the Cu catalyst with SiO_2 support, the correlation of catalyst structure evolution and ethanol selectivity during the reaction process was systematically discussed [35]. Although we had somewhat understood the relation between copper species and performance of the supported Cu-based catalysts, the direct comparison of Al_2O_3 and SiO_2 supported Cu catalysts and the effects of supports, treated at similar conditions, on the direct synthesis of higher alcohols from CO hydrogenation had not been sufficiently discussed.

Therefore, this work was mainly to clarify the reason of difference in reaction behaviors over the Cu catalysts supported on Al_2O_3 and SiO_2 for CO hydrogenation into higher alcohols. Considering that alkali addition strongly affected the selectivity towards higher alcohols [13,15,36–41], herein, the present study also put forth effort to explore the effects of potassium addition on the structure and performance of Al_2O_3 and SiO_2 supported Cu catalysts. Moreover, the physicochemical properties of the prepared catalysts were characterized via various techniques, including X-ray diffraction (XRD), N_2 absorption-desorption, H_2-temperature-programmed reduction (H_2-TPR), temperature-programmed desorption of ammonia (NH_3-TPD), X-ray photoelectron spectroscopy (XPS), and in situ Fourier-transform infrared spectroscopy (FTIR), and the characterization results were discussed alongside with the catalytic data in detail.

2. Materials and Methods

2.1. Materials

Analytical-grade chemicals, including $Cu(NO_3)_2 \cdot \times 3H_2O$, NaOH, $AlCl_3 \cdot \times 6H_2O$, and K_2CO_3, were purchased from the Beijing Chemical Co. Ltd. (Beijing, China) and used directly without further purification. The employed SiO_2 was purchased from Aladdin Industrial Co. Ltd. (Los Angeles, US). The γ-Al_2O_3, as a support was synthesized using a hydrothermal route, which was similar to the procedure described by Yang et al. [42]. Typically, the ammonia solution (28% NH_3), $AlCl_3 \cdot \times 6H_2O$ solution, and NaOH solution were mixed under hydrothermal treatment. Then, the mixture was dried and calcined to obtain the γ-Al_2O_3.

2.2. The Preparation of Cu/Al_2O_3, K-Cu/Al_2O_3 and Cu/SiO_2, K-Cu/SiO_2 Catalysts

The Al_2O_3 and SiO_2 supports with a near Brunauer-Emmett-Teller (BET) surface area (157–160 m^2/g) were chosen in this study. The catalysts were prepared using a sequential impregnation method. Typically, for the Cu/Al_2O_3 and K-Cu/Al_2O_3 catalysts, 6.74 g of $Cu(NO_3)_2 \times 3H_2O$ (10 wt % CuO) was dissolved in 20 mL of deionized water. Twenty grams of Al_2O_3 were added into the above copper nitrate solution and impregnated by the ultrasonic treatment for 1 h at room temperature. Afterward, the mixture was dried at 120 °C for 10 h and calcined at 900 °C for 5 h in air. The obtained solid was Cu/Al_2O_3 catalyst. K-Cu/Al_2O_3 catalyst was prepared through the second impregnation of Cu/Al_2O_3 in K_2CO_3 aqueous solution. Simply, 0.61 g of K_2CO_3 (4 wt % K_2O loading) was dissolved in another 10 mL of deionized water. The desired amount of the Cu/Al_2O_3 catalyst obtained above was impregnated in K_2CO_3 aqueous solution, along with the ultrasonic treatment, for 1 h at room temperature. Then, the resulting mixture was dried at 120 °C for 10 h and calcined at 500 °C for 5 h in air. The Cu/SiO_2 and K-Cu/SiO_2 catalysts were also prepared using a method similar to one above. For comparison, the Al_2O_3 and SiO_2 supports were also calcined at 900 °C for 5 h in air and denoted as Al_2O_3-900 and SiO_2-900, respectively.

2.3. Catalyst Characterization

The textural properties of the as-prepared catalysts were measured with N_2 absorption-desorption at −196 °C on a Tristar 3000 Micromeritics (Atlanta, GA, US) instrument. The specific surface area (S_{BET}) was calculated by the BET method. The micropore volume was obtained from the t-plot method. The pore size distributions were evaluated by using the density functional theory (DFT) method applied to the nitrogen adsorption data. The measurement of textural properties is accuracy (±1%). The experiments were repeated three times.

Powder XRD patterns of the catalysts were collected on a Rigaku MiniFlex II X-ray diffractometer (Tokyo, Japan), using Ni-filtered Cu-Kα radiation (k = 0.15418 nm) with a scanning angle (2θ) of 10–90°.

H_2-TPR was carried out on an automatic temperature-programmed chemisorption analyzer (TP-5080, Tianjin Xianquan Industrial Trade and Develpment Co. Ltd, Tianjin, China) equipped with a thermal conductivity detector. The catalyst with 100 mg was pretreated at 300 °C under a flow of N_2 (32 mL/min) for 1 h to remove traces of water and then cooled to 50 °C. Subsequently, the gas flow was switched to a 10% H_2/N_2 (v/v, 35 mL/min). The sample was heated to 900 °C at a rate of 10 °C/min.

NH_3-TPD was carried out on a TP-5080 chemisorption instrument in order to evaluate the acidity of the catalysts. The catalyst (100 mg) was pretreated at 400 °C under a flow of N_2 (32 mL/min) for 1 h and then cooled down to 100 °C. After that, sample was exposed on NH_3 flow for 15 min. The TPD spectra were recorded from 100 to 600 °C, using a heating rate of 10 °C/min.

Characterizations of XPS and Auger electron spectroscopy (XAES) were conducted on an AXIS ULTRA DLD instrument (Kratos, Manchester, UK) equipped with Al Kα (hν = 1486.6 eV). The binding energy values were corrected for charging effects by referring to the adventitious C1s line at 284.5 eV.

In situ FTIR spectra of CO adsorption and desorption were obtained with a TENSOR-27 in the range from 4000 to 1000 cm^{-1} with 4 cm^{-1} resolution. Before CO adsorption, all catalysts were reduced at 400 °C for 0.5 h in a 10% H_2/N_2 (v/v, 15 mL/min). CO adsorption was taken at 400 °C and after 30 min of pure Ar flow at the same temperature. IR spectra were collected after evacuation for 30 min.

2.4. Catalytic Performance Evaluation

The catalyst test of CO hydrogenation was performed in a stainless fixed-bed reactor. In a typical run, 5 mL of the prepared catalyst (30–40 meshes) was placed in the center of the reactor. The catalyst was reduced according to the designed temperature program, i.e., from room temperature to 400 °C in a 10% H_2/N_2 (v/v, 35 mL/min) mixture at 400 °C for 4 h. The reaction was conducted at 400 °C, 10 MPa and 5000 h^{-1}. The flow rate of fed syngas (with CO/H_2 ratio of 1 to 2.7) was controlled by a mass flow controller, and the exit gases were measured using a wet test meter. The products were analyzed using four chromatographs during the reaction. The organic gas products, consisting of hydrocarbons and methanol, were detected online on GC4000A (EastWest, Beijing, China) equipped with flame ionization detector and GDX-403 column (EastWest, Beijing, China) (3 mm, 1 m). The inorganic gas products were detected online by thermal conductivity measurements using an EastWest GC4000A (carbon molecular sieves column, 3 m, 3m). The H_2O and methanol products in the liquid phase were detected by thermal conductivity measurements using a GC4000A (Shimazduo, Kyoto, Japan) (GDX-401 column, Shimazduo, Kyoto, Japan, 3 mm, 3 m). The alcohol products in the liquid phase were detected by flame ionization measurements using a Shimazduo GC-7AG (Shimazduo, Kyoto, Japan) (Chromosorb 101, Shimazduo, Kyoto, Japan, 3 mm, 4 m).

3. Results and Discussion

3.1. Catalyst Characterization

3.1.1. XRD

The XRD patterns of Al_2O_3, Al_2O_3-900, Cu/Al_2O_3, K-Cu/Al_2O_3 catalysts (Figure 1a) and SiO_2, SiO_2-900, Cu/SiO_2, K-Cu/SiO_2 catalysts (Figure 1b) are shown in Figure 1. No obvious changes were observed in the XRD patterns of the supports (Al_2O_3 and SiO_2) before and after calcination, revealing that tuning calcination temperature did not influence the phases of the supports. In the case of the Cu/Al_2O_3 and Cu/SiO_2 catalysts, the XRD patterns were very different from that of the supports. Specifically, the diffraction peaks (2θ = 31.3, 39.4, 42.6, 52.5, and 55.7°) of the $CuAlO_2$ phase (JCPDS no. 39-0246) [43] and the peaks (2θ = 31.3, 36.9, 44.9, 55.7, 59.5, 65.3, 77.2, and 80.8°) of the $CuAl_2O_4$ phase (JCPDS no. 33-0448) [16,43] appeared in the Cu/Al_2O_3 catalysts (as shown in Figure 1a). Unlike the Cu/Al_2O_3 catalyst, the Cu/SiO_2 catalyst showed the peaks (2θ = 35.7 and 38.9°) of CuO phase (JCPDS no. 05-661) [22,44] and the peaks (2θ = 31.4, 57.5, and 62.4°) of copper phyllosilicate [33,34] (as displayed in Figure 1b). In addition, potassium introduction (such as the K-Cu/Al_2O_3 catalyst and the K-Cu/SiO_2 catalyst) did not seemingly induce obvious changes in the diffraction peaks. These findings clearly revealed that copper species reacted with the support to form new phases when calcined at 900 °C, and the supports strongly affected the forms of copper species, but the potassium had no obvious effect on the phases of the catalysts.

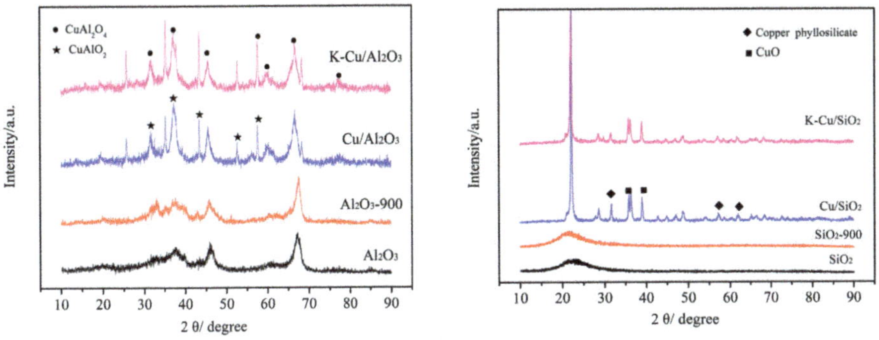

(**a**) (**b**)

Figure 1. X-ray diffraction (XRD) patterns of (**a**) Al_2O_3, Al_2O_3-900, Cu/Al_2O_3, K-Cu/Al_2O_3 catalysts and (**b**) SiO_2, SiO_2-900, Cu/SiO_2, K-Cu/SiO_2 catalysts.

3.1.2. N_2 Absorption-Desorption

The textural properties of the Al_2O_3, Al_2O_3-900, SiO_2, SiO_2-900, Cu/Al_2O_3, K-Cu/Al_2O_3 and Cu/SiO_2, K-Cu/SiO_2 catalysts were listed in Table 1. In comparison of the parent Al_2O_3 (157 m^2/g) and SiO_2 (160 m^2/g), the BET surface areas of Al_2O_3-900 and SiO_2-900 dramatically decreased to 87.0 and 25.8 m^2/g, respectively, via calcination at 900 °C, which was strongly associated with the collapse of porous structure during the high-temperature calcination process. In addition, when copper species were introduced into the uncalcined supports, the surface areas of the Cu/Al_2O_3 and Cu/SiO_2 catalysts sharply dropped to 41.6 and 6.69 m^2/g, which were much smaller than that of Al_2O_3-900 and SiO_2-900, respectively. The decrease in surface areas was probably due to both the formation of interfacial composite phases and copper as a sintering agent. As also shown in Table 1, compared with the Cu/Al_2O_3 and K-Cu/Al_2O_3 catalysts, the Cu/SiO_2 and K-Cu/SiO_2 catalysts showed considerably lower values of the surface area (4.63–6.69 m^2/g), smaller pore volume (0.006–0.008 cm^3/g), and average pore diameter (5.26–5.66 nm). When potassium was added, the surface area, pore volume, and average pore diameter of Al_2O_3 supported catalysts further decreased (from 41.6 to 40.7 m^2/g, 0.20 to 0.18 cm^3/g, and 19.1 to 17.8 nm, respectively). The surface areas and pore volume of SiO_2 supported catalysts also showed a decreasing trend (from 6.69 to 4.63 m^2/g and 0.008 to 0.006 cm^3/g, respectively), but the value of average pore diameter slightly increased from 5.26 to 5.66 nm, which was probably related to the corrosion of potassium to SiO_2. The results indicated that the textural parameters of the samples were greatly affected by both supports (Al_2O_3, SiO_2) and potassium.

Table 1. Textural properties of the representative samples.

Catalyst	S_{BET} (m^2/g)	V_{Pore} (cm^3/g)	d_{Pore} (nm)
Al_2O_3	157	0.43	10.8
Al_2O_3-900	87.0	0.39	18.1
Cu/Al_2O_3	41.6	0.20	19.1
K-Cu/Al_2O_3	40.7	0.18	17.8
SiO_2	160	0.54	13.7
SiO_2-900	25.8	0.07	10.6
Cu/SiO_2	6.69	0.008	5.26
K-Cu/SiO_2	4.63	0.006	5.66

The N_2 adsorption-desorption isotherms of the Cu/Al_2O_3, K-Cu/Al_2O_3 and Cu/SiO_2, K-Cu/SiO_2 catalysts were shown in Figure 2a. As observed, Cu/Al_2O_3 catalyst showed a type IV adsorption isotherm [9]. When potassium was added, the shape of the isotherms of Al_2O_3 supported catalysts did not change significantly. Unlike Al_2O_3 suppored catalysts, SiO_2 supported catalysts had no N_2 adsorption-desorption isotherms.

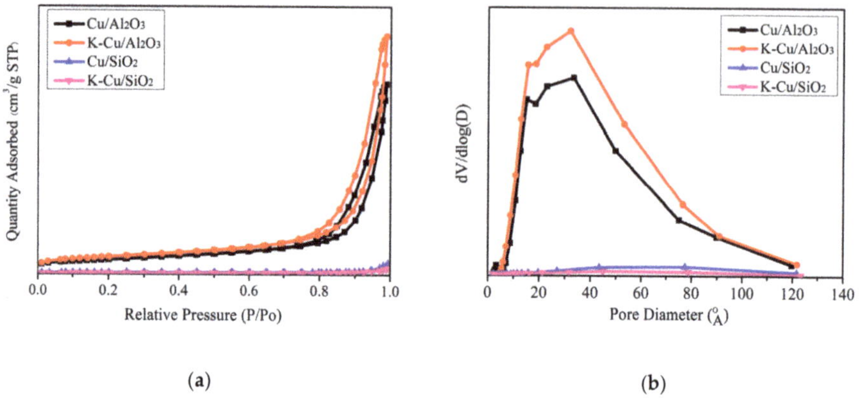

Figure 2. (**a**) Nitrogen adsorption-desorption isotherms and (**b**) pore size distribution of Cu/Al_2O_3, K-Cu/Al_2O_3 and Cu/SiO_2, K-Cu/SiO_2 catalysts.

Figure 2b presented the pore size distribution curves of the Cu/Al_2O_3, K-Cu/Al_2O_3 and Cu/SiO_2, K-Cu/SiO_2 catalysts. It was clearly observed that the Cu/Al_2O_3 catalyst had a wide range of 10–120 Å, while the addition of potassium, such as the K-Cu/Al_2O_3 catalyst, led to no obvious change in pore size distribution. In Figure 2b, note that no pore size distribution existed in the SiO_2 supported catalysts.

3.1.3. H_2-TPR

The reduction behaviors of the Al_2O_3, SiO_2, CuO, Cu/Al_2O_3, K-Cu/Al_2O_3 and Cu/SiO_2, K-Cu/SiO_2 catalysts were studied by H_2-TPR, and the results were presented in Figure 3. No reduction peak was observed in the Al_2O_3 and SiO_2, and one reduction peak at 299 °C was clearly detected in the CuO phase. As displayed in Figure 3, the H_2-TPR profile of Cu/Al_2O_3 catalyst showed three reduction peaks at around 280, 540, and 800 °C, which corresponded to the reduction of CuO [45], $CuAl_2O_4$ [16], and $CuAlO_2$ [43], respectively. When the potassium was introduced into the catalyst, only the reduction temperature of the CuO phase in the Cu/Al_2O_3 catalyst shifted towards a higher temperature. The observed shift could be attributed to that the chemical interaction between the copper species and alumina, which was somewhat affected by the addition of potassium, in agreement with the observations of Tien-Thao et al. [40], who reported an increase in the reduction temperature of copper in Co-Cu catalysts with increasing amounts of alkali additives. The above XRD results revealed that the diffraction peaks ascribed to CuO were not observed factually for Al_2O_3 supported catalysts. It was thought that CuO particles with small size were probably dispersed on Al_2O_3 support. From Figure 3, four reduction peaks at 435, 540, 700, and 770 °C were clearly found in the Cu/SiO_2 catalyst, suggesting that four types of copper species formed on the catalyst [46]. Apparently, the addition of potassium to the Cu/SiO_2 catalyst led to no obvious change in the position of all the reduction peaks, suggesting a weak influence of potassium on the interactions between Cu and Si. In comparison of two kinds of the catalyst with different supports (in Figure 3), SiO_2 supported catalysts (Cu/SiO_2, K-Cu/SiO_2) showed a much narrower reduction temperature range than Al_2O_3 supported catalysts (Cu/Al_2O_3, K-Cu/Al_2O_3). It was easily understood that the copper oxide interacted with Al_2O_3 or SiO_2, and different supports always led to different interactions, which implied different reaction behaviors on these catalysts.

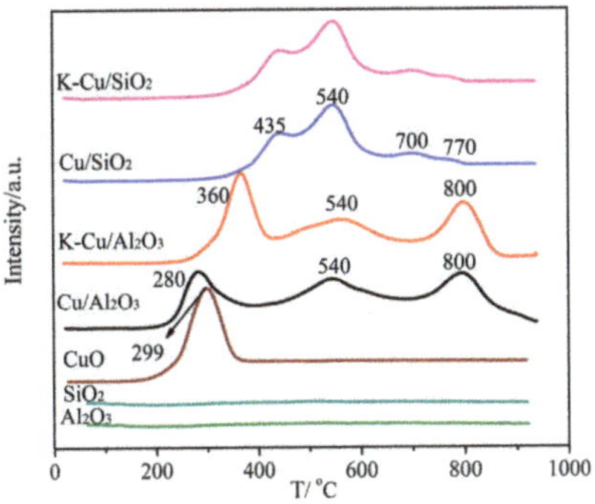

Figure 3. H_2-temperature programmed reduction (H_2-TPR) profiles of Al_2O_3, SiO_2, CuO, Cu/Al_2O_3, K-Cu/Al_2O_3 and Cu/SiO_2, K-Cu/SiO_2 catalysts.

3.1.4. NH_3-TPD

The acidity of Cu/Al_2O_3, K-Cu/Al_2O_3 and Cu/SiO_2, K-Cu/SiO_2 catalysts was studied by NH_3-TPD measurements, and the results were shown in Figure 4. No NH_3 desorption peak was found over the CuO phase, revealing that the acid of the CuO phase was very weak. It was clearly observed that the NH_3-TPD profiles of the Al_2O_3 and Cu/Al_2O_3 catalysts were exactly the same. Specifically, two peaks at 270 and 500 °C, ascribed to the weak acidic sites and the strong acidic sites, respectively, were obviously observed in the Al_2O_3 and Cu/Al_2O_3 catalysts. These results indicated that the acid stemmed mainly from the Al_2O_3 support. When the potassium was introduced, the peak of weak acidic sites shifted to a lower temperature, and yet that of strong acidic sites slightly shifted towards a higher temperature, revealing that the strength of weak acidic sites decreased and the strength of strong acidic sites increased slightly. In comparison of the area for NH_3 desorption, it was apparent that the ammonia amounts of both weak acidic sites and strong acidic sites obviously decreased when adding the potassium, due to the partial neutralization of the surface acidity by alkali compounds [36,39,40]. As known, SiO_2 possesses remarkably weak acidity. Therefore, the present SiO_2 supported Cu catalysts (eg., Cu/SiO_2, K-Cu/SiO_2) showed no NH_3 desorption peak, with or without the potassium addition [47]. These findings indicated that the acid-base property of the prepared catalysts was closely related to the support employed, such as SiO_2 and Al_2O_3, wherein, the difference in acid-base property easily resulted in obviously different reaction behaviors.

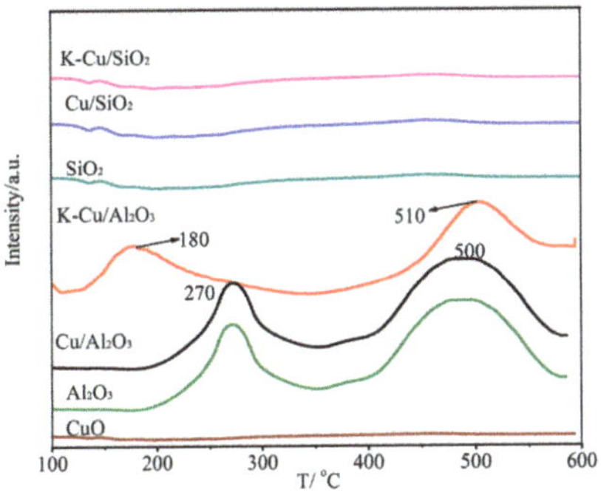

Figure 4. The temperature-programmed desorption of ammonia (NH_3-TPD) profiles of Al_2O_3, SiO_2, CuO, Cu/Al_2O_3, K-Cu/Al_2O_3 and Cu/SiO_2, K-Cu/SiO_2 catalysts.

3.1.5. XPS

XPS measurements of the representative catalysts were carried out in order to investigate the chemical state of the elements at the catalyst surface and the results were shown in Figures 5–8.

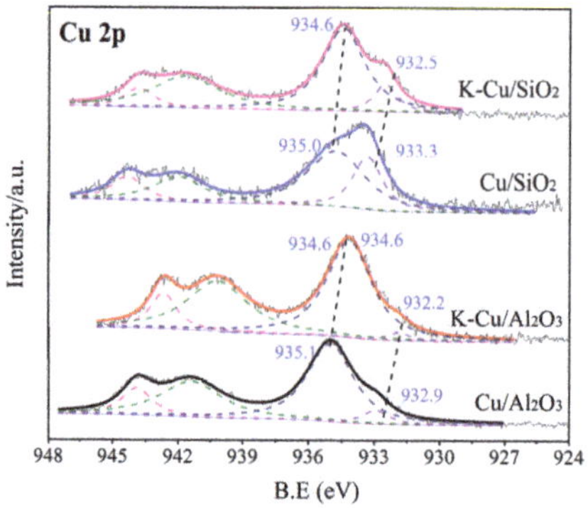

Figure 5. $Cu2p_{3/2}$ X-ray photoelectron spectroscopy (XPS) spectra of Cu/Al_2O_3, $K\text{-}Cu/Al_2O_3$ and Cu/SiO_2, $K\text{-}Cu/SiO_2$ catalysts.

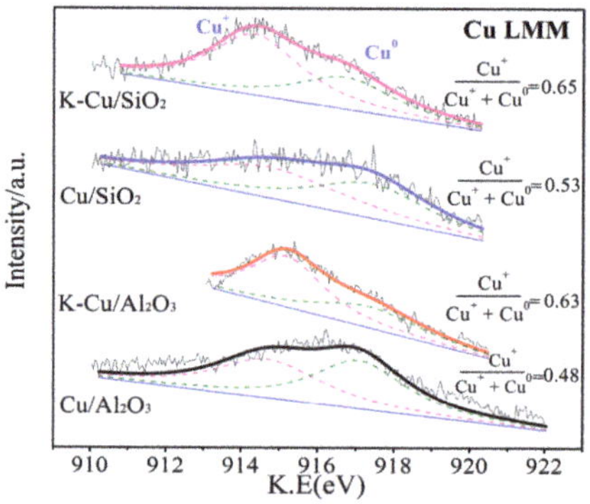

Figure 6. Cu LMM spectra of the reduced Cu/Al_2O_3, $K\text{-}Cu/Al_2O_3$ and Cu/SiO_2, $K\text{-}Cu/SiO_2$ catalysts.

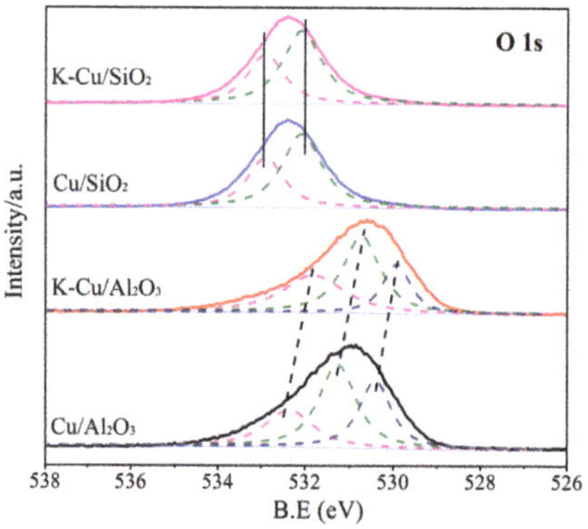

Figure 7. O1s XPS spectra of Cu/Al_2O_3, $K\text{-}Cu/Al_2O_3$ and Cu/SiO_2, $K\text{-}Cu/SiO_2$ catalysts.

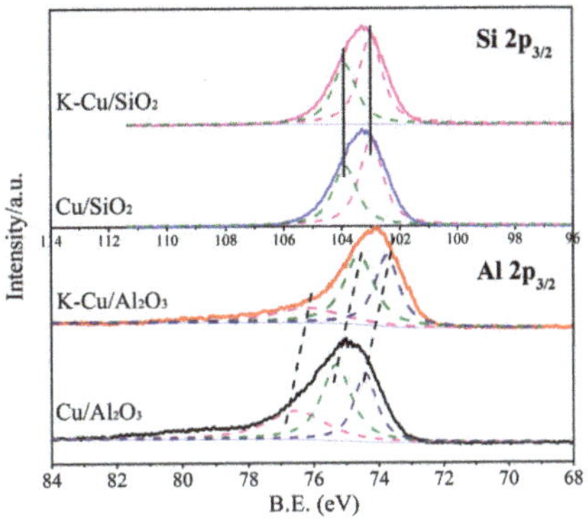

Figure 8. Al2p XPS spectra of Cu/Al_2O_3, $K\text{-}Cu/Al_2O_3$ and Si 2p XPS spectra of Cu/SiO_2, $K\text{-}Cu/SiO_2$ catalysts.

The $Cu2p_{3/2}$ binding energies (BEs) spectra of the catalysts were displayed in Figure 5. For Cu/Al_2O_3 sample, the peak appearing at ~934.0 eV, along with the shakeup satellites (940–945 eV), suggested the presence of Cu^{2+} species [30]. The asymmetry of the $Cu2p_{3/2}$ envelope could be deconvoluted into two peaks centered at around 933.0 and 935.0 eV, which were ascribed to Cu^{2+} in CuO and Cu^{2+} in $CuAl_2O_4$ respectively [16,17]. It indicated that copper oxides reacted with Al_2O_3 to form interfacial composite phases, consistent with the XRD results. When potassium was added into Cu/Al_2O_3, the peaks of the Cu^{2+} species shifted to lower BEs values. In the case of the Cu/SiO_2 catalyst, the peak (933.3 eV) of CuO and the peak (935.0 eV) of copper phyllosilicate [33,34,48] were clearly observed from Figure 5. Further, the BEs values of Cu^{2+} species shifted to the lower position with the addition of potassium. The results revealed that the chemical states of the Cu element were strongly affected by the supports employed, as well as potassium addition.

The distinction between the Cu^+ and Cu^0 species is feasible through the examination of Cu LMM XAES spectra. From the Cu LMM XAES of the reduced Cu/Al_2O_3, $K\text{-}Cu/Al_2O_3$ and Cu/SiO_2, $K\text{-}Cu/SiO_2$ catalysts shown in Figure 6, each Cu LMM spectrum contained two peaks centered at about 914–915 eV and 917–918 eV, with respect to the Cu^+ and Cu^0 species [34,48]. From the deconvolution results (inset), SiO_2 supported catalysts showed a slightly higher ratio of Cu^+ than Al_2O_3 supported catalysts. However, it was very obvious that the potassium addition induced an increase in Cu^+ species on two kinds of catalysts, which is in agreement with the observations of Lopez et al. [41].

O1s XPS spectra of Cu/Al_2O_3, $K\text{-}Cu/Al_2O_3$ and Cu/SiO_2, $K\text{-}Cu/SiO_2$ catalysts were given in Figure 7. The O1s signal of Cu/Al_2O_3 catalyst showed three overlapping peaks at around 530.1, 531.2, and 532.4 eV, indicating that three oxygen compounds formed on the catalyst surface [43]. After adding the potassium, a remarkable decrease in the BEs values was observed. However, different than the O1s XPS patterns of Al_2O_3 supported catalysts, two forms of oxygen compounds with higher BEs values were monitored over SiO_2 supported catalysts. Furthermore, the BEs values did not change with the addition of potassium. These results clearly revealed that the electronic environments of the O element on Al_2O_3 and SiO_2 supported catalysts were significantly different, and only the chemical states of the O element on the Cu/Al_2O_3 catalyst changed, apparently, by adding potassium.

Figure 8 displayed Al2p XPS spectra of Cu/Al_2O_3, $K\text{-}Cu/Al_2O_3$ and Si2p XPS spectra of Cu/SiO_2, $K\text{-}Cu/SiO_2$ catalysts. Three peaks attributed to aluminum species were obviously observed in the Cu/Al_2O_3 catalyst, revealing that three forms of aluminum species were present in this catalyst, which were related to the Al_2O_3, which formed $CuAl_2O_4$ and $CuAlO_2$, respectively [43]. The BEs values of Al2p obviously decreased when potassium was added, suggesting that the three chemical states of the Al element were affected by potassium. It was noted that two peaks centered at 103.1 and 103.9 eV were found on the Cu/SiO_2 catalyst, indicating that the Si element possessed two chemical states in the catalyst [34,48]. Also, adding potassium did not change the BEs values of Si2p. The results revealed that the addition of the potassium promoter altered the chemical states of the Al element, while it had no influence on that of the Si element.

3.2. Catalyst Evaluation

The performances of the Cu/Al_2O_3, K-Cu/Al_2O_3 and Cu/SiO_2, K-Cu/SiO_2 catalysts for higher alcohols synthesis from syngas were presented in Table 2. It was observed that the supports, such as Al_2O_3 and SiO_2, could profoundly influence catalytic behaviors. CO conversion of 84.6% and total alcohol selectivity of 7.7%, wherein the percentages of methanol and C_{2+} alcohols were 44.0 and 56.0 wt %, respectively, were achieved over the Cu/Al_2O_3 catalyst corresponding to CO_2 selectivity of 23.0%. Conversely, the Cu/SiO_2 catalyst showed relatively low CO conversion (18.2%) and CO_2 selectivity (2.4%), but much higher total alcohol selectivity (26.7%) in spite of slight lower percentage of C_{2+} alcohols (40.3 wt %). These results indicated that the supports had obvious effects on CO conversion, CO_2 selectivity, total alcohol selectivity, and alcohol distribution.

Table 2. The performances of Cu/Al_2O_3, K-Cu/Al_2O_3 and Cu/SiO_2, K-Cu/SiO_2 catalysts.

Samples	CO Conversion (%)	STY (mg/mlcath)	Carbon Selectivity (%)				Alcohol Distribution (wt %)				
			CH_4	C_{2-5}	CO_2	ROH	MeOH	EtOH	PrOH	BuOH	$C_{5+}OH$
Cu/Al_2O_3	84.6	93.7	42.2	27.1	23.0	7.7	44.0	42.5	8.0	4.6	0.9
K-Cu/Al_2O_3	48.5	141.4	23.2	19.5	32.5	24.1	34.9	38.3	16.2	8.5	2.0
Cu/SiO_2	18.2	49.5	42.3	28.3	2.4	26.7	59.8	34.6	4.5	1.1	0.1
K-Cu/SiO_2	16.8	55.0	27.4	27.7	16.1	28.8	51.8	32.6	9.8	4.4	1.4

Reaction conditions: 10 MPa, 400 °C, 5000 h^{-1}.

Further, one could observe from Table 2 that the potassium addition induced obviously different reaction behaviors over the Al_2O_3 and SiO_2 supported catalysts. Seemingly, for Al_2O_3 supported catalysts, the potassium addition resulted in a dramatic decrease in CO conversion, CH_4 selectivity, and C_{2-5} hydrocarbons selectivity (from 84.6 to 48.5%, from 42.2 to 23.2%, and from 27.1 to 19.5%, respectively) but an obvious increase in CO_2 selectivity and total alcohol selectivity (from 23.0 to 32.5% and from 7.7 to 24.1%, respectively). Besides, the K-CuO/Al_2O_3 showed a relatively lower selectivity to methanol but a higher one to C_{2+} alcohols, indicating that alcohol chain-growth was enhanced. Chain-growth probabilities (α) of alcohols were calculated, as shown in Figure 9. In the case of SiO_2 supported catalysts, the change trend in the CH_4 selectivity, CO_2 selectivity, and alcohol chain-growth, induced by potassium addition, was similar with that in Al_2O_3 supported catalysts. Whereas, different than the Cu/Al_2O_3 catalyst, adding the potassium into the Cu/SiO_2 catalyst did not significantly change the values of CO conversion (~17.0%), total alcohol selectivity (~27.0%), and C_{2-5} hydrocarbons selectivity (~28.0%). Conclusively, potassium introduction did not only promote the formation of CO_2 and inhibit the CH_4 formation, but also enhanced the carbon chain growth probability of products; moreover, the potassium had a greater impact on the CO conversion, total alcohol selectivity, and C_{2-5} hydrocarbons selectivity over the Al_2O_3 supported catalysts than that over SiO_2 supported catalysts.

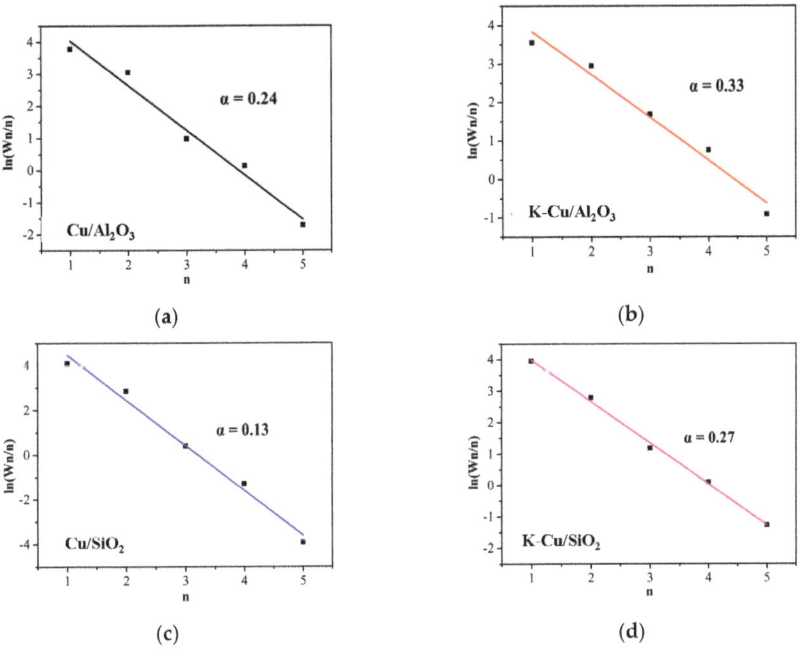

Figure 9. Anderson-Schulz-Flory (A-S-F) plots for the distribution of alcohols for catalysts: (**a**) Cu/Al_2O_3, (**b**) K-Cu/Al_2O_3 and (**c**) Cu/SiO_2, (**d**) K-Cu/SiO_2. $Wn = n \times (1 - \alpha)^2 \times \alpha^{n-1}$, where Wn stands for the mass fraction of alcohols containing n carbon atoms.

3.3. *In Situ FTIR*

To further obtain detailed information on the molecular events that occur on the surface of the catalyst, CO adsorption over reduced Cu/Al_2O_3, K-Cu/Al_2O_3 and Cu/SiO_2, K-Cu/SiO_2 catalysts was monitored using in situ FTIR. As displayed in Figure 10, there were two absorption bands in the region of 3200–2850 cm^{-1} (ν C-H) and 1650–1300 cm^{-1} (ν COO) observed, assigned to the adsorbed hydrocarbons and formate species (C1 oxygenate species), respectively [49,50]. For the Cu/Al_2O_3 catalyst, very weak peaks at 1650–1300 cm^{-1} were detected, revealing the presence of only a trace of C1 oxygenate species on the catalyst surface, which were approved to contribute to the formation of higher alcohols [2,41,50]. It was also noted that the peaks at around 3200–2850 cm^{-1}, ascribed to the hydrogenation, was strong on the Cu/Al_2O_3 catalyst. When potassium was added, the band peaks of the C1 oxygenate species obviously increased, whereas the intensity of hydrocarbons decreased. As Santos et al. [2] reported, potassium, in close vicinity to an adsorbed methyl group, stabilized oxygenate species that were found to play an important role in the syngas to alcohol route. Thus, the reaction shifted towards alcohols, rather than hydrocarbons, when potassium modified the Cu/Al_2O_3 catalyst. When SiO_2 was used as support, it was noted that the peak at 1650–1300 cm^{-1} ascribed to the C1 oxygenate species was enhanced on the Cu/SiO_2 catalyst, which was very different with that of Cu/Al_2O_3 catalyst, on which only trace amounts of the C1 oxygenate species were formed (Figure 10); moreover, it seemed that adding potassium had no effect on the amount of C1 oxygenate species formed. This explained why the Cu/SiO_2 and K-Cu/SiO_2 catalysts exhibited the similar alcohol selectivity of ~27% (Table 2).

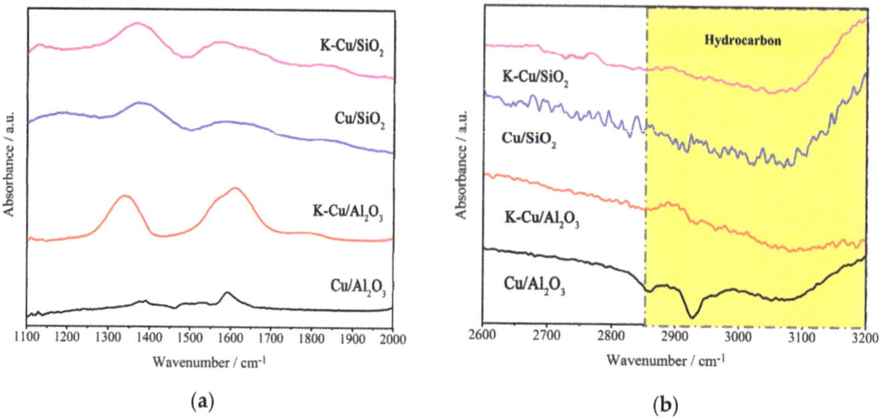

Figure 10. In situ Fourier-transform infrared (FTIR) spectra of CO adsorption over reduced Cu/Al_2O_3, $K\text{-}Cu/Al_2O_3$ and Cu/SiO_2, $K\text{-}Cu/SiO_2$ catalysts during CO flow at 400 °C for 30 min and then under Ar at 400 °C for 30 min.

3.4. Discussion

The results obtained by the test and characterizations of Cu/Al_2O_3 and Cu/SiO_2 samples clearly demonstrated that the physicochemical properties and catalytic performances of the Cu-based catalyst were strongly affected by the types of supports employed (Al_2O_3, SiO_2). In comparison to the Cu/Al_2O_3 catalyst with only 7.7% of total alcohol selectivity, the Cu/SiO_2 catalyst possessed a much higher total alcohol selectivity of 26.7%. As confirmed by the FTIR result, only a trace of the C1 oxygenate species was detected on the Cu/Al_2O_3 catalyst, but a relatively large amount of the C1 oxygenate species existed on the surface of the Cu/SiO_2 catalyst. Some authors have pointed out that the oxygenate species played an essential role in directing the synthesis toward alcohols, rather than hydrocarbons [39–41,50,51]. Zhang et al. [52,53] conducted a series of DFT studies to assess the mechanism of CO hydrogenation to higher alcohols on Cu (110) [52] and Cu (211) [53], and pointed out that the CH_xO species as key intermediates for higher alcohols synthesis could give the CH_x species through the C-O cleavage, and CH_x monomers subsequently combined with CO or CHO to form alcohol. Based on the above, it was apparent that the more adsorbed C1 oxygenate species on the Cu/SiO_2 catalyst inevitably resulted in an increase in the concentration of CH_x species via the C-O cleavage, and, finally, promoted the formation of alcohol. Hence, the Cu/SiO_2 catalyst with a higher amount of the C1 oxygenate species exhibited a higher total alcohols selectivity than the Cu/Al_2O_3 catalyst. Further, by comparing to the Cu/SiO_2 catalyst, the Cu/Al_2O_3 catalyst showed higher selectivities towards C_{2+} alcohols. As reported [54–56], aldol condensation as one of the key steps for carbon-chain growth of alcohol products easily proceeded on basic oxide catalysts or acid oxide catalysts. The NH_3-TPD result revealed that the surface acidity of the Cu/Al_2O_3 catalyst was much stronger than that of the Cu/SiO_2 catalyst. Therefore, the possibility of carbon-chain growth occurred on the acid Cu/Al_2O_3 catalyst more easily. Despite that the ethanol selectivities were both 32.6 wt % above on Al_2O_3 and SiO_2-supported catalysts, which was probably due to high $Cu^+/(Cu^++Cu^0)$ values, Cu/SiO_2 with a relatively high $Cu^+/(Cu^++Cu^0)$ value did not give higher ethanol selectivity than Cu/Al_2O_3. Our previous work [16,17] has demonstrated that the amounts of Cu^+ species were somewhat responsible for the formation of ethanol. These results suggested that the ethanol formation was affected by many factors, such as $Cu^+/(Cu^++Cu^0)$ value, the physical properties, the reduction behaviors, acidity, and electronic properties on the catalyst, which are synergetic.

In addition, CO conversion and CO_2 selectivity of the Cu/Al_2O_3 catalyst also showed significant differences from that of the Cu/SiO_2 catalyst. As revealed by the XRD and H_2-TPR results, although the refractory phases formed over Al_2O_3 or SiO_2 supported Cu catalysts under 900 °C, the two kinds of the catalysts with different supports presented differences in the types of copper species and the reduction behaviors of copper species. N_2 absorption-desorption results indicated that the textural parameters of the Cu/Al_2O_3 and Cu/SiO_2 catalysts, despite uncalcined supports with similar surface areas (157–160 m^2/g), were significantly different. Furthermore, the Cu/Al_2O_3 catalyst with a type IV adsorption isotherm showed a wide pore size distribution, but the Cu/SiO_2 catalyst had no N_2 adsorption-desorption isotherms and pore size distribution. The NH_3-TPD result suggested that the acidity was obviously detected in the Cu/Al_2O_3 catalyst, but no acidity was observed in the Cu/SiO_2 catalyst. Moreover, the XPS results showed that the Cu, O elements of the Cu/Al_2O_3 catalyst differed from that of the Cu/SiO_2 catalyst in electronic environments. As confirmed by the FTIR result, only a trace of C1 oxygenate species, contributing to alcohols formation, was detected on the Cu/Al_2O_3 catalyst, while a relatively large amount of C1 oxygenate species existed on the surface of the Cu/SiO_2 catalyst. These characterization results indicated that the structural and chemical properties of Cu/Al_2O_3 and Cu/SiO_2 catalysts showed obvious differences and thus affected the catalytic behaviors of the catalysts synergistically.

The performances of two kinds of catalysts with the support employed, such as SiO_2 and Al_2O_3, were also significantly affected by the potassium introduction. Doping potassium into the Al_2O_3 and SiO_2 supported catalysts improved the carbon chain growth probability of alcohol products. This was probably because potassium introduction provided new basic sites for the aldol condensation of lower alcohols to higher alcohols [40,57]. Moreover, alkali elements are known to be good promoters for the WGS reaction ($CO + H_2O \rightarrow CO_2 + H_2$) when introduced at optimum content [9,15,39,50]. As a result, the addition of potassium to Cu/Al_2O_3 and Cu/SiO_2 catalysts resulted in the improvement of CO_2 formation.

Additionally, according to the test results, doping potassium into Cu/Al_2O_3 and Cu/SiO_2 catalysts induced distinct differences in the total alcohol selectivity. For Al_2O_3 supported catalysts, total alcohol selectivity increased up to 24.1 from previous 7.7%, clearly, when the potassium was added. It indicated that the presence of potassium promoted the formation of alcohols. As confirmed by the FTIR result, by adding the K promoter, the relative amount of adsorbed C1 oxygenate species, as intermediates in higher alcohols synthesis [41,50], increased obviously by tuning the reduction behavior, neutralizing the surface acidity, and altering of the electronic properties, whereas the formation of hydrocarbons (mainly CH_4 and C_{2-5} hydrocarbons) was severely inhibited [39,50]. Anton et al. [39–41] reported that alkali (K/Cs/Rb) modified Cu-based catalysts can enhance the stability of CHx intermediate species. Therefore, total alcohols selectivity on the K-Cu/Al_2O_3 catalyst increased dramatically with potassium introduction. However, for the SiO_2 supported catalyst, the potassium addition hardly had an obvious effect on the relative amount of C1 oxygenate species, thus the total alcohol selectivity on K-Cu/SiO_2 catalyst remained almost unchanged. Due to very weak acidity of the SiO_2 supported catalyst, only CH_4 selectivity decreased sharply after potassium introduction, implying that potassium addition more preferentially inhibited CH_4 formation, compared with that of C_{2-5} hydrocarbons. Obviously, potassium addition could somehow modify the structural and chemical properties of the Cu/Al_2O_3 catalyst and enhance the amount of C1 oxygenate species, which narrowed the gap in performances of the two Al_2O_3 and SiO_2 supported catalysts.

4. Conclusions

In this work, the Al_2O_3 and SiO_2 supported Cu catalysts prepared by a facile impregnation method were used for higher alcohols synthesis from CO hydrogenation. Based on the remarkably different reaction behaviors over the Cu catalysts supported on Al_2O_3 and SiO_2, some systematical investigations were carried out to understand the main reasons. The Cu/SiO_2 catalyst possessed a higher amount of the C1 oxygenate species and showed higher total alcohols selectivity than the Cu/Al_2O_3 catalyst. Compared to the very weak acidity of the SiO_2 supported catalyst, the carbon chain growth probability of alcohol products occurred on the acid Cu/Al_2O_3 catalyst more easily.

Further, the Cu/Al_2O_3 and Cu/SiO_2 catalysts showed obvious differences in the structural and physicochemical properties, such as the types of copper species, the reduction behaviors, acidity, and electronic properties. As a result, the CO conversion, alcohol distribution, and CO_2 selectivity of the Cu/Al_2O_3 catalyst were different from that of the Cu/SiO_2 catalyst. Additionally, the performances of the Cu catalysts supported on Al_2O_3 and SiO_2 became more similar when the potassium was introduced. Wherein, the potassium was approved to modify the structural and chemical properties of the catalysts to some extent.

Author Contributions: The idea was conceived by Y.T. and X.L.; X.L. performed the experiments and drafted the paper under the supervision of Y.T. and J.Z.; M.Z. (Min Zhang), W.Z., M.Z. (Meng Zhang), H.X. and Y.W. helped to collect and analyse some characterization data. The manuscript was revised and checked through the comments of all authors. All authors have given approval for the final version of the manuscript.

Funding: This research was funded by the National Natural Science Foundation of China (No. 21573269), the Key Research and Development Program of Shanxi Province (No. MD2014-10).

Acknowledgments: The authors thank the financial support of National Natural Science Foundation of China (No. 21573269), the Key Research and Development Program of Shanxi Province (No. MD2014-10).

Conflicts of Interest: The authors declare no conflict of interest.

References

1. Yang, Y.; Qi, X.; Wang, X.; Lv, D.; Yu, F.; Zhong, L.; Wang, H.; Sun, Y.H. Deactivation study of CuCo catalyst for higher alcohol synthesis via syngas. *Catal. Today* **2016**, *270*, 101–107. [CrossRef]
2. Santos, V.P.; van der Linden, B.; Chojecki, A.; Budroni, G.; Corthals, S.; Shibata, H.; Meima, G.R.; Kapteijn, F.; Makkee, M.; Gascon, J. Mechanistic Insight into the Synthesis of Higher Alcohols from Syngas: The Role of K Promotion on MoS_2 Catalysts. *ACS Catal.* **2013**, *3*, 1634–1637. [CrossRef]
3. Goldemberg, J. Ethanol for a sustainable energy future. *Science* **2007**, *315*, 808–810. [CrossRef]
4. Luk, H.T.; Mondelli, C.; Ferré, D.C.; Stewart, J.A.; Pérez-Ramírez, J. Status and prospects in higher alcohols synthesis from syngas. *Chem. Soc. Rev.* **2017**, *46*, 1358–1426. [CrossRef]
5. Zhang, R.; Duan, T.; Wang, B.J.; Ling, L.X. Unraveling the role of support surface hydroxyls and its effect on the selectivity of C2 species over $Rh/\gamma\text{-}Al_2O_3$ catalyst in syngas conversion: A theoretical study. *Appl. Surf. Sci.* **2016**, *379*, 384–394. [CrossRef]
6. Choi, Y.; Liu, P. Mechanism of ethanol synthesis from syngas on Rh (111). *J. Am. Chem. Soc.* **2009**, *131*, 13054–13061. [CrossRef]
7. Gao, W.; Zhao, Y.F.; Liu, J.M.; Huang, Q.W.; He, S.; Li, C.M.; Zhao, J.W.; Wei, M. Catalytic conversion of syngas to mixed alcohols over CuFe-based catalysts derived from layered double hydroxides. *Catal. Sci. Technol.* **2013**, *3*, 1324–1332. [CrossRef]
8. Xiao, K.; Qi, X.Z.; Bao, Z.H.; Wang, X.X.; Zhong, L.S.; Fang, K.G.; Lin, M.G.; Sun, Y.H. CuFe, CuCo and CuNi nanoparticles as catalysts for higher alcohol synthesis from syngas: a comparative study. *Catal. Sci. Technol.* **2013**, *3*, 1591–1602. [CrossRef]
9. Zaman, S.F.; Pasupulety, N.; Al-Zahrani, A.A.; Daous, M.A.; Al-Shahrani, S.; Driss, H.; Petrov, L.A. Carbon monoxide hydrogenation on potassium promoted Mo_2N catalysts. *Appl. Catal. A Gen.* **2017**, *532*, 133–145. [CrossRef]
10. Wang, P.; Bai, Y.X.; Xiao, H.; Tian, S.P.; Zhang, Z.Z.; Wu, Y.Q.; Xie, H.J.; Yang, G.H.; Han, Y.Z.; Tan, Y.S. Effect of the dimensions of carbon nanotube channels on copper-cobalt-cerium catalysts for higher alcohols synthesis. *Catal. Commun.* **2016**, *75*, 92–97. [CrossRef]
11. Wang, P.; Zhang, J.F.; Bai, Y.X.; Xiao, H.; Tian, S.P.; Xie, H.J.; Yang, G.H.; Tsubaki, N.; Han, Y.Z.; Tan, Y.S. Ternary copper-cobalt-cerium catalyst for the production of ethanol and higher alcohols through CO hydrogenation. *Appl. Catal. A Gen.* **2016**, *514*, 14–23. [CrossRef]
12. Wang, P.; Chen, S.Y.; Bai, Y.X.; Gao, X.F.; Li, X.L.; Sun, K.; Xie, H.J.; Yang, G.H.; Han, Y.Z.; Tan, Y.S. Effect of the promoter and support on cobalt-based catalysts for higher alcohols synthesis through CO hydrogenation. *Fuel* **2017**, *195*, 69–81. [CrossRef]

13. Sun, J.; Cai, Q.X.; Wan, Y.; Wan, S.L.; Wang, L.; Lin, J.D.; Mei, D.H.; Wang, Y. Promotional effects of cesium promoter on higher alcohol synthesis from syngas over cesium-promoted $Cu/ZnO/Al_2O_3$ catalysts. *ACS Catal.* **2016**, *6*, 5771–5785. [CrossRef]
14. Liu, G.L.; Niu, T.; Cao, A.; Geng, Y.X.; Zhang, Y.; Liu, Y. The deactivation of Cu-Co alloy nanoparticles supported on ZrO_2 for higher alcohols synthesis from syngas. *Fuel* **2016**, *176*, 1–10. [CrossRef]
15. Gupta, M.; Smith, M.L.; Spivey, J.J. Heterogeneous catalytic conversion of dry syngas to ethanol and higher alcohols on Cu-based catalysts. *ACS Catal.* **2011**, *1*, 641–656. [CrossRef]
16. Li, X.L.; Zhang, Q.D.; Xie, H.J.; Gao, X.F.; Wu, Y.Q.; Yang, G.H.; Wang, P.; Tian, S.P.; Tan, Y.S. Facile preparation of Cu-Al oxide catalysts and their application in the direct synthesis of ethanol from syngas. *Chem. Select* **2017**, *2*, 10365–10370. [CrossRef]
17. Li, X.L.; Xie, H.J.; Gao, X.F.; Wu, Y.Q.; Wang, P.; Tian, S.P.; Zhang, T.; Tan, Y.S. Effects of calcination temperature on structure-activity of $K\text{-}ZrO_2/Cu/Al_2O_3$ catalysts for ethanol and isobutanol synthesis from CO hydrogenation. *Fuel* **2018**, *227*, 199–207. [CrossRef]
18. Sun, K.; Gao, X.F.; Bai, Y.X.; Tan, M.H.; Yang, G.H.; Tan, Y.S. Synergetic catalysis of bimetallic copper-cobalt nanosheets for direct synthesis of ethanol and higher alcohols from syngas. *Catal. Sci. Technol.* **2018**, *8*, 3936–3947. [CrossRef]
19. Liakakou, E.T.; Heracleous, E.; Triantafyllidis, K.S.; Lemonidou, A.A. K-promoted NiMo catalysts supported on activated carbon for the hydrogenation reaction of CO to higher alcohols: Effect of support and active metal. *Appl. Catal. B Environ.* **2015**, *165*, 296–305. [CrossRef]
20. Wang, J.J.; Chernavskii, P.A.; Khodakov, A.Y. Influence of the support and promotion on the structure and catalytic performance of copper-cobalt catalysts for carbon monoxide hydrogenation. *Fuel* **2013**, *103*, 1111–1122. [CrossRef]
21. Wang, J.J.; Chernavskii, P.A.; Khodakov, A.Y.; Wang, Y. Structure and catalytic performance of alumina-supported copper–cobalt catalysts for carbon monoxide hydrogenation. *J. Catal.* **2012**, *286*, 51–61. [CrossRef]
22. Ding, J.; Popa, T.; Tang, J.; Gasem, K.A.M.; Fan, M.; Zhong, Q. Highly selective and stable Cu/SiO_2 catalysts prepared with a green method for hydrogenation of diethyl oxalate into ethylene glycol. *Appl. Catal. B Environ.* **2017**, *209*, 530–542. [CrossRef]
23. Lee, J.H.; Reddy, K.H.; Jung, J.S.; Yang, E.; Moon, D.J. Role of support on higher alcohol synthesis from syngas. *Appl. Catal. A Gen.* **2014**, *480*, 128–133. [CrossRef]
24. Schumann, J.; Eichelbaum, M.; Lunkenbein, T.; Thomas, N.; Álvarez Galván, M.C.; Schlögl, R.; Behrens, M. Promoting strong metal support interaction: Doping ZnO for enhanced activity of Cu/ZnO:M (M = Al, Ga, Mg) catalysts. *ACS Catal.* **2015**, *5*, 3260–3270. [CrossRef]
25. Schneider, J.; Struve, M.; Trommler, U.; Schlüter, M.; Seidel, L.; Dietrich, S.; Rönsch, S. Performance of supported and unsupported Fe and Co catalysts for the direct synthesis of light alkenes from synthesis gas. *Fuel Process. Technol.* **2018**, *170*, 64–78. [CrossRef]
26. Kim, T.W.; Kleitz, F.; Jun, J.W.; Chae, H.J.; Kim, C.U. Catalytic conversion of syngas to higher alcohols over mesoporous perovskite catalysts. *J. Ind. And Eng. Chem.* **2017**, *51*, 196–205. [CrossRef]
27. Liu, Y.J.; Deng, X.; Han, P.D.; Huang, W. CO hydrogenation to higher alcohols over CuZnAl catalysts without promoters: Effect of pH value in catalyst preparation. *Fuel Process. Technol.* **2017**, *167*, 575–581. [CrossRef]
28. París, R.S.; Montes, V.; Boutonnet, M.; Järås, S. Higher alcohol synthesis over nickel-modified alkali-doped molybdenum sulfide catalysts prepared by conventional coprecipitation and coprecipitation in microemulsions. *Catal. Today* **2015**, *258*, 294–303. [CrossRef]
29. Song, Z.Y.; Shi, X.P.; Ning, H.Y.; Liu, G.L.; Zhong, H.X.; Liu, Y. Loading clusters composed of nanoparticles on ZrO_2 support via a perovskite-type oxide of $La_{0.95}Ce_{0.05}Co_{0.7}Cu_{0.3}O_3$ for ethanol synthesis from syngas and its structure variation with reaction time. *Appl. Surf. Sci.* **2017**, *405*, 1–12. [CrossRef]
30. Choi, S.M.; Kang, Y.J.; Kim, S.W. Effect of γ-alumina nanorods on CO hydrogenation to higher alcohols over lithium-promoted CuZn-based catalysts. *Appl. Catal. A Gen.* **2018**, *549*, 188–196. [CrossRef]
31. Jiang, Y.; Yang, H.Y.; Gao, P.; Li, X.P.; Zhang, J.M.; Liu, H.J.; Wang, H.; Wei, W.; Sun, Y.H. Slurry methanol synthesis from CO_2 hydrogenation over micro-spherical SiO_2 support Cu/ZnO catalysts. *J. CO_2 Util.* **2018**, *26*, 642–651. [CrossRef]

32. Su, J.J.; Zhang, Z.P.; Fu, D.L.; Liu, D.; Xu, X.C.; Shi, B.F.; Wang, X.; Si, R.; Jiang, Z.; Xu, J.; Han, Y.F. Higher alcohols synthesis from syngas over $CoCu/SiO_2$ catalysts: Dynamic structure and the role of Cu. *J. Catal.* **2016**, *336*, 94–106. [CrossRef]
33. Huang, X.M.; Ma, M.; Miao, S.; Zheng, Y.P.; Chen, M.S.; Shen, W.J. Hydrogenation of methyl acetate to ethanol over a highly stable Cu/SiO_2 catalyst: Reaction mechanism and structural evolution. *Appl. Catal. A Gen.* **2017**, *531*, 79–88. [CrossRef]
34. Ye, R.P.; Lin, L.; Yang, J.X.; Sun, M.L.; Li, F.; Li, B.; Yao, Y.G. A new low-cost and effective method for enhancing the catalytic performance of $Cu\text{-}SiO_2$ catalysts for the synthesis of ethylene glycol via the vapor-phase hydrogenation of dimethyl oxalate by coating the catalysts with dextrin. *J. Catal.* **2017**, *350*, 122–132. [CrossRef]
35. Li, X.L.; Yang, G.H.; Zhang, M.; Gao, X.F.; Xie, H.J.; Bai, Y.X.; Wu, Y.Q.; Pan, J.X.; Tan, Y.S. Insight into the correlation between Cu species evolution and ethanol selectivity in the direct ethanol synthesis from CO hydrogenation. *ChemCatChem* **2018**, *11*, 1123–1130. [CrossRef]
36. Cosultchi, A.; Pérez-Luna, M.; Morales-Serna, J.A.; Salmón, M. Characterization of modified Fischer-Tropsch catalysts promoted with alkaline metals for higher alcohol synthesis. *Catal. Lett.* **2012**, *142*, 368–377. [CrossRef]
37. Tan, L.; Yang, G.H.; Yoneyama, Y.; Kou, Y.L.; Tan, Y.S.; Vitidsant, T. Iso-butanol direct synthesis from syngas over the alkali metals modified Cr/ZnO catalysts. *Appl. Catal. A Gen.* **2015**, *505*, 141–149. [CrossRef]
38. Sun, J.; Wan, S.; Wang, F.; Lin, J.D.; Wang, Y. Selective synthesis of methanol and higher alcohols over $Cs/Cu/ZnO/Al_2O_3$ catalysts. *Ind. Eng. Chem. Res.* **2015**, *54*, 7841–7851. [CrossRef]
39. Anton, J.; Nebel, J.; Song, H.Q.; Froese, C.; Weide, P.; Ruland, H.; Muhler, M.; Kaluza, S. The effect of sodium on the structure-activity relationships of cobalt-modified $Cu/ZnO/Al_2O_3$ catalysts applied in the hydrogenation of carbon monoxide to higher alcohols. *J. Catal.* **2016**, *335*, 175–186. [CrossRef]
40. Tien-Thao, N.; Zahedi-Niaki, M.H.; Alamdari, H.; Kaliaguine, S. Effect of alkali additives over nanocrystalline Co-Cu-based perovskites as catalysts for higher-alcohol synthesis. *J. Catal.* **2007**, *245*, 348–357. [CrossRef]
41. Lopez, L.; Montes, V.; Kusar, H.; Cabrera, S.; Boutonnet, M.; Järås, S. Syngas conversion to ethanol over a mesoporous Cu/MCM-41 catalyst: Effect of K and Fe promoters. *Appl. Catal. A Gen.* **2016**, *526*, 77–83. [CrossRef]
42. Yang, Q. Synthesis of $\gamma\text{-}Al_2O_3$ nanowires through a boehmite precursor route. *Bull. Mater. Sci.* **2011**, *34*, 239. [CrossRef]
43. Kato, S.; Fujimaki, R.; Ogasawara, M.; Wakabayashi, T.; Nakahara, Y.; Nakata, S. Oxygen storage capacity of $CuMO_2$ (M = Al, Fe, Mn, Ga) with a delafossite-type structure. *Appl. Catal. B Environ.* **2009**, *89*, 183–188. [CrossRef]
44. Tada, S.H.; Watanabe, F.; Kiyota, K.; Shimoda, N.; Hayashi, R.; Takahashi, M.; Nariyuki, A.; Igarashi, A.; Satokawa, S. Ag addition to $CuO\text{-}ZrO_2$ catalysts promotes methanol synthesis via CO_2 hydrogenation. *J. Catal.* **2017**, *351*, 107–118. [CrossRef]
45. Li, M.M.J.; Zeng, Z.Y.; Liao, F.L.; Hong, X.L.; Tsang, S.C.E. Enhanced CO_2 hydrogenation to methanol over CuZn nanoalloy in Ga modified Cu/ZnO catalysts. *J. Catal.* **2016**, *343*, 157–167. [CrossRef]
46. Wang, B.; Cui, Y.Y.; Wen, C.; Chen, X.; Dong, Y.; Dai, W.L. Role of copper content and calcination temperature in the structural evolution and catalytic performance of Cu/P25 catalysts in the selective hydrogenation of dimethyl oxalate. *Appl. Catal. A Gen.* **2016**, *509*, 66–74. [CrossRef]
47. Soled, S. Silica-supported catalysts get a new breath of life. *Science* **2015**, *350*, 1171–1172. [CrossRef] [PubMed]
48. Liu, Y.T.; Ding, J.; Bi, J.C.; Sun, Y.P.; Zhang, J.; Liu, K.F.; Kong, F.H.; Xiao, H.C.; Chen, J.G. Effect of Cu-doping on the structure and performance of molybdenum carbide catalyst for low-temperature hydrogenation of dimethyl oxalate to ethanol. *Appl. Catal. A Gen.* **2017**, *529*, 143–155. [CrossRef]
49. Kwak, G.; Woo, M.H.; Kang, S.C.; Park, H.G.; Lee, Y.J.; Jun, K.W.; Ha, K.S. In situ monitoring during the transition of cobalt carbide to metal state and its application as Fischer-Tropsch catalyst in slurry phase. *J. Catal.* **2013**, *307*, 27–36. [CrossRef]
50. Claure, M.T.; Chai, S.H.; Dai, S.; Unocic, K.A.; Alamgir, F.M.; Agrawal, P.K.; Jones, C.W. Tuning of higher alcohol selectivity and productivity in CO hydrogenation reactions over K/MoS_2 domains supported on mesoporous activated carbon and mixed MgAl oxide. *J. Catal.* **2015**, *324*, 88–97. [CrossRef]

51. Tian, S.P.; Wang, S.C.; Wu, Y.Q.; Gao, J.W.; Wang, P.; Xie, H.J.; Yang, G.H.; Han, Y.Z.; Tan, Y.S. The role of potassium promoter in isobutanol synthesis over Zn-Cr based catalysts. *Catal. Sci. Technol.* **2016**, *6*, 4105–4115. [CrossRef]
52. Zhang, R.; Sun, X.; Wang, B. Insight into the Preference Mechanism of CHx (x = 1-3) and C-C Chain Formation Involved in C2 Oxygenate Formation from Syngas on the Cu (110) Surface. *J. Phys. Chem. C* **2013**, *117*, 6594–6606. [CrossRef]
53. Zhang, R.; Wang, G.; Wang, B. Insights into the mechanism of ethanol formation from syngas on Cu and an expanded prediction of improved Cu-based catalyst. *J. Catal.* **2013**, *305*, 238–255. [CrossRef]
54. Zhang, J.F.; Zhang, M.; Wang, X.X.; Zhang, Q.D.; Song, F.E.; Tan, Y.S.; Han, Y.Z. Direct synthesis of isobutyraldehyde from methanol and ethanol on Cu-Mg/Ti-SBA-15 catalysts: the role of Ti. *New J. Chem.* **2017**, *41*, 9639–9648. [CrossRef]
55. Zhang, J.F.; Wu, Y.Q.; Li, L.; Wang, X.X.; Zhang, Q.D.; Zhang, T.; Tan, Y.S.; Han, Y.Z. Ti-SBA-15 supported Cu-MgO catalyst for synthesis of isobutyraldehyde from methanol and ethanol. *RSC Adv.* **2016**, *6*, 85940–85950. [CrossRef]
56. Wu, Y.Q.; Xie, H.J.; Tian, S.P.; Tsubaki, N.; Han, Y.Z.; Tan, Y.S. Isobutanol synthesis from syngas over K-Cu/ZrO_2-$La_2O_{3(x)}$ catalysts: effect of La-loading. *J. Mol. Catal. A Chem.* **2015**, *396*, 254–260. [CrossRef]
57. Heracleous, E.; Liakakou, E.T.; Lappas, A.A.; Lemonidou, A.A. Investigation of K-promoted Cu-Zn-Al, Cu-X-Al and Cu-Zn-X (X= Cr, Mn) catalysts for carbon monoxide hydrogenation to higher alcohols. *Appl. Catal. A Gen.* **2013**, *455*, 145–154. [CrossRef]

catalysts

MDPI

Article

Study on Nanofibrous Catalysts Prepared by Electrospinning for Methane Partial Oxidation

Yuyao Ma, Yuxia Ma, Min Liu, Yang Chen, Xun Hu, Zhengmao Ye and Dehua Dong *

School of Materials Science and Engineering, University of Jinan, Jinan 250022, China; snsdyoona530@163.com (Y.M.); jndx_yxma@163.com (Y.M.); lm970222@163.com (M.L.); cy0204Yyqx@163.com (Y.C.); Xun.Hu@outlook.com (X.H.); mse_yezm@ujn.edu.cn (Z.Y.)

* Correspondence: mse_dongdh@ujn.edu.cn; Tel.: +86-531-8973-6011

Received: 4 April 2019; Accepted: 17 May 2019; Published: 23 May 2019

Abstract: Electrospinning is a simple and efficient technique for fabricating fibrous catalysts. The effects of preparation parameters on catalyst performance were investigated on fibrous Ni/Al_2O_3 catalysts. The catalyst prepared with H_2O/C_2H_5OH solvent showed higher catalytic activity than that with DMF/C_2H_5OH solvent because of the presence of NiO in the catalyst prepared with DMF/C_2H_5OH solvent. The metal ion content of the precursor also influences catalyst properties. In this work, the Ni/Al_2O_3 catalyst prepared with a solution containing the metal ion content of 30 wt % demonstrated the highest Ni dispersion and therefore the highest catalytic performance. Additionally, the Ni dispersion decreased as calcination temperature was enhanced from 700 to 900 °C due to the increased Ni particle sizes, which also caused a high reduction temperature and low catalytic activity in methane partial oxidation. Finally, the fibrous Ni/Al_2O_3 catalysts can achieve high syngas yields at high reaction temperatures and high gas flow rates.

Keywords: electrospinning; fibrous catalysts; metal ion content; calcination temperature; methane partial oxidation

---

1. Introduction

Electrospinning has been developed to fabricate one-dimensional materials with controllable fiber diameters, morphologies and compositions. Electrospun nanofibers have special features, such as hierarchically porous structure and high surface area, which have been successfully applied in various fields such as nanocatalysts, filtration, biomedical, optical electronics and electrodes for energy conversion or storage devices [1–5].

Ni/Al_2O_3 fibrous catalysts prepared by electrospinning have applied in catalytic methane reforming [1,6,7]. Ni nanoparticles can be in situ formed on the nanofiber surface via reducing catalyst precursor $NiAl_2O_4$. The fibrous structure of the catalysts is stable up to 1000 °C [6]. Moreover, the fibrous structure has a large void fraction (about 95%), which enables operation at high gas hourly space velocities through catalyst bed. It matches with the fast reaction of methane partial oxidation, which can be completed within a contact time of sub-milliseconds [8,9]. Therefore, the fibrous catalysts can produce high syngas yields [1].

Catalyst precursor solution greatly affects the electrospinning process via viscosity and evaporation [2]. To the best of our knowledge, the effect of preparation parameters on electrospun catalyst has not been reported previously. This study has investigated the effects of solvent, metal ion content and calcination temperature on catalyst properties, including crystallinity, particle size, microstructure, reducibility and catalytic performance. The preparation parameters were optimized to achieve high performance of methane partial oxidation (POM). The effects of reaction parameters on catalyst properties was also studied to utilize the advantages of fibrous catalysts.

2. Experimental

2.1. Catalyst Preparation

The fibrous Ni/Al_2O_3 catalysts were prepared by electrospinning, and the electrospinning process was started with preparing a spinning solution. A certain amount of polyvinyl pyrrolidone (PVP, molecular weight = 1.3×10^6, Shanghai Dibo Chemical Technology Co., Ltd., Shanghai, China) was dissolved in 2.0 g C_2H_5OH (≥99.7 wt %, Sinopharm Chemical Reagent Co., Ltd., Shanghai, China) and 8.0 g H_2O to prepare a PVP solvent. $Al(NO_3)_3{\cdot}9H_2O$ (≥99.0 wt %, Sinopharm Chemical Reagent Co., Ltd., Shanghai, China) and $Ni(NO_3)_2{\cdot}6H_2O$ (≥99.0 wt %, Sinopharm Chemical Reagent Co., Ltd., Shanghai, China) were dissolved in the solvent to form the catalyst with the Ni content of 30 wt % in Ni/Al_2O_3, and the Ni content was same for all catalysts. The electrospinning solutions with different metal ion contents are denoted CX (X = 10, 20, 30 and 40), where X represents the metal ion content, defined as solute (nitrate) weight percentage in precursor solution (solute + solvent). The ratio between solvent and solute in the C10, C20, C30 and C40 catalysts was 8.7:1, 4:1, 2.2:1 and 1.5:1, respectively.

Electrospinning was conducted on a device (Ucalery ET-2535H, Beijing, China) with a spinning distance of 30cm driven by a applied voltage of 19 kV. The feeding rate was maintained at 0.05 mm min^{-1}. The sample was calcined at 800 °C for 1 h in air.

The effect of solvents was compared by using the catalysts with 20 wt % metal ion content. Only H_2O in the solvent was changed to DMF (≥99.5 wt %, Sinopharm Chemical Reagent Co., Ltd., Shanghai, China), and the weight ratio to distilled water was 4:1. Other preparation parameters are the same. In addition, the catalyst prepared with a metal ion content of 30% was calcined at 700, 800 and 900 °C, separately, to study the effect of calcination temperature on catalyst properties.

2.2. Catalyst Characterisation

Scanning electron microscopy (SEM) images of catalyst microstructure were acquired with a FEI QUANTA FEG 250 microscope. Crystal sizes were measured using an X-ray diffractometer (XRD, Bruker D8 Advance) with Cu-Kα radiation (λ = 0.15408 nm). Temperature-programmed reduction (TPR) was conducted on a Micrometric ChemiSorb 2720 using a 10 mg of catalyst and a feeding gas of 10 vol % H_2 in Ar with a gas flow rate of 30 mL min^{-1}. The TPR tests were operated from room temperature to 1000 °C at a heating rate of 10 °C min^{-1}. CO-chemisorption was performed on a Micrometric ChemiSorb 2720 using a 30 mg of catalyst. First, the catalyst was reduced by the TPR process. Next, the catalyst was cooled to room temperature for pulse chemisorption using 5 vol % CO in He.

2.3. Catalytic Reaction

The calcined catalysts were crushed into sheets about 900 μm in size to ensure the similar density of catalyst beds. Catalytic evaluation was tested in a fixed bed quartz tube reactor (inner diameter = 6 mm) with a central K-type thermocouple. 10 mg of the catalyst was pre-reduced in situ by 10 vol % H_2 in Ar at 750 °C for 1 h. The reactant gas of CH_4, O_2 and Ar with a molar ratio of 2:1:17 was introduced into the reactor at 750 °C at a gas flow rate of 800 mL min^{-1}. Reaction products were sampled by a gas chromatography (GC, Shimadzu GC-2014).

3. Results and Discussions

3.1. Effect of Solvent

A solvent is used to dissolve catalyst precursors and polymer, forming electrospinning solution. During the electrospinning, the solvent needs to be evaporated before electrospun fibrous composites reach collectors so as to retain fibrous morphology. Solvent properties affect solution viscosity and solvent evaporation. Therefore, two common solvents, H_2O/C_2H_5OH and DMF/C_2H_5OH, are employed to investigate the solvent effect. Both catalysts had a metal ion content of 20 wt % and were calcined at 800 °C for 1 h.

3.1.1. XRD

Figure 1 shows the XRD patterns of the catalysts prepared with different solvents. The catalyst prepared with the H_2O/C_2H_5OH solvent has no NiO diffraction peaks while the catalyst prepared with the DMF/C_2H_5OH solvent shows NiO peaks, indicating Ni segregation during electrospinning. The segregation might be caused by the solubility difference of two metal ions in the two solvents. $Ni(NO_3)_2{\cdot}6H_2O$ has a lower solubility in the DMF/C_2H_5OH than the H_2O/C_2H_5OH, while $Al(NO_3)_3{\cdot}9H_2O$ has the similar solubility in the two solvents (Table 1). The lower solubility of $Ni(NO_3)_2{\cdot}6H_2O$ in the DMF/C_2H_5OH causes the segregation during drying electrospun fibrous composites. The phase segregation also resulted in the higher crystallinity of $NiAl_2O_4$ and Al_2O_3. After reduction, Ni presents in both catalysts while the catalyst prepared with the DMF/C_2H_5OH solvent shows the larger Ni crystal sizes due to NiO reduction.

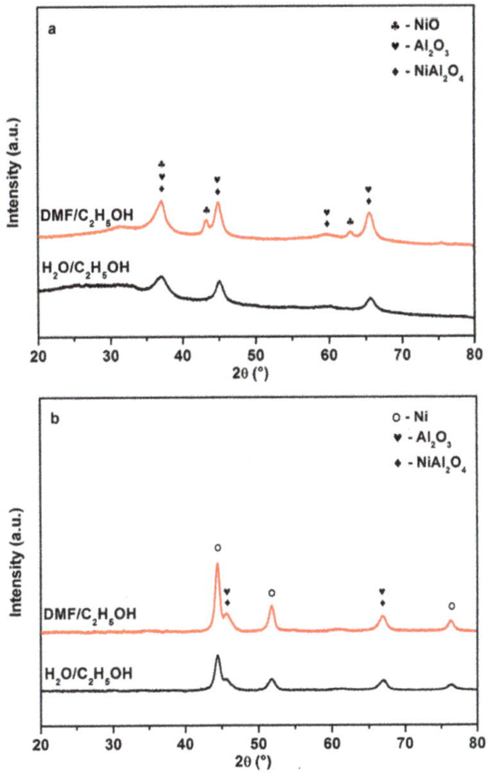

Figure 1. XRD patterns of the catalysts with different solvents: (**a**) Before reduction; (**b**) after reduction at 750 °C for 1 h.

Table 1. Solubility of nitrates in two solvents.

Solvent (8 g/2 g)	$Ni(NO_3)_2{\cdot}6H_2O$ (g)	$Al(NO_3)_3{\cdot}9H_2O$ (g)
DMF/C_2H_5OH	24.5	18.5
H_2O/C_2H_5OH	35.5	19.5

3.1.2. SEM

The morphologies of the reduced catalysts prepared with different solvents are presented in Figure 2. The fibrous structure has high void fraction and therefore achieves fast mass transfer [10].

Figure 2a shows the morphology of the catalyst prepared with the H_2O/C_2H_5OH solvent, uniform Ni particles anchored on the surface of fibrous support. As shown in Figure 2b, some large Ni particles appeared on the catalyst surface prepared with the DMF/C_2H_5OH solvent, which was attributed to NiO reduction.

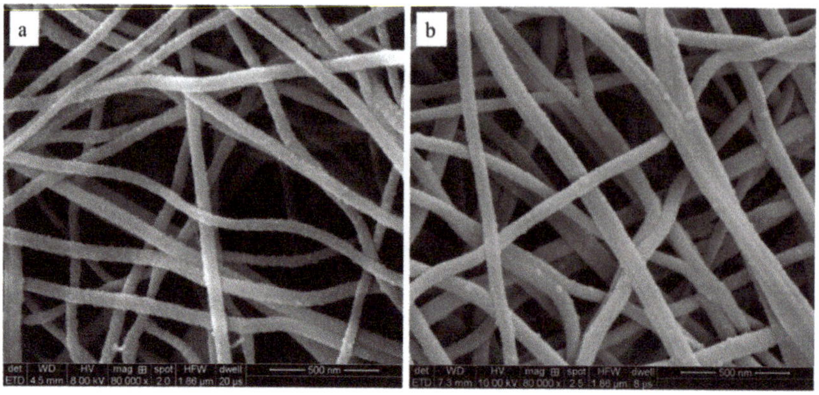

Figure 2. SEM images of the reduced catalysts prepared with different solvents: (**a**) H_2O/C_2H_5OH; (**b**) DMF/C_2H_5OH.

3.1.3. TPR and CO-Chemisorption

TPR was carried out on fibrous catalysts to investigate reducibility. Figure 3 shows that the TPR profiles of the catalysts consist of two reduction peaks centered at 500 and 800 °C, respectively. NiO reduction occurs at low temperatures (400–600 °C) while $NiAl_2O_4$ reduction takes place at temperatures above 600 °C [6,11]. For the catalyst prepared with the DMF/C_2H_5OH solvent, the NiO reduction peak is stronger than the catalyst prepared with the H_2O/C_2H_5OH solvent. The H_2 consumption peak areas and reducibilities are compared in Table 2. The reducibility of the catalyst prepared with the DMF/C_2H_5OH solvent is lower than that of the catalyst prepared with the H_2O/C_2H_5OH solvent. It might be attributed to Ni segregation because the Ni segregation causes NiO aggregation on the fiber surface, and the formed large NiO particles cause a decrease in reducibility. In addition, the Ni dispersion of the catalyst prepared with the DMF/C_2H_5OH solvent is smaller than that of the catalyst prepared with the H_2O/C_2H_5OH solvent (Table 2) because the large Ni particles formed by NiO reduction decrease the Ni dispersion.

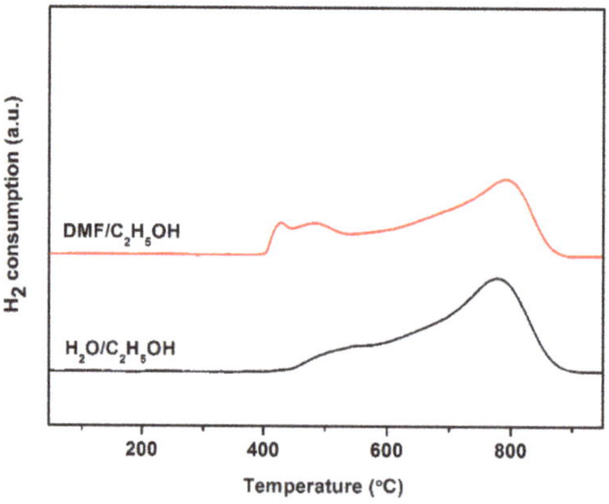

Figure 3. TPR profiles of the catalysts prepared with different solvents.

Table 2. Reducibility and Ni dispersion of Ni/Al_2O_3 catalysts.

Sample	Peak Area	Reducibility (%)	Ni Dispersion (%)
DMF/C_2H_5OH	48.3	76.6	0.05
H_2O/C_2H_5OH	51.6	83.4	0.27

3.1.4. Catalytic Performance

Figure 4 shows catalyst performance during methane partial oxidation at 750 °C and a gas flow rate of 800 mL·min^{-1}. The catalyst prepared with the DMF/C_2H_5OH solvent generated a low methane conversion of about 10%, which degraded rapidly. According to the TPR results of the spent catalysts in Figure 5, a substantial amount of Ni particles in the catalyst prepared with the DMF/C_2H_5OH solvent were oxidized into NiO during the POM. In contrast, there is no obvious NiO reduction peak in the catalyst prepared with the H_2O/C_2H_5OH solvent as the fresh catalyst. Our previous study shows the catalytic activity is mainly contributed by Ni-NiO_x particles formed from $NiAl_2O_4$ reduction rather than Ni particles formed from NiO reduction. Therefore, the catalyst prepared with the H_2O/C_2H_5OH solvent demonstrated a high and stable methane conversion of 30%, which is consistent with the results of Ni dispersion in Table 2.

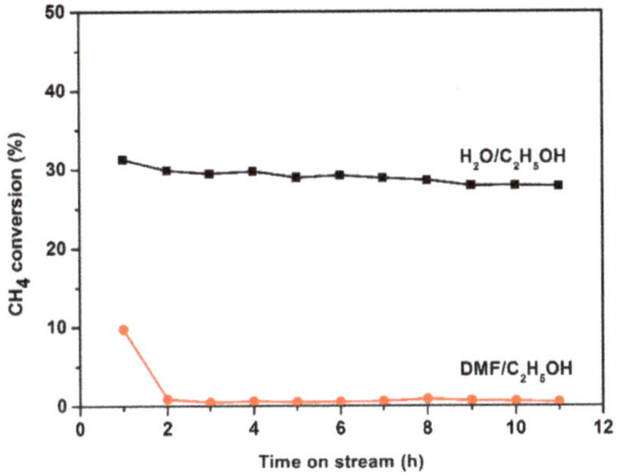

Figure 4. Methane conversion of the catalysts prepared with different solvents during the POM 10 h at 750 °C and a gas flow rate of 800 mL·min^{-1}.

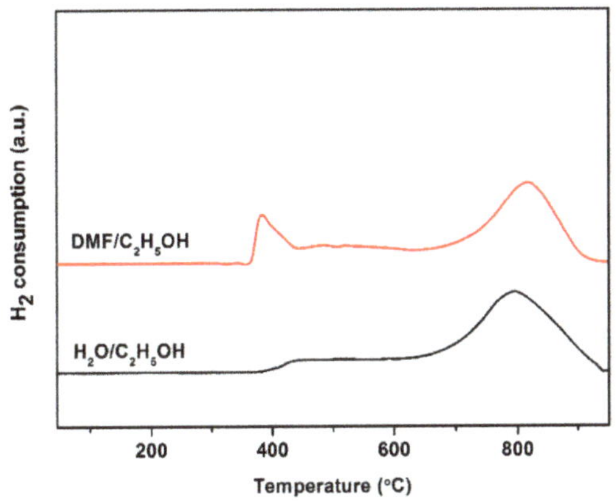

Figure 5. TPR profiles of the spent catalysts prepared with different solvents.

3.2. *Effect of Metal Ion Content*

During electrospinning, electrical filed force pulls solution drop from spinneret to form a jet. Metal ion content affects solution viscosity, electrical filed force and final ceramic fibers. Therefore, the effect of metal ion content was investigated. All catalysts were prepared with the H_2O/C_2H_5OH solvent and calcined at 800 °C for 1 h.

3.2.1. XRD

Figure 6 exhibits the XRD patterns of the catalyst prepared with different metal ion contents. As the metal ion content is increased, NiO phase started to present, indicating Ni segregation. It is because Ni prefers to accumulate on the surface in NiO-Al_2O_3 system [12,13]. As shown in Figure 6a, calculated using the Scherrer equation, the $NiAl_2O_4$ crystal sizes of the C10, C20, C30 and C40 catalysts are 9.7, 8.5, 8.0 and 7.5 nm, respectively. The $NiAl_2O_4$ crystal sizes decrease as the metal ion content is increased, which is because NiO presence dispersed $NiAl_2O_4$ phase, inhibiting $NiAl_2O_4$ growth. After reduction, Ni peaks are observed in addition to the $NiAl_2O_4$/Al_2O_3 peaks.

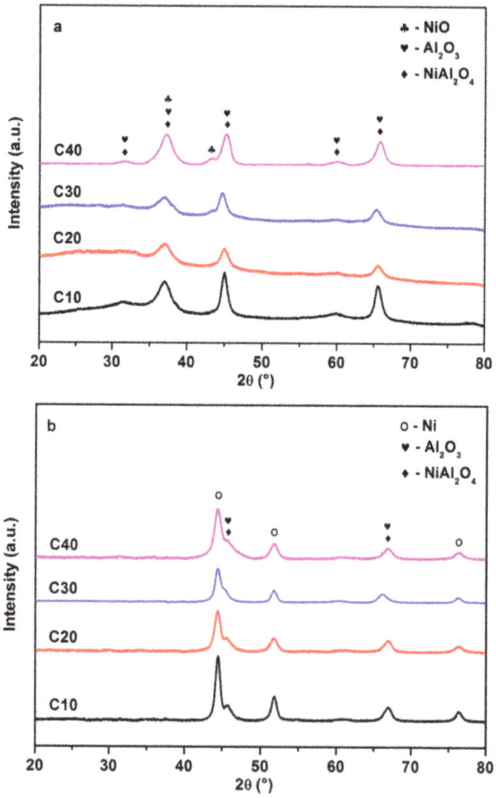

Figure 6. XRD patterns of the catalysts with different metal ion contents: (**a**) Before reduction; (**b**) after reduction at 750 °C for 1 h.

3.2.2. TPR and CO-Chemisorption

As shown in Figure 7, for the C10 catalyst, a single $NiAl_2O_4$ reduction peak was centred at 800 °C. When metal ion content was increased to 20 wt %, the reduction peak occurs around 400 °C, which is attributed to NiO reduction. The H_2 consumption peak areas and reducibilities are summarized in

Table 3. As the metal ion content is increased, the amount of reduced NiO increases, indicating more NiO segregated. Therefore, the reducibility of fibrous catalysts also increases. Furthermore, the Ni dispersion increases with metal ion content up to 30 wt %. However, the Ni dispersion of the C40 catalyst is reduced because of NiO aggregation [14]. Therefore, a certain amount of NiO can improve Ni dispersion.

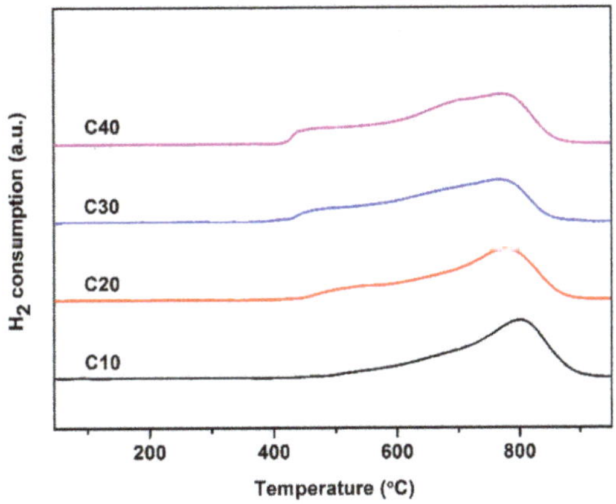

Figure 7. TPR profiles of the catalysts prepared with different metal ion contents.

Table 3. Reducibility and Ni dispersion of Ni/Al_2O_3 catalysts.

Sample	Peak Area	Reducibility (%)	Ni Dispersion (%)
C10	50.8	81.7	0.17
C20	51.6	83.4	0.27
C30	52.7	85.7	0.60
C40	57.4	95.4	0.39

3.2.3. SEM

The morphologies of reduced catalysts made from the solution prepared with different metal ion contents are shown in Figure 8. The fiber diameter increases with the increase of metal ion content from 50 to 300 nm due to the increased solution viscosity and metal ion loading. Ni particles present on fiber surface after reduction. The C10 catalyst shows the largest Ni particle sizes, and Ni particle sizes increase when metal ion content was increased from 20 to 40%. The change of Ni particle sizes are consistent with the change of crystal sizes in Figure 6a, and the large crystal sizes generate the big Ni particles.

3.2.4. Catalytic Performance

The fibrous catalyst prepared with different metal ion contents were tested for the POM at 750 °C and a gas flow rate of 800 mL min^{-1} to investigate catalytic activity. As shown in Figure 9, methane conversion improves with metal ion content up to 30 wt %, and further increasing metal ion content to 40 wt % causes a decline in CH_4 conversion. The catalytic performance is consistent with Ni dispersion in Table 3, and the higher dispersion contribute the higher catalytic performance. In addition, the Ni particles formed by NiO are easily oxidized during the POM, which has the limited contribution to catalytic performance [14].

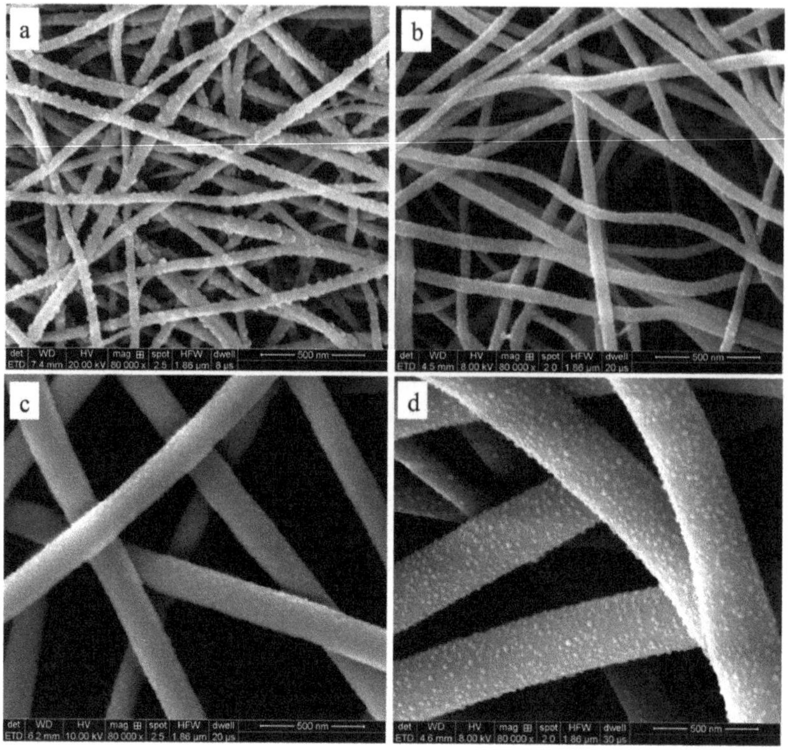

Figure 8. SEM images of the reduced catalysts made from the solution with different metal ion contents: (**a**) C10; (**b**) C20; (**c**) C30; (**d**) C40.

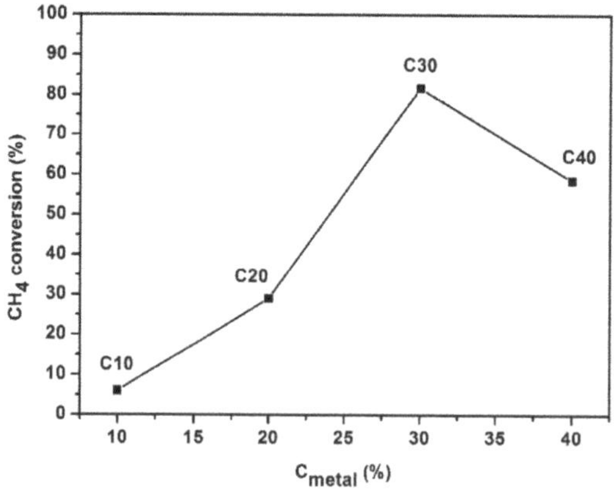

Figure 9. Methane conversion of the catalysts prepared with different metal ion contents during the POM for 10 h at 750 °C and a gas flow rate of 800 mL min^{-1}.

The morphologies of the catalysts after 10 h-POM test are shown in Figure 10, and the fibrous structure was stable during reactions. The Ni particles on the C10 and C20 catalyst surface disappeared after the reaction due to Ni oxidation. In contrast, the Ni particles retained for the C30 and C40 catalysts while carbon fibers could be found.

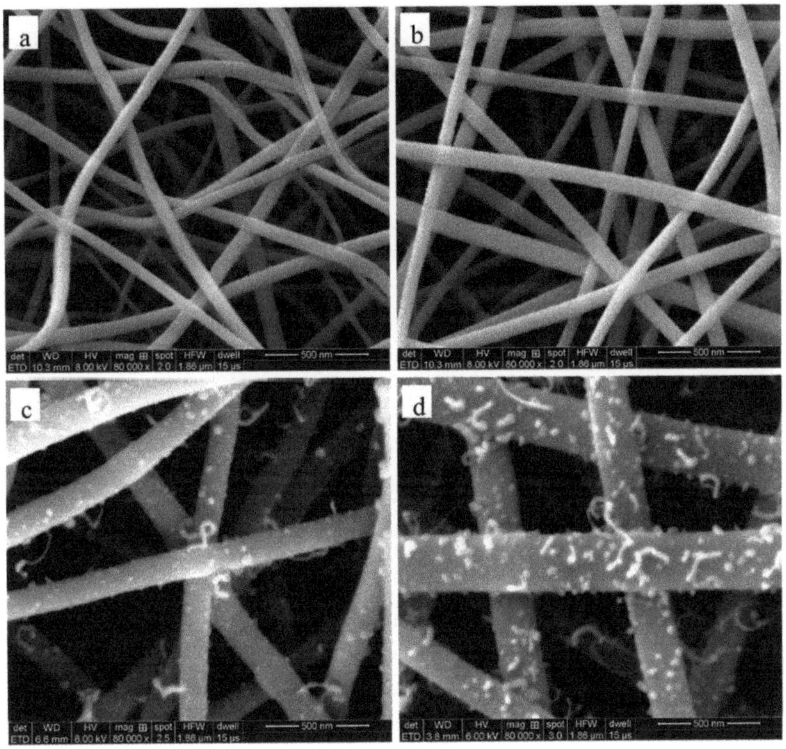

Figure 10. SEM images of the spent catalysts made from the solution with different metal ion contents: (**a**) C10; (**b**) C20; (**c**) C30; (**d**) C40.

3.3. *Effect of Calcination Temperature*

Ni catalysts are formed via reducing $NiAl_2O_4$, and catalytic activity is affected by catalyst crystallinity, which is determined by calcination temperature. To study the effect of calcination temperature on catalytic activity, the fibrous catalysts were calcined at temperatures ranging from 700 to 900 °C. All catalysts were prepared with a metal ion content of 30 wt % and the H_2O/C_2H_5OH solvent. As shown in Figure 11a, $NiAl_2O_4$ has a low crystallinity when calcined at 700 °C. With the increase of calcination temperature, the crystallinity is enhanced, resulting in the increase of crystal sizes. Accordingly, the reduced catalysts show the increased Ni crystal sizes with calcination temperature according to the diffraction intensity in Figure 11b. Calculated by the Scherrer equation, the Ni crystal sizes are 7.4, 8.1 and 9.2 nm at the calcination temperatures of 700, 800 and 900 °C, respectively.

TPR profiles are shown in Figure 12. According to the XRD results, $NiAl_2O_4$ crystallinity enhances with the increase of calcination temperature, resulting in the increase of reduction temperature. As shown in Figure 11a, Ni reducibility increases slightly due to the presence of NiO with the increase of calcination temperature. Additionally, the Ni dispersion of catalysts reduces with the increase of calcination temperature, which is attributed to the increase of Ni crystal size [6].

The POM was conducted at 750 °C and a gas flow rate of 800 mL min^{-1}, and the methane conversions are shown in Figure 13. According to the TPR results in Figure 12, the catalysts calcined at 700, 800 and 900 °C were reduced for 1 h at a reduction peak temperature of 615, 750 and 800 °C, respectively, which ensures that the catalysts were pre-reduced to the same extent. The catalyst calcined at the higher temperature showed the lower catalytic activity due to the decrease of Ni dispersion (Table 4).

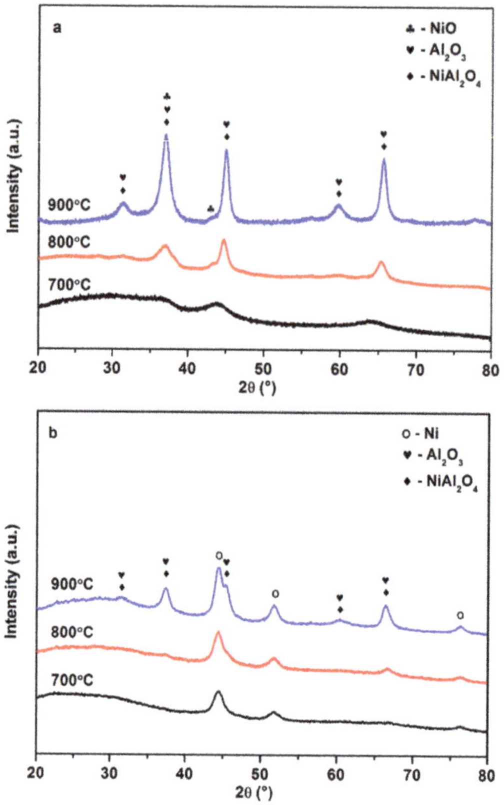

Figure 11. XRD patterns of the catalysts with different calcination temperatures: (**a**) Before reduction; (**b**) after reduction at 750 °C for 1 h.

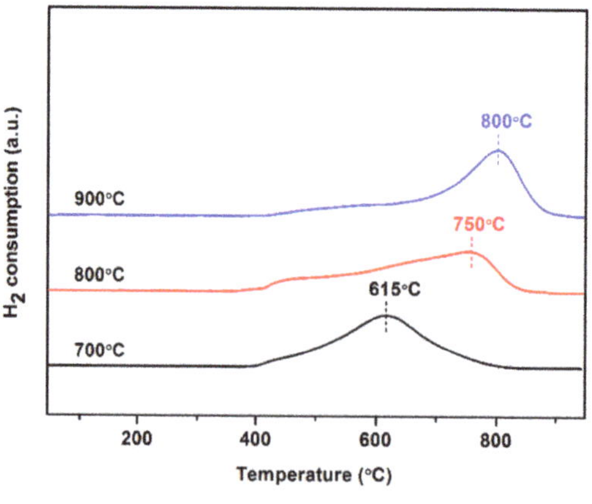

Figure 12. TPR profiles of the catalysts prepared at different calcination temperatures.

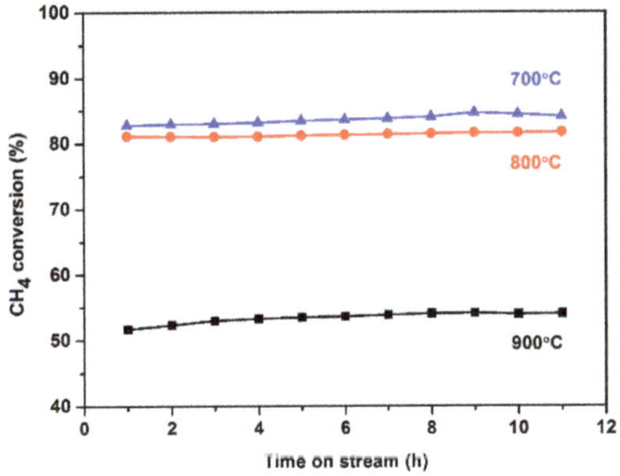

Figure 13. Methane conversion of the catalysts prepared at different calcination temperatures during the POM for 10 h at 750 °C and a gas flow rate of 800 mL min^{-1}.

Table 4. Reducibility and Ni dispersion of Ni/Al_2O_3 catalysts.

Temperature (°C)	Peak Area	Reducibility (%)	Ni Dispersion (%)
700	50.8	81.7	0.63
800	52.7	85.7	0.60
900	55.5	91.5	0.38

3.4. *Effect of Reaction Temperature and Gas Flow Rate*

The fibrous structure has high thermal stability and large void fraction, which makes it possible to operate at high temperatures and gas flow rates. The C30 catalyst was chosen to investigate the effect of reaction conditions on catalytic activity. Figure 14 shows that CH_4 conversion increases with reaction temperature at a gas flow rate of 1000 mL min^{-1}, and the catalytic reaction rate increases with gas flow rate at 750 °C. Under the reaction conditions of 800 °C and a flow rate of 1000 mL min^{-1}, the selectivity of H_2 and CO was 97% and 87%, respectively, and the yield was 9.8×10^5 L Kg^{-1} h^{-1} and 4.4×10^5 L Kg^{-1} h^{-1}, respectively. The H_2 and CO yields were calculated according to H_2 and CO amounts in the product gas. Therefore, the fibrous Ni/Al_2O_3 catalyst can generate high syngas yields owing to the fibrous structure.

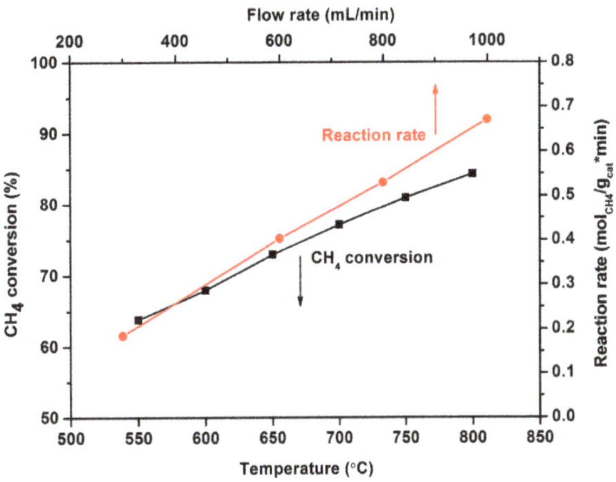

Figure 14. Catalytic performance of the C30 catalyst during the POM changing with operation temperature and gas flow rate.

4. Conclusions

In this study, the effects of preparation parameters on catalyst properties were investigated on electrospun fibrous Ni/Al_2O_3 catalysts. The catalyst prepared with the H_2O/C_2H_5OH solvent mainly consisted of $NiAl_2O_4$, while the catalyst prepared with the DMF/C_2H_5OH solvent formed NiO due to Ni segregation. The catalytic performance is mainly contributed by the Ni from $NiAl_2O_4$ reduction, and therefore the catalytic activity of the catalyst prepared with the H_2O/C_2H_5OH solvent was higher than that of the catalyst prepared with the DMF/C_2H_5OH solvent. The metal ion content affects catalyst composition, microstructure, reducibility and dispersion and therefore catalytic performance during the POM. The C30 catalyst had the highest catalytic performance. In addition, the higher calcination temperature produced the larger Ni particles due to the larger crystal size of $NiAl_2O_4$, which required a high reduction temperature. Therefore, the catalytic activity during the POM decreased with the increase of calcination temperature. Finally, it has been confirmed that the fibrous Ni/Al_2O_3 catalysts can achieve high syngas yields through the POM owing to the fibrous structure.

Author Contributions: Conceptualization, Y.M. (Yuyao Ma) and D.D.; methodology, Y.M. (Yuyao Ma), Y.M. (Yuxia Ma), M.L. and Y.C.; software, Y.M. (Yuyao Ma); validation, X.H., Z.Y. and D.D.; formal analysis, Y.M. (Yuyao Ma), Y.M. (Yuxia Ma) and D.D.; investigation, Y.M. (Yuyao Ma), Y.M. (Yuxia Ma), M.L. and Y.C.; resources, D.D.; data curation, Y.M. (Yuyao Ma), M.L. and Y.C.; writing—original draft preparation, Y.M. (Yuyao Ma); writing—review and editing, D.D.; visualization, X.H., Z.Y. and D.D.; supervision, X.H., Z.Y. and D.D.; project administration, D.D.; funding acquisition, D.D.

Funding: This research was funded by Natural Science Foundation of Shandong Province (ZR2017MEM022) and Shandong Province Key Research and Development Program (2018GGX102037).

Acknowledgments: D. H. Dong acknowledges the startup funding provided by the University of Jinan. The study is a part of the projects of Natural Science Foundation of Shandong Province (ZR2017MEM022) and Shandong Province Key Research and Development Program (2018GGX102037).

Conflicts of Interest: The authors declare no conflict of interest. The funder played a decisive role in the design of the study; in the collection, analyses or interpretation of data; in the writing of the manuscript and in the decision to publish the results.

References

1. Wang, Z.; Cheng, Y.; Shao, X.; Veder, J.P.; Hu, X.; Ma, Y.; Wang, J.; Xie, K.; Dong, D.; Jiang, S.P.; et al. Nanocatalysts anchored on nanofiber support for high syngas production via methane partial oxidation. *Appl. Catal. A* **2018**, *565*, 119–126. [CrossRef]
2. Dai, Y.; Liu, W.; Formo, E.; Sun, Y.; Xia, Y. Ceramic nanofibers fabricated by electrospinning and their applications in catalysis, environmental science, and energy technology. *Polym. Adv. Technol.* **2011**, *22*, 326–338. [CrossRef]
3. Lasprilla-Botero, J.; Álvarez-Láinez, M.; Lagaron, J.M. The influence of electrospinning parameters and solvent selection on the morphology and diameter of polyimide nanofibers. *Materi. Today Commun.* **2018**, *14*, 1–9. [CrossRef]
4. Bhardwaj, N.; Kundu, S.C. Electrospinning: A fascinating fiber fabrication technique. *Biotechnol. Adv.* **2010**, *28*, 325–347. [CrossRef] [PubMed]
5. Thavasi, V.; Singh, G.; Ramakrishna, S. Electrospun nanofibers in energy and environmental applications. *Energy Environ. Sci.* **2008**, *1*, 205–221. [CrossRef]
6. Wang, Z.; Hu, X.; Dong, D.; Parkinson, G.; Li, C.Z. Effects of calcination temperature of electrospun fibrous Ni/Al_2O_3 catalysts on the dry reforming of methane. *Fuel Process. Technol.* **2017**, *155*, 246–251. [CrossRef]
7. Liu, L.; Wang, S.; Guo, Y.; Wang, B.; Rukundo, P.; Wen, S.; Wang, Z.-J. Synthesis of a highly dispersed Ni/Al_2O_3 catalyst with enhanced catalytic performance for CO_2 reforming of methane by an electrospinning method. *Int. J. Hydrog. Energy* **2016**, *41*, 17361–17369. [CrossRef]
8. Goetsch, D.A.; Schmidt, L.D. Microsecond catalytic partial oxidation of alkanes. *Science* **1996**, *271*, 1560–1562. [CrossRef]
9. Hickman, D.A.; Schmidt, L.D. Production of syngas by direct catalytic-oxidation of methane. *Science* **1993**, *259*, 343–346. [CrossRef] [PubMed]

10. Reichelt, E.; Heddrich, M.P.; Jahn, M.; Michaelis, A. Fiber based structured materials for catalytic applications. *Appl. Catal. A* **2014**, *476*, 78–90. [CrossRef]
11. Sahli, N.; Petit, C.; Roger, A.C.; Kiennemann, A.; Libs, S.; Bettahar, M. Ni catalysts from $NiAl_2O_4$ spinel for CO_2 reforming of methane. *Catal. Today* **2006**, *113*, 187–193. [CrossRef]
12. Jiménez-González, C.; Boukha, Z.; De Rivas, B.; Delgado, J.J.; Cauqui, M.Á.; González-Velasco, J.R.; Gutiérrez-Ortiz, J.I.; López-Fonseca, R. Structural characterisation of Ni/alumina reforming catalysts activated at high temperatures. *Appl. Catal. A* **2013**, *466*, 9–20. [CrossRef]
13. Li, C.; Chen, Y.W. Temperature-programmed-reduction studies of nickel oxide/alumina catalysts: Effects of the preparation method. *Thermochim. Acta* **1995**, *256*, 457–465. [CrossRef]
14. Ma, Y.; Ma, Y.; Zhao, Z.; Hu, X.; Ye, Z.; Yao, J.; Buckley, C.E.; Dong, D. Comparison of fibrous catalysts and monolithic catalysts for catalytic methane partial oxidation. *Renew. Energy* **2019**, *138*, 1010–1017. [CrossRef]

catalysts

MDPI

Article

Insights over Titanium Modified $FeMgO_x$ Catalysts for Selective Catalytic Reduction of NO_x with NH_3: Influence of Precursors and Crystalline Structures

Liting Xu [1], Qilei Yang [2], Lihua Hu [1], Dong Wang [2,*], Yue Peng [2], Zheru Shao [1], Chunmei Lu [3] and Junhua Li [2]

1 Everbright Environmental Protection Technological Development (Nanjing) Limited, Nanjing 210000, China; xuliting@ebchinaintl.com.cn (L.X.); hulh@ebchinaintl.com.cn (L.H.); shaozr@ebchinaintl.com.cn (Z.S.)
2 State Key Joint Laboratory of Environment Simulation and Pollution Control, National Engineering Laboratory for Multi Flue Gas Pollution Control Technology and Equipment, School of Environment, Tsinghua University, Beijing 100084, China; yangqilei@tsinghua.edu.cn (Q.Y.); pengyue83@tsinghua.edu.cn (Y.P.); lijunhua@tsinghua.edu.cn (J.L.)
3 School of Energy and Power Engineering, Shandong University, Jinan 250061, China; cml@sdu.edu.cn
* Correspondence: sdu_wd@tsinghua.edu.cn; Tel.: +86-010-6278-2030

Received: 3 May 2019; Accepted: 17 June 2019; Published: 24 June 2019

Abstract: Titanium modified $FeMgO_x$ catalysts with different precursors were prepared by coprecipitation method with microwave thermal treatment. The iron precursor is a key factor affecting the surface active component. The catalyst using $FeSO_4$ and $Mg(NO_3)_2$ as precursors exhibited enhanced catalytic activity from 225 to 400 °C, with a maximum NO_x conversion of 100%. Iron oxides existed as γ-Fe_2O_3 in this catalyst. They exhibited highly enriched surface active oxygen and surface acidity, which were favorable for low-temperature selective catalytic reduction (SCR) reaction. Besides, it showed advantage in surface area, spherical particle distribution and pores connectivity. Amorphous iron-magnesium-titanium mixed oxides were the main phase of the catalysts using $Fe(NO_3)_3$ as a precursor. This catalyst exhibited a narrow T_{90} of 200/250–350 °C. Side reactions occurred after 300 °C producing NO_x, which reduced the NO_x conversion. The strong acid sites inhibited the side reactions, and thus improved the catalytic performance above 300 °C. The weak acid sites appeared below 200 °C, and had a great impact on the low-temperature catalytic performance. Nevertheless, amorphous iron-magnesium-titanium mixed oxides blocked the absorption and activation between NH_3 and the surface strong acid sites, which was strengthened on the γ-Fe_2O_3 surface.

Keywords: selective catalytic reduction (SCR); catalyst; precursor; NO_x conversion

1. Introduction

Nitrogen oxides (NO_x) mainly come from fossil fuel combustion [1–3] and have caused a series of environmental problems such as nitric acid rain, photochemical smog, ozone layer depletion and fine particle pollution [4–7]. As the severe NO_x emission situation and the rigorous emission legislation exhibit, many efforts have been made in NO_x reduction [8]. SCR (selective catalytic reduction) of NO_x with NH_3 [9–12] has been extensively proved to be the most efficient way for the removal of NO_x from stationary sources. A catalyst is critical to create an efficient SCR reaction and operating cost [13]. Commercial catalysts, such as V_2O_5/TiO_2, V_2O_5-WO_3/TiO_2 and V_2O_5-MoO_3/TiO_2 [8,14–18], were constrained for further development because of several inevitable drawbacks such as high cost, the bio-toxicity of the common vanadium compounds, etc. [19,20]. Furthermore, the commercial vanadium-titanium catalysts have been managed as hazardous waste. The study of novel high-efficiency catalysts is of great significance.

Many efforts have been made on iron oxides based catalysts, such as Fe-Ti [21–23], Fe/Ce-Ti [24,25], Fe-Ce-W [26,27], Fe-Sn-Mn [28], Ce-Fe/WMH [29], WO_3/Fe_2O_3 [30], Fe/WO_3-ZrO_2 [31], Mn-Fe/TiO_2 [32], FeMnTiO_x [33], etc. Iron oxides catalysts have low prices and free secondary pollution. They showed good catalytic performance and high N_2 selectivity for SCR reaction [8,34,35]. NO_x conversion of 60% was obtained over Fe_2O_3/TiO_2 prepared by Kato in 250–450 °C. In Fe/WO_3-ZrO_2 [31], Fe-Mn-Ce/γ-Al_2O_3 [36], and Fe-Er-V/TiO_2-WO_3-SiO_2 [37], the introduce of iron enlarged the surface area and pore volume, and meanwhile, it improved the Brönsted and Lewis acid sites. Zhu [38] studied Co-Fe/TiO_2 and Cu-Fe/TiO_2 and manifested that the reactants could be easily adsorbed on Co-Fe/TiO_2 because of its strong adsorption capacity, while Cu-Fe/TiO_2 showed better redox ability. The redox ability and surface acidity are the key factors that affect the catalytic performance. As far as we know, the type of precursors is the very first factor that could have great impact on the physicochemical properties of catalysts. It plays a decisive role on the surface active component, the surface area, the redox ability, the surface acidity, etc. Liu [39,40] et al prepared FeTiO_x via coprecipitation with $Fe(NO_3)_3$ and $Ti(SO_4)_2$ as precursors. The catalyst exhibited good NH_3-SCR activities and the NO_x conversion exceeded 90% in 200–350 °C. They found that there was strong interaction between iron and titanium. Ma [41] prepared $Fe_2(SO_4)_3/TiO_2$ by the impregnation method and found the NO_x conversion was up to 98% in 350–450 °C. Heterogeneous agglomeration of iron oxides could be weakened and the Brönsted acid sites could be improved by using $Fe_2(SO_4)_3$ as a precursor. In our previous work [42–44], we found that the magnesium-based catalyst showed good SCR activity and sulfur tolerance. Furthermore we studied on titanium modified FeMgO_x catalysts and found that titanium modified FeMgO_x catalysts exhibited excellent catalytic performance in SCR reaction. However, to the best of our knowledge, the catalytic performance could be further improvement via modifying the precursors.

In this work, a series of titanium modified FeMgO_x catalysts with different precursors were studied. The objective of this paper is to investigate the effect of precursor type on the physicochemical properties of titanium modified FeMgO_x catalysts and to reveal the optimization mechanism of catalytic performance.

2. Experimental

2.1. Catalyst Preparation

$FeSO_4 \cdot 7H_2O$, $FeCl_2 \cdot 7H_2O$ and $Fe(NO_3)_3 \cdot 9H_2O$ were used as iron precursors, $Mg(NO_3)_2 \cdot 6H_2O$ and $MgSO_4 \cdot 7H_2O$ (analytical pure, Tianjin Kermel Chemical Reagent Co., Ltd, Tianjin, China) were used as magnesium precursors and $NH_3 \cdot H_2O$ was used as precipitant in catalyst preparation. Titanium modified FeMgO_x catalysts were prepared via coprecipitation method with microwave thermal treatment. A certain amount of iron precursor, magnesium precursor and $TiSO_4$ (analytical pure, Sinopharm Group Co., Ltd, Shanghai, China) were dissolved in 250 mL deionized water and sufficiently stirred for 1 h. $NH_3 \cdot H_2O$ was titrated into the mixed solution with continuous stirring until the pH of the mixed solution was 9–10. Then the precipitate was filtered and washed by deionized water several times until neutral to remove the foreign ions. The precipitate was first impregnated by 1 mol/L Na_2CO_3 solution and then disposed of by microwave thermal treatment. The impregnated precipitate was washed by deionized water to be neutral and then dried at 105 °C. After calcined at 400 °C for 5 h, the obtained sample was crushed and sieved into 40–60 mesh (0.28 nm–0.45 nm) for the test. The catalysts prepared were denoted as Ti modified FeMgO_x with label SN, SS, CN, CS, NN and NS to represent different combination of precursors (S represents for sulfates, N represents for nitrates and C represents for chlorides), as seen in Table 1.

Table 1. Titanium modified $FeMgO_x$ catalysts with different precursors.

Label	Catalyst	Precursor
SN	Ti modified $FeMgO_x$	$FeSO_4{\cdot}7H_2O$, $Mg(NO_3)_2{\cdot}6H_2O$
SS		$FeSO_4{\cdot}7H_2O$, $MgSO_4{\cdot}7H_2O$
CN		$FeCl_2{\cdot}7H_2O$, $Mg(NO_3)_2{\cdot}6H_2O$
CS		$FeCl_2{\cdot}7H_2O$, $MgSO_4{\cdot}7H_2O$
NN		$Fe(NO_3)_3{\cdot}9H_2O$, $Mg(NO_3)_2{\cdot}6H_2O$
NS		$Fe(NO_3)_3{\cdot}9H_2O$, $MgSO_4{\cdot}7H_2O$

2.2. Activity Test

NH_3-SCR activity was completed in a quartz fixed-bed tube reactor at atmosphere pressure. The simulated flue gas was provided with standard gases, including 0.1 Vol % NO, 0.1 Vol % NH_3, 3.5 Vol % O_2 and balanced N_2. The total flow rate of simulated gas was 2 L/min and the catalyst used in each experiment was 4 mL, thus the corresponding gas hourly space velocity (GHSV) was 30,000/h^{-1}. The concentration of NO and NO_2 was monitored and analyzed by the MGA5 Flue Gas Analyzer (MRU Instruments, Inc. Emission Monitoring Systems, Neckarsulm-Obereisesheim, Germany). Before entering the flue gas analyzer, the flue gas should be washed by phosphoric acid (100% pure) to absorb ammonia and avoid the impact of ammonia on the analyzer. Data were recorded every 25 °C from 100 °C to 400 °C. NO_x conversion was calculated as follows:

$$\eta = \frac{C[NO_x(\text{inlet})] - C[NO_x(\text{outlet})]}{C[NO_x(\text{inlet})]} \times 100\% \tag{1}$$

where C[NO_x(inlet)] and C[NO_x(outlet)] meant the concentration of NO_x in the inlet and outlet of the reactor, μL/L. NO_x represented the sum of NO and NO_2.

2.3. Catalyst Characterization

A Rigaku D/max 2500 PC diffractometer (50 kV × 150 mA) with Cu Kα radiation was used to complete the X-ray Diffraction. The data of 2θ were collected from 10° to 90° by 4 °/min with the step size 0.1°.

N_2-adsorption-desorption was obtained by using an ASAP2020 Surface Area and Porosity Analyzer (Micromeritics Instrument Corp., Norcross, Georgia, USA) at −196 °C. The specific surface area and the average pore diameter were calculated by the Brunauer–Emmett–Teller (BET) method, and the specific pore volume and pore diameter distribution were calculated by Barrett–Joyner–Halenda (BJH) method.

Microstructure of the catalysts was conducted on a Japan JSM-6700F cold field emission scanning electron microscope. The elements on the surface of the catalysts were analyzed on an Oxford INCA X sight energy dispersive spectrometer (Be4-U92) with 5.9 KeV, UK.

To analyze the surface atomic concentration and distinguish the chemical states of the elements, a Thermo ESCALAB 250XI surface analyze system with Al Kα radiation (1486.6 eV, 150 W) was used to complete X-ray Photoelectron Spectroscopy. Prior to the measurement, each sample was degassed in vacuum to eliminate surface contamination.

Temperature-programmed Desorption of NH_3 (NH_3-TPD) was performed on a TP-5080 instrument using a 100 mg sample. The sample was pretreated in flowing He at 300 °C for 1 h before the measurement. Then, the sample was He-cooled to 100 °C, then treated with 5% NH_3/Ar at a flow rate of 30 mL/min for 0.5 h and flushed with He at 100 °C for 1 h. The desorption process was carried out by heating the sample from 100 °C to 700 °C at a rate of 10 °C/min.

3. Results and Discussion

3.1. Effect of Different Precursors on Catalytic Performance Over Titanium Modified $FeMgO_x$ Catalysts

The catalytic performance of titanium modified $FeMgO_x$ catalysts with different precursors is shown in Figure 1. It was obvious that temperature had a strong effect on titanium modified $FeMgO_x$ catalysts with different precursors at the temperature range from 100 to 400 °C. Generally speaking, it was necessary to reach a certain temperature for the catalyst to exhibit good catalytic activity, while an excessively high reaction temperature would lead to a decrease in catalytic activity due to the secondary reaction.

Titanium modified $FeMgO_x$ catalysts with different precursors revealed excellent catalytic activity from 100 to 400 °C and had wide temperature windows. Among all the catalysts, catalysts SN, SS, CN and CS showed similar catalytic performances, which was distinctly evident with catalysts NN and NS. When the temperature was blow 200 °C, catalyst NN and NS showed better catalytic activity than other catalysts, especially catalyst NN, whose NO_x conversion could exceed 50% and 90% when the reaction temperature was close to 150 °C and 200 °C, respectively. However, when the temperature exceeded 350 °C, the NO_x conversion of catalysts NN and NS apparently decreased to about 50–60% due to the oxidation and decomposition of NH_3. The NO_x conversion of catalysts SN, SS, CS and CN could be stable at 90% at high temperature range. Among all the catalysts, catalyst SN with precursors of $FeSO_4$ and $Mg(NO_3)_2$ exhibited excellent catalytic activity and N_2 selectivity in a wide temperature range, with NO_x conversion above 90% from 225 to 400 °C and N_2 selectivity above 90% in the whole temperature range. The NO_x conversion of which approached 100% from 250 to 375 °C.

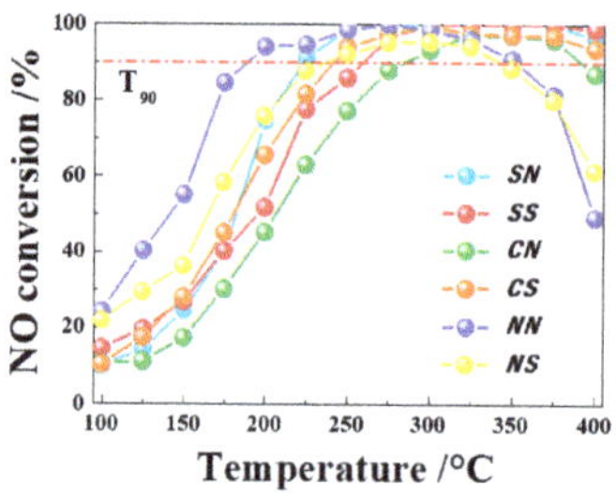

Figure 1. Catalytic performance of titanium modified $FeMgO_x$ catalysts with different precursors.

3.2. X-ray Diffraction (XRD) Patterns

Figure 2 shows the XRD patterns of titanium modified $FeMgO_x$ catalysts with different precursors. There were no diffraction peaks of magnesium or titanium in XRD patterns of all the catalysts according to JCPDF standard; it could be inferred that magnesium and titanium existed in a highly dispersed state or an amorphous state in the catalyst, or maybe that the crystallites formed were less than 5 nm. There were obvious sharp diffraction peaks of the catalysts SN, SS, CN and CS at $2\theta = 30.2°$, 35.5°, 43.2°, 53.7°, 53.7° and 62.8°, corresponding to maghemite (γ-Fe_2O_3) crystallite according to JCPDS PDF#39-1346 [39,40]. It could be inferred that maghemite crystallite was the main active component in these catalysts.

However, in catalyst NN and NS, the diffraction peaks were ascribe to amorphous oxides, comparing with that in Figure S1. It could be concluded that the active component was directly affected by the precursors. When $Fe(NO_3)_3$ was used as precursor, the active components of the titanium modified $FeMgO_x$ catalysts prepared were iron-magnesium-titanium mixed oxides; however, when $FeSO_4$ and $FeCl_2$ were used as precursors, the active component of the titanium modified $FeMgO_x$ catalysts prepared was γ-Fe_2O_3. It was reported that γ-Fe_2O_3 was an octahedral structure with vacancies and was in a metastable state with a lower activation energy, which resulted in better denitration activity. In general, different active components led to diversity between different catalysts.

Considering the result of the activity test, iron-magnesium-titanium mixed oxides and γ-Fe_2O_3 as active components were the substantial cause of different catalytic performance. The mixed oxides made the temperature window of the catalysts NN and NS obviously shift to a low temperature range and was the main cause of the secondary reaction at high temperature range.

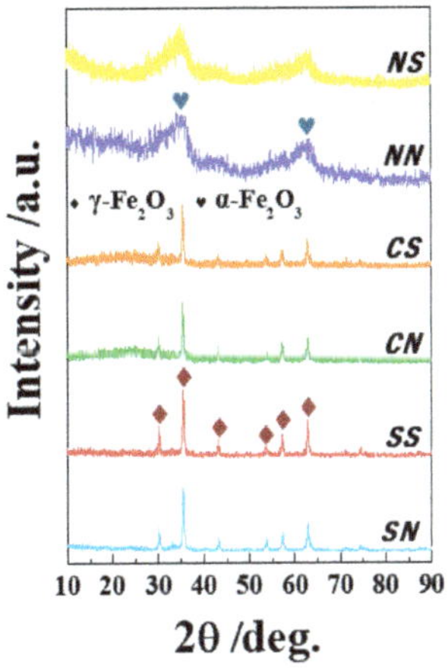

Figure 2. Powder X-ray diffraction (XRD) patterns of titanium modified $FeMgO_x$ catalysts with different precursors.

3.3. N_2-Adsorption-Desorption

BET surface area, average pore volume and average pore diameter of titanium modified $FeMgO_x$ catalysts with different precursors are enumerated in Table 2.

Table 2. Surface characterization of titanium modified $FeMgO_x$ catalysts with different precursors.

Catalysts	$S_{BET}/m^2 \cdot g^{-1}$	$V_{BJH}/cm^3 \cdot g^{-1}$	Average Pore Diameter/nm
SN	55.1245	0.2291	16.2670
SS	46.6441	0.1951	18.6213
CN	58.6066	0.2413	15.5461
CS	51.8948	0.2243	19.2990
NN	181.3934	0.2279	4.3710
NS	180.4130	0.2340	4.4512

The BET surface area, BJH pore volume and average pore diameter of the catalysts SN, SS, CN and CS were similar to each other. The BET surface area of catalyst SN, which exhibited the best catalytic activity, was 55.1245 m^2/g, the pore volume was 0.2291 cm^3/g and the average pore diameter was 16.2670 nm. Nevertheless, the BET surface area of the catalysts NN and NS using $Fe(NO_3)_3$ as a precursor were 181.3934 m^2/g and 180.4130 m^2/g, respectively, which were three times larger than the other four catalysts. The average pore diameter of the catalysts NN and NS were only 4.3710 nm and 4.4512 nm, which were almost four times smaller than the other four catalysts, but the pore volumes were close to each other. Generally speaking, more active sites could be provided by large surface area, pore volume and relatively small average pore diameter, which was beneficial for SCR reaction. For the catalysts NN and NS, large surface area could provide more active sites, but the small pore diameter would lead to the increment of diffusion resistance during the gas-solid reaction and would be bad for the adsorption-desorption process. Appropriate surface area, pore volume and average pore diameter in the catalyst SN could provide enough active sites and guarantee the diffusion and mass transfer processes, which were in favor of SCR reaction.

t-Plot microporous area and volume are enumerated in Table 3. It was obvious that the t-Plot microporous area was much smaller than t-Plot external surface area, and the microporous volume of all the catalysts was close to zero. It could be concluded that the t-Plot microporous area had less contribution to the surface area, and the t-Plot external surface area was much more important to produce large surface area. Meanwhile, mesopore (pore diameter ranging from 2 to 50 nm) was the main pore type in titanium modified $FeMgO_x$ catalysts with different precursors, and there were substantially fewer micropores (pore diameter less than 2 nm).

Table 3. t-Plot properties of titanium modified $FeMgO_x$ catalysts with different precursors.

Samples	t-Plot Microporous Area/$m^2 \cdot g^{-1}$	t-Plot External Surface Area/$m^2 \cdot g^{-1}$	t-Plot Microporous Volume/$cm^3 \cdot g^{-1}$
SN	0.5467	54.5777	6.5000×10^{-5}
SS	5.1114	41.5327	2.5290×10^{-3}
CN	4.3249	54.2817	2.0670×10^{-3}
CS	6.5865	45.3083	3.3030×10^{-3}
NN	2.7968	178.5966	3.5550×10^{-4}
NS	3.1251	177.2879	2.54×10^{-3}

Distribution characterization of pore structures over titanium modified $FeMgO_x$ catalysts with different precursors is shown in Figure 3. In Figure 3a,b, pore diameter of all the catalysts is distributed mainly from 2 nm to 10 nm, which also means that mesopores were the major pore type in the catalysts. In addition, the pore diameter distribution of the catalysts NN and NS is obviously distinguished from the other catalysts, and the intensities of the distribution peaks are a great deal stronger than other catalysts, indicating that the pores ranging from 2 to 10 nm made a greater contribution to surface area and pore volume.

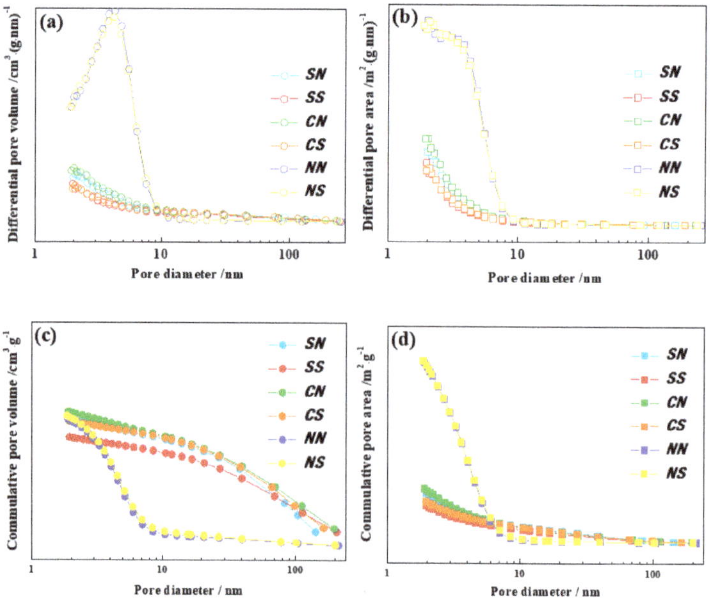

Figure 3. Distribution characterization of pore structures over titanium modified $FeMgO_x$ catalysts with different precursors: (**a**) Pore volume, (**b**) Pore area, (**c**) Cumulate pore volume and (**d**) Cumulate pore area.

In Figure 3c,d, it can be seen that the cumulative pore area and the cumulative pore volume of the catalysts SN, SS, CN and CS gently declined with pore diameter. However, the cumulative pore area and the cumulative pore volume of the catalysts NN and NS from 2 to 10 nm declined rapidly with pore diameter, and the cumulative pore area and the cumulative pore volume above 10 nm were pretty small compared with that from 2 to 10 nm. The intensive distribution of pore diameter from 2 to 10 nm could provide a large surface area, but the narrow distribution of pore diameter would lead to the hysteresis of the diffusion and mass transfer processes. For the catalyst SN, the reasonable distribution of pore diameter could guarantee enough surface area and also the diffusion and mass transfer processes.

The N_2-adsorption-desorption isotherms of titanium modified $FeMgO_x$ catalysts are shown in Figure 4. According to the International Union of Pure and Applied Chemistry classification, the N_2-adsorption-desorption isotherms of the catalysts SN, SS, CN and CS were classified as V-shaped isotherms with an H3 hysteresis loop. The absorbed volume was really small when the pressure was low. Only when the pressure was approaching the saturated vapor pressure did the absorbed volume increase rapidly due to capillary condensation. The absorption characteristic was often observed by the weak solid-gas interaction in mesopores on the surface of catalysts. It could be concluded from the type of isotherm and hysteresis loop that disorderly wedge-shaped mesopores were formed by particles accumulated loosely on the surface of the catalysts [44].

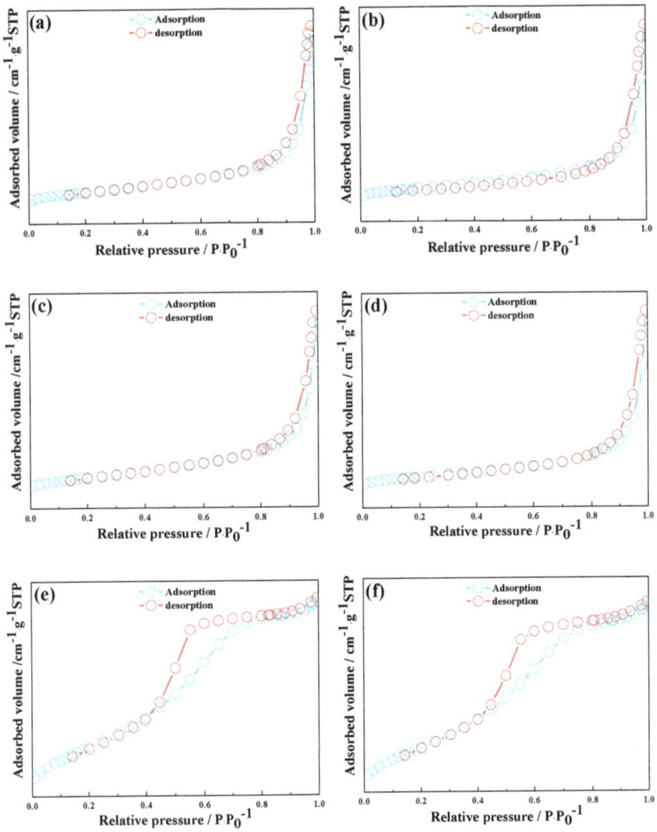

Figure 4. N_2-adsorption-desorption isotherms of titanium modified $FeMgO_x$ catalysts with different precursors: (**a**) SN, (**b**) SS, (**c**) CN, (**d**) CS, (**e**) NN and (**f**) NS.

However, the N_2-adsorption-desorption isotherms of the catalysts NN and NS were classified as IV-shaped isotherms with an H4 hysteresis loop. The absorbed volume gradually increased when the pressure was low. Only when the relative pressure was about 0.5 to 0.8 did the absorbed volume increase rapidly, and then the absorbing capacity became almost invariable, which meant there were mainly mesopores in the catalysts and less macropores were obtained. It could be concluded from the type of isotherm and hysteresis loop that wedge-shaped mesopores were formed by lamellar structures accumulated tightly on the surface of the catalysts.

3.4. SEM and Energy Dispersive Spectrometer (EDS)

The SEM images of titanium modified $FeMgO_x$ catalysts are shown in Figure 5. The surface of the catalysts SN, SS, CN and CS presented spherical particle distribution and the particles were significantly more independent and regular. There was hardly any accumulation of particles occurring on the surface of these catalysts, and the regular distribution of particles benefitted the formation of intergranular pores, which were good for the mass transfer process.

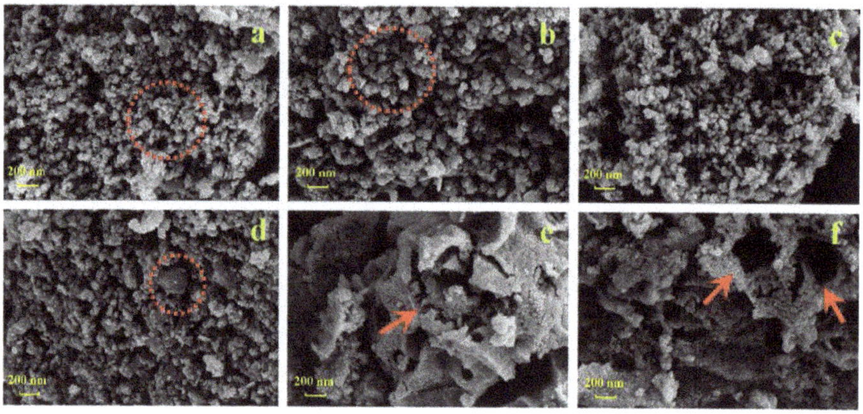

Figure 5. SEM images of titanium modified $FeMgO_x$ catalysts with different precursors: (**a**) SN, (**b**) SS, (**c**) CN, (**d**) CS, (**e**) NN and (**f**) NS.

However, the surface of the catalysts NN and NS exhibited stratiform and nubby distribution. There were a huge number of pyknotic fine intergranular pores on the stratiform and nubby structure. Fine intergranular pores were favorable for large surface area and provided more active sites, but they would resist the diffusion and mass transfer process. Meanwhile the results implied that the choice of an iron precursor had significant influence on the surface morphology of the catalyst. The catalyst SN exhibited regular and distributed spherical particles with good pores connectivity which were favor of SCR reaction.

The EDS composition analysis of titanium modified $FeMgO_x$ catalysts with different precursors is shown in Table 4.

Table 4. EDS composition analysis data of titanium modified $FeMgO_x$ catalysts with different precursors.

Catalyst	Percentage by Weight/wt %				Percentage by Atomicity/at %			
	Fe	Mg	Ti	O	Fe	Mg	Ti	O
SN	59.06	1.68	4.79	34.47	30.82	2.85	3.21	63.12
SS	59.25	3.10	4.73	32.92	31.72	3.81	2.95	61.52
CN	60.66	3.64	4.60	31.10	33.16	4.57	2.93	59.34
CS	59.33	3.18	4.40	33.09	31.68	3.90	2.74	61.68
NN	68.39	2.90	5.30	23.41	41.97	4.09	3.79	50.16
NS	67.93	3.47	4.88	23.72	41.53	4.45	3.17	50.85

The catalysts prepared all consisted of iron, magnesium, titanium and oxygen, as expected. From Table 4, it can be seen that the percentage of oxygen by atomicity of the catalysts NN and NS was obviously lower than that of the catalysts SN, SS, CN and CS, which means that it was difficult for oxygen atoms to enrich on the surface of the catalysts NN and NS. The percentage of oxygen by atomicity of the catalyst SN could reach 63.12%. In general, the surface oxidation ability could be improved by the abundant lattice oxygen on the surface of the catalyst. Strong oxidation ability could promote NO oxidation to NO_2 and then induce the rapid SCR reaction, which effectively guarantees the catalytic performance of the catalyst. The results illustrated that using $Fe(NO_3)_3$ as a precursor was not conductive to the enrichment of oxygen on the surface of the catalyst.

3.5. X-ray Photoelectron Spectroscopy (XPS)

An XPS test was used to better elucidate the spices, concentration and valency of different elements on the surface of titanium modified $FeMgO_x$ catalysts with different precursors, and the element surface concentration calculated is shown as Table 5. It can be seen that the main elements of titanium modified $FeMgO_x$ catalysts are iron, magnesium, titanium and oxygen, which is in agreement with EDS results.

Table 5. X-ray photoelectron spectroscopy (XPS) elementary surface concentration of titanium modified $FeMgO_x$ catalysts with different precursors.

Catalyst	Fe 2p/%	Mg 1s/%	O 1s/%	Ti 2p/%
SN	38.88	6.92	46.85	7.35
SS	46.35	7.89	37.20	8.57
CN	45.75	9.54	37.75	6.95
CS	47.49	7.47	37.48	7.55
NN	56.74	6.43	32.92	3.91
NS	56.65	6.20	32.86	4.29

The concentration of O 1s over the catalyst SN could reach 46.85%, which illustrates that using $FeSO_4$ and $Mg(NO_3)_2$ as precursors was in favor of the enrichment of oxygen on the surface of the catalyst and enhanced the surface oxidation ability.

Mg 1s and Ti 2p spectra of titanium modified $FeMgO_x$ catalysts with different precursors are shown in Figure 6. The Mg 1s peaks at a binding energy of about 1303 eV in, and coincides exactly with Mg^{2+}. For the catalysts SN, SS, CN and CS, the Ti $2p_{3/2}$ peaks at 458.23 eV and the Ti $2p_{1/2}$ peaks at 464.03 eV were attributed to Ti^{4+}. The binding energy of Ti 2p shifted higher compared with those of the catalysts NN and NS. The Ti $2p_{3/2}$ peaks appeared at 458.23 eV and the Ti $2p_{1/2}$ peaks appeared at 464.03 eV.

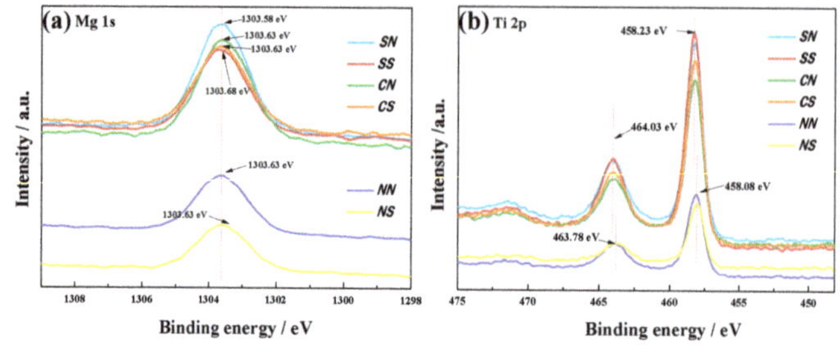

Figure 6. Mg 1s and Ti 2p spectra over titanium modified $FeMgO_x$ catalysts with different precursors.

The Fe 2p spectra of titanium modified $FeMgO_x$ catalysts with different precursors are shown in Figure 7. The Fe 2p spectra consisted of three overlapping peaks, and the binding energy of the Fe species were further analyzed by peak-fitting. The Fe $2p_{3/2}$ peaks at binding energy around 710.0 eV and 712.0 eV, the $Fe2p_{3/2,sat}$ peaks at binding energy about 718.8 eV and the Fe $2p_{1/2}$ peaks at binding energy of 724.2 eV were all ascribed to Fe^{3+} [21].

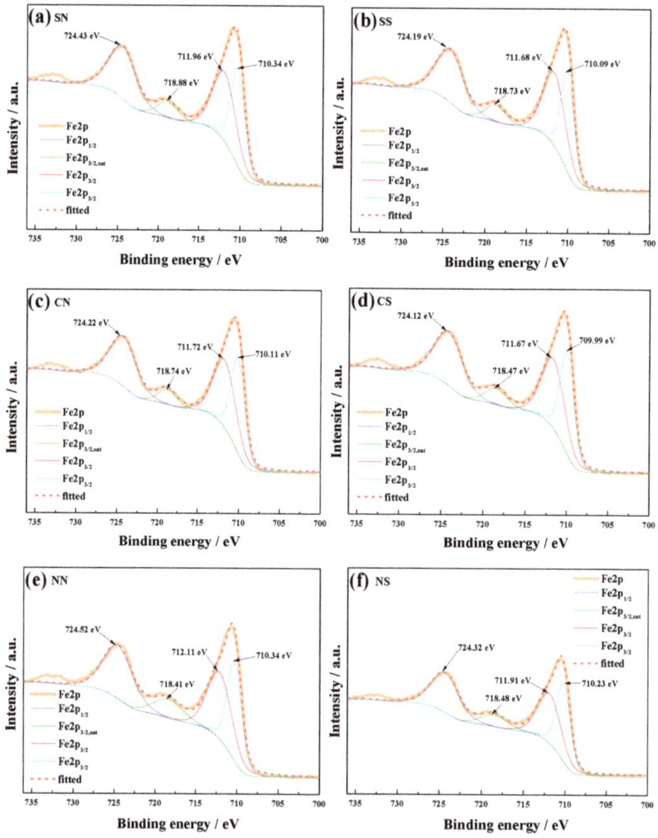

Figure 7. Fe 2p spectra of titanium modified $FeMgO_x$ catalysts with different precursors.

The O 1s spectra of titanium modified $FeMgO_x$ catalysts with different precursors are shown on Figure 8. The O 1s peaks at the binding energy of about 529.8 eV and 531.5 eV were the lattice oxygen (denoted as O_β) and the chemisorbed oxygen (denotes as O_α), respectively, which were both ascribed to O^{2-}. The intensity of the O 1s peaks of the catalysts NN and NS using $Fe(NO_3)_3$ as precursor was apparently lower than those of the other catalysts, demonstrating that the concentration of oxygen on the surface of the catalysts NN and NS was low. Analyzed by peak-fitting, it could be seen that the chemisorbed oxygen, which was reported most active for the catalysts NN and NS, was significantly poorer, and the concentration of chemisorbed oxygen of the catalyst SN could be up to 21.2% on the surface, which was in accordance with the results above. Meanwhile, the shifting of the O 1s binding energy means that the oxygen deficit was on the surface of the catalyst, which was in favor of the SCR reaction.

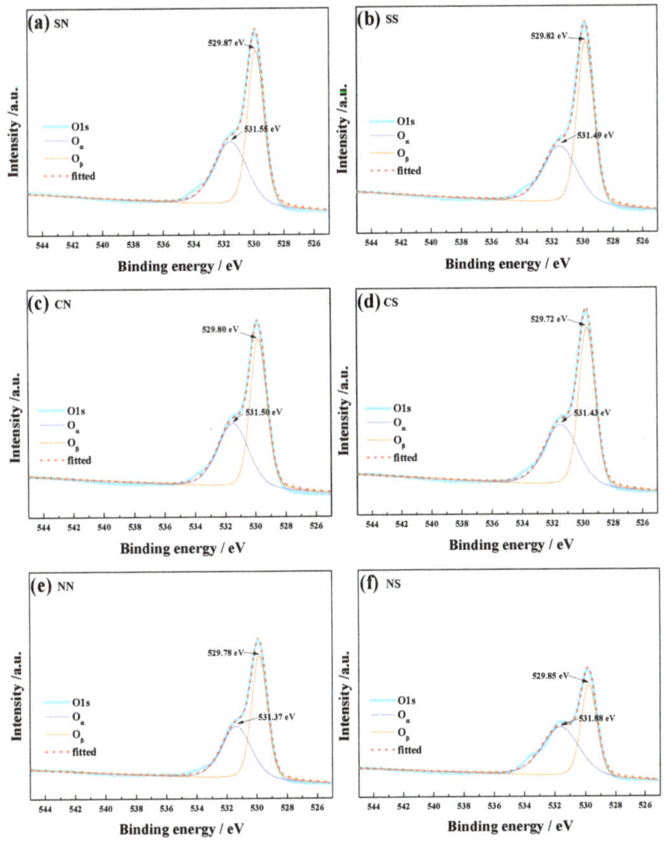

Figure 8. O 1s spectra of titanium modified $FeMgO_x$ catalysts with different precursors.

3.6. NH_3-TPD

In the SCR reaction, NH_3 should have firstly adsorbed and activated on the active sites and then reacted with NO, such as the Eley–Rideal mechanism, or reacted with both NO and adsorbed NO, such as the Langmuir–Hinshelwood mechanism. The presence of acid sites was of great importance for catalytic performance [45]. The surface acidity and acid species of titanium modified $FeMgO_x$ catalysts were investigated by NH_3-TPD and the results are shown in Figure 9. Usually the NH_3 desorption peaks below 300 °C were attributed to weak acid sites, and the NH_3 desorption peaks above 300 °C were ascribed to strong acid sites. The NH_3-TPD profiles of all the catalysts exhibited a desorption peaks close to 170 °C, and the intensity of desorption peaks over the catalysts NN and

NS was apparently stronger than that of the catalysts SN, SS, CN and CS. The desorption peaks ascribe to strong acid sites of the catalysts SN, SS, CN and CS appeared at about 380 °C, while those of the catalysts NN and NS appeared above 450 °C. The amount of total acidity of all the catalysts was calculated as 1362.03 μmol/g, 883.81μmol/g, 933.96 μmol/g, 525.86 μmol/g, 1915.58 μmol/g and 1977.91 μmol/g, with the order from catalyst SN to NS. Considering the activity results, it could be deduced that the weak acid sites that appeared at 170 °C had impact on the catalytic performance below 200 °C, and the strong acid sites that appeared at 380 °C made a great contribution to the catalytic activity and inhibited the secondary reaction at a higher temperature range. However, it was difficult for NH_3 to absorb and activate on the strong acid site of the catalysts NN and NS, and appeared to exceed 450 °C. The total acidity of the catalyst SN reached 1362.03 μmol/g, and the improvement of the total acidity of the catalyst SN could provide more adsorption sites. The enhancement of strong acid sites around 380 °C could inhibit the secondary reaction above 300 °C, which was the reason that the catalyst SN exhibited excellent catalytic performance and had a wide temperature range.

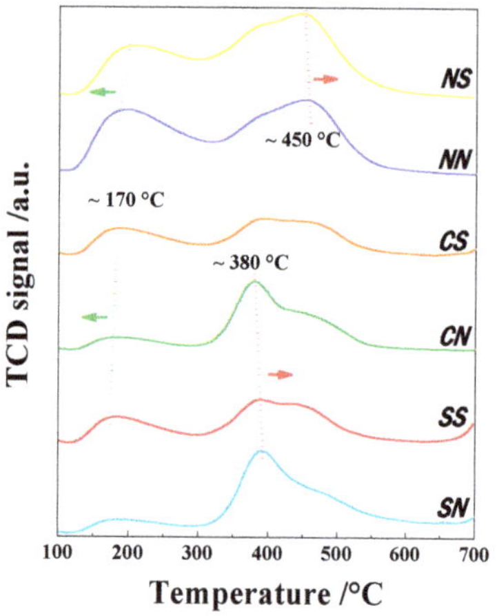

Figure 9. NH_3-TPD profiles over titanium modified $FeMgO_x$ catalysts with different precursors.

4. Conclusions

The influence of different precursors on titanium modified $FeMgO_x$ catalysts was investigated and characterized. The results show that the crystalline phase of the active component was directly affected by the iron precursors. γ-Fe_2O_3 formed as the main crystalline phase when $FeSO_4$ and $FeCl_2$ were used as precursors. The main crystalline phase would be amorphous iron-magnesium-titanium mixed oxides when $Fe(NO_3)_3$ was used as precursor. The catalyst using $FeSO_4$ and $Mg(NO_3)_2$ as precursors exhibited NO_x conversion above 90% from 225 to 400 °C, while approaching 100% from 250 to 375 °C. The temperature window of the catalysts using $Fe(NO_3)_3$ as a precursor shifted to lower temperature range, on which the secondary reaction occurred, leading to the decline of NO_x conversion at a high temperature. The regular spherical particle distribution and the good pores connectivity were advantageous to the mass transfer process. The acid sites that appeared at 170 °C played an important role for catalytic performance below 200 °C. The acid sites that appeared at 380 °C inhibited the secondary reaction at high temperature range. However, it was difficult for NH_3 to absorb and activate on the strong acid sites using $Fe(NO_3)_3$ as a precursor when exceeding 450 °C. The total acidity of the catalyst using $FeSO_4$ and $Mg(NO_3)_2$ as precursors could reach 1362.03 μmol/g, and the surface oxygen concentration was also enhanced, thereby SCR reaction was improved correspondingly.

Supplementary Materials: The following are available online at http://www.mdpi.com/2073-4344/9/6/560/s1, Figure S1: Powder XRD patterns of Fe_2O_3.

Author Contributions: Conceptualization, D.W. and L.X.; methodology, D.W., Y.P. and L.X.; validation, J.L., Y.P. and D.W.; formal analysis, L.H. and Z.S.; investigation, L.X., Q.Y. and L.H.; resources, D.W. and C.L.; data curation, L.X. and Z.S.; writing—original draft preparation, L.X.; writing—review and editing, D.W.; visualization, L.X., L.H. and Z.S.; supervision, J.L. and C.L.; project administration, D.W., Y.P. and C.L.; funding acquisition, D.W., Y.P. and C.L.

Funding: This research was funded by the National Key Research and Development Program, grant number 2017YFC0212800 and 2017YFC0210200, the National Natural Science Foundation of China, grant number 21777081 and 51576117.

Conflicts of Interest: The authors declare no conflict of interest.

References

1. Li, J.; Chang, H.; Ma, L.; Hao, J.; Yang, R.T. Low-temperature selective catalytic reduction of NO_x with NH_3 over metal oxide and zeolite catalysts—A review. *Catal. Today* **2011**, *175*, 147–156. [CrossRef]
2. Baek, J.; Lee, S.; Park, J.; Jeong, J.; Hwang, R.; Ko, C.; Jeon, S.; Choi, T.; Yi, K. Effects of steam introduction on deactivation of Fe-BEA catalyst in NH_3-SCR of N_2O and NO. *J. Ind. Eng. Chem.* **2017**, *48*, 194–201. [CrossRef]
3. Boningari, T.; Smirniotis, P.G. Impact of nitrogen oxides on the environment and human health: Mn-based materials for the NO_x abatement. *Curr. Opin. Chem. Eng.* **2016**, *13*, 133–141. [CrossRef]
4. Wang, D.; Luo, J.; Yang, Q.; Yan, J.; Zhang, K.; Zhang, W.; Peng, Y.; Li, J.; Crittenden, J. Deactivation mechanism of multipoisons in cement furnace flue gas on selective catalytic reduction catalysts. *Environ. Sci. Technol.* **2019**, *53*, 6937–6944. [CrossRef] [PubMed]
5. Skalska, K.; Miller, J.S.; Ledakowicz, S. Trends in NO_x abatement: A review. *Sci. Total Environ.* **2010**, *408*, 3976–3989. [CrossRef]
6. Wang, D.; Chen, J.; Peng, Y.; Si, W.; Li, X.; Li, B.; Li, J. Dechlorination of chlorobenzene on vanadium-based catalysts for low-temperature SCR. *Chem. Commun.* **2018**, *54*, 2032–2035. [CrossRef] [PubMed]
7. Qi, G.; Yang, R.T. Performance and kinetics study for low-temperature SCR of NO with NH_3 over MnO_x–CeO_2 catalyst. *J. Catal.* **2003**, *217*, 434–441. [CrossRef]
8. France, L.J.; Yang, Q.; Li, W.; Chen, Z.; Guang, J.; Guo, D.; Wang, L.; Li, X. Ceria modified $FeMnO_x$—Enhanced performance and sulphur resistance for low-temperature SCR of NO_x. *Appl. Catal. B Environ.* **2017**, *206*, 203–215. [CrossRef]
9. Shelef, M. Selective catalytic reduction of NOx with N-free reductants. *Chem. Rev.* **1995**, *95*, 209–225. [CrossRef]
10. Wang, D.; Peng, Y.; Yang, Q.; Xiong, S.; Li, J.; Crittenden, J. Performance of Modified LaxSr1–xMnO$_3$ Perovskite Catalysts for NH_3 Oxidation: TPD, DFT, and Kinetic Studies. *Environ. Sci. Technol.* **2018**, *52*, 7443–7449. [CrossRef]
11. Shi, X.; Liu, F.; Xie, L.; Shan, W.; He, H. NH_3-SCR performance of fresh and hydrothermally aged Fe-ZSM-5 in standard and fast selective catalytic reduction reactions. *Environ. Sci. Technol.* **2013**, *47*, 3293–3298. [CrossRef] [PubMed]
12. Ko, J.; Park, R.; Jeon, J.; Kim, D.; Jung, S.; Kim, S.; Park, Y. Effect of surfactant, HCl and NH_3 treatments on the regeneration of waste activated carbon used in selective catalytic reduction unit. *J. Ind. Eng. Chem.* **2015**, *32*, 109–112. [CrossRef]
13. Fan, Y.; Ling, W.; Huang, B.; Dong, L.; Yu, C.; Xi, H. The synergistic effects of cerium presence in the framework and the surface resistance to SO_2 and H_2O in NH_3-SCR. *J. Ind. Eng. Chem.* **2017**, *56*, 108–119. [CrossRef]
14. Boningari, T.; Ettireddy, P.R.; Somogyvari, A.; Liu, Y.; Vorontsov, A.; McDonald, C.A.; Smirniotis, P.G. Influence of elevated surface texture hydrated titania on Ce-doped Mn/TiO_2 catalysts for the low-temperature SCR of NO_x under oxygen-rich conditions. *J. Catal.* **2016**, *325*, 145–155. [CrossRef]
15. Aguilar-Romero, M.; Camposeco, R.; Castillo, S.; Marín, J.; Rodríguez-González, V.; García-Serrano, L.; Mejía-Centeno, I. Acidity, surface species, and catalytic activity study on V_2O_5-WO_3/TiO_2 nanotube catalysts for selective NO reduction by NH_3. *Fuel* **2017**, *198*, 123–133. [CrossRef]
16. Pappas, D.K.; Boningari, T.; Boolchand, P.; Smirniotis, P.G. Novel manganese oxide confined interweaved titania nanotubes for the low-temperature Selective Catalytic Reduction (SCR) of NOx by NH_3. *J. Catal.* **2016**, *334*, 1–13. [CrossRef]

17. Song, I.; Youn, S.; Lee, H.; Lee, S.G.; Cho, S.; Kim, D. Effects of microporous TiO_2 support on the catalytic and structural properties of V_2O_5/microporous TiO_2 for the selective catalytic reduction of NO by NH_3. *Appl. Catal. B Environ.* **2017**, *210*, 421–431. [CrossRef]
18. Thirupathi, B.; Smirniotis, P.G. Nickel-doped Mn/TiO_2 as an efficient catalyst for the low-temperature SCR of NO with NH_3: Catalytic evaluation and characterizations. *J. Catal.* **2012**, *288*, 74–83. [CrossRef]
19. He, Y.; Ford, M.E.; Zhu, M.; Liu, Q.; Tumuluri, U.; Wu, Z.; Wachs, I.E. Influence of catalyst synthesis method on selective catalytic reduction (SCR) of NO by NH_3 with V_2O_5-WO_3/TiO_2 catalysts. *Appl. Catal. B* **2016**, *193*, 141–150. [CrossRef]
20. Xu, H.; Li, Y.; Xu, B.; Cao, Y.; Feng, X.; Sun, M.; Gong, M.; Chen, Y. Effectively promote catalytic performance by adjusting W/Fe molar ratio of FeW_x/Ce0. 68Zr0. 32O_2 monolithic catalyst for NH_3-SCR. *J. Ind. Eng. Chem.* **2016**, *36*, 334–345. [CrossRef]
21. Wang, D.; Peng, Y.; Xiong, S.; Li, B.; Gan, L.; Lu, C.; Chen, J.; Ma, Y.; Li, J. De-reducibility mechanism of titanium on maghemite catalysts for the SCR reaction: An in situ DRIFTS and quantitative kinetics study. *Appl. Catal. B* **2018**, *221*, 556–564. [CrossRef]
22. Liu, F.; He, H.; Zhang, C.; Feng, Z.; Zheng, L.; Xie, Y.; Hu, T. Selective catalytic reduction of NO with NH_3 over iron titanate catalyst: Catalytic performance and characterization. *Appl. Catal. B Environ.* **2010**, *96*, 408–420. [CrossRef]
23. Yang, S.; Li, J.; Wang, C.; Chen, J.; Ma, L.; Chang, H.; Chen, L.; Peng, Y.; Yan, N. Fe–Ti spinel for the selective catalytic reduction of NO with NH_3: Mechanism and structure–activity relationship. *Appl. Catal. B Environ.* **2012**, *117*, 73–80. [CrossRef]
24. Liu, Z.; Liu, Y.; Chen, B.; Zhu, T.; Ma, L. Novel Fe–Ce–Ti catalyst with remarkable performance for the selective catalytic reduction of NO_x by NH_3. *Catal. Sci. Technol.* **2016**, *6*, 6688–6696. [CrossRef]
25. Du, X.; Wang, X.; Chen, Y.; Gao, X.; Zhang, L. Supported metal sulfates on Ce–TiO_x as catalysts for NH_3–SCR of NO: High resistances to SO_2 and potassium. *J. Ind. Eng. Chem.* **2016**, *36*, 271–278. [CrossRef]
26. Xiong, Z.; Peng, B.; Zhou, F.; Wu, C.; Lu, W.; Jin, J.; Ding, S. Magnetic iron-cerium-tungsten mixed oxide pellets prepared through critic acid sol-gel process assisted by microwave irradiation for selective catalytic reduction of NO_x with NH_3. *Powder Technol.* **2017**, *319*, 19–25. [CrossRef]
27. Xiong, Z.; Liu, J.; Zhou, F.; Liu, D.; Lu, W.; Jin, J.; Ding, S. Selective catalytic reduction of NO_x with NH_3 over iron-cerium-tungsten mixed oxide catalyst prepared by different methods. *Appl. Surf. Sci.* **2017**, *406*, 218–225. [CrossRef]
28. Xu, H.; Xie, J.; Ma, Y.; Qu, Z.; Zhao, S.; Chen, W.; Huang, W.; Yan, N. The cooperation of FeSn in a MnO_x complex sorbent used for capturing elemental mercury. *Fuel* **2015**, *140*, 803–809. [CrossRef]
29. Shu, Y.; Aikebaier, T.; Quan, X.; Chen, S.; Yu, H. Selective catalytic reaction of NO_x with NH_3 over Ce–Fe/TiO_2-loaded wire-mesh honeycomb: Resistance to SO_2 poisoning. *Appl. Catal. B Environ.* **2014**, *150*, 630–635. [CrossRef]
30. Liu, Z.; Su, H.; Chen, B.; Li, J.; Woo, S. Activity enhancement of WO_3 modified Fe_2O_3 catalyst for the selective catalytic reduction of NO_x by NH_3. *Chem. Eng. J.* **2016**, *299*, 255–262. [CrossRef]
31. Foo, R.; Vazhnova, T.; Lukyanov, D.B.; Millington, P.; Collier, J.; Rajaram, R.; Golunski, S. Formation of reactive Lewis acid sites on Fe/WO_3–ZrO_2 catalysts for higher temperature SCR applications. *Appl. Catal. B Environ.* **2015**, *162*, 174–179. [CrossRef]
32. Putluru, S.S.R.; Schill, L.; Jensen, A.D.; Siret, B.; Tabaries, F.; Fehrmann, R. Mn/TiO_2 and Mn–Fe/TiO_2 catalysts synthesized by deposition precipitation—Promising for selective catalytic reduction of NO with NH_3 at low temperatures. *Appl. Catal. B Environ.* **2015**, *165*, 628–635. [CrossRef]
33. Wu, S.; Zhang, L.; Wang, X.; Zou, W.; Cao, Y.; Sun, J.; Tang, C.; Gao, F.; Deng, Y.; Dong, L. Synthesis, characterization and catalytic performance of $FeMnTiO_x$ mixed oxides catalyst prepared by a CTAB-assisted process for mid-low temperature NH_3-SCR. *Appl. Catal. A Gen.* **2015**, *505*, 235–242. [CrossRef]
34. Qi, G.; Yang, R.T.; Chang, R. MnO_x-CeO_2 mixed oxides prepared by co-precipitation for selective catalytic reduction of NO with NH_3 at low temperatures. *Appl. Catal. B Environ.* **2004**, *51*, 93–106. [CrossRef]
35. Xiong, Z.; Wu, C.; Hu, Q.; Wang, Y.; Jin, J.; Lu, C.; Guo, D. Promotional effect of microwave hydrothermal treatment on the low-temperature NH_3-SCR activity over iron-based catalyst. *Chem. Eng. J.* **2016**, *286*, 459–466. [CrossRef]

36. Cao, F.; Su, S.; Xiang, J.; Wang, P.; Hu, S.; Sun, L.; Zhang, A. The activity and mechanism study of Fe–Mn–Ce/γ-Al_2O_3 catalyst for low temperature selective catalytic reduction of NO with NH_3. *Fuel* **2015**, *139*, 232–239. [CrossRef]
37. Casanova, M.; Llorca, J.; Sagar, A.; Schermanz, K.; Trovarelli, A. Mixed iron–erbium vanadate NH_3-SCR catalysts. *Catal. Today* **2015**, *241*, 159–168. [CrossRef]
38. Zhu, L.; Zhong, Z.; Yang, H.; Wang, C. Comparison study of Cu-Fe-Ti and Co-Fe-Ti oxide catalysts for selective catalytic reduction of NO with NH_3 at low temperature. *J. Collid Interface Sci.* **2016**, *478*, 11–21. [CrossRef]
39. Liu, F.; He, H.; Zhang, C.; Shan, W.; Shi, X. Mechanism of the selective catalytic reduction of NO_x with NH_3 over environmental-friendly iron titanate catalyst. *Catal. Today* **2011**, *175*, 18–25. [CrossRef]
40. Liu, F.; He, H. Structure−Activity Relationship of Iron Titanate Catalysts in the Selective Catalytic Reduction of NO_x with NH_3. *J. Phys. Chem. C* **2010**, *114*, 16929–16936. [CrossRef]
41. Ma, L.; Li, J.; Ke, R.; Fu, L. Catalytic Performance, Characterization, and Mechanism Study of $Fe_2(SO_4)_3/TiO_2$ Catalyst for Selective Catalytic Reduction of NO_x by Ammonia. *J. Phys. Chem. C* **2011**, *115*, 7603–7612. [CrossRef]
42. Xu, L.; Niu, S.; Lu, C.; Wang, D.; Zhang, K.; Li, J. NH_3-SCR performance and characterization over magnetic iron-magnesium mixed oxide catalysts. *Korean J. Chem. Eng.* **2017**, *34*, 1576–1583. [CrossRef]
43. Xu, L.; Niu, S.; Wang, D.; Lu, C.; Zhang, Q.; Zhang, K.; Li, J. Selective catalytic reduction of NO_x with NH_3 over titanium modified $Fe_xMg_yO_z$ catalysts: Performance and characterization. *J. Ind. Eng. Chem.* **2018**, *63*, 391–404. [CrossRef]
44. Xu, L.; Niu, S.; Lu, C.; Zhang, Q.; Li, J. Influence of calcination temperature on $Fe_{0.8}Mg_{0.2}O_z$ catalyst for selective catalytic reduction of NO_x with NH_3. *Fuel* **2018**, *219*, 248–258. [CrossRef]
45. Wang, D.; Peng, Y.; Yang, Q.; Hu, F.; Li, J.; Crittenden, J. NH_3-SCR performance of WO_3 blanketed CeO_2 with different morphology: Balance of surface reducibility and acidity. *Catal. Today* **2019**, *332*, 42–48. [CrossRef]

catalysts

MDPI

Article

A Characterization Study of Reactive Sites in ALD-Synthesized WO_x/ZrO_2 Catalysts

Cong Wang [1], Xinyu Mao [1], Jennifer D. Lee [2], Tzia Ming Onn [1], Yu-Hao Yeh [1], Christopher B. Murray [2,3] and Raymond J. Gorte [2,3,*]

1 Chemical & Biomolecular Engineering, University of Pennsylvania, Philadelphia, PA 19104, USA; wangcong@seas.upenn.edu (C.W.); xinyumao@seas.upenn.edu (X.M.); tonn@seas.upenn.edu (T.M.O.); yeh12@seas.upenn.edu (Y.-H.Y.)
2 Department of Chemistry, University of Pennsylvania, Philadelphia, PA 19104, USA; jleed@sas.upenn.edu (J.D.L.); cbmurray@sas.upenn.edu (C.B.M.)
3 Material Science & Engineering, University of Pennsylvania, Philadelphia, PA 19104, USA
* Correspondence: gorte@seas.upenn.edu; Tel.: +1-215-898-4439

Received: 20 June 2018; Accepted: 17 July 2018; Published: 19 July 2018

Abstract: A series of ZrO_2-supported WO_x catalysts were prepared using atomic layer deposition (ALD) with $W(CO)_6$, and were then compared to a WO_x/ZrO_2 catalyst prepared via conventional impregnation. The types of sites present in these samples were characterized using temperature-programmed desorption/thermogravimetric analysis (TPD-TGA) measurements with 2-propanol and 2-propanamine. Weight changes showed that the WO_x catalysts grew at a rate of 8.8×10^{17} W atoms/m^2 per cycle. Scanning transmission electron microscopy/energy-dispersive spectroscopy (STEM-EDS) indicated that WO_x was deposited uniformly, as did the 2-propanol TPD-TGA results, which showed that ZrO_2 was completely covered after five ALD cycles. Furthermore, 2-propanamine TPD-TGA demonstrated the presence of three types of catalytic sites, the concentrations of which changed with the number of ALD cycles: dehydrogenation sites associated with ZrO_2, Brønsted-acid sites associated with monolayer WO_x clusters, and oxidation sites associated with higher WO_x coverages. The Brønsted sites were not formed via ALD of WO_x on SiO_2. The reaction rates for 2-propanol dehydration were correlated with the concentration of Brønsted sites. While TPD-TGA of 2-propanamine did not differentiate the strength of Brønsted-acid sites, H–D exchange between D_2O and either toluene or chlorobenzene indicated that the Brønsted sites in tungstated zirconia were much weaker than those in H-ZSM-5 zeolites.

Keywords: solid acids; metal-oxide catalysts; tungstated zirconia; atomic layer deposition; Brønsted-acid strength; temperature-programmed desorption/thermogravimetric analysis; H–D exchange

1. Introduction

In the 1980s, it was reported that tungstated zirconia (WO_x/ZrO_2) could be used as a solid acid catalyst; however, there are still many questions and apparent contradictions regarding the nature and strength of the acid sites in this material, as demonstrated in the recent review by Zhou et al. [1]. A major difficulty with tungstated-zirconia catalysts is that their properties depend on how they are made. It is likely, for this reason, that some studies indicated that the structure of the underlying ZrO_2, whether amorphous or crystalline (tetragonal or monoclinic), influences the activity of the sites [2], while other workers reported that the crystallographic structure of the zirconia is not important [3,4]. It was also suggested that the sites may be of varying strength, depending on the WO_x cluster size, with one theoretical study reporting that the Brønsted sites not only depend on cluster size, but can also approach super-acid strength [1]. However, the observations that it is necessary to add Pt in order for tungstated zirconia to exhibit alkane-isomerization activity [5,6] would suggest that the sites are less strong, in addition to tungstated zirconia being selective to ether formation in the dehydration of alcohols [7], without forming olefins, unlike with protonic zeolites.

Most researchers prepared their tungstated-zirconia catalysts via the aqueous impregnation of ammonium metatungstate ($(NH_4)_{10}W_{12}O_{41}\cdot 5H_2O$), followed by calcination [1]. In addition to the fact that this precursor consists of a relatively large number of tungsten atoms, the drying process itself can result in multiple clusters coming together before decomposing upon calcination [8]. While the WO_x species may spread over the support surface during calcination, the implications of there being so many tungsten atoms together in the initial state implies that there will likely be large clusters on the support in all cases. This is important because mono-tungstate species are not believed to be the active component [1]. Furthermore, the fact that the catalytic properties are a function of the tungsten-oxide coverage and pretreatment conditions [9,10] implies that discrepancies in the results from different studies are likely due to differences in the detailed nature of the clusters that are formed in the catalysts.

In the presented work, we set out to investigate the catalytic sites in WO_x/ZrO_2 catalysts prepared using atomic layer deposition (ALD). ALD is a self-limiting process in which WO_x is deposited through repeated, cyclic exposures of $W(CO)_6$ and oxidants [11–13]. Because the amount of WO_x deposited per cycle in ALD is low and the deposition is uniform, ALD ensures the formation of uniform, atomic-scale layers, which maximize the interfacial contact between WO_x and ZrO_2. This allows the formation of isolated WO_x species, as well as much more control over the tungsten cluster sizes by increasing tungsten coverage. A complicating factor in the characterization is that zirconia itself exhibits activity in the reaction of amines and alcohols, and contributes to the activity of catalysts for some reactions. Fortunately, as demonstrated in this work, the chemistries on bare zirconia are oftentimes distinguishable from those that occur on WO_x/ZrO_2 interfaces, and thus, provide additional information on coverage and dispersion.

We also set out to test claims that tungstated-zirconia could show super-acidic properties. The quantification of Brønsted-acid strength in solid acids is difficult, and some common measures, such as heats of adsorption for ammonia or pyridine, were shown to be uninformative [14]. A simple reaction that appears to depend only on the ability of the solid to protonate a weak base involves the H–D exchange between an aromatic molecule and deuterated acid sites [15]. In our case, we examined light-off curves for H–D exchange between D_2O and either toluene (C_7H_8) or chlorobenzene (C_6H_5Cl) [15]. Essentially, all hydroxides on solids exchange readily with D_2O [14,16], meaning that the presence of D_2O ensures a high concentration of deuterated acid sites. However, the deuteration of toluene (proton affinity = 784 kJ/mol) or chlorobenzene (proton affinity = 753 kJ/mol) requires the formation of a carbenium ion, so that the temperature at which exchange becomes rapid should be a reasonable measure of the acid strength. In agreement with expectations based on proton affinities, higher temperatures are required for H–D exchange with chlorobenzene compared to that with toluene; however, the Brønsted sites in tungstated zirconia appear to be significantly weaker than those in H-ZSM-5.

2. Results

The ZrO_2 substrate prepared for this study had an initial BET surface area of 65 m^2/g. As shown by the X-ray diffraction (XRD) pattern in Figure 1a, its phase was primarily monoclinic with a small amount of the tetragonal phase. The WO_x loadings were then determined as a function of the number of ALD cycles by measuring the sample weights. Table 1 shows these loadings, together with BET surface areas. In the first five ALD cycles, the sample weight increased almost linearly at 22 mg WO_x/g ZrO_2 per cycle. Assuming an O:W stoichiometry of three and uniform film growth over the entire ZrO_2 surface, this corresponds to a growth rate of 0.048 nm/cycle, a value that is reasonable for the size of the $W(CO)_6$ precursor, but that is somewhat larger than the value reported in the literature for the growth of WO_3 films on flat surfaces (0.023 nm/cycle) for similar growth conditions [11,13]. It is interesting to notice that the growth rate of WO_x on the 140-m^2/g SiO_2 support, calculated from the amount deposited after five ALD cycles, was only 0.025 nm/cycle, implying that the substrate does influence the initial deposition rate. Finally, it is useful to consider that deposition of 22 mg of WO_3 onto the 65-m^2/g ZrO_2 sample corresponds to 8.8×10^{17} W atoms/m^2. This is a fraction of an oxide monolayer, which implies that, in the absence of surface migration, WO_x species are likely spatially isolated after a single ALD cycle.

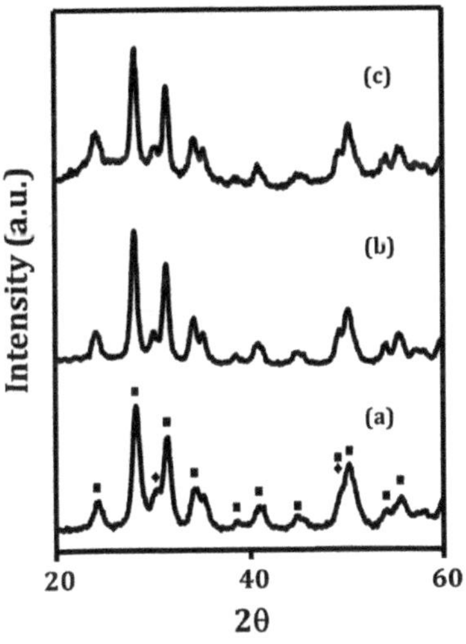

Figure 1. XRD patterns for (a) ZrO_2, (b) 5W-Zr (5 ALD cycles), and (c) 40W-Zr (40 ALD cycles). Monoclinic phase (■); tetragonal phase (♦).

Table 1. Weight gain, surface area, and calculated layer thickness for W-Zr and W-Si materials. The layer thickness was calculated using a WO_3 density of 7.16 g/cm^3. The nomenclature *x*W-Zr is used to refer to a sample exposed to *x* ALD cycles.

SAMPLE	Weight Gain (mg/g of Substrate)	Surface Area (m^2/g)	Layer Thickness (nm)
ZrO_2	0	65	0
1W-Zr	21	62	0.045
2W-Zr	47	60	0.10
3W-Zr	63	53	0.14
4W-Zr	87	47	0.19
5W-Zr	112	46	0.24
20W-Zr	430	36	0.92
40W-Zr	720	19	1.6
impW-Zr (10-wt% WO_x)	111	58	-
5W-Si	120	135	0.12

After five ALD cycles, the BET surface area decreased to 46 m^2/g. Some of the decrease was due to the increase in sample mass; however, the majority of the loss in surface area per mass of sample must be associated with the narrowing or blocking of pores. The WO_x added with five ALD cycles did not cause any changes in the XRD pattern (Figure 1b). Because of the decreasing surface area of the sample, the rate at which the weight changed decreased somewhat with the number of ALD cycles. After 40 ALD cycles, the WO_x loading was 42 wt%, or 0.72 g WO_x/g ZrO_2. Assuming that the film was uniform with a stoichiometry and density of bulk WO_3, the film thickness after 40 cycles was 1.5 nm, a thickness significantly greater than that of a monolayer. However, even with this relatively thick film, the XRD pattern (Figure 1c) showed no evidence of a well-defined crystalline WO_x phase. The only change in the XRD pattern with the addition of this large amount of WO_x was a slightly elevated baseline between 25 and 35 degrees 2θ, and again above 50 degrees 2θ. These broad features may be associated with a very small fraction of mixed WO_x phases. However, if large, three-dimensional, crystalline clusters were being formed, it would have been apparent in the diffraction pattern.

Scanning transmission electron microscopy (STEM) imaging and energy-dispersive spectroscopy (EDS) elemental mapping on the 40W-Zr sample (Figure 2) demonstrate that the WO_x ALD films deposited uniformly over the ZrO_2 surface. While there were no obvious features in the image, despite the high WO_x coverage, the EDS mapping of W and Zr indicates the co-existence of W and Zr elements in the sample. The overlap in the signals reveals their uniform distribution.

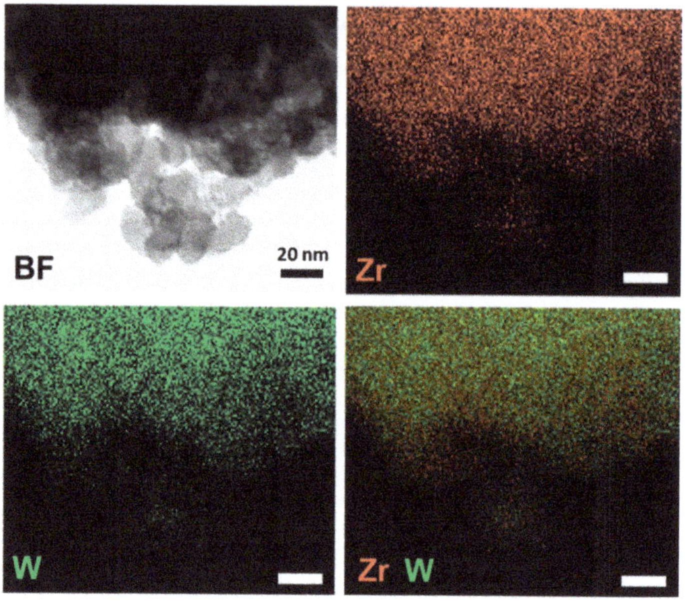

Figure 2. BF (Bright-Field) scanning transmission electron microscopy (STEM) image and energy-dispersive spectroscopy (EDS) elemental mapping of the 40W-Zr sample. The scale bars represent 20 nm.

To characterize the structure of the WO_x ALD films, Raman spectra of the ZrO_2 support and of the impW-Zr (10-wt% WO_x) and 5W-Zr samples were measured, both of which had WO_x loadings that were close to 10 wt%. The data for ZrO_2 (Figure 3a) show vibrational bands at ~347, 380, 478, 616, and 636 cm^{-1}, which are well known to be due to ZrO_2 [17]. The spectrum for the impW-Zr sample (Figure 3b) exhibits the same bands, but with a new, broad feature centered at ~962 cm^{-1}, which was previously assigned to the symmetric stretching mode of a terminal W=O bond [17]. This broad feature may also include a contribution from a bridging W–O–Zr bond at 915 cm^{-1}. The spectrum of the 5W-Zr sample (Figure 3c) is similar, except that the vibrational features associated with WO_x relative to those with ZrO_2 are significantly more intense, despite having the same WO_x loading. This is likely due to the WO_x being spread more uniformly over the ZrO_2 surface in the ALD-prepared sample.

A further indication that the WO_x layer was more uniform on the ALD-prepared sample is demonstrated in Figure 4, which shows the temperature-programmed desorption/thermogravimetric analysis (TPD-TGA) results for 2-propanol on the ZrO_2, 5W-Zr, and impW-Zr samples. The TGA data show that the initial coverages following room-temperature adsorption and 1-h evacuation were slightly higher on the ZrO_2 and impW-Zr samples (between 250 and 300 μmol/g) compared to the

5W-Zr (200 μmol/g). However, this is almost certainly due to the lower surface area of the 5W-Zr sample. The initial specific coverages were roughly 2.7×10^{18} molecules/m^2 for each of the samples. While some of the 2-propanol (m/e = 45) desorbed unreacted from all three samples, significant fractions of the 2-propanol desorbed from the samples as propene (m/e = 41) and H_2O due to reactions on either Lewis- or Brønsted-acid sites [18]. Water is not shown because it tends to desorb over a broad temperature range. What is more interesting is that the dehydration reaction occurred at very different temperatures on ZrO_2 and on WO_x/ZrO_x sites. On the ZrO_2 and 5W-Zr samples, the dehydration reaction occurred over narrow temperature ranges in peaks centered at 560 K on ZrO_2, and 405 K on 5W-Zr. The TPD result of impW-Zr shows similarly sized peaks at both temperatures, suggesting that impW-Zr has regions of bare ZrO_2 and regions covered by WO_x. Even though the WO_x loadings were the same for 5W-Zr and impW-Zr, the distributions of WO_x were clearly different, with tungsten distributed much more uniformly on the sample prepared using ALD.

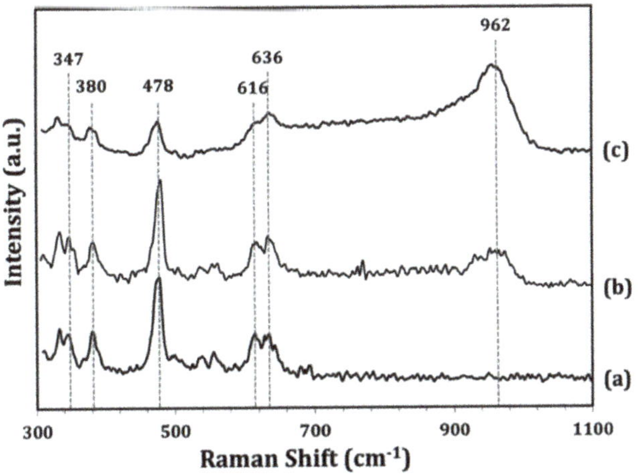

Figure 3. Raman spectra of (a) ZrO_2, (b) impW-Zr (10-wt% WO_x), and (c) 5W-Zr samples under ambient conditions.

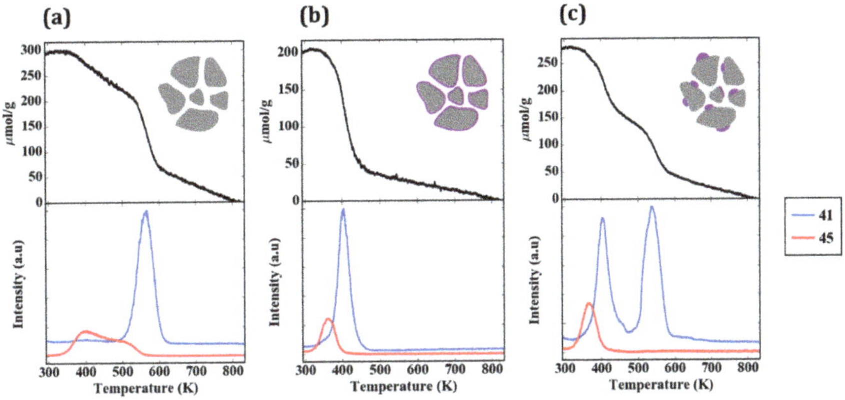

Figure 4. TPD-TGA of 2-propanol over (**a**) ZrO_2, (**b**) 5W-Zr, and (**c**) impW-Zr. The desorption features correspond to propene (m/e = 41) and unreacted 2-propanol (m/e = 45). The graphic symbols are a schematic of the WO_x (purple) over the ZrO_2 support (gray).

To qualitatively assess the nature of the sites on the ZrO_2 and 5W-Zr samples, FTIR measurements were performed following adsorption of pyridine, with results shown in Figure 5. The spectrum for ZrO_2 (Figure 5a) shows only bands at 1440 and 1460 cm^{-1}, which are characteristic of adsorption at Lewis sites. In contrast to this, the spectrum of pyridine on the ALD-prepared 5W-Zr (Figure 5b) also exhibits a band near 1540 cm^{-1}, which can be assigned to adsorbed pyridinium ions, implying 5W-Zr contains a significant concentration of Brønsted sites.

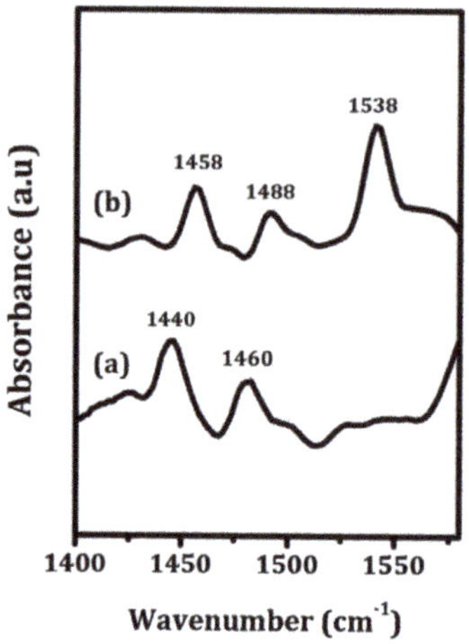

Figure 5. FTIR spectra of pyridine adsorbed on (**a**) ZrO_2 and (**b**) 5W-Zr. The samples were pretreated with helium at 373 K before pyridine was adsorbed onto the samples at room temperature.

To quantify the concentration of the sites, TPD-TGA measurements were performed using 2-propanamine as the probe molecule. On Brønsted sites, 2-propanamine forms 2-propylammonium which decompose during TPD to form ammonia and propylene between 573 and 650 K due to a Hoffman Elimination reaction [19,20], allowing site concentrations to be determined from this decomposition feature. Reaction of the amines can occur on catalytic sites other than Brønsted sites; however, the products that are formed and the temperature range in which the reactions occur are different and depend on the nature of the site [21]. The TPD-TGA results for the WO_x/ZrO_2 samples show evidence for the presence of three different types of reactive sites, with concentrations of those sites dependent on the WO_x coverages. A summary of the three reaction pathways is illustrated in Scheme 1.

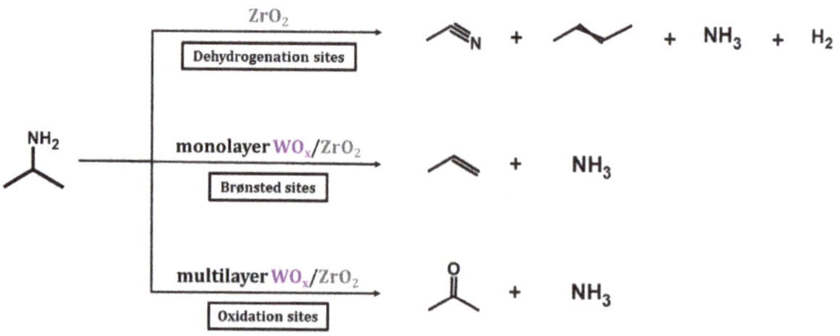

Scheme 1. A summary of reaction pathways for the conversion of 2-propanamine on the WO_x/ZrO_2 samples observed in the TPD-TGA measurements.

Figure 6a provides the TPD-TGA result for 2-propanamine with pure ZrO_2. Approximately half of the 300 μmol/g that remained on the sample after evacuation desorbed as unreacted amine (m/e = 42, 43, 44) below 500 K. Above that temperature, most of the amine molecules reacted in a feature centered at approximately 565 K to form H_2 (m/e = 2), a mixture of acetonitrile and butenes (m/e = 41), and ammonia (m/e = 17). The absence of a peak at m/e = 42 is particularly important for demonstrating that the high temperature peak at m/e = 41 is not propene, the product that would be formed on Brønsted sites. The identification of acetonitrile and propene as products was achieved by a more complete analysis of the mass spectra and was confirmed in steady-state flow-reactor measurements over ZrO_2 at 673 K. The same products were observed in those measurements, with the addition of small amounts of dipropylamine. Apparently, ZrO_2 catalyzes the dehydrogenation of 2-propanamine to form the imine, which is unstable and reacts to form the smaller nitrile and butenes. The reaction of adsorbed 1-propanamine on ZrO_2, shown in Figure S1, was simpler, forming primarily propionitrile and H_2. However, because the products formed by 1-propanamine on sites formed by WO_x were more difficult to distinguish from those formed on ZrO_2, most of our work focused on using 2-propanamine to characterize the samples. It is noteworthy that ZrO_2 was previously demonstrated to exhibit dehydrogenation chemistry under some conditions, and it is this functionality that is apparently responsible for the amine reactions [22].

The TPD-TGA result for the 5W-Zr sample, shown in Figure 6b, differs significantly from that obtained on ZrO_2. As in the case of 2-propanol, the initial coverage was slightly lower due to the lower specific surface area; similar to the case of ZrO_2, unreacted amine desorbed from the sample below 500 K. However, on the 5W-Zr sample, there are two distinct reaction features at ~585 K and ~620 K, and the products formed in both peaks differ from that observed on ZrO_2. The 620-K peak is characteristic of Brønsted-acid sites. The major products are propene (m/e = 41 and 42, in the correct ratio for propene) and ammonia (m/e = 17), and these were formed via the Hoffman elimination at exactly the same temperature reported for the reaction of adsorbed 2-propanamine on acidic zeolites [19]. The lower-temperature peak at ~585 K may have a small contribution from the bare ZrO_2, since the temperature is similar to the reaction temperature on ZrO_2; however, the major products formed in this case were very different. The major products determined from a complete analysis of the mass spectra in this temperature range were acetone (m/e = 43) and ammonia. Minimal amounts of acetonitrile and much less H_2 were formed, suggesting that most of the ZrO_2 was covered. The formation of acetone implies that partial oxidation took place. The total amount of 2-propanamine that reacted on the 5W-Z sample was ~100 μmol/g. By integrating the product peaks, we estimate that this sample had less than 10 μmol/g of dehydrogenation sites (from the amount of H_2 and nitrile that formed), 80 μmol/g of Brønsted sites (from propene and ammonia), and 15 μmol/g of oxidation sites (acetone). These results are summarized in Table 2.

Table 2. Reaction site densities determined from the desorption features of 2-propanamine TPD, determined as described in the text. W/B = number of W atoms per Brønsted site.

SAMPLE	Reaction-Site Density (μmol/g)			W/B
	Dehydrogenation	Oxidation	Brønsted	
ZrO_2	123	0	0	-
1W-Zr	115	0	6	15
2W-Zr	77	0	24	8.5
3W-Zr	31	4	67	4.1
4W-Zr	20	5	74	5.2
5W-Zr	6	16	80	6.0
20W-Zr	5	32	41	32
40W-Zr	1	15	12	148
impW-Zr	63	9	40	13
5W-Si	0	24	9	30

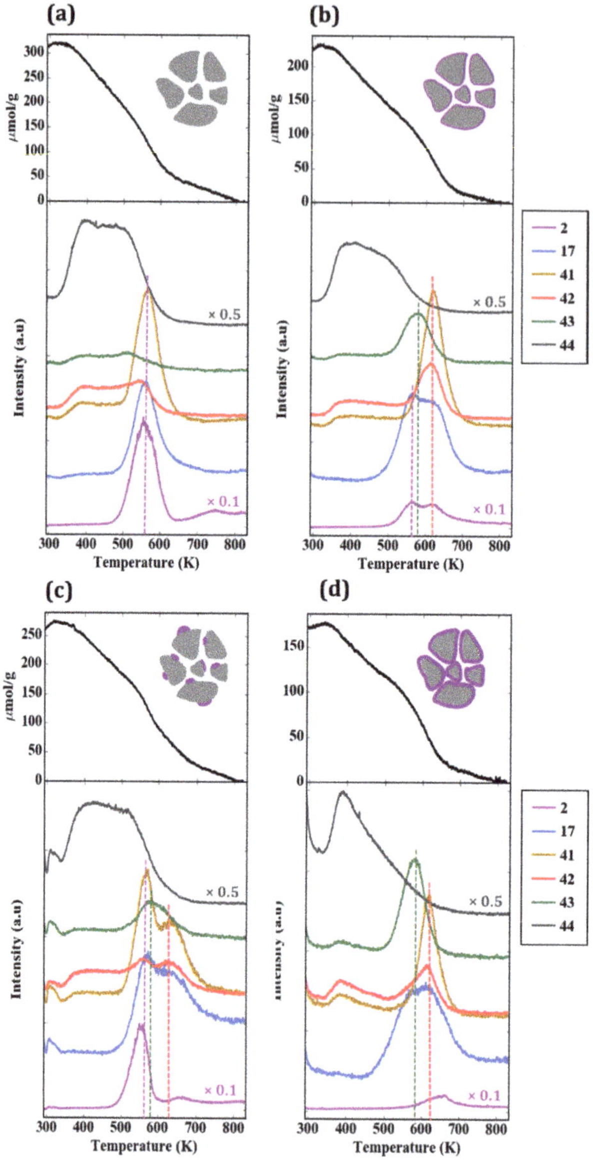

Figure 6. TPD-TGA of 2-propanamine over (**a**) ZrO_2, (**b**) 5W-Zr, (**c**) impW-Zr, and (**d**) 20W-Zr. The desorption features correspond to hydrogen (m/e = 2), ammonia (m/e = 17), a mixture of butenes and acetonitrile (m/e = 41), acetone (m/e = 43), propene (m/e = 41 and 42), and unreacted 2-propanamine (m/e = 44).

TPD-TGA measurements were also performed for the impW-Zr sample (10-wt% WO_x), with the result shown in Figure 6c. This sample exhibited all three types of reaction features, with a clearly defined H_2 and acetonitrile/butene feature at 565 K, associated with ZrO_2. An estimate of the site densities gave 60 μmol/g of dehydrogenation sites, 40 μmol/g of Brønsted sites, and 10 μmol/g of oxidation sites. Figure 6d is the TPD-TGA result obtained for 20W-Zr, the sample prepared using ALD which had a 30-wt% WO_x loading, corresponding to a uniform film of ~1 nm. In comparison to the near-monolayer 5W-Zr sample, the TPD trace shows more intense features associated with oxidation sites at ~585 K, and smaller features associated with Brønsted sites at 620 K.

To understand the effect of WO_x coverage, TPD-TGA measurements of 2-propanamine were performed on samples exposed to a varying number of ALD cycles. All of the samples exhibited the same three features in the TPD-TGA but at different concentrations. A summary of the calculated site concentrations is reported in Table 2 and Figure 7. The concentration of accessible dehydrogenation sites decreased dramatically with the number of ALD cycles, and essentially disappeared after five ALD cycles. Using the deposition rate of 8.8×10^{17} W atoms/m^2 per cycle reported earlier, five ALD cycles corresponds to 4.4×10^{18} W atoms/m^2, a reasonable value for an oxide monolayer. In agreement with previous reports that individual W atoms do not form Brønsted sites, the Brønsted-site concentration was negligible after one cycle, and increased non-linearly thereafter, converging to a maximum after about five cycles. At higher WO_x coverages, the Brønsted-site concentration decreased, implying that contact with the ZrO_2 substrate was essential for forming these sites. Confirmation of the importance of ZrO_2 came from the fact that the silica-supported sample, 5W-Si, showed a low Brønsted-site acidity. Finally, the concentration of oxidation sites, those responsible for forming acetone, were negligible below five ALD rounds, the point at which the dehydrogenation sites disappeared. This implies that the oxidation sites are associated with bulk WO_x that is not in contact with the ZrO_2 surface. It is also confirmed by the fact that oxidation-site density reached a maxima for 20W-Zr when multilayer WO_x was formed. However, the decrease in oxidation-site density for 40W-Zr was only due to a loss of surface area.

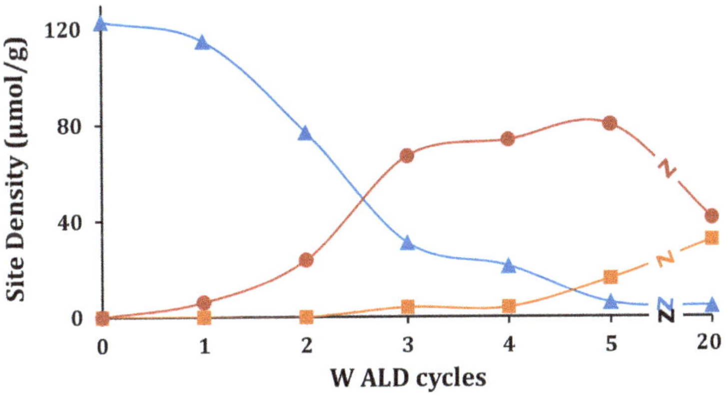

Figure 7. Reaction site densities of (▲) dehydrogenation sites, (●) Brønsted sites, and (■) oxidation sites as a function of the number of WO_x ALD cycles.

To determine the relationship between catalytic activity and the presence of Brønsted sites, the steady-state conversion of 2-propanol was performed on a series of ALD-prepared WO_x/ZrO_2 catalysts at 403 K and a Weight Hourly Space Velocity (WHSV) of 0.40 h^{-1}. Under these conditions, pure ZrO_2 was not catalytically active. As shown in Figure 8, there was no significant conversion over catalysts prepared with one or two WO_x ALD cycles. With higher WO_x concentrations, the propanol reacted to a mixture of water with dipropyl ether (~60%) and propene (40%). Conversions were maximized after five ALD rounds, and there was a reasonable correspondence between the concentration of Brønsted sites on these samples and the conversions. The fact that the 40W-Zr sample showed a higher conversion than would be expected based on its Brønsted-site density suggests that the bulk WO_x also has some activity for alcohol dehydration.

As discussed in the introduction, the quantification of Brønsted-acid strength in solid acids is difficult. Here, we measured light-off curves for H–D exchange between D_2O and either toluene (C_7H_8)

or chlorobenzene (C_6H_5Cl) on the 5W-Zr sample, and compared the results to those obtained on an H-ZSM-5 zeolite which had a similar Brønsted-site density of 80 μmol/g. Measurements began at room temperature, and were carried out in a steady-state reaction environment after each increment of 10 K. The results are shown in Figure 9. Firstly, on both H-ZSM-5 and 5W-Zr, the light-off temperatures for chlorobenzene occurred at higher temperatures, roughly 60 degrees higher on H-ZSM-5 and 75 degrees higher on 5W-Zr. This is consistent with a lower barrier for the reaction of toluene due to its significantly higher proton affinity. Secondly, the temperatures at which H–D exchange occurred were significantly higher on the tungstated zirconia. Sorption effects and other factors can play a role in the observed rates on solid acids [23]; however, the fact that the reaction occurred at higher temperatures for both adsorbates suggests that tungstated zirconia has much weaker acid sites than does H-ZSM-5.

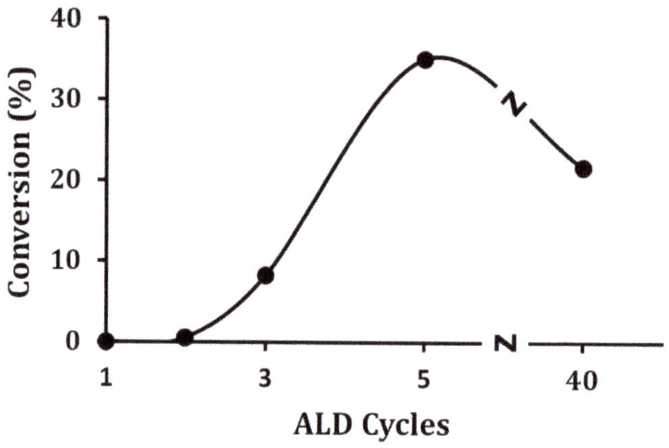

Figure 8. The 2-propanol conversions on *x*W-Zr (where *x* is the number of ALD cycles) samples for the steady-state dehydration reaction at 403 K. Reaction conditions: 2-propanol WHSV = 0.40 h^{-1}.

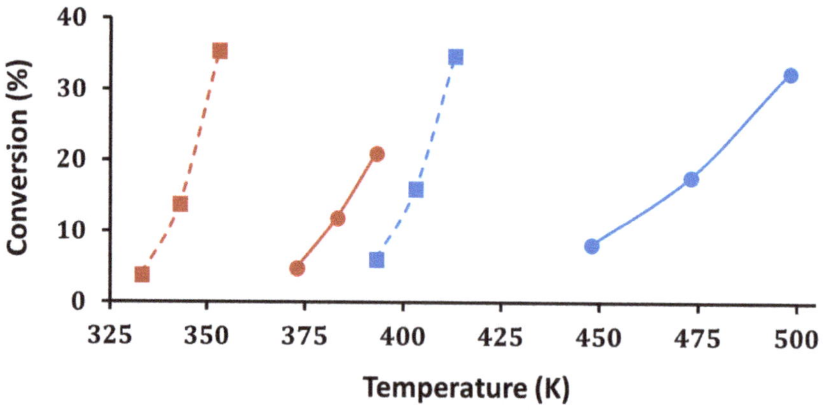

Figure 9. Conversions for H–D exchange of toluene (in red) and chlorobenzene (in blue) on (●) 5W-Zr and (■) H-ZSM-5 (280) as a function of temperature. The Brønsted-site densities for both materials were 80 μmol/g. Reaction conditions: 100 mg of catalyst, 1% toluene/chlorobenzene, and 1% D_2O in 20 mL/min He.

3. Discussion

In agreement with previous reports [1,17], our data here indicate that tungstated zirconia is a complex material that can exhibit Brønsted acidity, Lewis acidity, and oxidation activity, along with dehydrogenation activity on ZrO_2 [22]. The concentrations of the various types of sites depend on how the material is made, as shown by the differences we observed in materials synthesized via impregnation and using ALD. This complexity almost certainly accounts for at least some of the differences reported by the various groups who have worked on tungstated zirconia. In the presented study, we focused on controlling the composition and structure by synthesizing catalysts using ALD because of the uniformity that this approach provides.

The TPD results also demonstrate the power of using reactive probe molecules, rather than simple bases like ammonia or pyridine, for characterizing the types of sites that are present on these materials. The application of TPD in this manner is certainly not new, and 2-propanol [18] and 2-propanamine [19] were previously used to characterize site concentrations for Lewis and Brønsted acids. However, the results demonstrate that additional information can also be obtained. For example, the comparison of 2-propanol results for WO_x/ZrO_2 samples prepared via impregnation gave a strong indication that the WO_x did not uniformly cover the ZrO_2. The 2-propanamine results were also able to delineate concentrations of the various types of sites that were formed with increasing WO_x coverage.

Based on our reactivity studies, the Brønsted-acid sites formed in tungstated zirconia are relatively weak, certainly compared to the Brønsted-acid sites formed in zeolites. While site strengths may well depend on the detailed structure and method of synthesis, it is worth noting that others reported that WO_x/ZrO_2 is selective for the etherification of alcohols [7], without forming olefins as with zeolites, which would again imply the sites are relatively weak. We suggest that the high activity sometimes observed at low temperatures in tungstated zirconia is due to the combination of oxidation sites and/or dehydrogenation sites, together with Brønsted sites, similar to that which was proposed for sulfated zirconia [24]. With sulfated zirconia, the oxidation sites were shown to transform alkanes to olefins, which were then more easily activated by the acid sites. A similar combination of sites likely explains at least some of the catalytic properties of tungstated zirconia. The combination of sites allows isomerization reactions to occur at low temperatures without the need for especially strong Brønsted sites.

It is interesting that there were essentially no Brønsted sites formed after the deposition of a single WO_x ALD cycle on the ZrO_2. One would expect to form isolated WO_x species under these conditions, and the lack of Brønsted sites is interesting. The fact that more than one WO_x species must be present in order to form a Brønsted site was stated by others [1], who argued that the W:Brønsted-site ratio is about four. The value of four is indeed the approximate minimum value observed in Table 2, where the Brønsted-site concentrations are listed as a function of the number of ALD cycles added to the ZrO_2. While this ratio suggests that a $(WO_x)_n$ cluster is required for forming Brønsted sites, it is worth noting that the ratio of four occurs after three cycles, at which point it is expected that most of the tungsten is deposited in the first monolayer, implying that the clusters would be two-dimensional and still in contact with ZrO_2. The importance of zirconia is confirmed by the fact that we do not form significant quantities of Brønsted sites when WO_x is deposited on silica.

Finally, our H–D exchange results imply that the Brønsted sites in WO_x/ZrO_2 are relatively weak compared to the sites in high-silica zeolites. The early work by Hino and Arata on the isomerization of small alkanes at low temperatures argued that these materials are very strong Brønsted acids [25]; similarly, the calculated deprotonation energies that were reported imply that tungstated zirconia should have much stronger sites than those present in zeolites [1]. However, the recent publication on dodecanol etherification showed that WO_x/ZrO_2 was more selective than acidic zeolites for forming ethers, and produced less olefin and oligomerization products [7], a result that would imply weaker Brønsted sites. We suggest that the high activity reported for alkane activation at low temperatures could be the result of a combination of oxidation sites and Brønsted sites. For example, the low-temperature activity of sulfated zirconia was associated in part with oxidation sites which form olefins that then go on to react over the Brønsted sites [9]. Since the tungsten clusters on zirconia were shown to be reducible, both in this study and previously [26], the low-temperature, isomerization activity may be the result of a combination of acidic and oxidation sites. Similarly, the presence of Pt in low-temperature isomerization catalysts likely leads to the formation of alkenes, which are much more easily protonated [27,28].

Obviously, there is still much to learn about the nature and properties of the catalytic sites in tungstated zirconia. We believe that the combination of controlled synthesis using ALD with careful adsorption studies of the type performed here can help elucidate the nature of these sites.

4. Experimental

The ZrO_2 support was prepared by titrating (0.2 mL/s) 30% aqueous NH_4OH (Fischer) to a 5-wt% aqueous solution of $ZrO(NO_3)_2{\bullet}xH_2O$ (99%, Sigma Aldrich, St. Louis, MO, USA) with vigorous stirring. The precipitate was then dried at 333 K for 12 h before being calcined in a Muffle furnace at 773 K for 5 h. ALD was performed in a custom-built, static system that could be evacuated by a mechanical pump, and was described in previous studies [12,29]. In a typical ALD cycle, the evacuated 300 mg of zirconia powder was first exposed to the vapor of $W(CO)_6$ (99%, Strem, Newburyport, MA, USA) at 473 K for 3 min, followed by a 5-min evacuation. The samples were then exposed to air at the same temperature for 6 min, before once again being evacuated to complete the cycle. We used the nomenclature *x*W-Zr to refer to a sample exposed to *x* ALD cycles. For comparison purposes, a sample with 10-wt% WO_x/ZrO_2 (impW-Zr) was prepared via conventional aqueous incipient wetness of $(NH_4)_{10}W_{12}O_{41}{\bullet}5H_2O$ (99.999%, Alfa Aesar, Haverhill, MA, USA), followed by a 4-h calcination at 773 K. To determine the effect of the support, a WO_x/SiO_2 sample (5W-Si) was prepared by depositing five ALD cycles onto a stabilized SiO_2 (Degussa AG, Essen, Germany, Ultrasil VN 3 SP, 140 m^2/g).

The temperature-programmed desorption/thermogravimetric analysis (TPD-TGA) measurements were performed on samples held in a system consisting of an evacuated CAHN 2000 microbalance, equipped with an SRI quadrupole mass spectrometer (RGA100), that is described elsewhere [30]. The 50-mg samples were first heated in a vacuum to 823 K, before being cooled to room temperature in a vacuum, and then exposed to the vapor of the probe adsorbate, either 2-propanol (99.9%, Fisher, Hampton, NH, USA), or 1-propanamine (99+%, Alfa Aesar), or 2-propanamine (99%, Alfa Aesar). After 1 h of evacuation, the TPD and TGA measurements were obtained while ramping the temperature at 10 K/min.

Scanning transmission electron microscopy (STEM) and elemental mapping via energy-dispersive X-ray spectroscopy (EDS) were performed with a JEOL 2010F field-emission scanning transmission electron microscope (JEOL, Tokyo, Japan), operated at an accelerating voltage of 200 kV with a 0.7 nm STEM probe. Infrared spectra of adsorbed pyridine were performed on a Mattson Galaxy FTIR (Madison Instruments Inc., Middleton, WI, USA) with a diffuse-reflectance attachment (Collector II™) in order to confirm the presence of Brønsted sites [31]. In the FTIR cell, the samples were initially heated to 373 K in flowing He at 60 mL/min to remove any adsorbed water for 10 min. After cooling the samples to room temperature, pyridine vapors were exposed to the sample for 10 min. The samples were then flushed with flowing He at 60 mL/min for 10 min. Raman spectra were obtained with an NTEGRA Spectra system (NT-MDT) with an excitation laser wavelength of 532 nm. The experiments were carried out with a laser power of 15 mW (10% of 150 W from the natural-density filter setting) at the samples and a collection time of 60 s. Powder X-ray Diffraction (XRD) patterns were collected from a Rigaku Smartlab diffractometer equipped with a Cu Kα source (The Woodlands, TX, USA).

The steady-state reaction rates for various reactions (2-propanol dehydration, H–D exchange between toluene (C_7H_8) and D_2O, and H–D exchange between chlorobenzene (C_6H_5Cl) and D_2O) were measured in a flow reactor that consisted of a 200-mm long, 4.6-mm ID stainless-steel tube, packed with 100 mg of catalyst that was held in place by two quartz-wool plugs and an inert tube from the back end. Products were monitored using an online GC-MS (QP-5000, Shimadzu, Kyoto, Japan) and the results were compared to reaction measurements over an H-ZSM-5 catalyst (Si/Al_2 = 280;

Zeolyst, CBV 28014, Conshohocken, PA, USA). For the dehydration of 2-propanol, the inlet flow to the reactor was 5% 2-propanol, achieved by feeding 0.9 μL/min liquid 2-propanol with a syringe pump (Harvard Apparatus, PHD 2000, Holliston, MA, USA) in a 5 mL/min He flow. In the H–D exchange measurements, an equal molar fraction of 1% toluene (or chlorobenzene) and D_2O were co-fed with a syringe pump in a 20 mL/min He flow. Prior to the reaction, the catalysts were pretreated in flowing He at 523 K for 30 min before measuring steady-state reaction rates using an online GC-MS (QP-5000, Shimadzu). The conversions of H–D exchange were quantified via the deconvolution of mass fragmentations from *m*/*e* 91 to 97 for toluene, and from *m*/*e* 112 to 118 for chlorobenzene.

5. Conclusions

(1) The ALD of $W(CO)_6$ on ZrO_2 allows the deposition of uniform layers of WO_x, with coverages varying from sub-monolayer to multilayer.
(2) The TPD measurements with 2-propanamine on the WO_x/ZrO_2 samples identified three types of sites: dehydrogenation sites associated with uncovered ZrO_2, Brønsted sites formed by monolayer clusters of WO_x, and redox sites associated with multilayers of WO_x.
(3) The formation of Brønsted sites for WO_x/ZrO_2 requires the presence of multiple W atoms.
(4) The Brønsted sites in WO_x/ZrO_2 are active for alcohol dehydration, but are significantly weaker than the Brønsted sites in an H-ZSM-5 zeolite, as shown via H–D exchange with toluene and chlorobenzene.

Supplementary Materials: The following are available online at http://www.mdpi.com/2073-4344/8/7/292/s1, Figure S1. TPD-TGA of 1-propanamine over ZrO_2. The desorption features correspond to hydrogen (*m*/*e* = 2), unreacted 1-propanamine (*m*/*e* = 30) and propionitrile (*m*/*e* = 54). The graphic symbols represent schematic ZrO_2 support (gray).

Author Contributions: R.J.G. and C.B.M. initiated the concept; C.W. and J.D.L. designed and conceived the experiments; C.W., X.M., J.D.L., and T.M.O. performed the experiments; C.W., J.D.L., T.M.O., and Y.-H.Y. analyzed the data; C.W. and J.D.L. drafted the manuscript; R.J.G. and C.B.M. reviewed the manuscript prior to submission. All authors approved the final manuscript.

Funding: This research was funded by the US Department of Energy, Office of Science, Office of Basic Energy Sciences under Award no. DE-SC0001004.

Acknowledgments: We acknowledge support from the Catalysis Center for Energy Innovation, an Energy Frontier Research Center funded by the US Department of Energy, Office of Science, Office of Basic Energy Sciences under Award no. DE-SC0001004. We acknowledge the NSF Major Research Instrumentation Grant DMR-0923245 for the Raman spectroscopy measurement.

Conflicts of Interest: The authors declare no conflict of interest.

References

1. Zhou, W.; Soultanidis, N.; Xu, H.; Wong, M.S.; Neurock, M.; Kiely, C.J.; Wachs, I.E. Nature of Catalytically Active Sites in the Supported WO_3/ZrO_2 Solid Acid System: A Current Perspective. *ACS Catal.* **2017**, *7*, 2181–2198. [CrossRef]
2. Lebarbier, V.; Clet, G.; Houalla, M. A Comparative Study of the Surface Structure, Acidity, and Catalytic Performance of Tungstated Zirconia Prepared from Crystalline Zirconia or Amorphous Zirconium Oxyhydroxide. *J. Phys. Chem. B* **2006**, *110*, 13905–13911. [CrossRef] [PubMed]
3. Rossmedgaarden, E.; Knowles, W.; Kim, T.; Wong, M.; Zhou, W.; Kiely, C.J.; Wachs, I.E. New insights into the nature of the acidic catalytic active sites present in ZrO_2-supported tungsten oxide catalysts. *J. Catal.* **2008**, *256*, 108–125. [CrossRef]
4. Digregorio, F.; Keller, N.; Keller, V. Activation and isomerization of hydrocarbons over WO_3/ZrO_2 catalysts, II. Influence of tungsten loading on catalytic activity: Mechanistic studies and correlation with surface reducibility and tungsten surface species. *J. Catal.* **2008**, *256*, 159–171. [CrossRef]
5. Iglesia, E.; Barton, D.G.; Soled, S.L.; Miseo, S.; Baumgartner, J.E.; Gates, W.E.; Fuentes, G.A.; Meitzner, G.D. Selective isomerization of alkanes on supported tungsten oxide acids. In *Studies in Surface Science and Catalysis*; Elsevier: New York, NY, USA, 1996; Volume 101, pp. 533–542. ISBN 0167-2991.

6. Baertsch, C.D.; Soled, S.L.; Iglesia, E. Isotopic and Chemical Titration of Acid Sites in Tungsten Oxide Domains Supported on Zirconia. *J. Phys. Chem. B* **2001**, *105*, 1320–1330. [CrossRef]
7. Rorrer, J.; He, Y.; Toste, F.D.; Bell, A.T. Mechanism and kinetics of 1-dodecanol etherification over tungstated zirconia. *J. Catal.* **2017**, *354*, 13–23. [CrossRef]
8. Jung, S.; Lu, C.; He, H.; Ahn, K.; Gorte, R.J.; Vohs, J.M. Influence of composition and Cu impregnation method on the performance of Cu/CeO_2/YSZ SOFC anodes. *J. Power Sources* **2006**, *154*, 42–50. [CrossRef]
9. Barton, D.G.; Soled, S.L.; Meitzner, G.D.; Fuentes, G.A.; Iglesia, E. Structural and catalytic characterization of solid acids based on zirconia modified by tungsten oxide. *J. Catal.* **1999**, *181*, 57–72. [CrossRef]
10. Scheithauer, M.; Jentoft, R.; Gates, B.; Knözinger, H. *n*-Pentane isomerization catalyzed by Fe-and Mn-containing tungstated zirconia characterized by raman spectroscopy. *J. Catal.* **2000**, *191*, 271–274. [CrossRef]
11. Malm, J.; Sajavaara, T.; Karppinen, M. Atomic Layer Deposition of WO_3 Thin Films using $W(CO)_6$ and O_3 Precursors. *Chem. Vap. Depos.* **2012**, *18*, 245–248. [CrossRef]
12. Wang, C.; Lee, J.D.; Ji, Y.; Onn, T.M.; Luo, J.; Murray, C.B.; Gorte, R.J. A Study of Tetrahydrofurfuryl Alcohol to 1, 5-Pentanediol over Pt–WOx/C. *Catal. Lett.* **2018**, *148*, 1047–1054. [CrossRef]
13. Nandi, D.K.; Sarkar, S.K. Atomic Layer Deposition of Tungsten Oxide for Solar Cell Application. *Energy Procedia* **2014**, *54*, 782–788. [CrossRef]
14. Kresnawahjuesa, O.; Kühl, G.; Gorte, R.J.; Quierini, C. An examination of Brønsted acid sites in H-[Fe] ZSM-5 for olefin oligomerization and adsorption. *J. Catal.* **2002**, *210*, 106–115. [CrossRef]
15. Chen, K.; Damron, J.; Pearson, C.; Resasco, D.; Zhang, L.; White, J.L. Zeolite Catalysis: Water Can Dramatically Increase Or Suppress Alkane C–H Bond Activation. *ACS Catal.* **2014**, *4*, 3039–3044. [CrossRef]
16. Farneth, W.E.; Roe, D.C.; Kofke, T.G.; Tabak, C.J.; Gorte, R.J. Proton transfer to toluene in H-ZSM-5: TPD, IR, and NMR studies. *Langmuir* **1988**, *4*, 152–158. [CrossRef]
17. Ross-Medgaarden, E.I.; Wachs, I.E. Structural Determination of Bulk and Surface Tungsten Oxides with UV−vis Diffuse Reflectance Spectroscopy and Raman Spectroscopy. *J. Phys. Chem. C* **2007**, *111*, 15089–15099. [CrossRef]
18. Luo, J.; Yu, J.; Gorte, R.J.; Mahmoud, E.; Vlachos, D.G.; Smith, M.A. The effect of oxide acidity on HMF etherification. *Catal. Sci. Technol.* **2014**, *4*, 3074–3081. [CrossRef]
19. Gorte, R.J. What do we know about the acidity of solid acids? *Catal. Lett.* **1999**, *62*, 1–13. [CrossRef]
20. Kofke, T.G.; Gorte, R.J.; Farneth, W.E. Stoichiometric adsorption complexes in H-ZSM-5. *J. Catal.* **1988**, *114*, 34–45. [CrossRef]
21. Yeh, Y.-H.; Zhu, S.; Staiber, P.; Lobo, R.F.; Gorte, R.J. Zn-Promoted H-ZSM-5 for Endothermic Reforming of *n* -Hexane at High Pressures. *Ind. Eng. Chem. Res.* **2016**, *55*, 3930–3938. [CrossRef]
22. He, M.-Y.; Ekerdt, J.G. Temperature-programmed studies of the adsorption of synthesis gas on zirconium dioxide. *J. Catal.* **1984**, *87*, 238–254. [CrossRef]
23. Luo, J.; Gorte, R. High pressure cracking of *n*-hexane over H-ZSM-5. *Catal. Lett.* **2013**, *143*, 313–316. [CrossRef]
24. Wan, K.T.; Khouw, C.B.; Davis, M.E. Studies on the catalytic activity of zirconia promoted with sulfate, iron, and manganese. *J. Catal.* **1996**, *158*, 311–326. [CrossRef]
25. Hino, M.; Arata, K. Synthesis of solid superacid of tungsten oxide supported on zirconia and its catalytic action for reactions of butane and pentane. *J. Chem. Soc. Chem. Commun.* **1988**, 1259–1260. [CrossRef]
26. Occhiuzzi, M.; Cordischi, D.; Gazzoli, D.; Valigi, M.; Heydorn, P.C. WO_x/ZrO_2 catalysts. *Appl. Catal. Gen.* **2004**, *269*, 169–177. [CrossRef]
27. Filimonova, S.; Nosov, A.; Scheithauer, M.; Knözinger, H. *n*-Pentane Isomerization over Pt/WOx/ZrO_2 Catalysts: A ^1H and ^{13}C NMR Study. *J. Catal.* **2001**, *198*, 89–96. [CrossRef]
28. Kuba, S.; Che, M.; Grasselli, R.K.; Knözinger, H. Evidence for the formation of W^{5+} centers and OH groups upon hydrogen reduction of platinum-promoted tungstated zirconia catalysts. *J. Phys. Chem. B* **2003**, *107*, 3459–3463. [CrossRef]
29. Onn, T.M.; Monai, M.; Dai, S.; Fonda, E.; Montini, T.; Pan, X.; Graham, G.W.; Fornasiero, P.; Gorte, R.J. Smart Pd Catalyst with Improved Thermal Stability Supported on High-Surface-Area $LaFeO_3$ Prepared by Atomic Layer Deposition. *J. Am. Chem. Soc.* **2018**, *140*, 4841–4848. [CrossRef] [PubMed]

30. Yu, J.; Zhu, S.; Dauenhauer, P.J.; Cho, H.J.; Fan, W.; Gorte, R.J. Adsorption and reaction properties of SnBEA, ZrBEA and H-BEA for the formation of p-xylene from DMF and ethylene. *Catal. Sci. Technol.* **2016**, *6*, 5729–5736. [CrossRef]
31. Yu, J.; Luo, J.; Zhang, Y.; Cao, J.; Chang, C.-C.; Gorte, R.J.; Fan, W. An examination of alkali-exchanged BEA zeolites as possible Lewis-acid catalysts. *Microporous Mesoporous Mater.* **2016**, *225*, 472–481. [CrossRef]

catalysts

MDPI

Article

Efficient Multifunctional Catalytic and Sensing Properties of Synthesized Ruthenium Oxide Nanoparticles

Ruby Phul [1], Mohammad Perwez [2], Jahangeer Ahmed [3], Meryam Sardar [2], Saad M. Alshehri [3], Norah Alhokbany [3], Mohd A. Majeed Khan [4] and Tokeer Ahmad [1,*]

1 Nanochemistry Laboratory, Department of Chemistry, Jamia Millia Islamia, New Delhi 110025, India; rubyphul@gmail.com
2 Enzyme Technology Laboratory, Department of Biosciences, Jamia Millia Islamia, New Delhi 110025, India; perwezmohammad@gmail.com (M.P.); msardar@jmi.ac.in (M.S.)
3 Department of Chemistry, College of Science, King Saud University, Riyadh 11451, Saudi Arabia; jahmed@ksu.edu.sa (J.A.); alshehri@ksu.edu.sa (S.M.A.); nhokbany@ksu.edu.sa (N.A.)
4 King Abdullah Institute for Nanotechnology, King Saud University, Riyadh 11451, Saudi Arabia; mmkhan@ksu.edu.sa
* Correspondence: tahmad3@jmi.ac.in; Tel.: +91-11-26981717 (ext. 3261); Fax: +91-11-26980229

Received: 23 June 2020; Accepted: 7 July 2020; Published: 13 July 2020

Abstract: Ruthenium oxide is one of the most active electrocatalyst for oxygen evolution (OER) and oxygen reduction reaction (ORR). Herein, we report simple wet chemical route to synthesize RuO_2 nanoparticles at controlled temperature. The structural, morphological and surface area studies of the synthesized nanoparticles were conducted with X-ray diffraction, electron microscopy and BETsurface area studies. The bifunctional electrocatalytic performance of RuO_2 nanoparticles was studied under different atmospheric conditions for OER and ORR, respectively, versus reversible hydrogen electrode (RHE) in alkaline medium. Low Tafel slopes of RuO_2 nanoparticles were found to be ~47 and ~49 mV/dec for OER and ORR, respectively, in oxygen saturated 0.5 M KOH system. Moreover, the catalytic activity of RuO_2 nanoparticles was examined against the Horseradish peroxidase enzyme (HRP) at high temperature, and the nanoparticles were applied as a sensor for the detection of H_2O_2 in the solution.

Keywords: ruthenium oxide; nanoparticles; electrocatalysts; sensing

1. Introduction

The present generation is largely dependent on fossil fuelsto meet the present energy requirements—for instance, oil, coal, or natural gases. However, these energy demands fulfilled by these products directly affect the environment. The burning of fossil fuels leads up to the emissions of carbon dioxide gas (a greenhouse gas), which is affecting the world significantly through global warming, change in weather patterns and several other noteworthy geographical changes [1]. In addition, we know fossil fuels are nonrenewable resources, hence they will eventually deplete, so alternatives must be found. Therefore, the development of efficient, inexpensive and eco-friendly sources of energy has become a significant and crucial task for the researchers [2]. Scientists have investigated the use of renewable resources such as solar, wind, tidal, biomass, geothermal energy etc. The major research trend of today's era is the water-splitting phenomena for energy generation. Water splitting via electrocatalysis or photocatalysis is a clean, environmentally friendly and renewable source of energy for fuel cells, batteries and hydrogen generation [3–5]. In the presence of electro/photocatalyst, the water molecule splits into hydrogen and oxygen gas (i.e., $H_2O \rightarrow H_2 + 1/2\ O_2$). The evolved hydrogen gas is used in fuel cells as a fuel which further reacts with O_2 to produce an electric current. The evolved oxygen gas participates in the combustion reaction of fuel cells to generate power.

Forthe past few decades, researchers made headway in creating robust and efficient catalysts for the oxidation of water. Several noble metals viz. Pt, Ir, Ru and their oxide-based catalysts have been developed for oxidation–reduction reactions [6–8]. Various other earth-abundant metal-based catalysts such as Mn, Fe, Co, Cu and W were also reported for their role in water splitting reactions [9–13]. In a previous research era, stupendous research has been done in the fabrication of active electrode material for oxygen evolution reactions (OERs). The main limitation with this method is that the OER at the anode gives rise to a high energy loss. Therefore, the emphasis has been given on attaining a high oxygen evolution rate at a low overpotential by optimizing the overall water splitting reaction. Electrode corrosion and low current densities are also the major disadvantages of conventional anode materials for OER [14]. Currently, ruthenium oxide has been used in the fabrication of dimensionally stable anodes (DSA), which havebeen employed to yield chlorine [15]. In addition, it has been used as a heterogeneous catalyst for the low temperature dehydrogenation of NH_3 [16], HCl [17], and methanol [18], respectively. Further, ruthenium oxide was reported to work as anexcellent electrode material for OERs [19,20] and hydrogen evolution reactions (HERs) [21,22]. The metallic conductivity of RuO_2, along with IrO_2,is of the order of 10^4 $ohm^{-1}cm^{-1}$ [23]. IrO_2 shows high corrosion-resistance, whereas RuO_2 shows better OER activity [24]. Among all the other transition metal oxides, RuO_2 and IrO_2 are considered as the best electrocatalytic materials for electrolysis of water in acidic as well as an alkaline mediums [25,26].

In the past few years, nanocrystalline RuO_2 particles were synthesized by thermal evaporation [27], nanocasting [28], cryogenic decomposition of RuO_4 [29] and electro-spinning [30]. Recently, M. P. Browne et al. [31] synthesized a series of $Mn_xO_y/RuO_2/Ti$ mixed oxide anode materials via a thermal decomposition method for OERs in alkaline medium. They have shown that electrocatalysts were containing different concentrations of Mn viz. 10%, 25% and 90% show almost similar or improved OER activity as compared to pure RuO_2. Gustafson et al. [32] have synthesized RuO_2 nanocatalyst for chemical and photochemical oxidation of water, which showed better catalytic performance as reported in the literature.

Presently, the artificial enzymes, i.e., nanoenzymes, are receiving significant attention of researchers due to their low cost, high catalytic property and thermal stability as compared to the natural enzymes. Prototypically, Gao et al. discovered that magnetite nanoparticles exhibit intrinsic peroxidase-like activity similar to that of a natural peroxidase enzyme [33]. H_2O_2 plays a vital role as an intermediate in food, pharmaceutical, clinical, and environmental analysis [34–36]. So, the detection of H_2O_2 has been done by using different nanoparticles. Besides that, H_2O_2 possesses a strong oxidizing property which may lead to different types of disorders in the body [37–39]. Hence, the detection of hydrogen peroxide is of a great practical feature.

Herein, we report the redox reaction of water (OER/ORR) happening through as-synthesized ruthenium oxide nanoparticles in 0.5M KOH electrolytic solution at room temperature under different atmospheres (air, N_2 and O_2). Further, the as-synthesized ruthenium oxide nanoparticles were also used as a sensor for the detection of H_2O_2 in solution. Moreover, the synthesis of RuO_2 nanoparticles was carried out through a simple, environmentally friendly and cost-effective wet chemical method at 80 °C followed by annealing at 300 °C for 6h.

2. Results and Discussion

The powder X-ray diffractometry was used for the structural analysis of the as-synthesized nanoparticles. Figure 1a shows the X-ray diffraction pattern of the nanocrystalline RuO_2. The obtained diffraction peaks are as follows at Bragg's angles of 27.8°, 34.9°, 39.8°, 54.0°, 57.6°, 59.2°, 65.2°, 66.7°, 69.2°, 73.8°, 82.9° and 87.3° corresponds to the planes (110), (101), (200), (211), (220), (002), (310), (112), (301), (202), (321) and (222), respectively, which were correlated to a tetragonal unit cell of ruthenium oxide. The reflection pattern could be indexed to a pure tetragonal phase of RuO_2(JCPDS No. 065-2824).

No peaks from any impurity or other phase or metallic Ru were detected, which affirms the formation of monophasic RuO_2 nanoparticles.

The microstructure and surface texture of as-synthesized ruthenium oxide nanoparticles wereinvestigated through SEM studies. Figure 1b shows the SEM micrograph of the as-synthesized nanoparticles. Further analysis of the SEM micrograph depicted the highlydense and agglomerated RuO_2 nanoparticles. The nanoparticles aggregate randomly to form almost spherical shape with an average diameter of 28 nm, which is as per the TEM analysis.

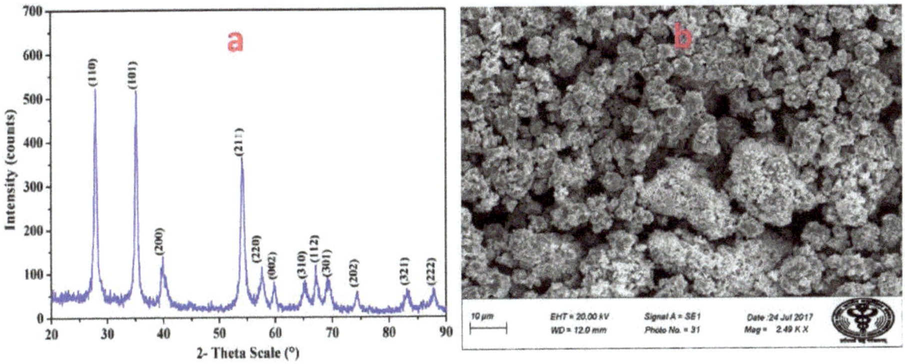

Figure 1. (**a**) X-ray diffraction pattern and (**b**) scanning electron microscope (SEM) micrograph of Ruthenium oxide nanoparticles.

The detailed structural analysis, shape and size distribution of RuO_2 nanoparticles was done with the help of TEM studies. The low magnification TEM micrograph is shown in Figure 2a, which indicates the formation of tiny sized tetragonal RuO_2 nanoparticles with slight agglomeration. Figure 2a also reveals that the small-sized nanoparticles tend to form large tetragonal structures, which are in accordance with X-ray diffraction studies. Furthermore, High Resolution Transmission Electron Microscope (HRTEM) analysis revealed the crystal structure, phase and growth direction of as-synthesized nanoparticles. Figure 2b shows the typical HRTEM image of ruthenium oxide nanoparticles, which depicts the well-resolved lattice fringes with an average lattice distance of 3.210 ± 0.05 Å and 2.560 ± 0.05 Å corresponding to (110) and (101) planes, respectively, of ruthenium oxide nanoparticles. The TEM average size distribution histogram of ruthenium oxide nanoparticles are shown in Figure 2c, which indicates that the particle size ranges from 5 nm to 35 nm, as the various small particles have combined to form a single massive particle. The average grain size was found to be ~20nm by using TEM micrograph as well as a size distribution plot.

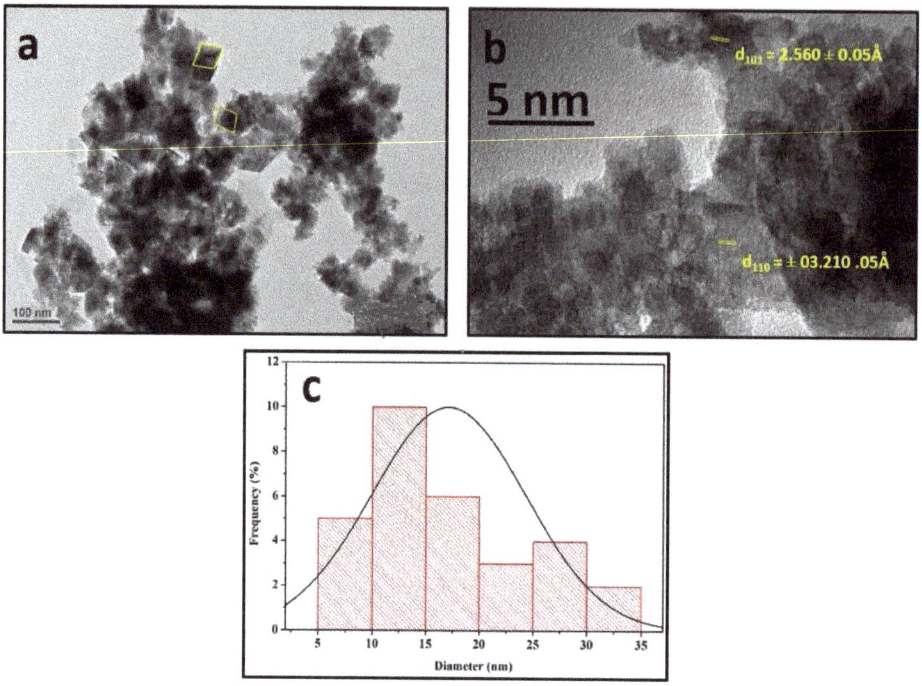

Figure 2. (**a**) TEM micrograph, (**b**) HR-TEM image and (**c**) size distribution histogram of Ruthenium oxide nanoparticles.

The specific surface area of as-prepared RuO_2 nanoparticles was estimated by using a multipoint BET equation that showed the linear relationship in the P/P_0 range of 0.05–0.35. Figure 3a shows the BET plot for RuO_2 nanoparticles. The specific surface area was found to be 64.5 m^2g^{-1}, which agrees with the earlier reported value [19]. The BJH (Barrett-Joyner-Halenda) model was used to determine the pore size. The pore size distribution plot of ruthenium oxide nanoparticles (Figure 3b) gives the pore radius value of 16 Å, which lies in the range of mesoporous materials. The pore radius was also determined by the DA (Dubinin-Astakhov) plot, as shown in Figure 3c, and it was found to be 13.5 Å, which is a bit smaller than the BJH results.

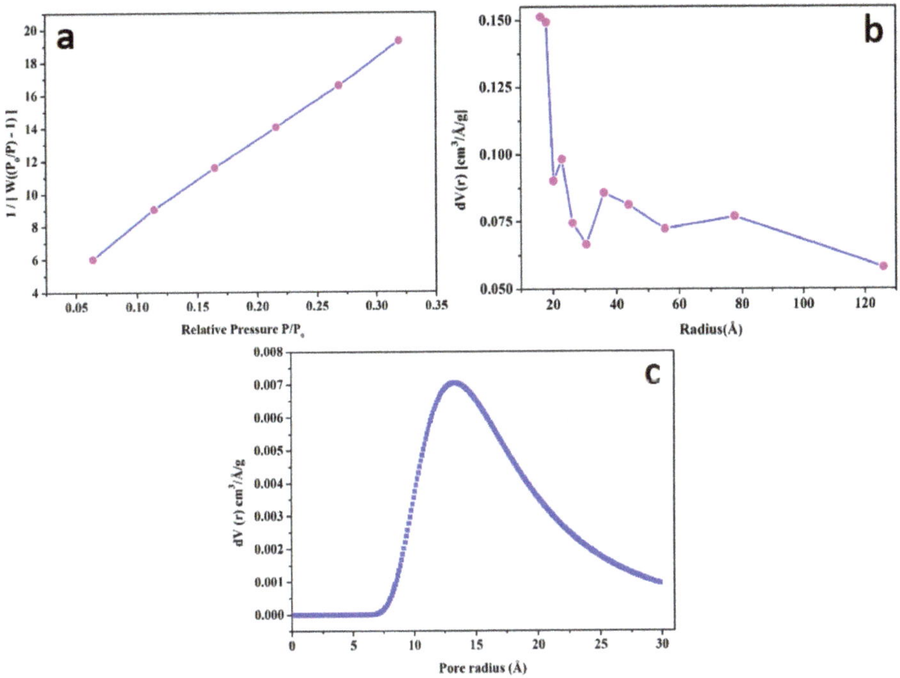

Figure 3. (**a**) BET, (**b**) BJH and (**c**) DA pore radius plots of ruthenium oxide nanoparticles.

The electrocatalytic activity of as-prepared nanoparticles for oxygen evolution reaction (OER) and oxygen reduction reaction (ORR) was evaluated by cyclic voltammetry (CV), linear sweep voltammetry (LSV) and Tafel polarization curves in 0.5M KOH electrolyte solution. The CV plots for OER (anodic sweep) and ORR (cathodic sweep) by ruthenium oxide electrode at the scan rate of 25 mVs^{-1} in the air (black), N_2 (red) and saturated O_2 (blue) atmosphere are shown in Figure 4a. The CV curves show that the OER starts from the low potential value of ~1.5 V vs. RHE (reversible hydrogen electrode). Figure 4a shows that the as-synthesized RuO_2 nanoparticles generate more current in O_2 saturated (17.5 $mAcm^{-2}$) as compared to air (11.5 $mAcm^{-2}$) and N_2 (10.5 $mAcm^{-2}$) atmosphere at 1.55 V vs. RHE at 25 mVs^{-1} for the oxygen evolution reaction. The ORR activity of ruthenium oxide nanoparticles in alkaline medium is also shown in Figure 4a. It was observed that as-synthesized RuO_2 nanoparticles show an almost comparable ORR reaction in all the atmospheric conditions. The LSV measurements optimized the electrocatalytic activity of ruthenium oxide nanoparticles in 0.5M KOH electrolyte at the scan rate of 25 mVs^{-1}. Figure 4b shows the LSV curves under air, nitrogen and O_2 saturated atmospheres. It was observed that the onset potential for OER was found to be ~1.5 (O_2) and ~1.61 V (air and N_2) vs. RHE. Notably, the resulting current density (current/area of the electrode) of RuO_2 electrode is directly related to the amount of oxygen evolved from the electrolysis of water. The geometric electro-active surface area of the working electrode could be estimated from the Randles–Sevik equation [40]. The current density of RuO_2 electrode at ~1.7 V versus RHE at 25 mVs^{-1} was found to be ~9.6 $mAcm^{-2}$ (in N_2), ~11.9 $mAcm^{-2}$ (in air) and ~15.5 $mAcm^{-2}$ (in O_2). The significance of LSV measurements is to find the onset potential, reaction kinetics and mechanism of the reaction, i.e., to identify the number of electrons taking part in the electrochemical reaction. Figure 4c shows the LSV curves of RuO_2 nanoparticles for the oxygen reduction reaction in air, N_2, and O_2 saturated 0.5M KOH at the scan rate of 25 mVs^{-1}. From this study, we clearly observed that the ORR activity performed significantly better in an O_2-saturated system compared to other systems as expected. The ORR activity in air could be due to the dissolved oxygen in system. Note that very weak ORR activity in N_2 was also observed, which could be due to the presence of oxygen content (5%) in N_2. The onset potential for ORR was found to be 0.7 V vs. RHE. Choronoamperometric (CA) measurements demonstrated the stability and the electrocatalyticactivity of the ruthenium oxide electrode at a fixed potential (1.5V vs. RHE) in O_2-saturated 0.5M KOH for 200 s. Figure 4d shows the CA curves, which demonstrate that the material is stable, and the constant current is generated with time. It was observed that the resulting current densities were consistent with time. The CA experiments also showed that on turning off the potential, the water redox reaction was stopped instantly, and therefore the current density dropped to zero. The current density obtained was directly proportional to the amount of gas evolved during the OER. The surface area of the electrode material used, and the Faradaic and Non-Faradaic processes were responsible for the resulting current density.

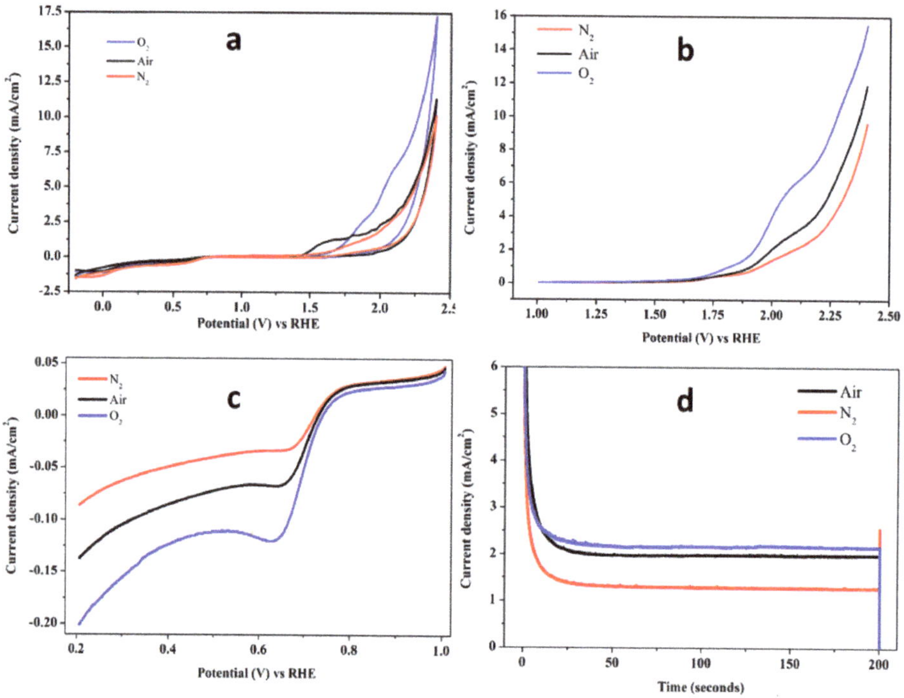

Figure 4. (**a**) CV, (**b**) LSV (OER), (**c**) LSV (ORR) curves of Ruthenium oxide photoanode in air, N_2 and O_2 saturated 0.5M KOH electrolyte vs. RHE at the scan rate of 25 mVs^{-1} and (**d**) CA curves of Ruthenium oxide nanoparticles for OER activity at 1.5 V vs. RHE in O_2-saturated 0.5M KOH.

The Tafel polarization studies determined the kinetics of the reaction, i.e., electrolysis of water. It was observed that reaction kinetics strongly depends upon the size, surface area, morphology and orientation of electrocatalysts [14,41]. Figure 5a,b shows the Tafel polarization plot of ruthenium oxide nanoparticles for the water redox reactions (OER and ORR) in air, and O_2, respectively. The linear curve fitting calculated the value of Tafel slopes of ruthenium oxide nanoparticles, and it comes out to be 76, and 47 $mVdec^{-1}$ in air, and O_2, respectively, for OER while for ORR, these values were found to be 48 (air), and 49 $mVdec^{-1}$ (O_2) with the experimental error of ±5. The effective electro-active catalysts for water splitting could lower the Tafel slope values to sustain the high activity, stability and to enhance the efficiency by reducing the loss of energy during the electrochemical reactions [42].

The comparison in the current electrocatalytic activities viz. OER/ORR of RuO_2 nanoparticles with the literature has been tabulated in Table 1.

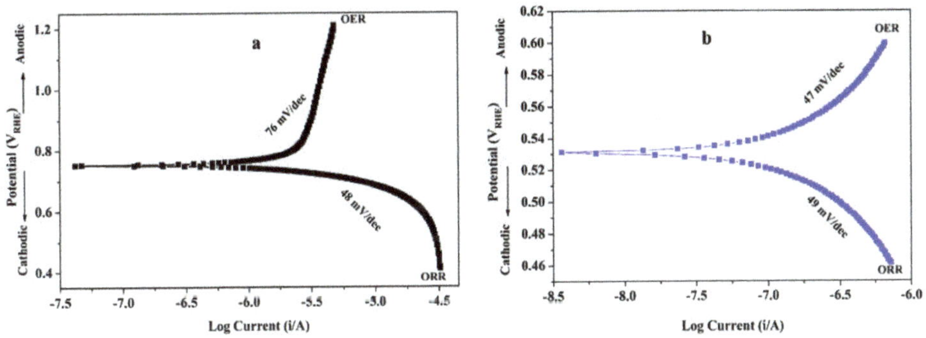

Figure 5. Tafel polarization curves of Ruthenium oxide nanoparticles for OER and ORR in (**a**) air, and (**b**) O_2 saturated atmosphere at 10 mVs^{-1}.

Table 1. Comparison of OER/ORR activity of RuO_2 nanoparticles with other reported literature.

Catalyst	Electrolyte	Scan Rate (mVs^{-1})	Onset Potential (V/RHE)	TafelSlope ($mVdec^{-1}$)		Ref.
				OER	ORR	
r-RuO_2	0.1M KOH	10	1.4	-	-	[25]
Ru@RuO_2	0.1M KOH	10	1.3	86	-	[43]
1D-RuO_2-CN_x	0.5M KOH	10	1.42	56	-	[44]
RuO_2	0.05M NaOH	10	1.27	-	-	[45]
$Mn_{25}Ru_{75}$@450	1M NaOH	10	1.4	66	-	[31]
RuO_2 nanoparticles	0.5M KOH	25	1.5	47	49	Present work

The catalytic activity of as-synthesized nanoparticles depends on the concentration of H_2O_2, and the reaction temperature just like that of HRP [33]. The catalytic oxidation of TMB substrate by RuO_2 nanoparticles at a range of H_2O_2 concentration and temperature is shown in Figure 6. It was observed that to attain maximum activity; the RuO_2 nanoparticles required a very high concentration of H_2O_2 (1M) as compared to HRP and on further increasing the concentration of H_2O_2 the catalytic activity was suppressed (Figure 6a). The effect of temperature on the catalytic activity of the RuO_2 nanoparticles was checked in the temperature range of 20–90 °C. Figure 6b indicates that the catalytic activity of RuO_2 nanoparticles increases with the increase in temperature till 60 °C and with further increase in temperature up to 90 °C the activity was quenched. Whereas, HRP shows maximum activity at 30 °C and it shows no activity at higher temperature, i.e., above 60 °C. The as-synthesized nanoparticles showed better activity at a wide range of temperature compared to HRP catalyst. The thermal stability of the RuO_2 nanoparticles and HRP was determined by incubating the reaction mixture with TMB substrate at 80 °C for 90 min, and the aliquots were taken at regular time intervals. It was found that the catalytic activity of RuO_2 nanoparticles was almost preserved until 90 min at 80 °C while the HRP showed no activity in similar reaction conditions (Figure 6c). We have also analyzed the sensitivity of RuO_2 nanoparticles for the detection of H_2O_2 in solution. The aliquots containing different concentration of H_2O_2 and colourimetric reagent were taken and analyzed via spectrophotometry. Figure 6d shows the absorbance at 505 nm, which increases nearly linearly as the concentration of H_2O_2 increased.

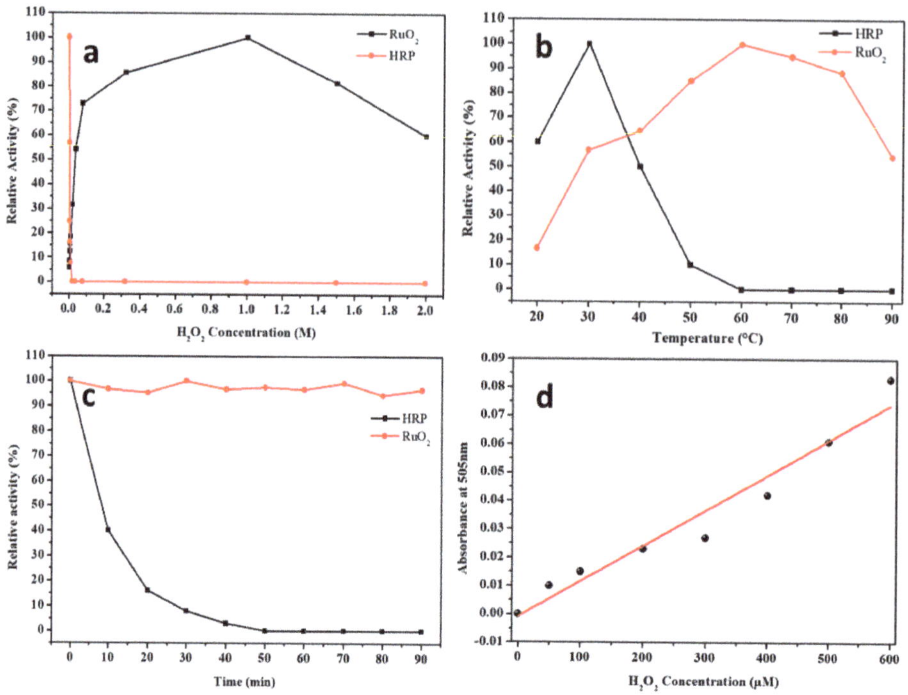

Figure 6. (**a**) H_2O_2 concentration optima and (**b**) temperature optima plots, (**c**) thermal stability (at 80 °C for 90 min under the standard reaction conditions) and (**d**) sensitivity (at 505 nm for the different concentrations of H_2O_2) of ruthenium oxide nanoparticles.

3. Experimental

Ruthenium (III) chloride ($RuCl_3.xH_2O$, Alfa Aesar, Haverhill, MA, USA, 99.9%), NaOH(Merck India Ltd., Mumbai, India), hydrogen peroxide (50% *w*/*v*, Merck India Ltd.), HRP and Tetramethylbenzidine (SRL, Mumbai, India) were used without further purification. 1.3 mmol of $RuCl_3.xH_2O$ was dissolved in 50 mL of water in a two-neck flask, followed by the dropwise addition of 1.5M aqueous NaOH to a pH ~8. The solution was continuously stirred and refluxed at 80 °C untilthe black precipitate was observed. The black precipitate was collected through centrifugation process and repeatedly washed with double-distilled water followed by ethanol to remove the Cl^- ions. The Cl^- ions present in the supernatant liquid were checked with the aqueous solution of silver nitrate. The black precipitate was dried at 100 °C in the hot-air oven and grounded to form a fine powder. Further, the black colour powder was annealed at 300 °C for 6 h in a high temperature furnace to obtained black colour ruthenium oxide nanoparticles. The as-synthesized nanoparticles were further employed to study the catalytic activity.

3.1. Physical Characterization

Powder X-ray diffraction (XRD) was performed using a RigakuUltima IV X-ray diffractometer with Ni-filtered Cu-Kα radiation (λ = 1.5416 Å). The morphological features of synthesized ruthenium oxide nanoparticles were determined by Zeiss EVO40 Scanning Electron Microscope (SEM) at an accelerating voltage of 20 kV. Transmission electron microscopic (TEM) analysis was carried out on FEI Technai G^2 20 HRTEM (High Resolution Transmission Electron Microscope) with an accelerating voltage of 200 kV. The surface area and the pore size of as-synthesized nanoparticles were determined using

Brunauer-Emmett-Teller (BET) surface area analyzer (Model: Nova 2000e, Quantachrome Instruments Limited, Boynton Beach, FL, USA) at liquid nitrogen temperature (77K).

3.2. Electrochemical Measurements

The electrolysis of water for OER and ORR was carried out with a three-electrode electrochemical work station (potentiostat/galvanostat, CHI 660E, Shenzhen, China) at room temperature in alkaline medium (0.5 M KOH) to investigate the redox behaviour of the synthesized RuO_2 nanoparticles. Pt wire, Ag/AgCl and glassy carbon electrodes were used as the counter, reference, and working electrodes in the electrochemical analyzer. The reference electrode (Ag/AgCl) was converted to the reversible hydrogen electrode (RHE) as per the conversion equation, i.e., Nernst Equation at 25 °C.

$$E_{RHE} = E_{Ag/AgCl} + 0.059\ pH + 0.197\ V$$

The slurry was prepared by sonicating 2.5 mg of catalysts in 0.5 mL of isopropanol with 0.1 mL of Nafion solution for 10 min. Then a drop of the slurry was cast on the surface of glassy carbon and dried at 60 °C in vacuum oven [46,47]. The loaded amount of the nanoparticles was of ~0.30 mg/cm^2 on the GC electrode, and the area of the working electrode was 0.07 cm^2. Freshly prepared electrodes were used for the electrochemical measurements. Cyclic Voltammetry (CV), linear sweep voltammetry (LSV) and Tafel measurements were done by applying redox potential versus Ag/AgCl electrode for OER and ORR at the scan rate of 25 mV s^{-1} in 0.5 M KOH electrolyte at room temperature in air, nitrogen and saturated oxygen atmosphere.

3.3. Catalytic Activity of RuO_2 Nanoparticles and HRP

The catalytic activity of as-synthesized nanoparticles and HRP was checked by peroxidase substrate Tetramethylbenzidine (TMB) [48]. In a typical procedure, 200 µL of RuO_2 nanoparticles (2 mg/mL), 1.5 mL TMB, and 5 µL H_2O_2 were added in 1mL of 0.02M acetate buffer (Ph~4.5) and incubated in a water-bath at 30 °C for 10 min and the progress of the reaction was monitored by A Shimadzu UV-2450 spectrophotometer. The leaching of the ions from the reaction mixture was checked by incubating a suspension of RuO_2 nanoparticles in acetate buffer (5 mg/mL, Ph~3.5) (marked as control) for 10 min at 45 °C; the nanoparticles were removed from the reaction mixture, and the catalytic activity of the reaction mixture was analyzed. No activity was found in control. In order to investigate the effect of H_2O_2 concentration on the activity of as-synthesized nanoparticles, it was examined by varying the concentrations of H_2O_2 (0.002–2M). The effect of change of temperature on the activity of RuO_2 nanoparticles was determined by incubating the reaction mixture from 20 to 90 °C. The obtained results were compared with the activity of HRP enzyme over the same range of parameters.

3.4. Detection of H_2O_2

In a typical experiment, 1 mL hydrogen peroxide (50–600 µM) was added to the 1mLcolourimetric reagent, i.e., 10 mg phenol, 10 mg of 4-aminoantipyrine, 50 mg of RuO_2 nanoparticles, dissolved in 20 mL of 100 mM acetic acid buffer (pH 5.6) [49]. The test tubes containing different concentrations of H_2O_2 and blank, i.e., without H_2O_2 were incubated in a water-bath for 10 min at 30 °C, and the progress of the reaction was monitored by spectrophotometer at 505 nm.

4. Conclusions

We have successfully synthesized the ultrafine ruthenium oxide nanoparticles via a simple co-precipitation method at 300 °C. The application of synthesized nanoparticles was successfully studied, and it was concluded that RuO_2 nanoparticles are an effective bifunctional and stable material for OER and ORR reactions in the air, N_2 and O_2 atmosphere. Also, RuO_2 nanoparticles were used as sensors for the detection of H_2O_2 in a solution. The RuO_2 nanoparticles have a comparable limit of detection and linear dynamic values ranging from 600 to 10 µM of H_2O_2. The economically viable as-synthesized nanoparticles could be used as an active nonenzymatic electrochemical sensor for the selective detection of H_2O_2. Further, these nanoparticles showed efficient electrocatalytic activity with low energy loss. Tafel slopes were found to be very low and the electrode material was stable as established by CA studies. Therefore, the ruthenium oxide nanoparticles consumed less energy during the water redox reaction (OER and ORR) and proved to be a better electrode material for OER/ORR reactions, showing its excellent potential for further applications in the future.

Author Contributions: R.P. is responsible for the experimental reactions, basic characterization and first draft of the manuscript, M.P. did the detection of H_2O_2 measurements, J.A. did the electrochemical measurement and its discussion, M.S. was the co-supervisor of first author and wrote the discussion part for the detection of H_2O_2 measurements, S.M.A. helps in the editing and electrochemical measurements, N.A. designed the electrochemical set for the water redox reaction and did the primary measurements, M.A.M.K.helps in the discussion and measurements electrochemical studies and T.A. is responsible for supervision, creation ideas, infrastructure, editing and final draft of the manuscript. All authors have read and agreed to the published version of the manuscript.

Funding: This research was funded by the scheme (SPARC/2018-2019/P843/SL) of MHRD, Government of India and King Saud University Research Project (RSP-2020/29).

Acknowledgments: T.A. thanks the MHRD-SPARC scheme of the Government of India for financial support. R.P. especially thanks to UGC, New Delhi, for the Senior Research Fellowship. Authors also acknowledge the measurement support provided through the DST PURSE program at CIF, Jamia Millia Islamia and AIIMS, New Delhi for electron microscopic studies. The authors extend their sincere appreciation to the Researchers Supporting Project at King Saud University for funding this Research.

Conflicts of Interest: The authors declare that there is no conflict of interest.

References

1. Solomonn, S.; Plattner, G.-K.; Knutti, R.; Friedlingstein, P. Irreversible climate change due to carbon dioxide emissions. *Proc. Natl. Acad. Sci. USA* **2009**, *106*, 1704–1709. [CrossRef] [PubMed]
2. Chhow, J.; Kopp, R.J.; Portney, P.R. Energy resources and global development. *Science* **2003**, *302*, 1528–1531. [CrossRef] [PubMed]
3. Ahmad, T.; Lone, I.H.; Ansari, S.G.; Ahmed, J.; Ahamad, T.; Alshehri, S.M. Multifunctional properties and applications of yttrium ferrite nanoparticles prepared by citrate precursor route. *Mater. Des.* **2017**, *126*, 331–338. [CrossRef]
4. Vondrák, J.; Klápště, B.; Velická, J.; Sedlaříková, M.; Ćerný, R. Hydrogen-oxygen fuel cells. *J. Solid State Electrochem.* **2003**, *8*, 44–47. [CrossRef]
5. Zhang, X.; Wang, X.-G.; Xie, Z.; Zhou, Z. Recent progress in rechargeable alkali metal-air batteries. *Green Energy Environ.* **2016**, *1*, 4–17. [CrossRef]
6. Gutsche, C.; Moeller, C.J.; Knipper, M.; Borchert, H.; Parisi, J.; Plaggenborg, T. Synthesis, structure, and electrochemical stability of Ir-decorated RuO_2 nanoparticles and Ptnanorods as oxygen catalysts. *J. Phys. Chem. C.* **2016**, *120*, 1137–1146. [CrossRef]
7. Tseng, H.-W.; Zong, R.; Muckerman, J.T.; Thummel, R. Mononuclear ruthenium (II) complexes that catalyze water oxidation. *Inorg. Chem.* **2008**, *47*, 11763–11773. [CrossRef]
8. Reier, T.; Oezaslan, M.; Strasser, P. Electrocatalytic oxygen evolution reaction (OER) on Ru, Ir, and Ptcatalysts: A comparative study of nanoparticles and bulk materials. *ACS Catal.* **2012**, *2*, 1765–1772. [CrossRef]
9. Karlsson, E.A.; Lee, B.-L.; Åkermark, T.; Johnston, E.V.; Kärkäs, M.D.; Sun, J.; Hansson, Ö.; Bäckvall, J.-E.; Åkermark, B. Photosensitized water oxidation by use of a bioinspired manganese catalyst. *Angew. Chem. Int. Ed.* **2011**, *50*, 11715–11718. [CrossRef]
10. Ellis, W.C.; McDaniel, N.D.; Bernhard, S.; Collins, T.J. Fast water oxidation using iron. *J. Am. Chem. Soc.* **2010**, *132*, 10990–10991. [CrossRef]
11. Alshehri, S.M.; Ahmed, J.; Alhabarah, A.N.; Ahamad, T.; Ahmad, T. Nitrogen doped cobalt ferrite/carbon nanocomposites for supercapacitor application. *ChemElectroChem* **2017**, *4*, 2952–2958. [CrossRef]
12. Coggins, M.; Zhang, K.M.T.; Chen, Z.; Song, N.; Meyer, T.J. Single-site copper (II) water oxidation electrocatalysis: Rate enhancements with HPO_4^{2-} as a proton acceptor at pH-8. *Angew. Chem. Int. Ed. Engl.* **2014**, *53*, 12226–12230. [CrossRef] [PubMed]

13. Alshehri, S.M.; Ahmed, J.; Ahamad, T.; Alhokbany, N.; Arunachalam, P.; Al-Mayouf, A.M.; Ahmad, T. Synthesis, characterization, multifunctional electrochemical (OGR/ORR/SCs) and photodegradable activities of $ZnWO_4$ nanobricks. *J. Sol-Gel Sci. Technol.* **2018**, *87*, 137–146. [CrossRef]
14. Fang, Y.-H.; Liu, Z.-P. Mechanism and tafel lines of electro-oxidation of water to oxygen on RuO_2 (110). *J. Am. Chem. Soc.* **2010**, *132*, 18214–18222. [CrossRef]
15. Michas, A.; Andolfatto, F.; Lyons, M.E.G.; Durand, R. Gas evolution reactions at conductive metallic oxide electrodes for solid polymer electrolyte water electrolysis. *Key Eng. Mater.* **1992**, *535*, 72–74. [CrossRef]
16. Cui, X.; Zhou, J.; Ye, Z.; Chen, H.; Li, L.; Ruan, M.; Shi, J. Selective catalytic oxidation of ammonia to nitrogen over mesoporous CuO/RuO_2 synthesized by co-nanocasting-replication method. *J. Catal.* **2010**, *270*, 310–317. [CrossRef]
17. Seki, K. Development of RuO_2/Rutile-TiO_2 catalyst for industrial HCl oxidation process. *Catal. Surv. Asia* **2010**, *14*, 168–175. [CrossRef]
18. Liu, H.; Iglesia, E. Selective oxidation of methanol and ethanol on supported ruthenium oxide clusters at low temperatures. *J. Phys. Chem. B* **2005**, *109*, 2155–2163. [CrossRef]
19. Ma, H.; Liu, C.; Liao, J.; Su, Y.; Xue, X.; Xing, W. Study of ruthenium oxide catalyst for electrocatalytic performance in oxygen evolution. *J. Mol. Catal. A-Chem.* **2006**, *247*, 7–13. [CrossRef]
20. Kiele, N.M.; Herrero, C.; Ranjbari, A.; Aukauloo, A.; Grigoriev, S.A.; Villagra, A.; Millet, P. Ruthenium-based molecular compounds for oxygen evolution in acidic media. *Int. J. Hydrogen Energy* **2013**, *38*, 8590–8596. [CrossRef]
21. Jeon, H.S.; Permana, A.D.C.; Kim, J.; Min, B.K. Water splitting for hydrogen production using a high surface area RuO_2 electrocatalyst synthesized in supercritical water. *Int. J. Hydrogen Energy* **2013**, *38*, 6092–6096. [CrossRef]
22. Tilley, S.D.; Schreier, M.; Azevedo, J.; Stefik, M.; Grätzel, M. Ruthenium oxide hydrogen evolution catalysis on composite cuprous oxide water-splitting photocathodes. *Adv. Funct. Mater.* **2014**, *24*, 303–311. [CrossRef]
23. Ryden, W.D.; Lawson, A.W.; Sartain, C.C. Temperature dependence of the resistivity of RuO_2 and IrO_2. *Phys. Lett.* **1968**, *26*, 209–210. [CrossRef]
24. Hu, J.-M.; Zhang, J.-Q.; Cao, C.-N. Oxygen evolution reaction on IrO_2-based DSA®type electrodes: Kinetics analysis of tafel lines and EIS. *Int. J. Hydrogen Energy* **2004**, *29*, 791–797. [CrossRef]
25. Lee, Y.; Suntivich, J.; May, K.J.; Perry, E.E.; Shao-Horn, Y. Synthesis and activities of rutile IrO_2 and RuO_2 nanoparticles for oxygen evolution in acid and alkaline solutions. *J. Phys. Chem. Lett.* **2012**, *3*, 399–404. [CrossRef]
26. Ahmed, J.; Mao, Y. Ultrafine iridium oxide nanorods synthesized by molten salt method toward electrocatalytic oxygen and hydrogen evolution reactions. *Electrochim. Acta* **2016**, *212*, 686–693. [CrossRef]
27. Chen, Y.M.; Korotcov, A.; Hsu, H.P.; Huang, Y.S.; Tsai, D.S. Raman scattering characterization of well-aligned RuO_2 nanocrystals grown on sapphire substrates. *New J. Phys.* **2007**, *9*, 1–11. [CrossRef]
28. Shen, W.; Shi, J.; Chen, H.; Gu, J.; Zhu, Y.; Dong, X. Synthesis and CO oxidation catalytic character of high surface area ruthenium dioxide replicated by cubic mesoporous silica. *Chem. Lett.* **2005**, *34*, 390–391. [CrossRef]
29. Ryan, J.V.; Berry, A.D.; Anderson, M.L.; Long, J.W.; Stroud, R.M.; Cepak, V.M.; Browning, V.M.; Rolison, D.R.; Merzbacher, C.I. Electronic connection to the interior of a mesoporous insulator with nanowires of crystalline RuO_2. *Nature* **2000**, *406*, 169–172. [CrossRef]
30. Viswanathamurthi, P.; Bhattarai, N.; Kim, C.K.; Kim, H.Y.; Lee, D.R. Ruthenium doped TiO_2 fibers by electrospinning. *Inorg. Chem. Commun.* **2004**, *7*, 679–682. [CrossRef]
31. Browne, M.P.; Nolan, H.; Duesberg, G.S.; Colavita, P.E.; Lyons, M.E.G. Low-overpotential high-activity mixed manganese and ruthenium oxide electrocatalysts for oxygen evolution reaction in alkaline media. *ACS Catal.* **2016**, *6*, 2408–2415. [CrossRef]
32. Gustafson, K.P.J.; Shatskiy, A.; Verho, O.; Kärkäs, M.D.; Schluschass, B.; Tai, C.-W.; Åkermark, B.; Bäckvall, J.-E.; Johnston, E.V. Water oxidation mediated by ruthenium oxide nanoparticles supported on siliceous mesocellular foam. *Catal. Sci. Technol.* **2017**, *7*, 293–299.
33. Gao, L.; Zhuang, J.; Nie, L.; Zhang, J.; Zhang, Y.; Gu, N.; Wang, T.; Feng, J.; Yang, D.; Perrett, S.; et al. Intrinsic peroxidase-like activity of ferromagnetic nanoparticles. *Nat. Nanotechnol.* **2007**, *2*, 577–583. [CrossRef] [PubMed]

34. Chen, X.; Zhou, X.; Hu, J. Pt-DNA complexes as peroxidase mimetics and their applications in colorimetric detection of H_2O_2 and glucose. *Anal. Methods* **2012**, *4*, 2183–2187. [CrossRef]
35. Chen, T.; Tian, L.; Chen, Y.; Liu, B.; Zhang, J. A facile one-pot synthesis of Au/Cu_2O nanocomposites for nonenzymatic detection of hydrogen peroxide. *Nanoscale Res. Lett.* **2015**, *10*, 935. [CrossRef] [PubMed]
36. Guan, J.; Peng, J.; Jin, X. Synthesis of copper sulfide nanorods as peroxidase mimics for the colorimetric detection of hydrogen peroxide. *Anal. Methods* **2015**, *7*, 5454–5461. [CrossRef]
37. Wei, Y.; Zhang, Y.; Liu, Z.; Guo, M. A Novel Profluorescent Probe for detecting oxidative stress induced by metal and H_2O_2 in living cells. *Chem. Commun.* **2010**, *46*, 4472–4474. [CrossRef]
38. Dickinson, B.C.; Chang, C.J. A targetable fluorescent probe for imaging hydrogen peroxide in the mitochondria of living cells. *J. Am. Chem. Soc.* **2008**, *130*, 9638–9639. [CrossRef]
39. Finkel, T.; Serrano, M.; Blasco, M.A. The common biology of cancer and ageing. *Nature* **2007**, *448*, 767–774. [CrossRef]
40. Bard, A.J.; Faulkner, L.R. *Electrochemical Methods: Fundamentals and Applications*, 2nd ed.; John Wiley & Sons: New York, NY, USA, 2001.
41. Stoerzinger, K.A.; Qiao, L.; Biegalski, M.D.; Shao-Horn, Y. Orientation-dependent oxygen evolution activities of rutile IrO_2 and RuO_2. *J. Phys. Chem. Lett.* **2014**, *5*, 1636–1641. [CrossRef]
42. AlShehri, S.M.; Ahmed, J.; Ahamad, T.; Arunachalam, P.; Ahmad, T.; Khan, A. Bifunctional electro-catalytic performances of $CoWO_4$ nanocubes for water redox reactions (OER/ORR). *RSC Adv.* **2017**, *7*, 45615–45623. [CrossRef]
43. Jiang, R.; Tran, D.T.; Li, J.; Chu, D. Ru@RuO_2 core-shell nanorods: A highly active and stable bifunctional catalyst for oxygen evolution and hydrogen evolution reactions. *Energy Environ. Mater.* **2019**, *2*, 201–208. [CrossRef]
44. Bhowmik, T.; Kundu, M.K.; Barman, S. Growth of one-dimensional RuO_2 nanowires on g-carbon nitride: An active and stable bifunctional electrocatalyst for hydrogen and oxygen evolution reactions at all pH values. *ACS Appl. Mater. Inter.* **2016**, *8*, 28678–28688. [CrossRef] [PubMed]
45. Cherevko, S.; Geiger, S.; Kasian, O.; Kulyk, N.; Grote, J.-P.; Savan, A.; Shrestha, B.R.; Merzlikin, S.; Breitbach, B.; Ludwig, A.; et al. Oxygen and hydrogen evolution reactions on Ru, RuO_2, Ir, and IrO_2 thin film electrodes in acidic and alkaline electrolytes: A comparative study on activity and stability. *Catal. Today* **2016**, *262*, 170–180. [CrossRef]
46. Farooq, U.; Phul, P.; Alshehri, S.M.; Ahmed, J.; Ahmad, T. Electrocatalytic and enhanced photocatalytic applications of sodium niobate nanoparticles developed by citrate precursor route. *Sci. Rep.* **2019**, *9*, 4488. [CrossRef]
47. Ahmed, J.; Ubaidullah, M.; Ahmad, T.; Alhokbany, N.; Alshehri, S.M. Synthesis of graphite oxide/cobalt molybdenum oxide hybrid nanosheets for enhanced electrochemical performances in supercapacitors and OER. *ChemElectroChem* **2019**, *6*, 2524–2530. [CrossRef]
48. Bos, E.S.; van der Doelen, A.A.; van Rooy, N.; Schuurs, A.H.W.M. 3,3′,5,5′-etramethylbenzidine as an ames test negative chromogen for horse-radish peroxidase in enzyme immunoassay. *J. Immunoass.* **1981**, *2*, 187–204. [CrossRef]
49. Zhou, B.; Wang, J.; Guo, Z.; Tan, H.; Zhu, X. A simple colorimetric method for determination of hydrogen peroxide in plant tissues. *Plant Growth Regul.* **2006**, *49*, 113–118. [CrossRef]

catalysts

MDPI

Article

Mesoporous Composite Networks of Linked $MnFe_2O_4$ and $ZnFe_2O_4$ Nanoparticles as Efficient Photocatalysts for the Reduction of Cr(VI)

Euaggelia Skliri [1], Ioannis Vamvasakis [1], Ioannis T. Papadas [2], Stelios A. Choulis [2] and Gerasimos S. Armatas [1,*]

1 Department of Materials Science and Technology, University of Crete, 70013 Heraklion, Greece; sklirieva@materials.uoc.gr (E.S.); j.vamvasakis@gmail.com (I.V.)
2 Department of Mechanical Engineering and Materials Science and Engineering, Cyprus University of Technology, Limassol 3041, Cyprus; ioannis.papadas@cut.ac.cy (I.T.P.); stelios.choulis@cut.ac.cy (S.A.C.)
* Correspondence: garmatas@materials.uoc.gr; Tel.: +30-2810-545004

Abstract: Semiconductor photocatalysis has recently emerged as an effective and eco-friendly approach that could meet the stringent requirements for sustainable environmental remediation. To this end, the fabrication of novel photocatalysts with unique electrochemical properties and high catalytic efficiency is of utmost importance and requires adequate attention. In this work, dual component mesoporous frameworks of spinel ferrite $ZnFe_2O_4$ (ZFO) and $MnFe_2O_4$ (MFO) nanoparticles are reported as efficient photocatalysts for detoxification of hexavalent chromium (Cr(VI)) and organic pollutants. The as-prepared materials, which are synthesized via a polymer-templated aggregating self-assembly method, consist of a continuous network of linked nanoparticles (ca. 6–7 nm) and exhibit large surface area (up to 91 $m^2\ g^{-1}$) arising from interstitial voids between the nanoparticles, according to electron microscopy and N_2 physisorption measurements. By tuning the composition, MFO-ZFO composite catalyst containing 6 wt.% MFO attains excellent photocatalytic Cr(VI) reduction activity in the presence of phenol. In-depth studies with UV-visible absorption, electrochemical and photoelectrochemical measurements show that the performance enhancement of this catalyst predominantly arises from the suitable band edge positions of constituent nanoparticles that efficiently separates and transports the charge carriers through the interface of the ZFO/MFO junctions. Besides, the open pore structure and large surface area of these ensembled networks also boost the reaction kinetics. The remarkable activity and durability of the MFO-ZFO heterostructures implies the great possibility of implementing these new nanocomposite catalysts into a realistic Cr(VI) detoxification of contaminated wastewater.

Keywords: zinc ferrite; manganese ferrite; mesoporous materials; nanoparticles; metal oxides; electronic band structure; photocatalysis; hexavalent chromium; organic pollutants; environmental remediation

Citation: Skliri, E.; Vamvasakis, I.; Papadas, I.T.; Choulis, S.A.; Armatas, G.S. Mesoporous Composite Networks of Linked $MnFe_2O_4$ and $ZnFe_2O_4$ Nanoparticles as Efficient Photocatalysts for the Reduction of Cr(VI). *Catalysts* **2021**, *11*, 199. https://doi.org/10.3390/catal11020199

Academic Editors: Michalis Konsolakis and Vassilis Stathopoulos

Received: 28 December 2020
Accepted: 28 January 2021
Published: 4 February 2021

1. Introduction

The rapid development of civilization and industrial activities has led to a large number of pollutants being disposed into the environment either intentionally or accidentally [1]. Hexavalent chromium (Cr(VI)) is a highly toxic and non-biodegradable pollutant that is discarded in water resources as a by-product of many industrial processes, like leather tanning, electroplating, metal finishing, and others [2–4]. Compared to trivalent chromium (Cr(III)), Cr(VI) oxyanions are far more toxic and mobile, and therefore, difficult to remove from water. The World Health Organization (WHO) recommended a maximum allowable concentration of 50 $\mu g\ L^{-1}$ for Cr in drinking water [5]. Moreover, as a consequence of its high toxicity, Cr(VI) has also been classified as a group I human carcinogen by the International Agency for Research on Cancer (IARC) [6]. Therefore, finding effective ways for detoxification of Cr(VI)-contaminated solutions is undoubted of high priority in the field of environmental and health protection.

Over the past years, a variety of techniques has emerged for remediation of Cr(VI) from aqueous solutions and industrial effluents, among them chemical precipitation [7], adsorption [8], ion exchange [9], reverse osmosis, and more recently chemical reduction with organic reducing substances or sulfate-based materials [10,11]. However, the effectiveness of these techniques is accompanied by high capital and operation cost as well as complex purification steps; for example, they need a large quantity of chemicals and usually generate secondary wastes as by-products [10]. Recently, semiconductor photocatalysis has been considered a viable and eco-friendly approach for the degradation of environmental pollutants. In this context, various semiconductor materials, such as TiO_2 [12], SnS_2 [13], ZnO [14], Bi_2O_3 [15], and CoO [16,17], have been applied to UV or visible light-induced photocatalytic reduction of toxic Cr(VI) to less harmful Cr(III). Unlike Cr(VI), aqueous Cr(III) can be easily precipitated as $Cr(OH)_3$ or Cr_2O_3 solids in alkaline solutions [18,19]. Although many of these catalysts demonstrated remarkable Cr(VI) photoreduction activity with good ability for wastewater remediation, their poor electron–hole separation, low solar light absorption, and limited structural stability are still important challenges to be overcome. In addition to heavy metals, wastewaters frequently contain recalcitrant dyes, pesticides, and phenolic contaminants that may increase the difficulty of pollutant abatement [20]. In such photocatalytic systems, the organic substances may compete with Cr(VI) ions for absorption on the catalyst surface, resulting in the activity decrease for Cr(VI) reduction. The absorbed organic compounds may block some of the surface-active sites of the catalyst and inhibit light absorption. Therefore, the simultaneous redox degradation of Cr(VI) and organic contaminants is an interesting task.

Zinc ferrite ($ZnFe_2O_4$, ZFO) is a kind of spinel-type oxide which possesses low cost, visible light responsiveness (it has a band gap of around 1.9–2.1 eV), and excellent photochemical stability [21–23]. Consequently, $ZnFe_2O_4$ based materials have been investigated as photocatalysts for photo-Fenton-like degradation of organic dyes [24–26], and photochemical hydrogen production [27,28]. Also, manganese spinel ferrite ($MnFe_2O_4$, MFO) with good magnetic responsivity and functional surface has been widely used as an adsorbent for removing heavy metals from water [29]. Recently, we reported the synthesis of mesoporous assemblies of spinel ferrite MFe_2O_4 (M = Zn^{2+}, Mn^{2+}, Ni^{2+}, Cd^{2+}, and Co^{2+}) nanoparticles (NPs) and demonstrated their functionality as catalysts in the reductive remediation of Cr(VI)-contaminated solution [30]. These materials exhibited very good Cr(VI) photoreduction performance and stability, due to their unique open porous structure and improved charge transfer along with the NP-linked framework. In this study, we present the first demonstration of chemically stable and robust mesoporous MFO/ZFO composite networks as effective photocatalysts for the detoxification of Cr(VI)-containing wastewaters under UV-visible light irradiation. We use a block copolymer-templated cross-linking aggregation of colloidal NPs to assemble dual component MFO-ZFO NP linked networks with different MFO content (i.e., 4, 6, 8, and 12 wt.%). Characterization with X-ray diffraction, high-resolution transmission electron microscopy, and N_2 porosimetry confirmed a highly porous structure consisting of connected small-sized (ca. 6–7 nm) spinel ferrite nanoparticles. Mechanistic studies with UV-visible optical absorption, electrochemical and photoelectrochemical measurements indicated that the enhanced reactivity of this catalytic system arises from the suitable band edge positions of constituent NPs which promotes the efficient separation and transport of photogenerated charges at the ZFO/MFO junctions. Finally, a plausible mechanistic scheme for the photocatalytic reduction of Cr(VI) over MFO-ZFO composite catalysts is proposed based on the experimental results.

2. Results and Discussion

2.1. Structure and Morphology of MFO-ZFO MNAs

Mesoporous assemblies from $MnFe_2O_4$ (MFO) and $ZnFe_2O_4$ (ZFO) spinel ferrite NPs were prepared by cross-linking polymerization of NP colloids in the presence of a block copolymer template. Briefly, the co-assembly of ZFO and MFO NPs with polymer template (Pluronic P123, BASF) occurs via a solvent evaporation-induced aggregating self-assembly process, in which slow evaporation of solvent promotes the arrangement of NPs into mesostructured NP/polymer composites [31]. The NP/polymer composites were then calcined at 350 °C in the air to give a continuous network of assembled NPs with an open pore structure. By tuning the ratio of the ZFO and MFO precursor NPs, we succeeded in preparing a series of samples with different compositions. The obtained materials are denoted as MFO-*n*-ZFO mesoporous nanoparticle assemblies (MNAs), where *n* refers to the weight percent of the MFO component, i.e., *n* = 4, 6, 8, and 12 wt.%. The crystallinity and phase purity of the resultant materials were determined by X-ray diffraction (XRD) measurements. Figure 1a shows the XRD patterns of the single-component ZFO and MFO MNAs and composite MFO-ZFO MNAs samples, where all the reflection peaks can be assigned to the spinel structure of metal ferrites (ZFO and MFO). Analysis of the (311) reflection with the Scherrer equation gives an average grain size of ~6.2–6.8 nm, which is very close to the size of starting NPs (ca. 6–7 nm) [30], see Table 1. This suggests that the size changes of the crystallites are limited after thermal annealing. Notably, no impurities (like Mn_xO_y, ZnO, or other metal oxide phases) were evident in the XRD patterns, suggesting the phase purity of the samples.

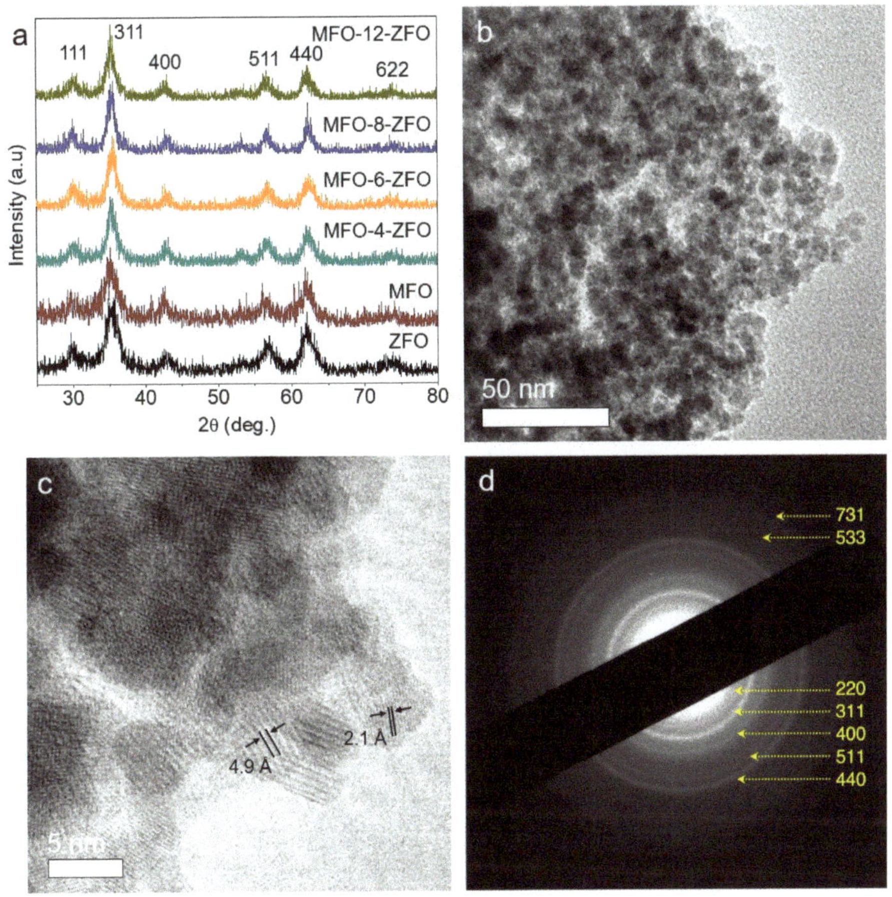

Figure 1. (**a**) Powder XRD patterns of ZFO, MFO, and MFO-ZFO MNAs. (**b**) Typical TEM, (**c**) high-resolution TEM (HRTEM), and (**d**) SAED pattern taken from a small area of mesoporous structure of the MFO-6-ZFO sample. In panel (**d**): All the diffraction rings are indexed to the cubic spinel structure of metal ferrites.

Table 1. Textural properties of mesoporous assemblies of ZFO and MFO NPs.

Sample	Surface Area ($m^2\ g^{-1}$)	Pore Volume ($cm^3\ g^{-1}$)	Pore Width (nm)	Crystallite Size [1] (nm)
ZFO	105	0.15	5.8	6.2
MFO	106	0.21	6.1	6.7
MFO-4-ZFO	91	0.14	5.9	6.5
MFO-6-ZFO	68	0.09	5.8	6.3
MFO-8-ZFO	79	0.13	6.2	6.8
MFO-12-ZFO	82	0.12	6.0	6.4

[1] Average crystallite size (d_p) of metal ferrite NPs calculated by the Scherrer equation: $d_p = 0.9\lambda/B\cos\theta$, where λ is the wavelength of Cu Kα radiation (λ = 1.5406 Å) and B is the full-width half-maximum of the diffraction peal centered at 2θ degrees.

The morphology and microstructure of the assembled materials were observed by transmission electron microscopy (TEM), and typical results for the MFO-6-ZFO MNAs, which is the most active catalyst of this work, are shown in Figure 1b,c. The images reveal that the sample consists of a porous network of aggregated small NPs. As shown in the high-resolution TEM (HRTEM) image in Figure 1c, the constituent NPs have an average diameter of around 6–7 nm, in agreement with XRD results, and are interlinked to a continuous structure. The direct NP-to-NP contact is advantageous to photocatalytic processes, since it efficiently transports and separates the photogenerated charge carriers within the assembled structure. HRTEM also gave further information on the single-crystalline nature of the constituent NPs, showing well-defined lattice fringes throughout the particles. The lattice fringes of 2.1 Å and 4.9 Å spacing in the NPs are indexed to the (004) and (111) crystal plane of spinel ferrite structure, respectively. Consistent with XRD and HRTEM analyses, the spinel structure of assembled NPs was evidenced by selected area electron diffraction (SAED) analysis. The SAED pattern in Figure 1d shows a series of diffuse Debye-Scherrer rings that can be readily assigned to the spinel ferrite phase of ZFO and MFO.

The porosity of the as-prepared materials was probed with N_2 physisorption measurements. Figure 2 shows N_2 adsorption–desorption isotherms and the corresponding pore size distribution plots for the MFO-ZFO MNAs. The corresponding plots for the ZFO and MFO MNAs are given in Figure S1. All the isotherms show typical type-IV curves, according to the IUPAC classification, with a distinct H_3-type hysteresis loop, being characteristic of porous solids with slit-like mesopores. The specific surface areas and total pore volumes of the composite materials were measured to be 68–91 $m^2\ g^{-1}$ and 0.09–0.14 $cm^3\ g^{-1}$, respectively, which are slightly lower than the surface area and pore volume of single component ZFO and MFO MNAs (105–106 $m^2\ g^{-1}$, 0.15–0.21 $cm^3\ g^{-1}$, see Figure S1). The pore size in these materials was derived from the adsorption branch of isotherms using the NLDFT method (based on slit-like pores). The NLDFT analysis indicated quite narrow size distributions of pores with an average pore diameter of ~5 to 6 nm (insets of Figure 2 and Figure S1). Table 1 summarizes the textural parameters of the prepared catalyst.

2.2. Photocatalytic Study of MFO-ZFO MNAs

The photocatalytic Cr(VI) reduction activity of the title materials was initially assessed in the presence of phenol (400 mg L^{-1}) as coexistent pollutant. Figure 3a displays the temporal concentration changes (C_t/C_o) of Cr(VI) during the photocatalytic process with different catalysts, namely, ZFO, MFO, and MFO-ZFO MNAs. All the catalytic reactions were performed at the same dose of catalyst (500 mg L^{-1}) dispersed in a Cr(VI) contaminated water (50 mg L^{-1}), under $\lambda > 360$ nm light irradiation. The comparison shows that the integration of MFO with the ZFO NPs has a profound effect on the Cr(VI) photoreduction performance. As shown in Figure 3a, MFO-6-ZFO exhibits the highest reactivity among the samples, achieving an almost complete (>99%) reduction of the Cr(VI) in 3 h. Under identical conditions, the 4, 8, and 12 wt.% MFO loaded samples reduce the Cr(VI) by ~70%,

~94%, and ~83%, respectively, while the single component MFO and ZFO MNAs reduce ~90% and ~88% of Cr(VI). Control experiments in the absence of catalyst or light irradiation showed almost no changes in the initial concentration of Cr(VI) (see Figure 3a), confirming that the reduction of Cr(VI) originated from the photoredox reactions on the catalyst under light illumination. Assuming that the reaction rate is proportional to the concentration of Cr(VI), the photocatalytic reaction can be expressed by the pseudo-first-order kinetics of the Langmuir–Hinshelwood model (Equation (1)).

$$\mathrm{Ln}(C_t/C_o) = -k_{app} \times t \tag{1}$$

where, C_o and C_t are the initial and at time t concentration of Cr(VI), respectively, and k_{app} is the apparent reaction rate constant.

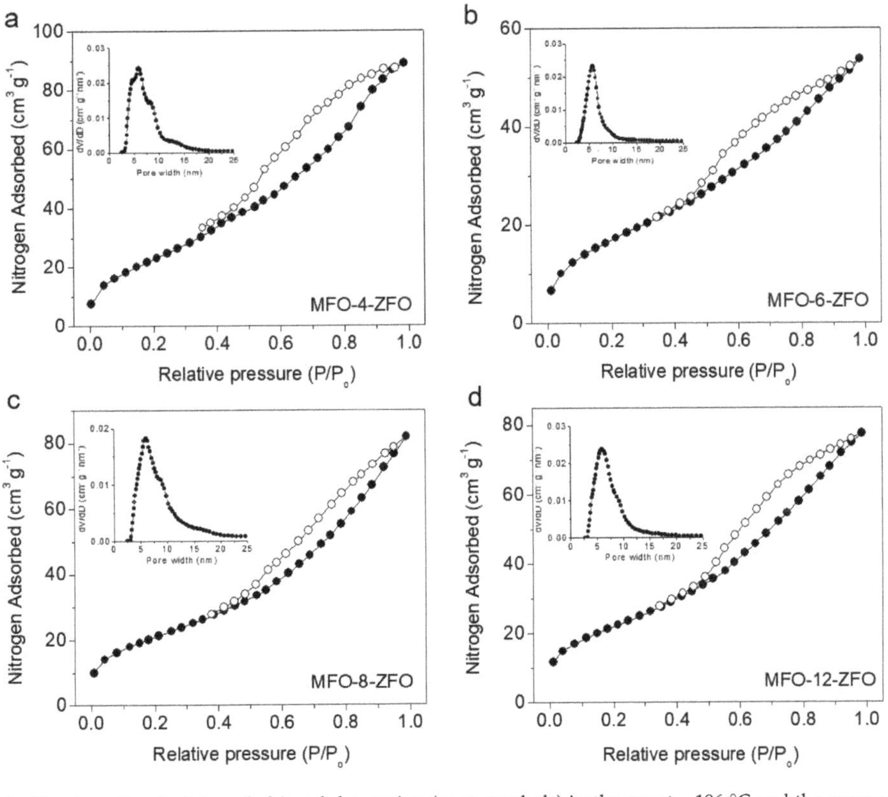

Figure 2. N_2 adsorption (solid symbols) and desorption (open symbols) isotherms at −196 °C and the corresponding NLFDT pore-size distribution plots calculated from the adsorption branch of the isotherms (inset) for the MFO-ZFO catalysts: (**a**) MFO-4-ZFO, (**b**) MFO-6-ZFO, (**c**) MFO-8-ZFO and (**d**) MFO-12-ZFO.

Thus, analysis of the temporal evolution of Cr(VI) concentration using Equation (1) reveals a rate constant k_{app} of 3.6×10^{-3} and 5.3×10^{-3} min^{-1} for ZFO and MFO MNAs, respectively, and in the range of 4.3–14.4 $\times 10^{-3}$ min^{-1} for dual component MFO-ZFO assemblies. The $\ln(C_t/C_o)$ versus time plots for different catalysts are shown in Figure S2. As shown in Figure 3b, the MFO-6-ZFO MNAs outperform the other catalysts, yielding faster reaction kinetics with k_{app} value of 14.41×10^{-3} min^{-1}, which is about 4 and 2.7 times higher than the k_{app} value obtained for the ZFO and MFO MNAs, respectively. We suggest that the high reactivity of MFO-6-ZFO is related to the open pore structure, which facilitates fast molecular diffusion, and the suitable electronic band structure of constituent NPs, which efficiently separates and transports the photogenerated electron-hole pairs (see below). Accordingly, we focused on Cr(VI) reduction reactions with this catalyst in our further studies.

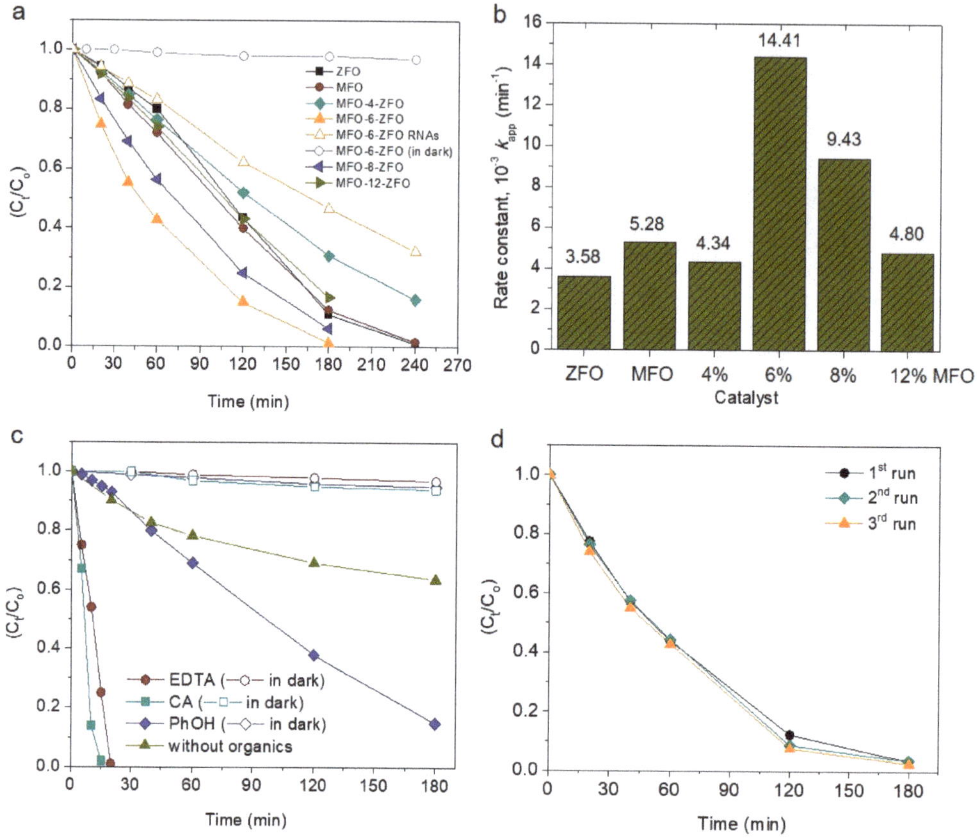

Figure 3. (**a**) Time courses of photocatalytic reduction of Cr(VI) in the presence of phenol (400 mg L^{-1}) and (**b**) kinetic rate constants (k_{app}) for different MFO, ZFO and MFO-ZFO MNAs catalysts. The photocatalytic reduction of Cr(VI) over the untemplated MFO-6-ZFO catalyst (MFO-6-ZFO RNAs) and over the MFO-6-ZFO MNAs in the dark is also given. (**c**) Photocatalytic reduction of Cr(VI) over MFO-6-ZFO catalyst in the absence and presence of three equivalents (ca. 1.28 mM) of phenol (PhOH), citric acid (CA), and ethylenediaminetetraacetic acid (EDTA) under $\lambda > 360$ nm light irradiation and in the dark. (**d**) Recycling study of the MFO-6-ZFO catalyst. The standard deviation of all measurements is about 3%. Reaction conditions: 50 mg L^{-1} Cr(VI) solution, 500 mg L^{-1} catalyst, pH = 2, UV-vis light ($\lambda > 360$ nm) irradiation, 20 °C.

Figure 3c shows comparative results of the Cr(VI) photoreduction over MFO-6-ZFO composite in the absence and presence of phenol (three equiv. compared to Cr(VI)). It can be observed that, without phenol, the Cr(VI) photoreduction proceeds, but at a lower reaction rate; MFO-6-ZFO MNAs show a ~25% Cr(VI) conversion in 2 h in pure water. Unlike oxidation of phenol, the photo-oxidation of water to dioxygen is a sluggish reaction that involves several uphill, multi-electron reaction steps, such as dissociation of –OH species and formation of O–O bonds. In a previous study, we showed that water oxidation to molecular oxygen over ZFO NP assemblies is a viable process [30]. Therefore, phenol enables more efficient utilization of the surface-reaching holes, leading to a significant improvement of the photo-oxidation efficiency. This study clearly suggests that the photocatalytic reduction of Cr(VI) and oxidation of phenol are collaborative over the mesoporous MFO-ZFO assemblies, and this process can enhance the photoreduction effect. To explore the possibility of MFO-6-ZFO MNAs in practical application, the Cr(VI) photoreduction activity of MFO-6-ZFO MNAs was examined in the presence of other pollutants, such as citric acid (CA) and ethylenediaminetetraacetic acid (EDTA). All these experiments were made by using the same dose of catalyst (0.5 g L^{-1}) in 50 mL of Cr(VI) aqueous solution (50 mg L^{-1}) containing three equiv. of organic pollutants that represent typical concentrations in industrial wastewaters [32,33]. The results showed that the Cr(VI) photoreduction in the presence of citric acid and EDTA is much faster than with phenol (Figure 3c); the photocatalytic reaction with citric acid and EDTA was complete within only ~15–20 min. Meanwhile, control experiments showed that the Cr(VI) reduction did not proceed in the dark, again confirming that this reaction is photocatalytic in nature (Figure 3c).The observed increased Cr(VI) reduction rate can be attributed to the fact that, compared to phenol, citric acid and EDTA can consume the photogenerated holes more effectively because of their favorable adsorption on the surface of catalyst, preventing multiple holes accumulation, and thus, electron-hole recombination at the MFO-ZFO surface. In particular, under acidic conditions (pH ~2), there are electrostatic attractions between the positively charged $\equiv M{-}OH_2^+$ catalyst surface and negatively charged EDTA (in the form of H_3EDTA^- ($pKa_1 = 1.99$) and H_2EDTA^{2-} ($pKa_2 = 2.67$)) and citric acid (in the form of H_2CA^- $pKa_1 = 3.1$) ions, which result in an increased concentration of these species near the catalyst's surface [34]. On the contrary, in these reaction conditions, phenol predominately exists in its neutral molecular form (it has a pKa of about 9.88), which is absorbed by the catalyst through weak van der Waals and hydrogen bonds.

In addition to chemical composition, morphological effects may also contribute to the photocatalytic activity of MFO-6-ZFO MNAs. To elucidate this possibility, we prepared random aggregates of MFO and ZFO NPs (denoted as MFO-6-ZFO RNAs) following a similar procedure as being used for the MFO-6-ZFO MNAs, but without a template. These two samples feature with ZFO and MFO NPs linked into a network structure (the weight content of MFO is ca. 6%), but with different porosity. We obtained an N_2 adsorption-desorption isotherm that is a type-II curve with an H_3 hysteresis loop for the MFO-6-ZFO RNAs, which is characteristic of nanoporous solids with slit-shaped pores (Figure S3). Analysis of the adsorption data indicated a surface area of 50 m^2 g^{-1} and a pore size of 2.9 nm. The catalytic results presented in Figure 3a show that the photocatalytic Cr(VI) reduction efficiency of the randomly aggregated NPs (MFO-6-ZFO RNAs) is significantly lower than that of polymer-templated analogs, giving a respective Cr(III) conversion yield of ~54% in 3 h. Furthermore, pseudo-first-order analysis of Cr(VI) photoreduction also reveals that the reaction proceeds at a lesser rate over the random aggregates MFO-6-ZFO RNAs ($k_{app} = 3.05 \times 10^{-3}$ min^{-1}) than the mesoporous MFO-6-ZFO MNAs ($k_{app} = 14.41 \times 10^{-3}$ min^{-1}), see Figure S4. The variance in catalytic activity between MFO-6-ZFO RNAs and MFO-6-ZFO MNAs should be ascribed to their different pore structures. The untemplated material contains a random distribution of interstitial pores between the NPs (ca. 2.9 nm, according to N_2 physisorption data), which may result in slow transport kinetics of Cr(VI) ions.

The reusability of the MFO-6-ZFO MNAs catalyst was investigated by conducting three recycling experiments in the presence of 50 mg L^{-1} Cr(VI) and 400 mg L^{-1} phenol. The catalyst was isolated by centrifugation after completion of the reaction, washed with water, and placed in a fresh Cr(VI)/phenol solution. As seen in Figure 3d, MFO-6-ZFO MNAs catalyst retains its initial activity after at least three 3-h recycling tests. Moreover, XRD and N_2 porosimetry data confirmed that the crystal and porous structure of the reused catalyst are well maintained after photocatalysis, indicating high durability, see Figure S5.

2.3. Mechanism of Photocatalytic Cr(VI) Reduction

To help explain the relationship between photochemical activity and electronic band structure, we investigated the electronic structure of as-prepared catalysts by combining electrochemical impedance (EIS) and optical absorption spectroscopy. Figure 4a and Figure S6a display the Mott–Schottky plots recorded at a frequency of 1 kHz and the corresponding linear fits on the inverse square capacitance ($1/C_{sc}^2$) versus applied potential (E) data for the mesoporous ZFO, MFO, and MFO-ZFO catalysts. Using extrapolation to $1/C_{sc}^2 = 0$, the flat band potential (E_{FB}) for MFO-ZFO MNAs is seen to be located at −0.19 V to −0.33 V versus NHE, while the E_{FB} position of ZFO and MFO MNAs locates at −0.17 V and −0.31 V, respectively (see Table 2). Obviously, the E_{FB} level of the composite materials undergoes a cathodic shift upon incorporating MFO NPs into the ZFO assembled structure, which is consistent with the more negative E_{FB} potential of MFO. The electron donor density (N_d) obtained from the slope of $1/Csc^2$ versus E curves ranges between 9.71×10^{15} and 1.88×10^{17} cm^{-3} for MFO-ZFO MNAs and appears around 3.63×10^{16} and 1.12×10^{17} cm^{-3} for ZFO and MFO MNAs, respectively (Table 2). It is apparent that all the Mott–Schottky plots show a positive linear slope, indicating n-type conductivity. The optical band gap (E_g) of the prepared materials was determined from UV-vis/NIR diffuse reflectance spectra, using Tauc plot analysis for direct band gap semiconductor (i.e., $(\alpha hv)^2$ versus energy (hv) plots), as shown in Figure 4b and Figure S6b. This analysis yields E_g values of 2.17 eV and 1.46 eV for ZFO and MFO MNAs, respectively, and from 2.13 to 2.16 eV for MFO-ZFO MNAs samples. Table 2 summarizes the results of electrochemical and optical absorption characteristics of ZFO, MFO, and MFO-ZFO catalysts.

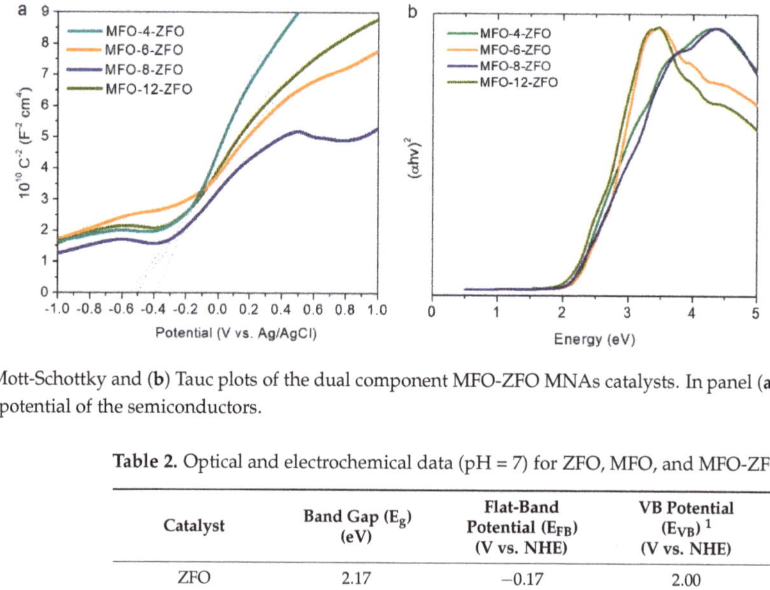

Figure 4. (**a**) Mott-Schottky and (**b**) Tauc plots of the dual component MFO-ZFO MNAs catalysts. In panel (**a**), the red lines show the E_{FB} potential of the semiconductors.

Table 2. Optical and electrochemical data (pH = 7) for ZFO, MFO, and MFO-ZFO MNAs.

Catalyst	Band Gap (E_g) (eV)	Flat-Band Potential (E_{FB}) (V vs. NHE)	VB Potential (E_{VB}) [1] (V vs. NHE)	Donor Density (N_d) (cm^{-3})
ZFO	2.17	−0.17	2.00	3.63×10^{16}
MFO	1.46	−0.31	1.15	1.10×10^{17}
MFO-4-ZFO	2.16	−0.19	1.97	9.71×10^{15}
MFO-6-ZFO	2.16	−0.33	1.83	1.79×10^{16}
MFO-8-ZFO	2.14	−0.32	1.82	1.88×10^{16}
MFO-12-ZFO	2.13	−0.30	1.83	1.46×10^{16}

[1] The E_{VB} potential of the semiconductor catalysts was estimated from E_{FB} + E_g.

Based on the measured E_{FB} potentials and optical band gaps, the energy band diagram for each catalyst can be obtained and is illustrated in Figure 5. Here, we assumed that the E_{FB} level is located very close to the CB edge of the catalysts, which is quite reasonable for heavily doped n-type semiconductors, such as $ZnFe_2O_4$; typically, the E_{FB} potential is about 0.1–0.3 V lower than the CB edge level [35]. Therefore, the valence band potential (E_{VB}) of the catalysts can be obtained by the difference between E_{FB} and E_g. From these results, it is clear that the CB edge of the MFO-ZFO MNAs is aligned well above the redox potential of Cr(VI)/Cr(III), thus demonstrating the ability of these materials for multi-electron reduction of Cr(VI) to Cr(III) under irradiation. Meanwhile, all the catalysts meet the electrochemical requirement for the oxidation of organic pollutants (such as phenol); that is, the VB position in these materials is below (more positive) the oxidation potential of phenol. The high photocatalytic activity obtained with the MFO-6-ZFO MNAs can be explained by the favorable alignment of band edges relative to the reduction potential of Cr(VI) and oxidation potential of organic pollutants. Namely, MFO-6-ZFO MNAs have a relatively high E_{FB} potential (−0.33 V), which reflects a better reducing ability of CB electrons, and a sufficiently deep VB potential (1.83 V), which favors the oxidation of organic pollutants (e.g., phenol). Moreover, MFO-6-ZFO MNAs possess a relatively high electron donor density and efficient electron transportability (see below). Taken together, these thermodynamic and kinetic effects definitely impact the overall photocatalytic performance of the MFO-6-ZFO MNAs catalyst.

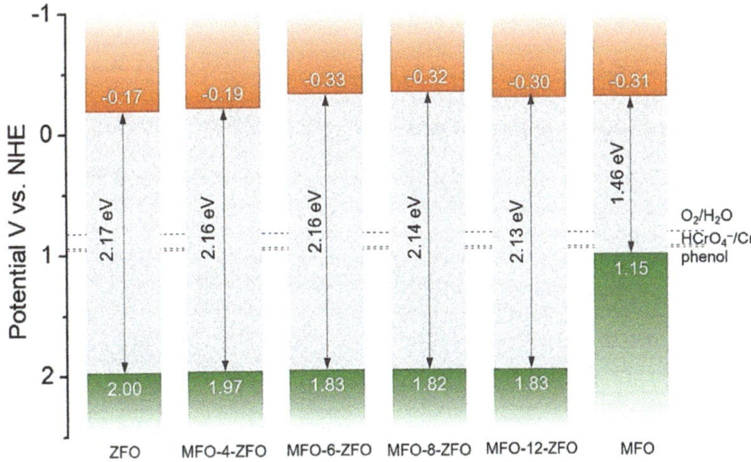

Figure 5. Energy band diagrams of ZFO, MFO, and MFO-ZFO MNAs catalysts. The redox potential levels of O_2/H_2O (0.82 V), $HCrO_4^-/Cr^{3+}$ (0.94 V), and phenol (0.96 V) are also presented. All the potentials are given versus the NHE scale at pH = 7.

Electrochemical impedance spectroscopy was also used to study the charge transfer events at the catalyst/electrolyte interface. Figure 6a shows the Nyquist plots of MFO-6-ZFO compared to that of single component ZFO and MFO samples. All measurements were carried out at open circuit potential from 1 Hz to 1 MHz using an alternating current amplitude of 10 mV in 0.5 $NaSO_4$ solution. A Randles circuit (inset of Figure 6a) was used to fit the impedance data and interpret the charge transfer resistance (R_{ct}) of the catalysts. From the fitting results, the R_{ct} value was determined to be 17.8, 15.6, and 10.4 kΩ for ZFO, MFO, and MFO-6-ZFO, respectively. This suggests that MFO-6-ZFO exhibits more efficient charge transport at the interface of electrode/electrolyte, as we will show below, due to the better electron-hole separation in the MFO/ZFO junction.

The improved charge transport properties of MFO-6-ZFO MNAs were further affirmed using photoelectrochemical measurements. Figure 6b shows the transient photocurrent responses of ZFO, MFO, and MFO-6-ZFO electrodes measured at a bias of 0.2 V (vs. Ag/AgCl) under visible light (380–780 nm) irradiation and dark conditions. Apparently, the MFO-6-ZFO composite generates the largest photocurrent as compared to ZFO and MFO samples, indicating a better electron conductivity and a higher charge separation efficiency. In addition, the efficient separation of photogenerated carriers in the MFO-ZFO composite structure was also confirmed using open-circuit photovoltage (OCP) decay analysis. The OCP decay technique is useful for determining the lifetime of photoinduced charge carriers of semiconducting materials. It consists of cutting off the illumination at the equilibrium state of the electrode/solution system and monitoring the decay of photovoltage (V_{oc}) with time. The V_{oc} decay rate conveys information on the transporting lifetime of photogenerated electrons (both free and trapped electrons) within the CB of semiconductors. The potential-dependent photoelectron lifetime (τ_n) could be calculated according to the following equation [36]:

$$\tau_n = -(k_B T/e)(dV_{oc}/dt)^{-1} \quad (2)$$

where, k_B is the Boltzmann's constant, e is the electron charge, T is the temperature, and V_{oc} is the open-circuit voltage at time t.

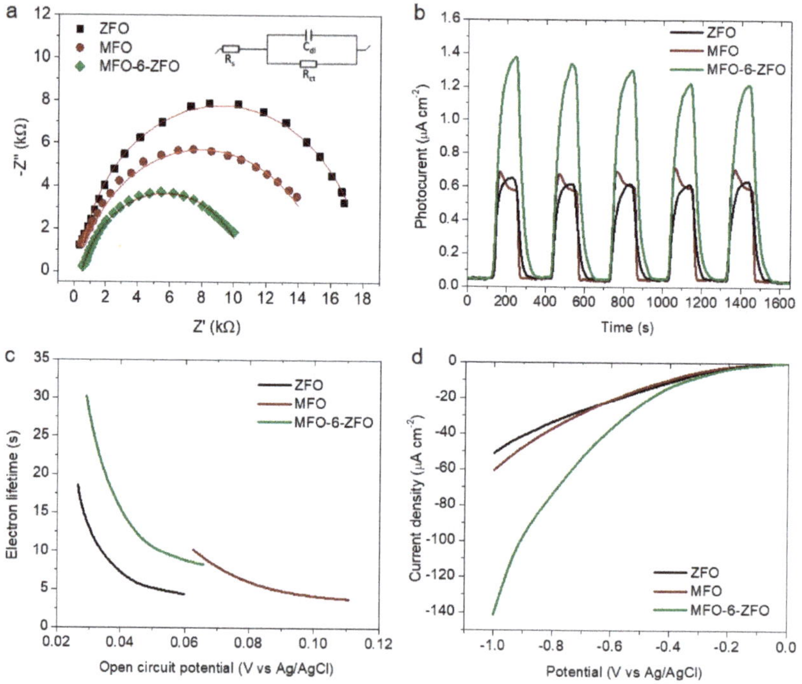

Figure 6. (**a**) Electrochemical impedance spectra (Inset: Equivalent circuit used for data analysis; C_{dl}—double layer capacitance, R_{ct}—charge-transfer resistance, R_s—electrolyte resistance), (**b**) transient photocurrent responses, (**c**) electron lifetime determined from the OCP decay curves, and (**d**) polarization curves of ZFO, MFO, and MFO-6-ZFO MNAs catalysts. In panel (**a**), the red lines are fit to the data.

The OCP decay profiles for different catalysts are shown in Figure S7, and the obtained τ_n versus V_{oc} plots are shown in Figure 6c. These results demonstrate that the composite structure of MFO-6-MFO promotes better separation of photoexcited electron-hole pairs, exhibiting a markedly prolonged electron lifetime; based on Equation (2), the τ_n is calculated to be ~30, ~18 and ~10 s for MFO-6-ZFO, ZFO, and MFO, respectively. This means that MFO-6-ZFO engages a more substantial portion of photogenerated carriers to redox reactions, instead of losing them to recombination, which is in line with its enhanced photocatalytic performance. In good agreement with the photoelectrochemical behaviour of ZFO, MFO, and MFO-6-ZFO samples, the polarization curves in Figure 6d show that the dual component MFO-ZFO structure obviously enhances the current density and reduces the over-potential for hydrogen evolution. For example, the MFO-6-ZFO generated a current density of 74 μA cm^{-2} at −0.8 V versus Ag/AgCl, while ZFO and MFO generated 34 and 37 μA cm^{-2}, respectively, at the same potential. It is unlikely that the increase in current density for MFO-6-ZFO is due to the change in majority carrier density because MFO-6-ZFO exhibits lower carrier density than single component ZFO and MFO MNAs, as revealed by the Mott-Schottky plots analysis (see Table 2). Therefore, the difference in current density should be attributed to the facile charge transport and separation of photoinduced charge carriers along the composite structure.

On the basis of the above discussion, a tentative reaction mechanism for the photocatalytic reduction of Cr(VI) over MFO-ZFO MNAs catalysts has been proposed (Scheme 1). Under UV-visible light irradiation, both ZFO and MFO components are excited and produce CB electrons and VB holes. The potential gradient created at the ZFO/MFO interface promotes the electron migration from the CB of MFO to ZFO, while the photogenerated holes transfer in the opposite direction and accumulate in the VB of MFO; MFO has more negative E_{FB} and E_{VB} potentials than those of ZFO according to electrochemical and UV-vis/NIR diffuse reflectance data (see Figure 5). The accumulated electrons on ZFO surface can readily transfer to Cr(VI) to produce Cr(III), while the photogenerated holes moving to the MFO surface can oxidize organic pollutants. As such, the photocatalytic performance of dual component MFO-ZFO catalysts is significantly enhanced by the unidirectional electron and hole flow on different components for reduction and oxidation reactions, respectively. This mechanism obviously reduces the recombination between photogenerated electrons and holes and agrees with electronic band structure characterizations obtained from electrochemical and photoelectrochemical experiments.

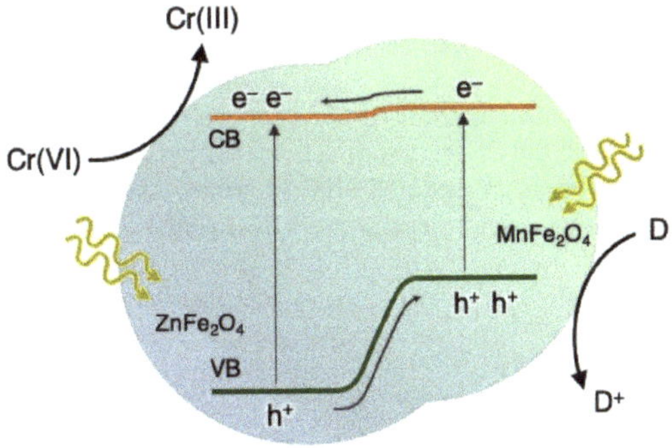

Scheme 1. Schematic illustration of the Cr(VI) photoreduction mechanism over dual component $MnFe_2O_4$–$ZnFe_2O_4$ catalyst (MFO-ZFO MNAs).

3. Materials and Methods

3.1. Materials

Block copolymer (Pluronic P123, Mn ~5800 g mol^{-1}), sodium dodecylbenzenesulfonate (NaDBS, technical grade), absolute ethanol, toluene (99.7%) and potassium dichromate ($K_2Cr_2O_7$ > 99.8%) were purchased from Aldrich Chemical Co. 3-aminopropanoic acid (3-APA, 99%) and sodium hydroxide (NaOH, 98%) were obtained from Acros Organics. Zinc (II) nitrate hexahydrate ($Zn(NO_3)_2{\cdot}6H_2O$, 98%), manganese (II) nitrate tetrahydrate ($Mn(NO_3)_2{\cdot}4H_2O$, 98%), iron (III) nitrate nonahydrate ($Fe(NO_3)_3{\cdot}9H_2O$, 98%), phenol (>99.5%), citric acid (>99.5%) and ethylenediaminetetraacetic acid (EDTA, >99%) were obtained from Sigma-Aldrich (Steinheim, Germany).

3.2. Preparetion of Spinel Ferrite NPs

Spinel ferrite MFe_2O_4 (M = Zn, Mn) NPs were synthesized according to a previously reported method [30,37]. For a typical synthesis of $ZnFe_2O_4$ NPs, $Zn(NO_3)_2{\cdot}6H_2O$ (5 mmol) and $Fe(NO_3)_3{\cdot}9H_2O$ (10 mmol) were added in 25 mL of deionized (DI) water at room temperature (RT). After obtaining a clear solution, 25 mL of 0.4 M NaDBS solution and 500 mL of toluene were added, and the resulting mixture was kept under stirring overnight. Next, 40 mL of NaOH solution (1 M) was dropwise added to the above mixture, and the solution was stirred for another 2 h. After that, the solution was concentrated by rotary evaporator and washed with water and ethanol several times to remove excess of surfactant. Finally, the product was collected with centrifugation and treated under a N_2 atmosphere at 350 °C for 12 h. $MnFe_2O_4$ NPs were synthesized using a similar procedure as described above, but using $Mn(NO_3)_2{\cdot}4H_2O$ as precursor.

3.3. Preparation of Colloidal Metal Ferrite NPs

The spinel ferrite surface was functionalized with 3-APA according to a previously reported method [30]. In a typical experiment, as-prepared Mfe_2O_4 (M = Zn, Mn) NPs (230 mg) were suspended in an aqueous solution of 3-APA (45 mg, 4 mL) and the pH was adjusted to 4 with by adding a HCl aqueous solution (2 M). The mixture was then stirred at RT until the NPs are completely transferred into the liquid phase, typically within 24 h. The dispersion was initially assisted with sonication for about 20 min. The 3-APA-stabilized NPs were collected with centrifugation, washed several times with water, and dispersed in ethanol to form a stable colloidal solution (120 mg mL^{-1}).

3.4. Preparetion of Mesoporous MFO-ZFO NP Assemblies

Mesoporous assemblies of $MnFe_2O_4$ and $ZnFe_2O_4$ NPs (denoted as MFO-*n*-ZFO MNAs, where *n* refers to the weight percent of MFO) were prepared as follows: 1 mL of colloidal ZFO NP solution (120 mg mL^{-1}) and appropriate amount of MFO NP solution (120 mg mL^{-1}) were slowly added to an ethanol solution (1 mL) of Pluronic P123 ($EO_{70}PO_{20}EO_{70}$) block copolymer (0.2 g). The mixture was sonicated for 10 min and then placed in an oven at 40 °C for about 2–3 days under static condition to form mesostructured NP/polymer composites. The removal of template and cross-linking of NPs were carried out by heating the dry gel product in air for 4 h at 350 °C, using a 0.5 °C min^{-1} heating rate. The amount of 3-APA-capped MFO NPs used in reactions was varied between 5, 8, 11 and 17 mg to give a series of mesoporous dual component MFO-*n*-ZFO materials with different MFO content, i.e., *n* = 4, 6, 8, and 12 wt.%, respectively. For comparison, mesoporous MFO and ZFO NP assemblies (denoted as MFO and ZFO MNAs) were also prepared using only MFO and ZFO NPs as starting materials. Also, random aggregates of MFO and ZFO NPs (denoted as MFO-6-ZFO RNAs) were synthesized through a similar procedure without adding polymer template.

3.5. Photocatalytic Experiments

The photocatalytic Cr(VI) reductions were carried out in a Pyrex glass cell (100 mL capacity) containing 50 mL of Cr(VI) and phenol (400 mg L^{-1}) aqueous solution and 0.5 g L^{-1} concentration of catalyst. A stock solution of Cr(VI) (50 mg L^{-1}) was prepared by dissolving $K_2Cr_2O_7$ in DI water, and the pH was adjusted to the desired value with dilute H_2SO_4. The reaction mixture was stirred in the dark for 30 min to establish adsorption-desorption equilibrium between the catalyst and pollutants, and then irradiated with UV-vis light using a 300-W Xe lamp with a 360 nm cutoff filter (Variac Cermax). All the experiments were performed at 20 ± 2 °C using a water bath cooling system. The residual concentration of Cr(VI) was analyzed using the 1,5-diphenylcarbazide (DPC) colorimetric method on a Perkin Elmer Lambda 25 UV-vis spectrometer (Perkin Elmer Inc., Waltham, MA, USA). The concentration (C_t/C_o) of Cr(VI) in solution at different illumination times was determined commensurate to the absorbance (A_t/A_o, where A_o and A_t are the absorbances at initial and at the time t, respectively) of the DPC-Cr(VI) complex at 540 nm.

3.6. Physical Characterization

Powder X-ray diffraction (XRD) was performed on a PANalytical X'Pert Pro MPD X-ray diffractometer (Malvern PANalytical Ltd., Almelo, The Netherlands) using Cu Kα (λ = 1.5406 Å) radiation (45 kV, 40 mA). Diffraction patterns were obtained in Bragg–Brentano geometry at a 2θ range of 20–80° using a step size (2θ) of 0.01° and a scanning speed of 0.1°/min. Transmission electron microscopy (TEM) images were obtained on a JEOL JEM-2100 electron microscope (LaB_6 filament) (JEOL Ltd., Tokyo, Japan) using an acceleration voltage of 200 kV. Samples were prepared by drop-casting a sample dispersion (0.5 mg mL^{-1} in ethanol) onto a carbon-coated copper grid. N_2 physisorption measurements were performed at −196 °C using a Quantachrome NOVA 3200*e* sorption analyzer (Quantachrome Co., Boynton Beach, FL, USA). Before analysis, samples were deaerated at 110 °C for 12 h under vacuum (<10^{-5} Torr). The specific surface areas were determined using the Brunauer–Emmett–Teller (BET) method [38] on the adsorption isotherm in the relative pressure (P/Po) range from 0.05 to 0.22. Total pore volumes were determined from the adsorbed amount at the P/Po = 0.98, and the pore size distributions were derived from the adsorption branches using the nonlocal density functional theory (NLDFT) method [39]. UV-vis/near-IR diffuse reflectance spectra were obtained on a Perkin Elmer Lambda 950 spectrophotometer (Perkin Elmer Inc., Waltham, MA, USA), using $BaSO_4$ powder as the reflectance reference. The energy band gaps (E_g) of the samples were determined from Tauc plots for direct allowed transition [40], i.e., $(f\text{hv})^2$ as a function of photon energy (hv), where f is the Kubelka–Munk function of the reflectance (R): $f(R) = (1 - R)^2/(2R)$ [41].

3.7. Electrochemical Measurements

Mott–Schottky plots were collected with a Metrohm Autolab PGSTAT 302N potentiostat (Metrohm AG, Herisau, Switzerland). A three-electrode set-up, with a Pt counter electrode and an Ag/AgCl (3M KCl) reference electrode was used to carry out all electrochemical studies. The capacitance of the semiconductor/electrolyte interface was obtained using 10 mV AC voltage amplitude at 1 kHz, in 0.5 M Na_2SO_4 solution (pH = 6.87). The measured flat-band potentials were converted to the normal hydrogen electrode (NHE) scale using the Nernst equation:

$$E_{NHE} = E_{Ag/AgCl} + 0.21 + 0.059 \times pH \quad (3)$$

The working electrodes were prepared as follows: 1 mL of aqueous dispersion of samples (10 mg) was sonicated in a water bath to obtain a uniform suspension. After that, 40 μL of the suspension was placed on a fluorine-doped tin oxide (FTO, 10 Ω/sq) substrate, which was covered with an epoxy resin to leave an effective area of 1 cm^2. Then the electrode was dried at 60 °C for 30 min. The donor density (N_d) of the samples was determined according to the Mott–Schottky equation:

$$N_d = \frac{2(E - E_{FB}) \cdot C_{SC}^2}{\varepsilon \varepsilon_o e} \quad (4)$$

where, ε is the relative dielectric constant of the material, ε_o is the permittivity of vacuum (8.8542×10^{-10} F cm^{-1}), e is the electron charge (1.602×10^{-19} C), E is the applied potential, and C_{sc} is the space charge capacitance of the semiconductor.

The electrochemical impedance spectroscopy, polarization, chronoamperometric and open-circuit photovoltage decay measurements were carried out in a 0.5 M Na_2SO_4 solution (pH = 6.87) using a VersaSTAT 4 electrochemical workstation (Princeton Applied Research, Oak Ridge, TN, USA) with an air-tight three-electrode cell, consisting of the samples as the working electrode, an Ag/AgCl (saturated KCl) as the reference electrode, and a Pt wire as the counter electrode. For Nyquist plots, the different current output was measured throughout a frequency range of 1 Hz to 1 MHz using a small AC perturbation of 20 mV, under open-circuit potential conditions. The electrochemical impedance data were fitted to an equivalent circuit model using ZView Software (Version 3.5h, Scribner Associates,

Southern Pines, NC, USA, 2020). Polarization curves were recorded at the sweep rate of 50 mV s^{-1}. Photochronoamperometric data were obtained at a bias voltage of 0.2 V (vs. Ag/AgCl).

4. Conclusions

In summary, high-surface-area dual component MFO-ZFO mesoporous networks were successfully synthesized via a polymer-assisted method that allows the co-assembly of the spinel ferrite ZFO and MFO colloidal NPs and amphiphilic block-copolymer aggregates. The resulting materials consist of small-sized (ca. 6–7 nm) MFO and ZFO NPs that form a continuous network-like structure with high porosity. This allows for a large surface area (ca. 68–91 m^2 g^{-1}) that is accessible for photochemical reactions, and a direct NP-to-NP contact for efficient charge separation. As a result, the MFO-ZFO composite materials exhibit excellent performance for photocatalytic reduction of Cr(VI) in aqueous solutions with coexisting organic pollutants (such as phenol, citric acid, and EDTA), under UV-vis light irradiation. Taken together, microscopic and spectroscopic characterization techniques revealed that the enhanced photocatalytic activity of dual component MFO-ZFO mesoporous networks is originated from the combined effect of accessible pore structure, which permits facile diffusion of reactants and products, and suitable electronic band structure, which efficiently separates and transports the charge carriers through the ZFO/MFO interface. Therefore, such mesoporous spinel ferrite NP-networks manifest improved photochemical activity and demonstrate great potential for applications in photocatalysis and environmental remediation.

Supplementary Materials: The following are available online at https://www.mdpi.com/2073-4344/11/2/199/s1, Figure S1: N_2 physisorption data for the ZFO and MFO MNAs, Figure S2: Kinetic plots for different ZFO and MFO based catalysts, Figure S3: N_2 physisorption data for the MFO-6-ZFO RNAs, Figure S4: Kinetic plot for the untemplated MFO-6-ZFO catalyst, Figure S5: Powder XRD and N_2 physisorption data for the reused MFO-6-ZFO catalyst, Figure S6: Mott-Schottky and Tauc plots of the ZFO and MFO MNAs, Figure S7: Open-circuit photovoltage decay curves for ZFO, MFO, and MFO-6-ZFO MNAs.

Author Contributions: Conceptualization and methodology, E.S. and G.S.A.; synthesis and physicochemical characterization, E.S.; electrochemical measurements, I.T.P. and S.A.C.; photoelectrochemical measurements, I.V.; writing—original draft preparation, E.S., I.V. and G.S.A.; writing—review and editing, G.S.A.; supervision, G.S.A. All authors have read and agreed to the published version of the manuscript.

Funding: This research was supported by the Hellenic Foundation for Research and Innovation (H.F.R.I.) under the "1st Call for H.F.R.I. Research Projects to support Faculty Members & Researchers and the Procurement of High-cost research equipment grant" (Project Number: 400). Funding for open access publication was supported by the Special Account for Research Funds (SARF) of the University of Crete (Project KA 3650).

Data Availability Statement: The data presented in this study are available in article and supplementary material.

Conflicts of Interest: The authors declare no conflict of interest.

References

1. Tchounwou, P.B.; Yedjou, C.G.; Patlolla, A.K.; Sutton, D.J. Heavy metal toxicity and the environment. *Exp. Suppl.* **2012**, *101*, 133–164. [CrossRef] [PubMed]
2. Bakshi, A.; Panigrahi, A.K. A comprehensive review on chromium induced alterations in fresh water fishes. *Toxicol. Rep.* **2018**, *5*, 440–447. [CrossRef] [PubMed]
3. Joutey, N.T.; Sayel, H.; Bahafid, W.; Ghachtouli, N.E. Mechanisms of hexavalent chromium resistance and removal by microorganisms. *Rev. Environ. Contam. Toxicol.* **2015**, *233*, 45–69. [CrossRef] [PubMed]
4. Laxmi, V.; Kaushik, G. Toxicity of hexavalent chromium in environment, health threats, and its bioremediation and detoxification from tannery wastewater for environmental safety. In *Bioremediation of Industrial Waste for Environmental Safety: Volume I: Industrial Waste and Its Management*; Saxena, G., Bharagava, R.N., Eds.; Springer: Singapore, 2020; pp. 223–243.

5. World Health Organization. *Guidelines for Drinking-Water Quality: Third Edition, Incorporating the First and Second Addenda*; World Health Organization: Geneva, Switzerland, 2008; Volume 1, pp. 334–335.
6. Siboni, M.S.; Samarghandi, M.R.; Azizian, S.; Kim, W.G.; Lee, S.M. The Removal of hexavalent chromium from aqueous solutions using modified holly sawdust: Equilibrium and kinetics studies. *Environ. Eng. Res.* **2011**, *16*, 55–60. [CrossRef]
7. Kongsricharoern, N.; Polprasert, C. Electrochemical precipitation of chromium (Cr^{6+}) from an electroplating wastewater. *Water Sci. Technol.* **1995**, *31*, 109–117. [CrossRef]
8. Etemadi, M.; Samadi, S.; Yazd, S.S.; Jafari, P.; Yousefi, N.; Aliabadi, M. Selective adsorption of Cr(VI) ions from aqueous solutions using Cr^{6+}-imprinted Pebax/chitosan/GO/APTES nanofibrous adsorbent. *Int. J. Biol. Macromol.* **2017**, *95*, 725–733. [CrossRef]
9. Xing, Y.; Chen, X.; Wang, D. Electrically Regenerated ion exchange for removal and recovery of Cr(VI) from wastewater. *Environ. Sci. Technol.* **2007**, *41*, 1439–1443. [CrossRef]
10. Khorramabadi, G.S.; Soltani, R.D.C.; Rezaee, A.; Khataee, A.R.; Jafari, A.J. Utilisation of immobilised activated sludge for the biosorption of chromium (VI). *Can. J. Chem. Eng.* **2012**, *90*, 1539–1546. [CrossRef]
11. Shirzad-Siboni, M.; Farrokhi, M.; Soltani, R.D.C.; Khataee, A.; Tajassosi, S. Photocatalytic reduction of hexavalent chromium over ZnO nanorods immobilized on kaolin. *Ind. Eng. Chem. Res.* **2014**, *53*, 1079–1087. [CrossRef]
12. Sun, B.; Reddy, E.P.; Smirniotis, P.G. Visible light Cr(VI) reduction and organic chemical oxidation by TiO_2 photocatalysis. *Environ. Sci. Technol.* **2005**, *39*, 6251–6259. [CrossRef]
13. Mondal, C.; Ganguly, M.; Pal, J.; Roy, A.; Jana, J.; Pal, T. Morphology controlled synthesis of SnS_2 nanomaterial for promoting photocatalytic reduction of aqueous Cr(VI) under visible light. *Langmuir* **2014**, *30*, 4157–4164. [CrossRef] [PubMed]
14. Yu, J.; Zhuang, S.; Xu, X.; Zhu, W.; Feng, B.; Hu, J. Photogenerated electron reservoir in hetero-p–n CuO–ZnO nanocomposite device for visible-light-driven photocatalytic reduction of aqueous Cr(VI). *J. Mater. Chem. A* **2015**, *3*, 1199–1207. [CrossRef]
15. Wang, Q.; Shi, X.; Liu, E.; Crittenden, J.C.; Ma, X.; Zhang, Y.; Cong, Y. Facile synthesis of AgI/BiOI-Bi_2O_3 multi-heterojunctions with high visible light activity for Cr(VI) reduction. *J. Hazard. Mater.* **2016**, *317*, 8–16. [CrossRef] [PubMed]
16. Velegraki, G.; Vamvasakis, I.; Papadas, I.T.; Tsatsos, S.; Pournara, A.; Manos, M.J.; Choulis, S.A.; Kennou, S.; Kopidakis, G.; Armatas, G.S. Boosting photochemical activity by Ni doping of mesoporous CoO nanoparticle assemblies. *Inorg. Chem. Front.* **2019**, *6*, 765–774. [CrossRef]
17. Velegraki, G.; Miao, J.; Drivas, C.; Liu, B.; Kennou, S.; Armatas, G.S. Fabrication of 3D mesoporous networks of assembled CoO nanoparticles for efficient photocatalytic reduction of aqueous Cr(VI). *Appl. Catal. B Environ.* **2018**, *221*, 635–644. [CrossRef]
18. Hernández-Gordillo, L.; Tzompantzi, F.J.; Gomez, R. Enhanced photoreduction of Cr(VI) using $ZnS(en)_{0.5}$ hybrid semiconductor. *Catal. Commun.* **2012**, *19*, 51–55. [CrossRef]
19. Wang, D.; Ye, Y.; Liu, H.; Ma, H.; Zhang, W. Effect of alkaline precipitation on Cr species of Cr(III)-bearing complexes typically used in the tannery industry. *Chemosphere* **2018**, *193*, 42–49. [CrossRef]
20. Srivastava, S.; Ahmad, A.H.; Thakur, I.S. Removal of chromium and pentachlorophenol from tannery effluents. *Bioresour. Technol.* **2007**, *98*, 1128–1132. [CrossRef]
21. Kulkarni, A.M.; Desai, U.V.; Pandit, K.S.; Kulkarni, M.A.; Wadgaonkar, P.P. Nickel ferrite nanoparticles–hydrogen peroxide: A green catalyst-oxidant combination in chemoselective oxidation of thiols to disulfides and sulfides to sulfoxides. *RSC Adv.* **2014**, *4*, 36702–36707. [CrossRef]
22. Guo, X.; Zhu, H.; Li, Q. Visible-light-driven photocatalytic properties of ZnO/$ZnFe_2O_4$ core/shell nanocable arrays. *Appl. Catal. B Environ.* **2014**, *160–161*, 408–414. [CrossRef]
23. Hou, Y.; Li, X.; Zhao, Q.; Chen, G. $ZnFe_2O_4$ multi-porous microbricks/graphene hybrid photocatalyst: Facile synthesis, improved activity and photocatalytic mechanism. *Appl. Catal. B Environ.* **2013**, *142–143*, 80–88. [CrossRef]
24. Cai, C.; Zhang, Z.; Liu, J.; Shan, N.; Zhang, H.; Dionysiou, D.D. Visible light-assisted heterogeneous Fenton with $ZnFe_2O_4$ for the degradation of Orange II in water. *Appl. Catal. B Environ.* **2016**, *182*, 456–468. [CrossRef]
25. Kong, L.; Jiang, Z.; Xiao, T.; Lu, L.; Jones, M.O.; Edwards, P.P. Exceptional visible-light-driven photocatalytic activity over BiOBr–$ZnFe_2O_4$ heterojunctions. *Chem. Commun.* **2011**, *47*, 5512–5514. [CrossRef]
26. Lee, K.-T.; Chuah, X.-F.; Cheng, Y.-C.; Lu, S.-Y. Pt coupled $ZnFe_2O_4$ nanocrystals as a breakthrough photocatalyst for Fenton-like processes—Photodegradation treatments from hours to seconds. *J. Mater. Chem. A* **2015**, *3*, 18578–18585. [CrossRef]
27. Yu, T.-H.; Cheng, W.-Y.; Chao, K.-J.; Lu, S.-Y. $ZnFe_2O_4$ decorated CdS nanorods as a highly efficient, visible light responsive, photochemically stable, magnetically recyclable photocatalyst for hydrogen generation. *Nanoscale* **2013**, *5*, 7356–7360. [CrossRef] [PubMed]
28. Kim, J.H.; Jang, Y.J.; Kim, J.H.; Jang, J.-W.; Choi, S.H.; Lee, J.S. Defective $ZnFe_2O_4$ nanorods with oxygen vacancy for photoelectrochemical water splitting. *Nanoscale* **2015**, *7*, 19144–19151. [CrossRef]
29. Ren, Y.; Li, N.; Feng, J.; Luan, T.; Wen, Q.; Li, Z.; Zhang, M. Adsorption of Pb(II) and Cu(II) from aqueous solution on magnetic porous ferrospinel $MnFe_2O_4$. *J. Colloid Interface Sci.* **2012**, *367*, 415–421. [CrossRef]
30. Skliri, E.; Miao, J.; Xie, J.; Liu, G.; Salim, T.; Liu, B.; Zhang, Q.; Armatas, G.S. Assembly and photochemical properties of mesoporous networks of spinel ferrite nanoparticles for environmental photocatalytic remediation. *Appl. Catal. B Environ.* **2018**, *227*, 330–339. [CrossRef]
31. Papadas, I.T.; Vamvasakis, I.; Tamiolakis, I.; Armatas, G.S. Templated self-assembly of colloidal nanocrystals into three-dimensional mesoscopic structures: A perspective on synthesis and catalytic prospects. *Chem. Mater.* **2016**, *28*, 2886–2896. [CrossRef]

32. Kong, X.K.; Zhou, Y.; Xu, T.; Hu, B.; Lei, X.; Chen, H.; Yu, G. A novel technique of COD removal from electroplating wastewater by Fenton-alternating current electrocoagulation. *Environ. Sci. Pollut. Res.* **2020**, *27*, 15198–15210. [CrossRef]
33. Dsikowitzky, L.; Schwarzbauer, J. Organic contaminants from industrial wastewaters: Identification, toxicity and fate in the environment. In *Pollutant Diseases, Remediation and Recycling. Environmental Chemistry for a Sustainable World*; Lichtfouse, E., Schwarzbauer, J., Robert, D., Eds.; Springer: Cham, Switzerland, 2013; Volume 4, pp. 45–101. [CrossRef]
34. Hsu, H.T.; Chen, S.S.; Tang, Y.F.; Hsi, H.C. Enhanced photocatalytic activity of chromium(VI) reduction and EDTA oxidization by photoelectrocatalysis combining cationic exchange membrane processes. *J. Hazard. Mater.* **2013**, *248–249*, 97–106. [CrossRef] [PubMed]
35. Wang, J.; Yu, Y.; Zhang, L. Highly efficient photocatalytic removal of sodium pentachlorophenate with Bi_3O_4Br under visible light. *Appl. Catal. B Environ.* **2013**, *136–137*, 112–121. [CrossRef]
36. Zaban, A.; Greenshtein, M.; Bisquert, J. Determination of the electron lifetime in nanocrystalline dye solar cells by open-circuit voltage decay measurements. *ChemPhysChem* **2003**, *4*, 859–864. [CrossRef] [PubMed]
37. Liu, C.; Zou, B.; Rondinone, A.J.; Zhang, Z.J. Reverse micelle synthesis and characterization of superparamagnetic $MnFe_2O_4$ spinel ferrite nanocrystallites. *J. Phys. Chem. B* **2000**, *104*, 1141–1145. [CrossRef]
38. Brunauer, S.; Deming, L.S.; Deming, W.E.; Teller, E. On a theory of the van der Waals adsorption of gases. *J. Am. Chem. Soc.* **1940**, *62*, 1723–1732. [CrossRef]
39. Ravikovitch, P.I.; Wei, D.; Chueh, W.T.; Haller, G.L.; Neimark, A.V. Evaluation of pore structure parameters of MCM-41 catalyst supports and catalysts by means of nitrogen and argon adsorption. *J. Phys. Chem. B* **1997**, *101*, 3671–3679. [CrossRef]
40. Tauc, J. Optical properties of amorphous semiconductors. In *Amorphous and Liquid Semiconductors*; Tauc, J., Ed.; Springer: Boston, MA, USA, 1974; pp. 159–220.
41. Kubelka, P. New contributions to the optics of intensely light-scattering materials. Part I. *J. Opt. Soc. Am.* **1948**, *38*, 448–457. [CrossRef]

catalysts

MDPI

Article

Development of a New Arylamination Reaction Catalyzed by Polymer Bound 1,3-(Bisbenzimidazolyl) Benzene Co(II) Complex and Generation of Bioactive Adamanate Amines

Baburajeev Chumadathil Pookunoth [1,†], Shilpa Eshwar Rao [1,2,†], Suresha Nayakanahundi Deveshegowda [3], Prashant Kashinath Metri [3], Kashifa Fazl-Ur-Rahman [1], Ganga Periyasamy [1], Gayathri Virupaiah [1,4], Babu Shubha Priya [3], Vijay Pandey [5], Peter E. Lobie [5,6,*], Rangappa Knchugarakoppal Subbegowda [7,*] and Basappa [1,3,*]

1 Department of Chemistry, Central College Campus, Bangalore University, Bangalore 560001, Karnataka, India; baburajeevnambiar@gmail.com (B.C.P.); shilpa.gaikwad1989@gmail.com (S.E.R.); kashifaf07@gmail.com (K.F.-U.-R.); ganga.periyasamy@gmail.com (G.P.); gayathritvr@yahoo.co.in (G.V.)
2 Department of Inorganic and Physical Chemistry, Indian Institute of Science, Bangalore 560012, Karnataka, India
3 Laboratory of Chemical Biology, Department of Studies in Organic Chemistry, University of Mysore, Manasagangotri, Mysore 570006, Karnataka, India; sureshand92@gmail.com (S.N.D.); prashant.metri@gmail.com (P.K.M.); priyabs_chem@yahoo.com (B.S.P.)
4 Department of Chemistry, Bengaluru City University, Bangalore 560001, Karnataka, India
5 Tsinghua Berkeley Shenzhen Institute, Tsinghua Shenzhen International Graduate School, Shenzhen 518055, China; vijay.pandey@sz.tsinghua.edu.cn
6 Shenzhen Bay Laboratory, Shenzhen 518055, China
7 Institution of Excellence, Vijnana Bhavan, University of Mysore, Mysore 570005, Karnataka, India
* Correspondence: pelobie@sz.tsinghua.edu.cn (P.E.L.); rangappaks@chemistry.uni-mysore.ac.in (R.K.S.); salundibasappa@gmail.com (B.); Tel.: +91-821-241-9428 (R.K.S.); +91-9481200076 (B.)
† These authors contributed equally.

Received: 10 October 2020; Accepted: 29 October 2020; Published: 13 November 2020

Abstract: We herein report the preparation and characterization of an inexpensive polymer supported 1,3-bis(benzimidazolyl)benzeneCo(II) complex [PS-Co(BBZN)Cl_2] as a catalyst by using the polymer (divinylbenzene cross-linked chloromethylated polystyrene), on which 1,3-bis(benzimidazolyl) benzeneCo(II) complex (PS-Co(BBZN)Cl_2) has been immobilized. This'catalyst was employed to develop arylamination reaction and robustness of the same reaction was demonstrated by synthesizing various bioactive adamantanyl-tethered-biphenylamines. Our synthetic methodology was much improved than reported methods due to the use of an inexpensive and recyclable catalyst.

Keywords: arylamination reactions; adamantanyl-tethered-biphenylamines; polymer-supported catalyst; cobalt complex; Buchwald–Hartwig reaction

1. Introduction

Transition metal-catalyzed cross-coupling reactions between aryl halides and primary/secondary amines to obtain aminated aryl compounds has been an area of interest due to the wide applications of arylamines in the synthetics and pharmaceutical industries [1–5]. In this direction, the Buchwald–Hartwig cross-coupling reaction was performed by using transition metal catalysts, ligands and bases with substrates to obtain the desired arylamine products [6–8]. The disadvantage of this reaction is the use of expensive catalysts, which offers the chemist the opportunity to discover cheaper, reusable catalysts to drive the arylamination reactions. Inspired by major developments in cobalt-catalyzed arylamination reactions, we developed a complementary method to perform an arylamination reaction using cobalt as a metal catalyst [9–11]

In addition, benzimidazole ligand coordinated metal complexes are widely used as catalysts in arylamination reactions [12]. Since these catalysts were found to be less hydrophobic, immobilization of such metal complexes with polymer support was observed to be stable, selective, and recyclable, attributed to the steric, electrostatic, hydrophobic and conformational effects of the polymer support [13]. Hence, several reports pertaining to the synthesis of arylamines using polymer-supported transition metal complexes are found [14–16]. Specifically, chloromethylated polystyrene cross-linked with divinyl-benzene was employed as a macromolecular support to perform the arylamination reactions [17–22].

In medicinal chemistry, an adamantane-coupled bicyclical core structure was used as an important pharmacophore, which was inserted in many drugs [23]. Hence, the adamantane structure was recognized as a readily available "liphophilic bullet" for providing critical liphophilicity to known pharmacophoric units. Given the remarkable importance of adamantane chemistry, we recently reported the synthesis and biology of adamantyl-tethered biphenylic compounds as potent anticancer agents [24]. In our continued efforts to synthesize newer bioactive agents [25–31], we herein report a practical, economically feasible and efficient arylamination reaction using polymer-supported 1,3-bis(benzimidazolyl)benzeneCo(II) complex (PS-Co(BBZN)Cl_2) as a catalyst. Interestingly, the recovered (PS-Co(BBZN)Cl_2) could be reused three times without a significant loss of activity.

2. Results

2.1. Chemistry of Catalyst Design and Method Development

We initially synthesized polymer-supported 1,3-bis(benzimidazolyl)benzeneCo(II) complex [PS-Co(BBZN)Cl_2] as shown in Scheme 1.

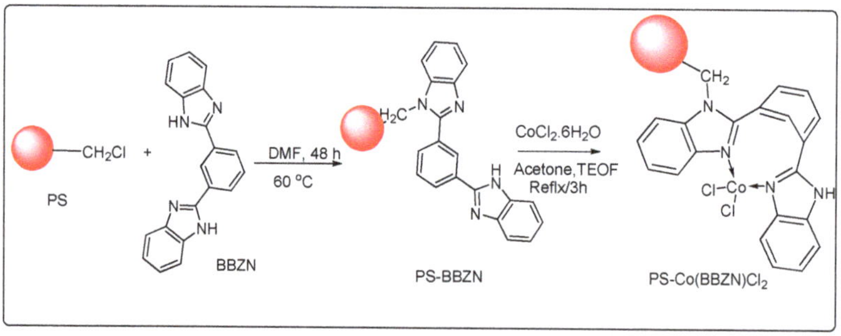

Scheme 1. Schematic representation to show synthesis of PS-Co(BBZN)Cl_2.

For this, 1, 3-bis(benzimidazolyl)benzene was treated with chloromethylated polystyrene divinylbenzene and followed by the addition of cobalt chloride. The obtained PS-Co (BBZN)Cl_2 was characterized by analytical techniques including CHNS, UV-Vis, FT-IR, SEM-EDX and TGA as presented in supporting information (Figure 1, Supplementary SI-02). Based on N% and Co% obtained through elemental and metal ion analysis, the complex formed on the polymer support was about 0.0053 moles per 1 g of the polymer support which corresponded to 7.16% of Co intake. This further confirmed the formation of the complex on the polymer support.

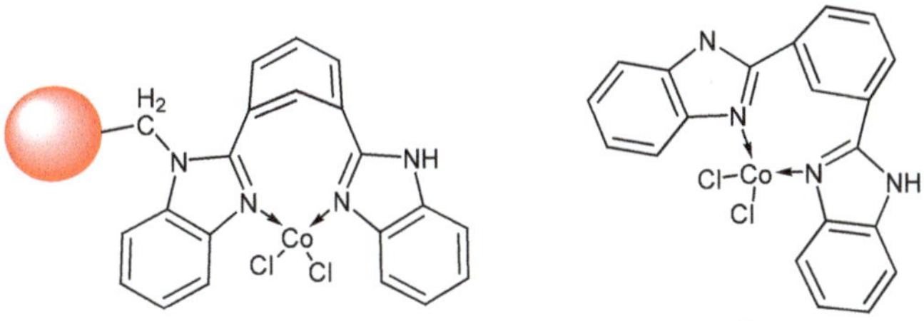

(A) (B)

Figure 1. Structure of (**A**) PS-Co (BBZN)Cl_2 and (**B**) unbound Co(BBZN)Cl_2.

Motivated by the increased understanding of the Co-catalyzed amination reaction, we next investigated the applicability of (PS-Co (BBZN)Cl_2) in the arylamination reaction. To examine this hypothesis, 1-(5-bromo-2-methoxyphenyl)adamantine (**1a**) and 4-chloro aniline (**2a**) were selected as model substrates and reagents for the reaction in 1,4-dioxane media and Cs_2CO_3 as a base (Scheme 2).

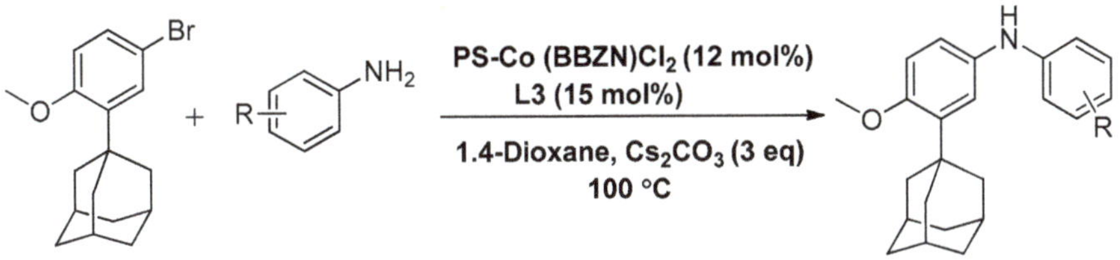

1a **2a-l** **3a-l**

Scheme 2. General scheme of arylamination reaction between adamantane bromide and various amines using PS-Co(BBZN)Cl_2 as a catalyst.

Control experiments established the importance of both PS-Co(BBZN)Cl_2 and ligand, as no product was obtained (Table 1, entry **1**). Gratifyingly, the substrate was transformed into the desired product 3-(adamantan-1-yl)-N-(4-chlorophenyl)-4-methoxyaniline (**3a**) with 51% yield in the presence of catalyst (PS-Co(BBZN)Cl_2) (10 mol%) and ligand **L3** (Table 2, entry **10**). Screening of various classes of ligands (Figure 2) to improve the yield revealed that the use of phosphine based ligand BINAP (**L3**) or Xphose (**L4**) gave improved yields at different catalyst concentrations (Table 1, entry **10, 11, 14, 15**), whereas the other ligands such as bidentate ligands (**L1, L2**) and N-heterocyclic carbine ligands (**L5, L6**) yielded no products indicating the high role of selectivity of ligands in the forward reaction. The most robust reaction was achieved by the use of 12 mol% of PS-Co(BBZN)Cl_2 in the presence of BINAP with an 86% yield at 10 h reaction condition (Table 1, entry **14**). Further investigation revealed that there was no considerable improvement in yield when the catalyst load was increased to 15 mol% (Table 1, entry **18, 19**) whereas the yield dropped to 69% when the reaction time was reduced to 6 h with 15 mol% catalyst (Table 1, entry **20**). Using the above better protocol, we further synthesized ABTAs by reacting adamantine bromo compounds (**1a**) and various amines (Table 2). It was observed that all amine partners productively coupled with good yields of around 70–86%.

All novel compounds exhibited spectral properties consistent with the assigned structures and were fully characterized by their spectroscopic data (mass, elemental, ^{1}H and ^{13}C NMR analysis).

The majority of reactions were done by keeping time point for 16 h and when the concentration of the catalyst was increased to 12%, the reaction was completed in 12 h and in many cases pure product was produced with excellent yield. The above developed method tolerated the presence of substituent in the aromatic amino-compounds. Specifically, we observed that the electron-donating para-substituted aromatic amine partners were well-tolerated to produce corresponding products in good to excellent yields (Table 2, entries **1–12**). However, ortho-substituted and electron-withdrawing group bearing compounds were not productive giving lower yields (Table 2, entries **5**, **11**, **12**, **13**).

Table 1. PS-Co(BBZN)Cl_2-catalyzed coupling of 1-(5-bromo-2-methoxyphenyl)adamantane with 4-Chloro aniline [a].

Entry	PS-Co(BBZN)Cl_2	Ligand [b]	Time	Yield (%) [c]
1	5 mol%	—	16	NR
2	5 mol%	L1	16	NR
3	5 mol%	L2	16	NR
4	5 mol%	L3	16	NR
5	5 mol%	L4	16	NR
6	5 mol%	L5	16	NR
7	5 mol%	L6	16	NR
8	10 mol%	L1	16	NR
9	10 mol%	L2	16	NR
10	10 mol%	L3	16	51
11	10 mol%	L4	16	42
12	10 mol%	L5	16	20
13	10 mol%	L6	16	26
14	12 mol%	L3	10	86
15	12 mol%	L4	12	78
16	12 mol%	L5	16	36
17	12 mol%	L6	16	41
18	15 mol%	L3	10	86
19	15 mol%	L4	12	79
20	15 mol%	L3	6	69

[a] Conditions: admantane-bromo compounds (1 mmol), 4-chloro aniline (1 mmol) (PS-Co (BBZN)Cl_2) (12 mol%); Cs_2CO_3 (3 eq); 1, 4 dioxane (10 mL); N_2 atmosphere: 100 °C. [b] ligands (15 mol%): L1 = 2, 2′-bipyridine, L2 = 1,10-phenanthroline; L3 = 2,2′-bis(diphenylphosphino)-1,1′-binaphthalene, L4 = dicyclohexyl(2-(2,4,6-trisopropylphenyl)cycohexyl)phosphine, L5 = 2,6-bis(3-methylimidazoline-1yl)pyridine, L6 = 1,3-dimessityl-4,5-dihydro-1H-imidazole-3-ium chloride; [c] isolated yield; NR = no reaction.

Table 2. PS-Co(BBZN)Cl_2 composite-catalyzed coupling of various substituted halo aromatic compounds with various substituted aromatic amines [a].

Entry	Amine	Product [b] and Yield (%) [c]
1	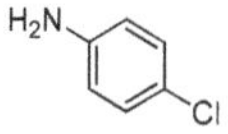 2a	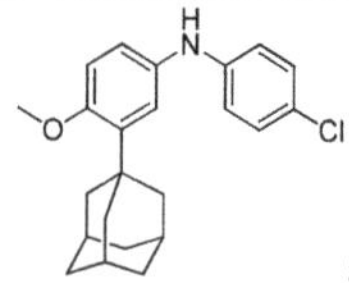 3a(86)
2	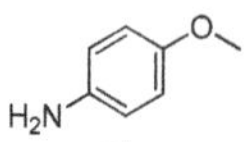 2b	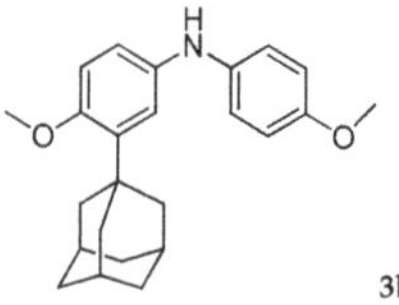 3b(82)
3	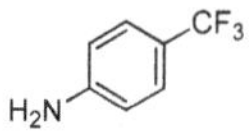 2g	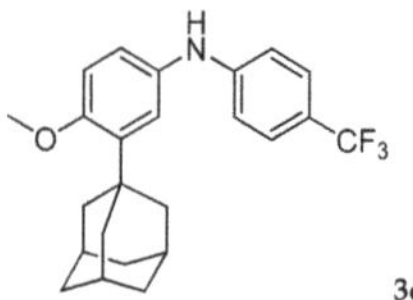 3c(79)

Table 2. *Cont.*

Entry	Amine	Product [b] and Yield (%) [c]
4	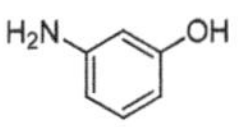 2h	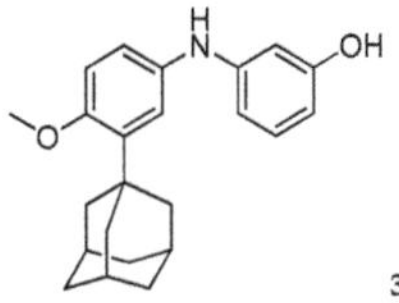 3d(84)
5	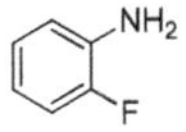 2i	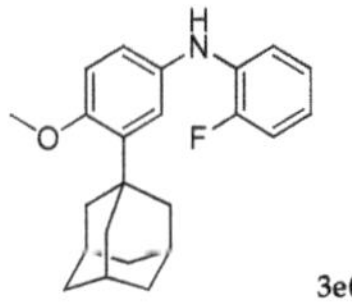 3e(62)
6	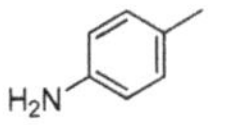 2j	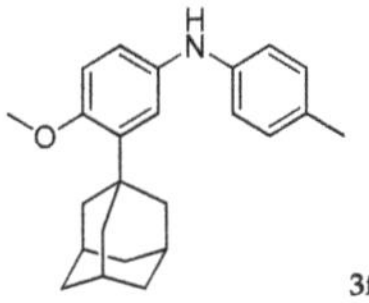 3f(87)
7	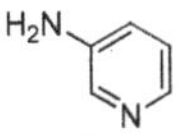 2k	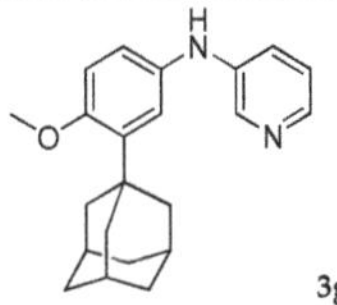 3g(84)
8	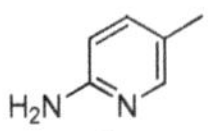 2l	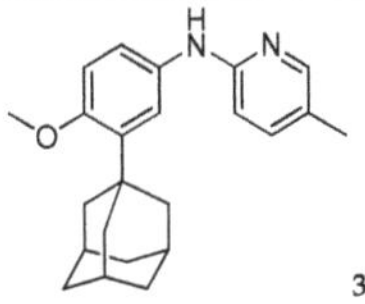 3h(80)
9	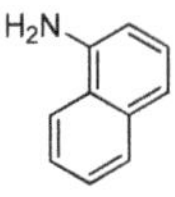 2m	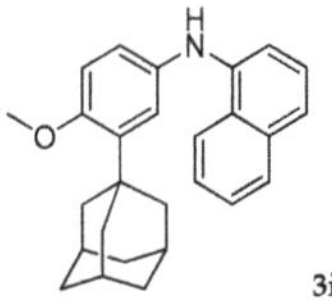 3i(83)
10	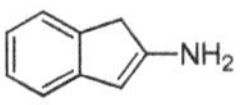 2n	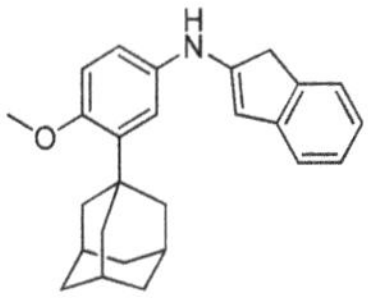 3j(76)
11	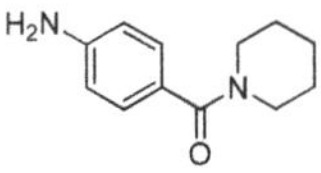 2o	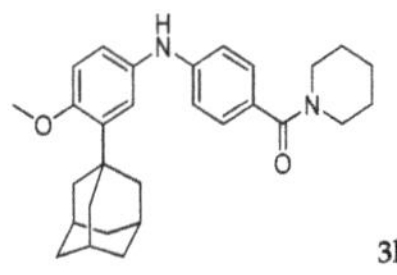 3k(64)

Table 2. *Cont.*

Entry	Amine	Product [b] and Yield (%) [c]
12	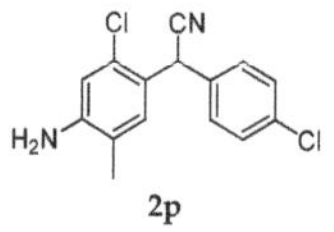 2p	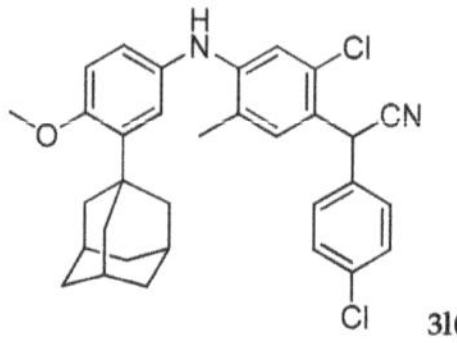 31(61)
13	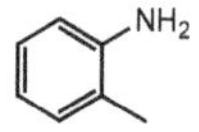 2q	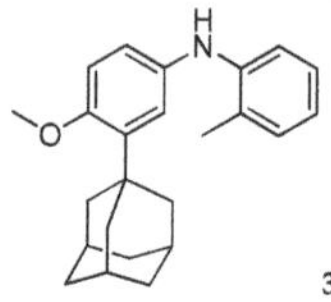 3m (64)

[a] Reaction conditions—Aromatic halo compounds (1 mmol), aromatic amine (1 mmol), BINAP(15 mol%), PS-Co(BBZN)Cl_2 (12 mol%), CS_2CO_3(3 mmol), 1,4-dioxane (5 mL), N_2 atmosphere 10 h, 100 °C. [b] All new compounds were characterized by their spectroscopic data shown in supporting information; [c] isolated yield.

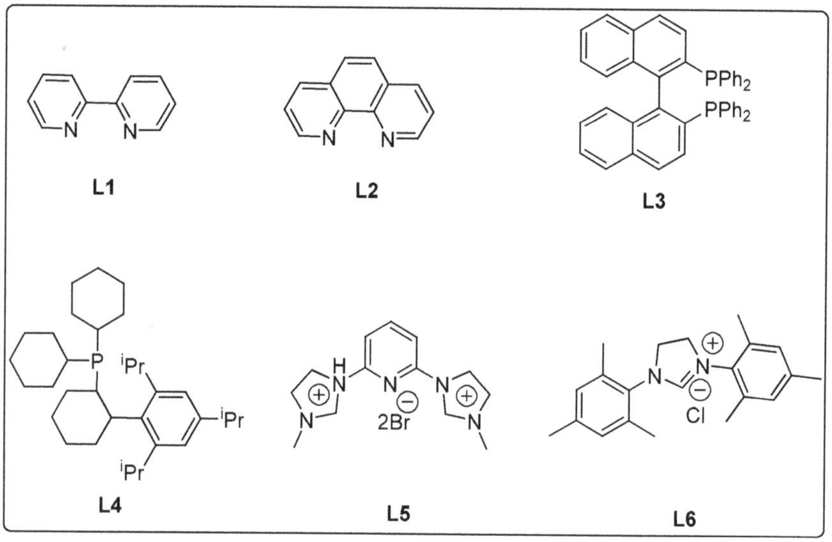

Figure 2. Various classes of ligands used in this study.

With the reaction conditions established we tried to investigate the scope of the new protocol on different substituted aromatic bromo compounds by treating with various amines (Table 3). We found that electron donating para-substituted on aromatic halo partner was tolerated well to give corresponding products in good to excellent yields (entries **1**, **2**, **3** and **5**), but with ortho-substituted and electron-withdrawing group bearing aromatic bromo compounds observed a loss in yield (entries **4**, **6** and **7**) with no improvement in the reaction conversion on prolonged reaction.

Table 3. Various substrates and reagents used to optimization of arylamination reaction.

Entry	Aromatic Halo Compounds	Amine	Product [a] and Yield (%)
1	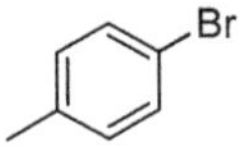 4 a	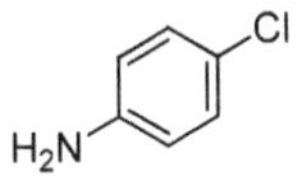 2 a	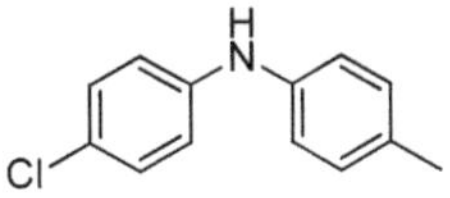 5 a [a] (81)
2	4 a	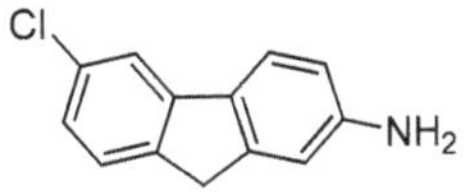 2 r	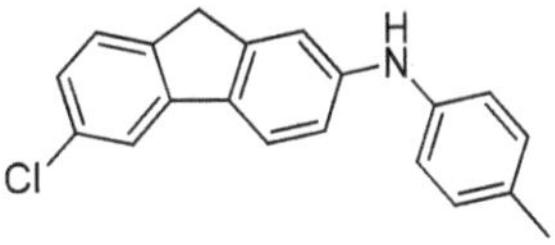 5 b (80)
3	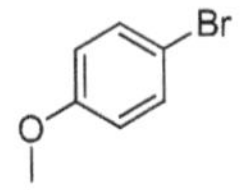 4 b	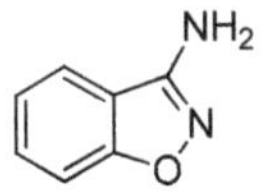 2 s	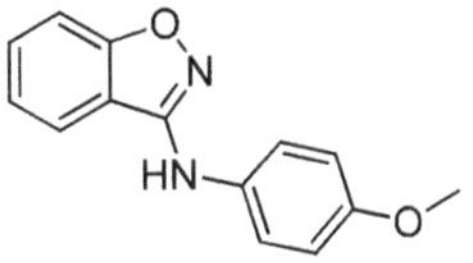 5 c (79)
4	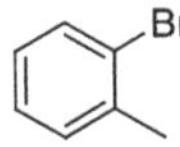 4 c	2 a	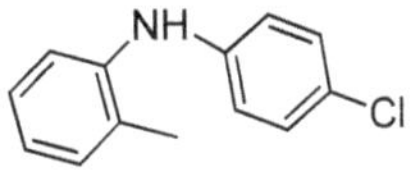 5 d [a] (61)
5	4 b	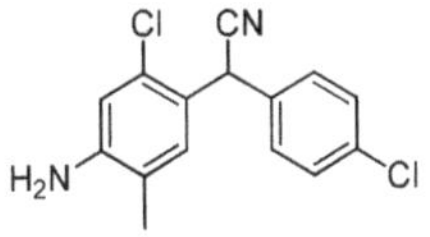 2 p	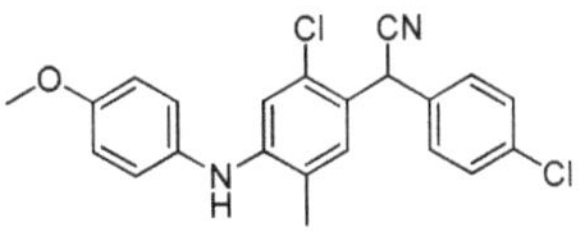 5 e(78)
6	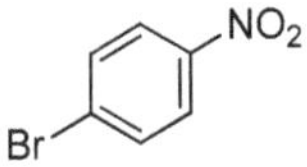 4 d	2 a	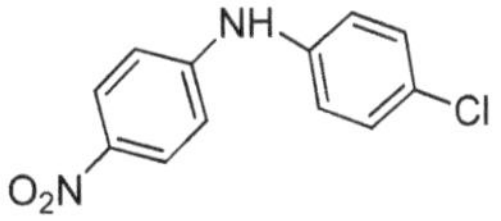 5 f [a] (59)
7	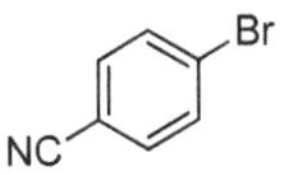	2 a	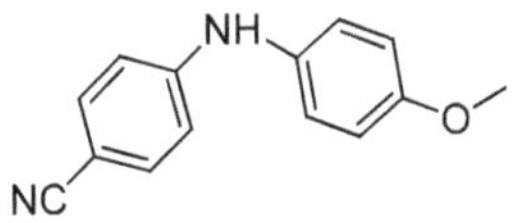 5 g [a] (58)

[a] Reported compounds.

All novel compounds exhibited spectral properties consistent with the assigned structures and were fully characterized by their spectroscopic data (mass, elemental, ^{1}H and ^{13}C NMR analysis). It was found that the use of a catalyst PS-Co(BBZN)Cl_2, in combination with some ligands provided a robust catalytic system. On the basis of previous mechanistic studies in cobalt-catalyzed C–N bond formation reactions, it was possible to propose a mechanism for the conversion of 3-(adamantan-1-yl)-N-(4-chlorophenyl)-4-methoxyaniline (**3 a**) as shown in Figure 3 [32–34].

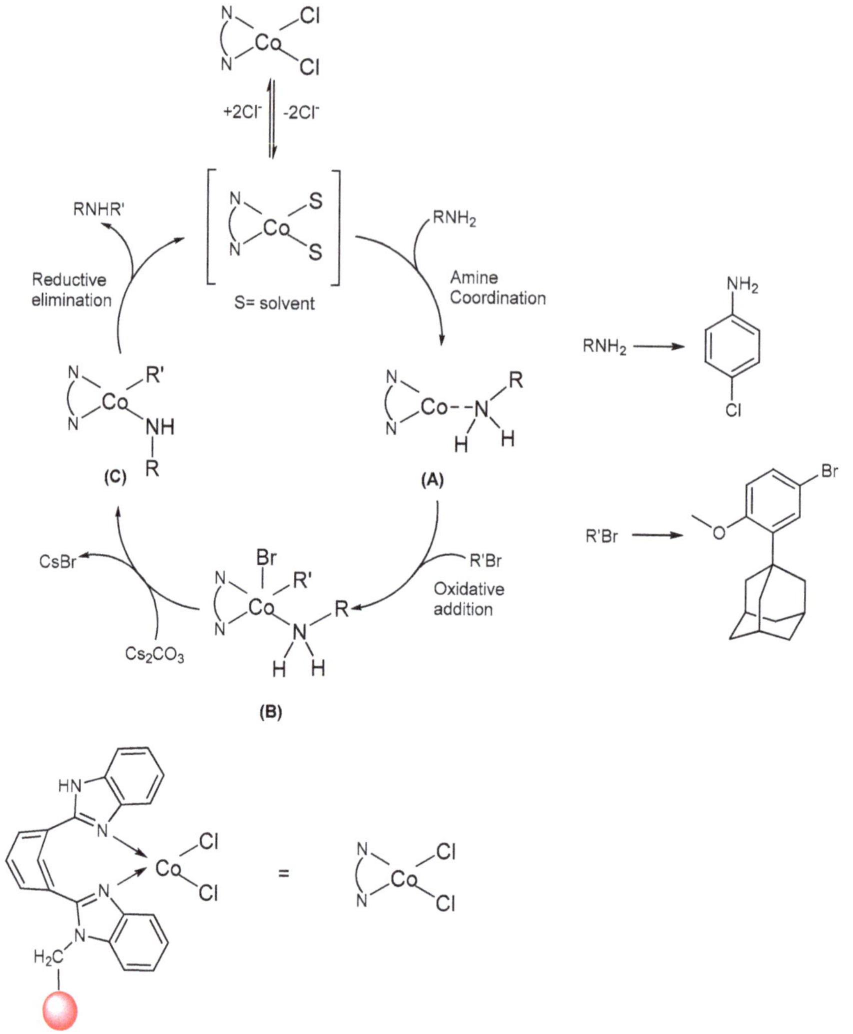

Figure 3. Plausible mechanism for the generation of arylamines using PS-Co(BBZN)Cl_2 as a catalyst.

Initially, the catalyst makes a complex with amine to form a catalyst-amine complex **A**, which undergoes an oxidative addition reaction with 1-(5-bromo-2-methoxyphenyl)adamantane and complex **B** formation occurs. Complex **B** reacts with cesium carbonate base and undergoes metathesis step, which gave complex **C**. Finally, the reductive elimination reaction complex **C** takes place and thereby catalyst regeneration and the desired product formation occur in the last step (Figure 3).

Further, we performed density-functional theory calculations using dispersion corrected CAM-B3 LYP functional and 6–31+G method [35]. All electron basis set as implemented in the Gaussian 09 package [36]. The minima nature of the structures has been confirmed based on computed real harmonic vibrational analysis at the same level of theory. Gibbs free energy calculations for four intermediate cobalt complexes were chosen for our mechanistic elucidation. Initially $CoCl_2$ makes the coordination complex with the ligand and reacts with aromatic amine and forms Co-NH bond quickly [intermediate (**a**); ΔE = −6.03 kcal/mole], which in turn gets stabilized by releasing HCl and attains a lower energy intermediate with a ΔE of −9.62 kcal/mole. Alkyl bromide adds to the intermediate (**b**) quickly and attains still lower energy of ΔE of −17.18 kcal/mole where the bindentate ligand detachment takes place and immediate loss of HCl takes place and again attains lowest energy intermediate (**d**) of ΔE = −19.62 kcal/mole, which gives the product immediately. The optimized geometries and the energy profile diagram of intermediates (**a**–**d**) are shown in Figures 4 and 5, respectively. On the basis of lower Gibbs free energy of intermediates across (**a**) to (**d**), we can conclude that the reaction occurs naturally upon cobalt chloride coordination complex formation occurring with the bidentate ligands.

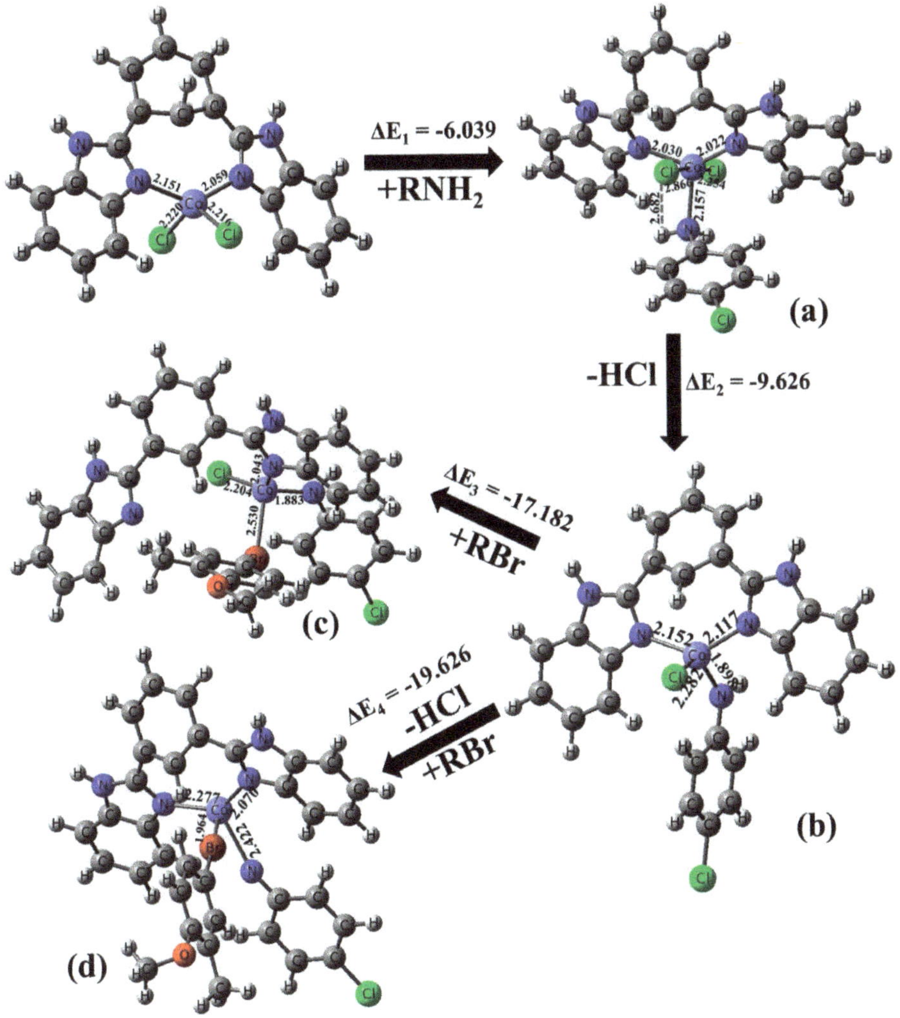

Figure 4. Computed intermediate structures (**a**–**d**) and reaction path. Energy difference ΔE are given in kcal/mole.

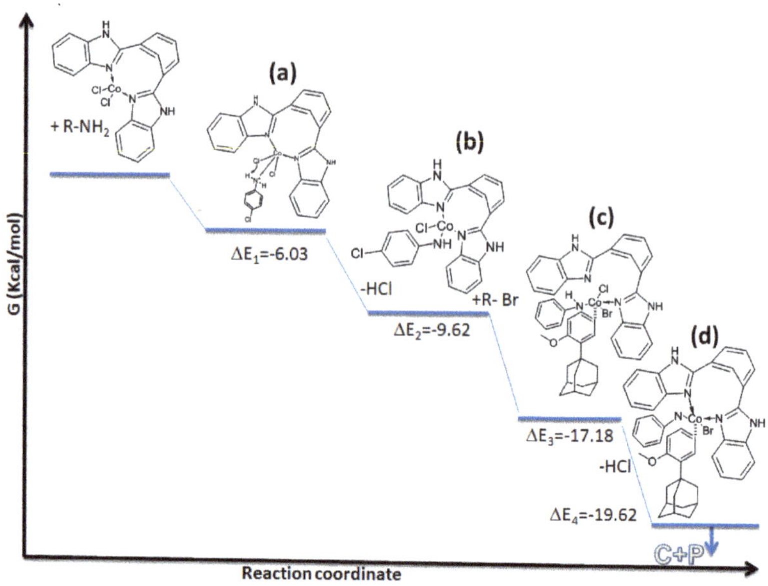

Figure 5. Energy profile diagram or arylamination reaction. C= catalyst; P = product.

2.2. Recyclability of the Catalyst

Further, the superiority of PS-Co(BBZN)Cl_2 catalyst was its recyclability, which was investigated by using the compound **1 a** and **2 b** as a model reaction. After each run, the catalyst was filtered off and washed with water followed by methanol, it was then dried in an oven at 120 °C for 15 min and used directly for the next reaction. The results were summarized (Table 4). We recorded that the catalyst could be used thrice and isolated yields achieved were above 70%.

Table 4. The recycling of the catalyst [a].

Run	1	2	3
Yield [b] (%)	86	81	75

[a] Reaction conditions—1 a (1 mmol), 2 b (1 mmol), BINAP (15 mol%), (PS-Co (BBZN)Cl_2) (12 mol%), CS_2 CO_3 (3 mmol), 1,4-dioxane (5 mL), N_2 atmosphere 10 h, 100 °C. [b] Isolated yield.

3. Materials and Methods

3.1. Procedure for the Synthesis of PS-Co(BBZN)Cl_2 Complex

3.1.1. Preparation of BBZN Functionalized Polymer Support

The chloromethylated polystyrene beads cross-linked with 6.5% divinylbenzene were first washed with a mixture of THF and water in the ratio 4:1 using Soxhlet extractor for 48 h. The beads were then vacuum dried. The chloromethylated polystyrene beads (3 g) were allowed to swell in DMF solution of BBZN ligand (5.2 g) was added to the above suspension followed by the addition of triethylamine (12 mL) in ethylacetate (105 mL) and was heated at 60 °C for 45 h in a water bath. It was cooled to room temperature, filtered, and washed with DMF. The beads were then Soxhlet extracted with ethanol to remove any unreacted BBZN and dried in an oven at 60 °C overnight.

3.1.2. Preparation of PS-Co(BBZN)Cl_2 Complex

The functionalized beads (1.0 g) were allowed to swell in 50 mL acetonitrile and toluene mixture in the ratio 1:1 for 1 h. Then the solvent was decanted. To this, 1.426 g of $CoCl_2$.6 H_2 O dissolved in methanol (100 mL) was added at intervals (4 times) and heated at 60 °C for 48 h. It was filtered, washed with alcohol and Soxhlet extracted to remove any unreacted $CoCl_2$.6 H_2 O. It was filtered and dried in an oven at 60 °C for 10 h and vacuum dried.

3.2. General Procedure for (PS-Co(BBZN)Cl_2) Complex Catalyzed C–N Bond-Formation Reaction

A dried Schlenk tube was charged with substrate **1 a** (320 mg, 1 mmol), **2 a** (127.6 mg, 1 mmol), BINAP (48 mg, 15 mol%), (PS-Co(BBZN)Cl_2](38 mg, 12 mol%). The tube was evacuated and backfilled with N_2, and Cs_2 CO_3 (975 mg, 3 mmol) followed by reagent grade 1, 4-dioxane (5 mL). The reaction mixture was heated to 100 °C for 10 h. After completion of reaction the mass was cooled to room temperature, filtered off the catalyst, the solvent quenched with water and diluted with ethyl acetate (10 mL). The layers were separated, and the aqueous layer was extracted with (5 mL) ethyl acetate. The combined organic layer was washed with water (10 mL), dried over anhydrous sodium sulphate and the solvent was removed in vacuum. The crude product was purified using silica gel column chromatography.

3.2.1. 3-(Adamantan-1-yl)-N-(4-chlorophenyl)-4-methoxyaniline (**3 a**)

Pale Yellow colored solid; mp 140–142 °C: ^{1}H NMR (400 MHz,$CDCl_3$) 7.14–7.12 (d, J = 8.0 Hz, 2 H), 6.95–6.91 (m, 2 H), 6.81–6.78 (m, 3 H), 5.46 (s, 1 H), 3.80 (s, 3 H), 2.05 (m, 9 H), 1.74 (m, 6 H); ^{13}C NMR (100 MHz,CDCl3) 155.0, 144.2, 139.9, 134.8, 129.2, 123.6, 120.6, 119.1, 116.4, 112.6, 55.4, 40.6, 37.1, 29.1; LCMS (MM : ES + APCI) 368.4 $(M + H)^+$; Anal.Calcd for C_{23} H_{26} ClNO: C, 75.08; H, 7.12; N, 3.81. Found: C, 75.01; H, 7.15; N, 3.88.

3.2.2. 3-(Adamantan-1-yl)-4-methoxy-N-(4-methoxyphenyl)aniline (**3 b**)

Brown colored solid; mp 117–119 °C: ^{1}H NMR (400 MHz,$CDCl_3$) 7.48–7.46 (d, J = 8.0 Hz, 2 H), 7.24 (m, 1 H), 6.88–6.86 (d, J = 8.0 Hz, 2 H), 6.72–6.70 (d, J = 8.0 Hz, 2 H), 5.39 (s, 1 H), 3.89 (s, 3 H), 3.83 (s, 3 H), 2.06–2.03 (m, 9 H), 1.75 (m, 6 H); ^{13}C NMR (100 MHz, $CDCl_3$) 153.8, 142.2, 138.9, 130.4, 128.2, 122.5, 118.1, 115.5, 111.8, 55.0, 53.7, 40.3, 36.96, 28.9; LCMS (MM : ES + APCI) 364.4 $(M + H)^+$; Anal.Calcd for C_{24} H_{29} NO_2: C, 79.30; H, 8.04; N, 3.85. Found: C, 79.26; H, 8.11; N, 3.79.

3.2.3. 3-(Adamantan-1-yl)-4-methoxy-N-(4-(trifluoromethyl)phenyl)aniline (**3 c**)

Off-white colored solid; mp 124–126 °C: ^{1}H NMR (400 MHz,$CDCl_3$) 7.64–7.61 (m, 2 H), 7.45–7.39 (m, 3 H), 7.24 (s, 1 H), 6.96–6.94 (d, J = 8.0 Hz, 1 H), 5.36 (s, 1 H), 3.87 (s, 3 H), 2.13–2.04 (m, 9 H), 1.77 (m, 6 H); ^{13}C NMR (100 MHz,$CDCl_3$) 159.1, 145.1, 139.1, 131.8, 127.0, 125.7 (JCF = 25.7 Hz), 112.1, 55.2, 40.6, 37.2 (JCF = 7.6 Hz), 29.7, 29.1; LCMS (MM : ES + APCI) 402.2 $(M + H)^+$; Anal.Calcd for C_{24} H_{26} F_3 NO: C, 71.80; H, 6.53; N, 3.49. Found: C, 71.76; H, 6.59; N, 3.41.

3.2.4. 3-((3-(Adamantan-1-yl)-4-methoxyphenyl)amino)phenol (**3 d**)

Off-white colored solid; mp 98–100 °C; ^{1}H NMR (400 MHz,CDC13) 7.43 (s, 1 H), 7.08 (s, 1 H), 6.99–6.96 (m, 4 H), 6.84–6.82 (d, J = 8.0 Hz, 1 H), 5.62 (s, 1 H), 4.80(s, 1 H), 3.82 (s, 3 H), 2.05 (m, 9 H), 1.74 (m, 6 H); ^{13}C NMR (100 MHz,CDC13) 159.2, 154.3, 146.2, 139.9, 129.7, 121.3, 119.9, 117.7, 115.3, 112.6, 111.2, 55.4, 40.6, 37.1, 29.1; HRMS Calcd 372.1934 Found: 372.1938 $(M + H)^+$; Anal.Calcd for C_{24} H_{29} NO_2: C, 79.30; H, 8.04; N, 3.85. Found: C, 79.26; H, 8.11; N, 3.79.

3.2.5. 3-(Adamantan-1-yl)-N-(2-fluorophenyl)-4-methoxyaniline (**3 e**)

Yellow colored solid; mp 106–108 °C: ^{1}H NMR (400 MHz, $CDCl_3$) 7.10–7.06 (m, 1 H), 6.98–6.96 (dd, J1 = 2.7 Hz, J2 = 2.2 Hz, 2 H), 6.84–6.83 (d, J = 4.0 Hz, 1 H), 6.81 (s, 1 H), 5.50 (s, 1 H), 3.81 (s, 3 H), 2.05 (m, 9 H), 1.75 (m, 6 H); ^{13}C NMR (100 MHz,$CDCl_3$) 158.4, 155.4, 147.1, 139.9, 134.1, 130.3, 121.5,

120.0, 118.8, 114.6, 112.8 (JCF = 53.4 Hz), 55.4, 40.5, 37.2, 29.0; LCMS (MM : ES + APCI) 352.4 $(M + H)^+$; Anal.Calcd for C_{23} H_{26} FNO: C, 78.60; H, 7.46; N, 3.99. Found: C, 78.71; H, 7.39; N, 3.91.

3.2.6. 3-(Adamantan-1-yl)-4-methoxy-N-(p-tolyl)aniline (**3 f**)

Off-white colored solid; mp 108–110 °C: ^{1}H NMR (400 MHz,$CDCl_3$) 7.03–7.01 (d, J = 8.0 Hz, 2 H), 6.95–6.89 (dd, J1 = 4.0 Hz, J2 = 4.0 Hz, 2 H), 6.85–6.83 (d, J = 8.0 Hz, 2 H), 6.80–6.78(d, J = 8.0 Hz, 1 H), 5.39 (s, 1 H), 3.80 (s, 3 H), 2.26 (s, 3 H), 2.05 (m, 9 H), 1.74 (m, 6 H); ^{13}C NMR (100 MHz,$CDCl_3$) 154.2, 142.6, 139.7, 129.8, 128.9, 119.4, 117.6, 116.3, 112.7, 55.5, 40.6, 37.1, 37.0, 29.1, 20.6; LCMS (MM : ES + APCI) 348.4 $(M + H)^+$; Anal. Calcd for C_{24} H_{29} NO: C, 82.95; H, 8.41; N, 4.03. Found: C, 82.90; H, 8.46; N, 3.99.

3.2.7. N-(3-Adamantan-1-yl)-4-methoxyphenyl)pyridin-3-amine (**3 g**)

Pale yellow colored solid; mp 103–104 °C; ^{1}H NMR (400 MHz, $CDCl_3$) 8.42–8.41 (d, J = 4.0 Hz, 1 H), 7.47–7.43 (m, 2 H), 7.29–7.27 (m, 2 H), 7.23–7.22 (m, 2 H), 5.58 (s, 1 H), 3.85 (s, 3 H), 2.14–207 (m, 9 H), 1.79–1.73 (m, 6 H); ^{13}C NMR (100 MHz, $CDCl_3$) 161.9, 153.0, 143.7, 136.1, 126.8, 125.9, 124.0, 123.0, 121.9, 117.6, 115.1, 56.1, 40.8, 38.0, 28.6; LCMS (MM : ES + APCI) 335.4 $(M + H)^+$; Anal. Calcd for C_{22} H_{26} N_2 O: C, 79.00; H, 7.84; N, 8.38. Found: C, 79.08; H, 7.79; N, 8.33.

3.2.8. N-(3-Adamantan-1-yl)-4-methoxyphenyl)-5-methylpyridin-2-amine (**3 h**)

Off-white colored solid; mp 101–102 °C: ^{1}H NMR (400 MHz, $CDCl_3$); 8.42 (s, 1 H), 7.53–7.47 (m, 3 H), 7.03–7.01 (d, J = 8.0 Hz, 2 H), 5.60 (s, 1 H), 3.83 (s, 3 H), 2.30 (s, 3 H), 2.09 (m, 9 H), 1.78 (m, 6 H); ^{13}C NMR (100 MHz, CDCl3) 160.2, 155.2, 145.0, 139.2, 128.9, 126.4, 121.4, 119.5, 117.4, 115.7, 54.8, 39.9, 36.4, 27.5, 27.0, 21.9; HRMS Calcd: 371.2094; Found: 371.2098 $(M + H)^+$; Anal. Calcd for C_{23} H_{28} N_2 O: C, 79.27; H, 8.10; N, 8.04. Found: C, 79.32; H, 7.99; N, 8.09.

3.2.9. N-(3-Adamantan-1-yl)-4-methoxyphenyl)naphthalen-1-amine (**3 i**)

Pale yellow colored solid; mp 105–106 °C; ^{1}H NMR (400 MHz, $CDCl_3$) 8.09–8.07 (d, J = 8.0 Hz, 1 H), 8.03–8.01 (d, J = 8.0 Hz, 1 H), 7.96–7.94 (d, J= 8 Hz, 1 H), 7.60–7.28 (m, 6 H), 5.51 (s, 1 H), 3.83 (s, 3 H), 2.13–2.05 (m, 9 H), 1.76–1.73 (m, 6 H); ^{13}C NMR (100 MHz,$CDCl_3$) 153.9, 145.7, 138.3, 133.9, 132.7, 131.9, 128.7, 128.3, 128.2, 127.1, 126.9, 126.3, 125.9, 125.5, 111.4, 55.2, 40.4, 37.2, 29.2; LCMS (MM : ES + APCI) 384.4 $(M + H)^+$; Anal.Calcd for C_{27} H_{29} NO: C, 84.55; H, 7.62; N, 3.65. Found: C, 84.61; H, 7.59; N, 3.69.

3.2.10. N-(3-Adamantan-1-yl)-4-methoxyphenyl)-1 H-inden-2-amine (**3 j**)

Pale yellow colored solid; mp 116–118 °C: ^{1}H NMR (400 MHz, $CDCl_3$) 7.64–7.61 (m, 2 H), 7.45–7.39 (m, 3 H), 6.96–7.94 (d, J = 8 Hz, 2 H), 6.19 (s, 1 H), 5.34 (s, 1 H), 3.83 (s, 3 H), 3.29 (s, 2 H), 2.13–2.07 (m, 9 H), 1.77 (m, 6 H); ^{13}C NMR (100 MHz, $CDCl_3$) 154.0, 144.4, 138.9, 132.0, 130.0, 126.0, 125.9, 120.1, 119.9, 115.5, 104.4, 55.4, 44.3, 40.8, 37.3, 29.7, 29.2; LCMS (MM : ES + APCI) 372.2 $(M + H)^+$ Anal.Calcd for C_{27} H_{29} NO: C, 84.06; H, 7.87; N, 3.77. Found: C, 84.11; H, 7.94; N, 3.72.

3.2.11. 4.((3-(Adamantan-1-yl)-4-methoxyphenyl)amino)phenyl)(piperidin-1-yl)methanone (**3 k**)

Off-white colored solid; mp 121–122 °C; ^{1}H NMR (400 MHz, $CDCl_3$) 7.65–7.62 (m, 2 H), 7.45–7.39 (m, 3 H), 6.97–6.95 (d, J = 8.0 Hz, 1 H), 5.37 (s, 1 H), 3.87 (s, 3 H), 3.47–3.39 (m, 4 H), 2.14–2.13 (m, 9 H), 2.08–2.04 (m, 3 H), 1.78 (m, 6 H), 1.45 (m, 6 H); ^{13}C NMR (100 MHz, $CDCl_3$) 170.4, 155.2, 145.6, 138.9, 131.8, 127.0, 125.8, 125.6, 112.5, 55.5, 46.2, 40.5, 37.2, 37.1, 29.7, 29.1, 24.4; LCMS (MM : ES + APCI) 445.2 $(M + H)^+$; Anal.Calcd for C_{29} H_{36} N_2 O_2: C 78.34; H 8.16; N 6.30; Found: C 78.39; H 8.10; N 6.24.

3.2.12. 2.(4-((3-Adamantan-1-yl)-4-methoxyphenyl)amino)-2-chloro-5-methylphenyl)-2-(4-chlorophenyl)acetonitrile (**3 l**)

Yellow colored solid; mp 129–132 °C; ^{1}H NMR (300 MHz, $CDCl_3$) 7.62–7.51 (m, 3 H), 7.32–7.26 (m, 2 H), 6.84–6.81 (d, J = 12.0 Hz, 2 H), 6.35–6.25 (m, 2 H), 5.66 (s, 1 H), 5.27 (s, 1 H), 3.82 (s, 3 H), 2.16–1.66

(m, 18 H); ^{13}C NMR (75 MHz,$CDCl_3$) 156.0, 1139.8, 136.6, 133.3, 126.8, 125.8, 125.6, 123.3, 121.7, 117.6, 55.3, 42.5, 41.6, 41.1, 40.9, 36.9, 36.5, 36.1, 35.6, 28.5, 27.9, 18.13; HRMS Calcd: 553.1784; Found: 553.1892 $(M + H)^+$; Anal.Calcd for C_{32} H_{32} Cl_2 N_2 O: C 72.31; H 6.07; N 5.27; Found: C 72.39; H 6.01; N 5.24.

Off-white colored solid; mp 118–120 °C; 3-(adamantan-1-yl)-4-methoxy-N-(o-tolyl)aniline (**3 m**): ^{1}H NMR (400 MHz, CDC13) 7.40 (s, 1 H), 7.36 (d, 1 H), 7.17–7.10 (m, 4 H), 6.94 (d, 1 H), 5.37 (s, 1 H), 3.87 (s, 3 H), 2.41 (s, 3 H), 2.14 (m, 6 H), 2.06 (m, 3 H), 1.77 (m, 6 H); HRMS Calcd 370.214. Found: 370.212 $(M + H)^+$.

6-Chloro-N-(p-tolyl)-9 H-fluoren-2-amine (**5 b**)

Yellow colored solid; mp 131–132 °C ^{1}H NMR (400 MHz, DMSO-d_6); 8.10 (s, 1 H), 8.05–8.03 (d, J = 8 Hz, 1 H), 7.76 (s, 1 H), 7.53–7.51 (d, J = 8.0 Hz, 1 H), 7.38–7.21 (m, 4 H), 7.16–7.12 (m, 2 H), 5.36 (s, 1 H), 4,37 (s, 2 H), 2.39 (s, 3 H); ^{13}C NMR (100 MHz, DMSO-d_6); 140.7, 140.5, 137.8, 135.5, 129.1, 127.8, 127.5, 127.0, 125.4, 123.2, 121.4, 119.9, 112.1, 110.6, 41.20, 23.5; HRMS Calcd 328.0863; Found: 328.0866 $(M + Na)^+$; Anal.Calcd for C_{20} H_{16} ClN: C, 78.55; H, 5.27; N, 4.58; Found: C, 78.58; H, 5.21; N, 4.55.

N-(4-Methoxyphenyl)benzo[d]isoxazol-3-amine (**5 c**)

White colored solid; mp 98–100 °C: ^{1}H NMR (400 MHz, DMSO-d_6); 8.44–8.42 (d, J = 8.0 Hz, 1 H), 8.08–8.06 (d, J = 8.0 Hz, 1 H), 8.02–8.00 (d, J = 8.0 Hz, 1 H), 7.95–7.91 (m, J = 8.0 Hz, 2 H), 7.71–7.67 (m, 1 H), 7.26–7.24 (d, J = 8.0 Hz, 2 H), 5.32 (s, 1 H), 3.88 (s, 3 H); ^{13}C NMR (100 MHz, DMSO-d_6); 164.1, 159.7, 152.0, 147.4, 132.2, 127.2,125.6, 123.4, 121.7, 118.9, 114.4, 113.8, 55.2; HRMS Calcd 263.0791; Found: 263.0794 $(M + Na^+)$; Anal.Calcd for C_{14} H_{12} N_2 O_2: C, 69.99; H, 5.03; N, 11.66; Found: C, 70.05; H, 5.08; N, 11.59.

2-(2-Chloro-4-((4-methoxyphenyl)amino)-5-methylphenyl)-2-(4-chlorophenyl)acetonitrile (**5 e**)

Off-white colored solid; mp 111–112 °C: ^{1}H NMR (400 MHz, DMSO-d_6); 7.57–7.53 (m, 3 H), 7.51–7.42 (m, 3 H), 7.35–7.31 (m, 2 H), 7.08–7.02 (m, 2 H), 5.72 (s, 1 H), 5.32 (s, 1 H), 3.83 (s, 3 H), 2.13 (s, 3 H); ^{13}C NMR (100 MHz, DMSO-d_6); 152.0, 149.8, 141.1, 140.0, 137.1, 132.4, 132.3, 132.1, 129.1, 128.6, 123.1, 120.8, 116.1, 54.99, 36.6, 17.9; HRMS Calcd 419.0688; Found: 419.0692 $(M + Na^+)$; Anal.Calcd for C_{22} H_{18} Cl_2 N_2 O: C, 66.51; H, 4.57; N, 7.05; Found: C, 66.59; H, 4.52; N, 7.11.

4. Conclusions

In conclusion, we prepared PS-Co $(BBZN)Cl_2$ catalyst and used it for the C−N bond formation reaction. A series of adamantyl-tethered-amino biphenylic compounds were synthesized by new protocol. Our synthetic methodology is much improved compared to existing methodologies as the catalyst is effective, inexpensive and recyclable.

Supplementary Materials: The following are available online at http://www.mdpi.com/2073-4344/10/11/1315/s1, SI-01: Experiment Section, SI-02: Spectral characterization $Co(BBZN)Cl_2$, SI-03 to14: Spectral characterization of compounds 4 a-4 l.

Author Contributions: B.C.P., S.E.R., S.N.D. performed the organic synthesis and material characterization experiments; P.K.M., G.V., B.S.P., V.P., P.E.L. and R.K.S. supported through suggestions and guidance; K.F.-U.-R. and G.P. provided DFT calculations; B. designed the research, provided resources and wrote the manuscript. All authors have read and agreed to the published version of the manuscript.

Funding: B. thanks Council of Scientific and Industrial Research (No. 02(0291)17/EMR-II), Department of Biotechnology (BT/PR24978/NER/95/938/2017), Vision Group on Science and Technology, Government of Karnataka for funding. PEL thanks Tsinghua Berkeley Shenzhen Institute Faculty Start-Up Fund, the Shenzhen Development and Reform Commission Subject Construction Project for funding.

Conflicts of Interest: The authors declare that the research was conducted in the absence of any commercial or financial relationships that could be construed as a potential conflict of interest.

References

1. Rasheed, S.; Rao, D.N.; Das, P. Copper-Catalyzed Inter- and Intramolecular C-N Bond Formation: Synthesis of Benzimidazole-Fused Heterocycles. *J. Org. Chem.* **2015**, *80*, 9321–9327. [CrossRef] [PubMed]
2. Wu, Z.; Huang, Q.; Zhou, X.; Yu, L.; Li, Z.; Wu, D. Synthesis of pyrido[1,2-a]benzimidazoles through a copper-catalyzed cascade C-N coupling process. *Eur. J. Org. Chem.* **2011**, *2011*, 5242–5245. [CrossRef]
3. Hesp, K.D.; Bergman, R.G.; Ellman, J.A. Rhodium-catalyzed synthesis of branched amines by direct addition of benzamides to imines. *Org. Lett.* **2012**, *14*, 2304–2307. [CrossRef] [PubMed]
4. Peng, J.; Ye, M.; Zong, C.; Hu, F.; Feng, L.; Wang, X.; Wang, Y.; Chen, C. Copper-catalyzed intramolecular C-N bond formation: A straightforward synthesis of benzimidazole derivatives in water. *J. Org. Chem.* **2011**, *76*, 716–719. [CrossRef]
5. Brain, C.T.; Steer, J.T. An improved procedure for the synthesis of benzimidazoles, using palladium-catalyzed aryl-amination chemistry. *J. Org. Chem.* **2003**, *68*, 6814–6816. [CrossRef] [PubMed]
6. Wolfe, J.P.; Wagaw, S.; Marcoux, J.; Buchwald, S.L. Rational Development of Practical Catalysts for Aromatic Carbon−nitrogen bond formation. *Acc. Chem. Res.* **1998**, *31*, 805–818. [CrossRef]
7. Tsang, W.C.P.; Zheng, N.; Buchwald, S.L. Combined C-H functionalization/C-N bond formation route to carbazoles. *J. Am. Chem. Soc.* **2005**, *127*, 14560–14561. [CrossRef] [PubMed]
8. Yang, Y.G.; Buchwald, S.L. Palladium-catalyzed amination of aryl halides and sulfonates. *J. Organomet. Chem.* **1999**, *576*, 125–146. [CrossRef]
9. Corcoran, E.B.; Pirnot, M.T.; Lin, S.; Dreher, S.D.; Dirocco, D.A.; Davies, I.W.; Buchwald, S.L.; Macmillan, D.W.C. Aryl amination using ligand-free Ni(II) salts and photoredox catalysis. *Science* **2016**, *353*, 279–283. [CrossRef]
10. Toma, G.; Yamaguchi, R. Cobalt-catalyzed C-N bond-forming reaction between chloronitrobenzenes and secondary amines. *Eur. J. Org. Chem.* **2010**, *2010*, 6404–6408. [CrossRef]
11. Ahmad, K.; Chang, C.R.; Li, J. Mechanistic investigations of Co(II)-Catalyzed C-N coupling reactions. *J. Organomet. Chem.* **2018**, *868*, 144–153. [CrossRef]
12. Ruiz-Castillo, P.; Buchwald, S.L. Applications of Palladium-Catalyzed C-N Cross-Coupling Reactions. *Chem. Rev.* **2016**, *116*, 12564–12649. [CrossRef] [PubMed]
13. Cahiez, G.; Moyeux, A. Cobalt-catalyzed cross-coupling reactions. *Chem. Rev.* **2010**, *110*, 1435–1462. [CrossRef]
14. Sherrington, D.C. Polymer-supported metal complex alkene epoxidation catalysts. *Catal. Today* **2000**, *57*, 87–104. [CrossRef]
15. Eshwar Rao, S.; Gayathri, V. Poly(styrene–divinyl benzene)-immobilized Fe(III) complex of 1,3-bis (benzimidazolyl)benzene: Efficient catalyst for the photocatalytic degradation of xylenol orange. *J. Appl. Polym. Sci.* **2018**, *135*, 1–13. [CrossRef]
16. Fan, Q.H.; Ren, C.Y.; Yeung, C.H.; Hu, W.H.; Chan, A.S.C. Highly effective soluble polymer-supported catalysts for asymmetric hydrogenation. *J. Am. Chem. Soc.* **1999**, *121*, 7407–7408. [CrossRef]
17. Howard, I.C.; Hammond, C.; Buchard, A. Polymer-supported metal catalysts for the heterogeneous polymerisation of lactones. *Polym. Chem.* **2019**, *10*, 5894–5904. [CrossRef]
18. Annis, D.A.; Jacobsen, E.N. Polymer-supported chiral Co(salen) complexes: Synthetic applications and mechanistic investigations in the hydrolytic kinetic resolution of terminal epoxides. *J. Am. Chem. Soc.* **1999**, *121*, 4147–4154. [CrossRef]
19. Karjalainen, J.K.; Hormi, O.E.O.; Sherrington, D.C. Efficient Polymer-Supported Sharpless Alkene Epoxidation Catalyst. *Molecules* **1998**, *3*, 51–59. [CrossRef]
20. Canali, L.; Sherrington, D.C. Utilisation of homogeneous and supported chiral metal(salen) complexes in asymmetric catalysis. *Chem. Soc. Rev.* **1999**, *28*, 85–93. [CrossRef]
21. Sherrington, D.C. Polymer-supported metal complex oxidation catalysts. *Pure Appl. Chem.* **1988**, *60*, 401–414. [CrossRef]
22. Conte, V.; Floris, B. Vanadium catalyzed oxidation with hydrogen peroxide. *Inorg. Chim. Acta* **2010**, *363*, 1935–1946. [CrossRef]
23. Farzaneh, S.; Zarghi, A. Estrogen receptor ligands: A review (2013–2015). *Sci. Pharm.* **2016**, *84*, 409–427. [CrossRef]

24. Sebastian, A.; Pandey, V.; Mohan, C.D.; Chia, Y.T.; Rangappa, S.; Mathai, J.; Baburajeev, C.P.; Paricharak, S.; Mervin, L.H.; Bulusu, K.C.; et al. Novel adamantanyl-based thiadiazolyl pyrazoles targeting EGFR in triple-negative breast cancer. *ACS Omega* **2016**, *1*, 1412–1424. [CrossRef]
25. Anusha, S.; Mohan, C.D.; Ananda, H.; Baburajeev, C.P.; Rangappa, S.; Mathai, J.; Fuchs, J.E.; Li, F.; Shanmugam, M.K.; Bender, A.; et al. Adamantyl-tethered-biphenylic compounds induce apoptosis in cancer cells by targeting Bcl homologs. *Bioorg. Med. Chem. Lett.* **2016**, *26*, 1056–1060. [CrossRef]
26. Sadashiva, M.P.; Nanjundaswamy, S.; Li, F.; Manu, K.A.; Sengottuvelan, M.; Prasanna, D.S.; Anilkumar, N.C.; Sethi, G.; Sugahara, K.; Subbegowda, K.; et al. Anti-cancer activity of novel dibenzo[b,f]azepine tethered isoxazoline derivatives. *BMC Chem. Biol.* **2012**, *12*. [CrossRef]
27. Ningegowda, R.; Shivananju, N.S.; Rajendran, P.; Basappa; Rangappa, K.S.; Chinnathambi, A.; Li, F.; Achar, R.R.; Shanmugam, M.K.; Bist, P.; et al. A novel 4,6-disubstituted-1,2,4-triazolo-1,3,4-thiadiazole derivative inhibits tumor cell invasion and potentiates the apoptotic effect of TNFα by abrogating NF-κB activation cascade. *Apoptosis* **2017**, *22*, 145–157. [CrossRef]
28. Priya, B.S.; Nanjunda Swamy, S.; Tejesvi, M.V.; Basappa; Sarala, G.; Gaonkar, S.L.; Naveen, S.; Shashidhara Prasad, J.; Rangappa, K.S. Synthesis, characterization, antimicrobial and single crystal X-ray crystallographic studies of some new sulfonyl, 4-chloro phenoxy benzene and dibenzoazepine substituted benzamides. *Eur. J. Med. Chem.* **2006**, *41*, 1262–1270. [CrossRef]
29. Anusha, S.; Anandakumar, B.S.; Mohan, C.D.; Nagabhushana, G.P.; Priya, B.S.; Rangappa, K.S. Preparation and use of combustion-derived Bi2O3 for the synthesis of heterocycles with anti-cancer properties by Suzuki-coupling reactions. *RSC Adv.* **2014**, *4*, 52181–52188. [CrossRef]
30. Rakesh, K.S.; Jagadish, S.; Vinayaka, A.C.; Hemshekhar, M.; Paul, M.; Thushara, R.M.; Sundaram, M.S.; Swaroop, T.R.; Mohan, C.D.; Sadashiva, M.P.; et al. A new ibuprofen derivative inhibits platelet aggregation and ros mediated platelet apoptosis. *PLoS ONE* **2014**, *9*, e2718. [CrossRef]
31. Nirvanappa, A.C.; Mohan, C.D.; Rangappa, S.; Ananda, H.; Sukhorukov, A.Y.; Shanmugam, M.K.; Sundaram, M.S.; Nayaka, S.C.; Girish, K.S.; Chinnathambi, A.; et al. Novel synthetic oxazines target NF-κB in colon cancer in vitro and inflammatory bowel disease in vivo. *PLoS ONE* **2016**, *11*, e0163209. [CrossRef] [PubMed]
32. Tan, B.Y.H.; Teo, Y.C. Efficient cobalt-catalyzed C-N cross-coupling reaction between benzamide and aryl iodide in water. *Org. Biomol. Chem.* **2014**, *12*, 7478–7481. [CrossRef] [PubMed]
33. Moselage, M.; Li, J.; Ackermann, L. Cobalt-Catalyzed C-H Activation. *ACS Catal.* **2016**, *6*, 498–525. [CrossRef]
34. Ibrahim, H.; Bala, M.D. Air stable pincer (CNC) N-heterocyclic carbene-cobalt complexes and their application as catalysts for C-N coupling reactions. *J. Organomet. Chem.* **2015**, *794*, 301–310. [CrossRef]
35. Yanai, T.; Tew, D.P.; Handy, N.C. A new hybrid exchange-correlation functional using the Coulomb-attenuating method (CAM-B3LYP). *Chem. Phys. Lett.* **2004**, *393*, 51–57. [CrossRef]
36. Frisch, M.J.; Trucks, G.W.; Schlegel, H.B.; Scuseria, G.E.; Robb, M.A.; Cheeseman, J.R.; Scalmani, G.; Barone, V.; Mennucci, B.; Petersson, G.A.; et al. *Gaussian 09, Revision D.01*; Gaussian Inc.: Wallingford, CT, USA, 2009.

www.ingramcontent.com/pod-product-compliance
Lightning Source LLC
LaVergne TN
LVHW070200100526
838202LV00015B/1971

MDPI
St. Alban-Anlage 66
4052 Basel
Switzerland
Tel. +41 61 683 77 34
Fax +41 61 302 89 18
www.mdpi.com

Catalysts Editorial Office
E-mail: catalysts@mdpi.com
www.mdpi.com/journal/catalysts